APPLIED ISOTOPE HYDROGEOLOGY
A CASE STUDY IN NORTHERN SWITZERLAND

# Nagra   Cédra   Cisra

Nationale Genossenschaft für die Lagerung radioaktiver Abfälle

Société coopérative nationale pour l'entreposage de déchets radioactifs

Società cooperativa nazionale per l'immagazzinamento di scorie radioattive

TECHNICAL REPORT 88-01

APPLIED ISOTOPE HYDROGEOLOGY – A CASE STUDY IN NORTHERN SWITZERLAND

F.J. Pearson Jr., W. Balderer, H.H. Loosli, B.E. Lehmann, A. Matter, Tj. Peters, H. Schmassmann, A. Gautschi

With contributions by J.N. Andrews, P. Baertschi, G. Däppen, J.-Ch. Fontes, M. Ivanovich, M. Kullin, J.-L. Michelot, K. Ramseyer, D. Rauber, W. Rauert, S. Soreau, W. Stichler

MAY 1991

Parkstrasse 23   5401 Baden/Switzerland   Tel. 056/20 55 11

Studies in Environmental Science 43

APPLIED ISOTOPE HYDROGEOLOGY
A CASE STUDY IN NORTHERN SWITZERLAND

F.J. Pearson Jr.[7], W. Balderer[2], H.H. Loosli[3],
B.E. Lehmann[3], A. Matter[4], Tj. Peters[5],
H. Schmassmann[6] and A. Gautschi[1]

With contributions by
J.N. Andrews[8], P. Baertschi[1], G. Däppen[3], J.-Ch. Fontes[9], M. Ivanovich[10],
M. Kullin[6], J.-L. Michelot[9], K. Ramseyer[4], D. Rauber[3], W. Rauert[11], S. Soreau[9],
W. Stichler[11]

1   NAGRA (National Cooperative for the Storage of Radioactive Waste), Parkstrasse 23, CH-5401 Baden, Switzerland
2   Ingenieurgeologie, ETH-Hönggerberg, CH-8093 Zürich, Switzerland
3   Physikalisches Institut, Universität Bern, Sidlerstrasse 5, CH-3012 Bern, Switzerland
4   Geologisches Institut, Universität Bern, Baltzerstrasse 1, CH-3012 Bern, Switzerland
5   Mineralogisch-petrographisches Institut, Universität Bern, Baltzerstrasse 1, CH-3012 Bern, Switzerland
6   Geologisches Institut Dr. Schmassmann AG, Langhagstrasse 7, CH-4410 Liestal, Switzerland
7   Ground-Water Geochemistry, 1304 Walnut Hill Lane, Suite 210, Irving, Texas 75038, USA
8   Research Institute for Sedimentology, The University, P.O. Box 227, Whiteknights, Reading RG6 2AB, UK
9   Laboratoire d'Hydrologie et de Géochimie Isotopique, Université de Paris-Sud, F-91405 Orsay Cedex, France
10 AEA Industrial Technology, Isotope Geoscience Section, Harwell Laboratory, Oxfordshire OX11 0RA, UK
11 GSF Forschungszentrum für Umwelt und Gesundheit, Institut für Hydrologie, D-8042 Neuherberg, Germany

ELSEVIER
Amsterdam — Oxford — New York — Tokyo   1991

ELSEVIER SCIENCE PUBLISHERS B.V.
Sara Burgerhartstraat 25
P.O. Box 211, 1000 AE Amsterdam, The Netherlands

Distributors for the United States and Canada:

ELSEVIER SCIENCE PUBLISHING COMPANY INC.
655, Avenue of the Americas
New York, NY 10010, U.S.A.

ISBN 0-444-88983-3

© Elsevier Science Publishers B.V., 1991

All rights reserved. No part of this publication may be reproduced, stored in a retrieval system or transmitted in any form or by any means, electronic, mechanical, photocopying, recording or otherwise, without the prior written permission of the Publisher, Elsevier Science Publishers B.V., P.O. Box 211, 1000 AE Amsterdam, The Netherlands.

Special regulations for readers in the USA – This publication has been registered with the Copyright Clearance Center Inc. (CCC), Salem, Massachusetts. Information can be obtained from the CCC about conditions under which photocopies of parts of this publication may be made in the USA. All other copyright questions, including photocopying outside of the USA, should be referred to the Publisher.

No responsibility is assumed by the Publisher for any injury and/or damage to persons or property as a matter of products liability, negligence or otherwise, or from any use or operation of any methods, products, instructions or ideas contained in the material herein.

This book is printed on acid-free paper.

Printed in The Netherlands

Studies in Environmental Science
Other volumes in this series

1 **Atmospheric Pollution 1978** edited by M.M. Benarie
2 **Air Pollution Reference Measurement Methods and Systems**
   edited by T. Schneider, H.W. de Koning and L.J. Brasser
3 **Biogeochemical Cycling of Mineral-Forming Elements**
   edited by P.A. Trudinger and D.J. Swaine
4 **Potential Industrial Carcinogens and Mutagens** by L. Fishbein
5 **Industrial Waste Management** by S.E. Jørgensen
6 **Trade and Environment: A Theoretical Enquiry** by H. Siebert, J. Eichberger, R. Gronych and R. Pethig
7 **Field Worker Exposure during Pesticide Application** edited by W.F. Tordoir and E.A.H. van Heemstra-Lequin
8 **Atmospheric Pollution 1980** edited by M.M. Benarie
9 **Energetics and Technology of Biological Elimination of Wastes**
   edited by G. Milazzo
10 **Bioengineering, Thermal Physiology and Comfort** edited by K. Cena and J.A. Clark
11 **Atmospheric Chemistry. Fundamental Aspects** by E. Mészáros
12 **Water Supply and Health** edited by H. van Lelyveld and B.C.J. Zoeteman
13 **Man under Vibration. Suffering and Protection** edited by G. Bianchi, K.V. Frolov and A. Oledzki
14 **Principles of Environmental Science and Technology** by S.E. Jørgensen and I. Johnsen
15 **Disposal of Radioactive Wastes** by Z. Dlouhý
16 **Mankind and Energy** edited by A. Blanc-Lapierre
17 **Quality of Groundwater** edited by W. van Duijvenbooden, P. Glasbergen and H. van Lelyveld
18 **Education and Safe Handling in Pesticide Application** edited by E.A.H. van Heemstra-Lequin and W.F. Tordoir
19 **Physicochemical Methods for Water and Wastewater Treatment** edited by L. Pawlowski
20 **Atmospheric Pollution 1982** edited by M.M. Benarie
21 **Air Pollution by Nitrogen Oxides** edited by T. Schneider and L. Grant
22 **Environmental Radioanalysis** by H.A. Das, A. Faanhof and H.A. van der Sloot
23 **Chemistry for Protection of the Environment** edited by L. Pawlowski, A.J. Verdier and W.J. Lacy
24 **Determination and Assessment of Pesticide Exposure** edited by M. Siewierski
25 **The Biosphere: Problems and Solutions** edited by T.N. Veziroğlu
26 **Chemical Events in the Atmosphere and their Impact on the Environment** edited by G.B. Marini-Bettòlo
27 **Fluoride Research 1985** edited by H. Tsunoda and Ming-Ho Yu
28 **Algal Biofouling** edited by L.V. Evans and K.D. Hoagland
29 **Chemistry for Protection of the Environment 1985** edited by L. Pawlowski, G. Alaerts and W.J. Lacy
30 **Acidification and its Policy Implications** edited by T. Schneider
31 **Teratogens: Chemicals which Cause Birth Defects** edited by V. Kolb Meyers
32 **Pesticide Chemistry** by G. Matolcsy, M. Nádasy and V. Andriska
33 **Principles of Environmental Science and Technology (second revised edition)** by S.E. Jørgensen and I. Johnson
34 **Chemistry for Protection of the Environment 1987** edited by L. Pawlowski, E. Mentasti, C. Sarzanini and W.J. Lacy

35 **Atmospheric Ozone Research and its Policy Implications** edited by T. Schneider, S.D. Lee, G.J.R. Wolters and L.D. Grant
36 **Valuation Methods and Policy Making in Environmental Economics** edited by H. Folmer and E. van Ierland
37 **Asbestos in the Natural Environment** by H. Schreier
38 **How to Conquer Air Pollution. A Japanese Experience** edited by H. Nishimura
39 **Aquatic Bioenvironmental Studies: The Hanford Experience, 1944–1984** by C.D. Becker
40 **Radon in the Environment** by M. Wilkening
41 **Evaluation of Environmental Data for Regulatory and Impact Assessment** by S. Ramamoorthy and E. Baddaloo
42 **Environmental Biotechnology** edited by A. Blazej and V. Prívarová

PREFACE

In 1980, Nagra, the Swiss National Cooperative for the Storage of Radioactive Waste, began a comprehensive field investigation programme to assess the feasibility and safety of a repository for the final disposal of high-level radioactive waste in northern Switzerland. The host rock of first priority was crystalline basement covered by a few hundred metres of sedimentary rocks. The various investigations cover an area of about 1200 km^2 and include a regional geophysical programme, a regional hydrogeological programme comprising hydrodynamic modelling and hydrogeochemical investigations, a neotectonic programme and a deep drilling programme. Seven deep boreholes with depths between 1306 and 2482 m have been drilled to date. As the study programme for the seventh borehole was still in progress during the final preparation of this report, only incomplete results were available for inclusion in this report. The scientific programmes were designed by Nagra and its geology consultants and carried out under the project management of Nagra by more than 200 scientists from over 50 companies and universities from 8 countries.

An understanding of the deep groundwater flow regime is needed to predict flow paths, travel times and dilution effects of groundwater moving from a repository location to the biosphere. To assess the flow field, regional and local hydrodynamic models have been prepared. Currently, extensive improvements to the input data are being made and anisotropic flow properties are being included for several formations. Subsequently, validation of the hydrodynamic models with the results of the hydrogeochemical investigations will be attempted. The isotopic investigations presented in this report will probably be the most important contribution to this validation.

The hydrogeochemical programme started in 1981 with the collection of waters from about 140 springs and wells of different aquifers in northern Switzerland and adjacent areas; this was called the "regional programme" (SCHMASSMANN AND OTHERS, 1984). Detailed hydrochemical and isotopic analyses have been carried out on these samples. About 60 waters from the Nagra deep boreholes were also sampled and analysed, of which some 40 gave useful results. Additional geochemical and isotopic data on rock material are also available from these boreholes. A literature study yielded complementary hydrochemical data, the so-called "Fremde Analysen", both from points sampled during the regional programme and from additional points.

The purpose of this report is to present all the isotope data collected within the Nagra programme in northern Switzerland. The report also includes the first comprehensive synthesis and interpretation of these data. It is anticipated that special aspects of the data interpretation will be discussed elsewhere in the reviewed scientific literature.

Marc Thury
Department Head, Geology
Nagra, Baden, Switzerland

CONTENTS

PREFACE vii
CONTENTS ix
TABLES xiv
ILLUSTRATIONS xviii

1.	INTRODUCTION	1
1.1	Overview of This Report	1
1.2	Acknowledgments	2
1.3	Scope of Data Presented	2
1.4	Reporting of Isotopic Data	4
1.5	Regional Geology and Stratigraphy	7
1.5.1	Introduction	7
1.5.2	Crystalline Basement	7
1.5.3	Sedimentary Cover	11
1.6	Hydrogeology and Hydrochemistry	15
1.6.1	General Situation	15
1.6.2	Data Base	17
1.6.3	Groundwaters in Tertiary and Upper Malm Aquifers	17
1.6.4	Groundwaters in the Dogger, Lias and Keuper Aquifers .	20
1.6.5	Groundwaters in the Upper Muschelkalk Aquifer	22
1.6.6	Groundwaters of Lower Triassic and Permian Aquifers ..	24
1.6.7	Groundwaters in the Crystalline Basement	26
2.	LIMITATIONS OF SAMPLING AND ANALYTICAL PROCEDURES .	31
2.1	Contamination of Samples	32
2.1.1	Large-Volume Gas Samples for ^{39}Ar and ^{85}Kr .	33
2.1.2	Samples for Carbon Isotopes	36
2.1.2.1	Samples from Test Borehole Programme	37
2.1.2.2	Other Samples	40
2.2	Comparison of Carbon Isotope Results From Conventional and AMS Measurements	43
2.3	Evaluation of Replicate Analytical Data	44
2.3.1	Dissolved Gases	45
2.3.1.1	Oxygen	45
2.3.1.2	Noble Gases	45
2.3.2	Uranium, Thorium and Daughter Elements	48
3.	INFILTRATION CONDITIONS	65
3.1	Isotopic Composition of Modern Recharge	65
3.1.1	Stable Isotope-Altitude Relationships	68
3.1.2	δ^{18}O-δ^{2}H Relationship	78
3.1.3	Local Altitude-Temperature Equations	80
3.1.4	Stable Isotope-Temperature Relationships	82
3.2	Isotopic Composition of Groundwater	90
3.2.1	Samples from Quaternary, Tertiary and Malm Aquifers ..	90

3.2.2		Samples from the Dogger and Lias	93
3.2.3		Samples from the Keuper	93
3.2.4		Samples from the Muschelkalk	94
3.2.5		Samples from the Buntsandstein, Permian and Crystalline	98
3.3	Noble Gases in Groundwater		116
3.3.1		Introduction	116
3.3.2		Calculation of Recharge Temperature Corrections	118
3.3.3		Correlation Between Recharge Temperature and δ^2H and $\delta^{18}O$ Values	125
3.3.4		Results	128
3.3.5		Discussion	129
3.3.6		Recharge Temperatures and ^{14}C Results	138
3.3.7		Summary	140
4.	DATING BY RADIONUCLIDES		153
4.1	Introduction		153
4.2	Overview of the Sources of the Individual Isotopes		154
4.3	Detection of Young Water Components With ^3H and ^{85}Kr		157
4.4	^{39}Ar Results for Waters from Sedimentary Formations		166
4.4.1		General Conclusions from the ^{39}Ar Results	167
4.4.2		Conclusions from Corrected ^{39}Ar Results about Residence Times of the "Old" Water Components	170
5.	CARBONATE ISOTOPES		175
5.1	Overview of Groundwater Carbonate Evolution		175
5.1.1		Isotope Equilibria	176
5.1.1.1		Oxygen Isotopes	177
5.1.1.2		Carbon Isotopes	177
5.1.2		Evolution of the Isotopic Composition of Dissolved Carbonate	180
5.1.2.1		Chemical Mass Balances	180
5.1.2.2		Isotope Evolution Model	183
5.1.2.3		Application of Model Equations	188
5.1.2.4		Uncertainties in Results of Modelling	191
5.1.2.5		Comparison with Other Models for ^{14}C Correction	200
5.2	Age Interpretations of Dissolved Carbonate Isotopes		204
5.2.1		Calculation of Adjusted ^{14}C Ages	204
5.2.2		Discussion of Results	207
5.2.2.1		Samples Containing Modern ^{14}C	208
5.2.2.2		Samples with Virtually No Modern ^{14}C	211
5.2.2.3		Samples Yielding Definite ^{14}C Ages Greater than c. 12 ka.	212
5.3	Isotopic Composition of Carbonate Minerals and Water		224
5.3.1		Böttstein Samples	225
5.3.2		Weiach Samples	228
5.3.3		Riniken Samples	229
5.3.4		Schafisheim Samples	230
5.3.5		Kaisten Samples	231
5.3.6		Leuggern Samples	234
5.3.7		Summary	237

6.	ISOTOPES FORMED BY UNDERGROUND PRODUCTION	239
6.1.	Introduction	239
6.2	Chlorine-36	250
6.2.1	Results	250
6.2.2	Origin of ^{36}Cl in Groundwater Samples	250
6.2.3	^{36}Cl in Recharge	252
6.2.4	Deep Subsurface Production of ^{36}Cl	256
6.2.5	Origin of Aqueous Chloride	258
6.2.6	^{36}Cl in Water from Sedimentary Formations	259
6.2.7	^{36}Cl in Water from the Crystalline	261
6.2.8	^{36}Cl Measurements on Rock Samples	263
6.2.9	Summary	264
6.3	Argon-39 and Argon-37	266
6.3.1	Introduction	266
6.3.2	^{39}Ar Results	266
6.3.3	Escape of Argon Atoms from Rock into Water	268
6.3.4	Fractional Loss Coefficients of ^{39}Ar for the Individual Formations	268
6.3.5	^{37}Ar Measurements	272
6.3.6	Comparison of ^{39}Ar and ^{37}Ar	273
6.4	^{3}He and ^{4}He	276
6.4.1	Introduction	276
6.4.2	Results and First Interpretation	276
6.4.3	Subsurface Production	278
6.4.4	Helium Accumulation Times and Water Residence Times	279
6.4.5	Helium Transport Models	281
6.4.6	Correlation Between Helium and Chloride Concentrations	284
6.4.7	Summary	286
6.5	^{40}Ar/^{36}Ar Ratios	288
6.5.1	Introduction	288
6.5.2	Results	288
6.5.3	Subsurface Production of ^{40}Ar	288
6.5.4	Correlations Between ^{40}Ar/^{36}Ar Ratios and the Potassium Contents of Rock and Water	292
6.5.5	Correlations Between He Concentrations and ^{40}Ar/^{36}Ar Ratios	292
6.5.6	Summary	294
7.	FORMATION-SPECIFIC CHARACTERISTICS OF GROUNDWATERS	297
7.1	Sulphur and Oxygen Isotopes in Sulphate and Sulphide	297
7.1.1	Comparison of Sulphur and Oxygen Isotope Analytical Results	298
7.1.2	Samples from the Tertiary and Jurassic	301
7.1.3	Samples from the Keuper	305
7.1.4	Samples from the Muschelkalk	305
7.1.5	Samples from the Buntsandstein and Permian	309
7.1.6	Samples from the Crystalline	312
7.1.7	Summary of Sulphur and Oxygen Isotope Results	316
7.2	Strontium Isotopes in Groundwaters and Minerals	323
7.2.1	Introduction	323
7.2.2	Analytical Procedures	324

7.2.3	Samples Analysed	324
7.2.4	Results	325
7.2.4.1	Groundwater Data	325
7.2.4.2	Crystalline Basement	327
7.2.4.3	Sedimentary Rocks	327
7.2.5	Discussion	328
7.3	Uranium and Thorium-Series Nuclides	336
7.3.1	Origin and Significance of Decay Series Disequilibria	336
7.3.2	Analysis of Concentrations and Activity Ratios	338
7.3.3	Interpretation of Groundwater Analyses	340
7.3.3.1	Uranium Content	341
7.3.3.2	Activity Ratios	341
7.3.3.3	Water from the Tertiary, Malm and Keuper	346
7.3.3.4	Water from the Muschelkalk	346
7.3.3.5	Water from the Buntsandstein	346
7.3.3.6	Water from the Permian	347
7.3.3.7	Water from the Crystalline	347
7.3.3.8	Conclusions	350
7.3.4	Interpretation of Rock Sample Analyses	351
7.3.4.1	Lias	353
7.3.4.2	Muschelkalk	353
7.3.4.3	Buntsandstein	354
7.3.4.4	Permian and Carboniferous	354
7.3.4.5	Crystalline	355
7.3.4.6	Fracture Infillings and Minerals	356
7.3.4.7	Conclusions	357
8.	SYNTHESIS OF ISOTOPE RESULTS	375
8.1	Tertiary and Malm	375
8.1.1	Calcium-Magnesium-Bicarbonate Groundwaters	376
8.1.2	Sodium-Bicarbonate Groundwaters	377
8.1.3	Sodium-Chloride Groundwaters	378
8.1.4	Conclusion	379
8.2	Dogger, Lias, and Keuper	380
8.3	Muschelkalk	381
8.3.1	Recharge Conditions and Residence Times	381
8.3.2	Geochemical Evolution	384
8.3.3	Conclusions about Muschelkalk Groundwater	387
8.4	Buntsandstein, Permian and Crystalline	387
8.4.1	Waters of Low Total Mineralisation from the Crystalline	390
8.4.2	Waters of High Mineralisation from Permian Sediments	392
8.4.3	Waters of Mixed Origin	394
8.4.4	Conclusions about Buntsandstein, Permian and Crystalline Water	396
8.5	Inter-Aquifer and Regional Flow	397
8.6	Recommended Further Studies	398

9.	SUMMARY	409
9.1	Analytical Results and Interpretative Methods	409
9.1.1	Quality of Sampling and Analysis	409
9.1.2	Water Origin and Recharge Conditions	410
9.1.3	Water Ages	412
9.1.4	Chemical Evolution	414
9.2	Hydrogeological and Hydrochemical Conclusions	416
9.2.1	Tertiary and Malm	416
9.2.2	Dogger, Lias and Keuper	417
9.2.3	Muschelkalk	417
9.2.4	Buntsandstein, Permian and Crystalline	417
9.2.5	Inter-Aquifer and Regional Flow	418
10.	LITERATURE CITED	421
	APPENDIX: Table of samples discussed in this report	A1

TABLES

Table 1.3.1:	List of analyses made as part of the Nagra programme showing substances analysed, and the source of the analyses	3
Table 1.6.1:	Chemistry of deep groundwaters from the Malm and Tertiary aquifers	19
Table 1.6.2:	Chemistry of deep groundwaters from the Dogger, Lias, and Keuper aquifers.	21
Table 1.6.3:	Chemistry of deep groundwaters from the Upper Muschelkalk aquifer	23
Table 1.6.4:	Chemistry of deep groundwaters from the Lower Triassic and Permian aquifers	25
Table 1.6.5:	Chemistry of deep groundwaters from the Crystalline aquifer	28
Table 2.1.1:	Evaluation of contamination of large-volume gas samples	51
Table 2.1.2:	Estimates of ^{14}C contamination of samples from Nagra deep boreholes	56
Table 2.1.3:	Young water contents of regional programme samples	60
Table 2.3.1:	Noble gas analyses on replicate samples made at Weizmann Institute, the University of Bath, and PSI	46
Table 2.3.2:	Dissolved helium concentrations measured on replicate samples at the Universities of Cambridge and Bath and at PSI	63
Table 2.3.3:	Radium concentrations measured on replicate samples at Fresenius and by AERE, Harwell	64
Table 3.1.1:	Sources of data on isotopic composition of modern recharge	85
Table 3.1.2:	Stable isotope-altitude equations	81
Table 3.2.1:	Isotopic composition of groundwaters from the Quaternary, Tertiary and Malm	105
Table 3.2.2:	Isotopic composition of groundwaters from the Dogger and Lias	107
Table 3.2.3:	Isotopic composition of groundwaters from the Keuper	107
Table 3.2.4:	Isotopic composition of groundwaters from the Muschelkalk	108

Table 3.2.5:	Isotopic composition of groundwaters from the Buntsandstein, Permian and Crystalline	113
Table 3.2.6:	Corrections to chemistry and isotopic composition of contaminated samples.	101
Table 3.3.1:	Summary of calculated recharge temperatures	144
Table 3.3.2:	Noble gas concentrations of samples from the deep boreholes and the regional programme	149
Table 4.3.1:	Samples with ^{3}H and ^{85}Kr concentrations indicating very young or young water contents below two to three per cent	161
Table 4.3.2:	Isotope results for discussion of a possible young water component (more than a few per cent)	165
Table 4.4.1:	Isotope results of waters from sedimentary rock; He analyses from Bath, except 1/1 from Weizmann	171
Table 4.4.2:	Summary of conclusions from ^{39}Ar results	173
Table 5.1.1:	Chemistry of modelled solutions used to illustrate the carbon isotope evolution of groundwaters	192
Table 5.2.1:	Adjusted ^{14}C ages of samples from Nagra deep boreholes	214
Table 5.2.2:	Adjusted ^{14}C ages of samples from the regional programme and other sources	219
Table 5.3.1:	Isotopic composition of water and carbonate minerals from the Böttstein borehole	225
Table 5.3.2:	Isotopic composition of water and carbonate minerals from the Weiach borehole	228
Table 5.3.3:	Isotopic composition of water and carbonate minerals from the Riniken borehole	229
Table 5.3.4:	Isotopic composition of water and carbonate minerals from the Schafisheim borehole	231
Table 5.3.5:	Isotopic composition of water and carbonate minerals from the Kaisten borehole	232
Table 5.3.6:	Isotopic composition of water and carbonate minerals from the Leuggern borehole	235

Table 6.1.1:	Elemental composition of rock used for the calculations of sub-surface production rates and calculated *in situ* neutron flux	241
Table 6.1.2:	Calculated production rates or equilibrium concentrations in rock	246
Table 6.2.1:	Results of ^{36}Cl analyses of groundwater samples	251
Table 6.2.2:	Maximum possible ^{36}Cl contamination from nuclear weapons test or drilling fluid	255
Table 6.2.3:	^{36}Cl measurements on rock samples	263
Table 6.3.1:	Results of ^{39}Ar analyses of groundwater samples from Nagra boreholes; Additional results from the regional programme are given in Table 4.4.1	267
Table 6.3.2:	Results of ^{37}Ar analyses of groundwater samples	273
Table 6.3.3:	^{39}Ar/^{37}Ar activity ratios calculated in rock and measured in water samples	274
Table 6.4.1:	^{3}He/^{4}He ratios and ^{4}He concentrations in waters from the Nagra boreholes	277
Table 7.1.1:	Range of δ^{34}S and δ^{18}O values of evaporite sulphate minerals from the literature and Nagra boreholes	299
Table 7.1.2:	δ^{34}S and δ^{18}O values and concentrations of sulphate and sulphide dissolved in groundwater from northern Switzerland and adjacent regions	318
Table 7.1.3:	δ^{34}S and δ^{18}O values of sulphate minerals from the sedimentary section of the Nagra deep boreholes	320
Table 7.1.4:	δ^{34}S and δ^{18}O values of barite from the crystalline of the Nagra deep boreholes	321
Table 7.1.5:	δ^{34}S values of sulphide minerals from the sediments and the crystalline of the Nagra deep boreholes	322
Table 7.2.1:	Results of strontium and rubidium analyses	333
Table 7.3.1:	Groundwater Samples: Uranium- and thorium-natural series nuclides abundance data, Böttstein borehole	359
Table 7.3.2:	Groundwater Samples: Uranium- and thorium-natural series nuclides abundance data, Weiach borehole	360

XVII

Table 7.3.3:	Groundwater Samples: Uranium- and thorium-natural series nuclides abundance data, Riniken borehole	362
Table 7.3.4:	Groundwater Samples: Uranium- and thorium-natural series nuclides abundance data, Schafisheim borehole	363
Table 7.3.5:	Groundwater Samples: Uranium- and thorium-natural series nuclides abundance data, Kaisten borehole	364
Table 7.3.6:	Groundwater Samples: Uranium- and thorium-natural series nuclides abundance data, Leuggern borehole	365
Table 7.3.7:	Groundwater Samples: Uranium- and thorium-natural series nuclides abundance data from the Nagra regional hydrochemical programme	367
Table 7.3.8:	Natural decay-series disequilibria in rocks useful for single event dating in rock-water interaction	351
Table 7.3.9:	Rock Samples: Uranium-and thorium-natural series nuclides abundance data, Böttstein borehole	369
Table 7.3.10:	Rock Samples: Uranium-and thorium-natural series nuclides abundance data, Weiach borehole	370
Table 7.3.11:	Rock Samples: Uranium-and thorium-natural series nuclides abundance data, Riniken and Schafisheim boreholes	371
Table 7.3.12:	Rock Samples: Uranium-and thorium-natural series nuclides abundance data, Kaisten borehole	372
Table 7.3.13:	Rock Samples: Uranium-and thorium-natural series nuclides abundance data, Leuggern borehole	373
Table 8.1.1:	Summary of isotope data on recharge temperatures and model ages of Tertiary and Malm aquifer waters	400
Table 8.2.1:	Summary of isotope data on recharge temperatures and model ages of Dogger, Lias and Keuper aquifer waters	402
Table 8.3.1:	Isotope data on recharge temperatures and model ages of Upper Muschelkalk aquifer waters	403
Table 8.4.1:	Isotope data on recharge temperatures and model ages of Buntsandstein, Permian and Crystalline waters	405

ILLUSTRATIONS

Plate I:	General Location Map	(Rear Pocket)
Plate II:	Tectonic map of northern Switzerland and adjacent areas	(Rear Pocket)
Plate III:	Stratigraphy of the Nagra deep boreholes	(Rear Pocket)

Figure 1.5.1: Map showing major structural units of northern Switzerland and adjacent areas and location of Nagra boreholes 8

Figure 1.5.2: Geological cross section through central and northern Switzerland (from THURY AND DIEBOLD, 1987) 9

Figure 1.5.3: Geological cross section showing Permo-Carboniferous Troughs, Tabular and Folded Jura, and the edge of the Molasse Basin (from DIEBOLD, 1986) 9

Figure 1.5.4: Schematic stratigraphic section showing hydrogeological characteristics 16

Figure 1.6.1: Distribution of water types in the Molasse Basin in northern Switzerland (from NAGRA, 1988, p. 104) 18

Figure 2.1.1: Graph of ^{3}H against ^{14}C of samples from northern Switzerland and adjacent areas 42

Figure 2.3.1: Graph comparing argon analyses made at the University of Bath with those made at PSI 47

Figure 2.3.2: Graph comparing helium analyses made at the University of Bath with those made at the University of Cambridge and at PSI 48

Figure 2.3.3: Graph comparing uranium concentrations measured at Fresenius, PSI and AERE, Harwell 50

Figure 2.3.4: Graph comparing radium concentrations measured at Fresenius and AERE, Harwell 50

Figure 3.1.1: Location map of stable isotope data (δ^2H or δ^{18}O) of modern recharge in Switzerland and adjacent regions. 66

Figure 3.1.2: Graph of stable isotope values against recharge altitudes, grouped according to climatic regions 67

Figure 3.1.3: Climatic regions of Switzerland (after MÜLLER, 1980) 68

Figure 3.1.4:	Relationships between stable isotope data (δ^2H or $\delta^{18}O$) of modern waters and mean recharge altitude for areas of Switzerland and adjacent regions	70
Figure 3.1.5:	Relationships between $\delta^{18}O$ and average recharge altitude for all areas	76
Figure 3.1.6:	Relationships between δ^2H and average recharge altitude for areas with δ^2H values and from conversion of $\delta^{18}O$ relationships	77
Figure 3.1.7:	Relationship between $\delta^{18}O$ and δ^2H for modern waters in Switzerland and adjacent areas	79
Figure 3.1.8:	Relationships between stable isotopes and mean annual air temperature relationships for northern Switzerland	84
Figure 3.2.1:	Graph of δ^2H against $\delta^{18}O$ for groundwaters from the Quaternary, Tertiary, and Malm	91
Figure 3.2.2:	$\delta^{18}O$ values of waters from the Tertiary and Malm *versus* their residues on evaporation	92
Figure 3.2.3:	Graph of δ^2H against $\delta^{18}O$ for groundwaters from the Dogger and Lias	94
Figure 3.2.4:	Graph of δ^2H against $\delta^{18}O$ for groundwaters from the Keuper	95
Figure 3.2.5:	Graph of δ^2H against $\delta^{18}O$ for groundwaters from the Muschelkalk with ^3H contents greater than 20 TU	96
Figure 3.2.6:	Graph of δ^2H against $\delta^{18}O$ for groundwaters from the Muschelkalk with ^3H contents less than 20 TU	97
Figure 3.2.7:	Graph of δ^2H against $\delta^{18}O$ for groundwaters from the Buntsandstein, Permian and Crystalline	99
Figure 3.2.8:	Graph of δ^2H against $\delta^{18}O$ for relatively enriched groundwaters from the Buntsandstein, Permian and Crystalline	102
Figure 3.2.9:	Graph of δ^2H against residue on evaporation for relatively enriched waters from the Buntsandstein, Permian and Crystalline	103
Figure 3.2.10:	Graph of $\delta^{18}O$ against residue on evaporation for relatively enriched waters from the Buntsandstein, Permian and Crystalline	104
Figure 3.3.1:	Solubilities of the noble gases He, Na, Ar, Kr, Xe and of N_2 in ccSTP/1H$_2$O · atm (IUPAC, 1979)	117

Figure 3.3.2:	The effect of excess air on the calculated recharge temperatures of three samples before (left) and after (right) correction	119
Figure 3.3.3:	Effect of $^{40}Ar/^{36}Ar$ correction on the calculated recharge temperatures	121
Figure 3.3.4:	Effect of the elevation (and pressure) correction on the calculated recharge temperatures	122
Figure 3.3.5:	Effect of a correction for gas losses of the water sample assuming equilibrium degassing	123
Figure 3.3.6:	Schematic representation of processes of potential influence on the location of points in the δ^2H/recharge temperature diagram	127
Figure 3.3.7:	Summary of measured δ^2H values *versus* noble gas recharge temperatures for all samples	130
Figure 3.3.8a:	Samples of Group 1	132
Figure 3.3.8b:	Samples of Group 2	133
Figure 3.3.8c:	Samples of Group 3	135
Figure 3.3.8d:	Samples of Groups 4 and 5	137
Figure 3.3.9:	Recharge temperatures calculated from noble gas contents plotted against measured ^{14}C of samples from all groups	138
Figure 3.3.10:	Calculated temperature plotted against measured ^{14}C of samples from the crystalline of Böttstein, Leuggern and the Black Forest	139
Figure 4.2.1:	Relative importance of various sources of isotopes in groundwaters	155
Figure 4.3.1:	Measured ^3H activity in precipitation of Switzerland (SIEGENTHALER AND OTHERS, 1983; LUDIN, 1989). Practically all activity originates from nuclear weapons tests	158
Figure 4.3.2:	Measured ^{85}Kr activity in atmospheric air (ZIMMERMANN AND OTHERS, 1987; LUDIN, 1989). All activity is man-made, mainly originating from nuclear fuel reprocessing plants	158
Figure 4.3.3:	Calculated $^3H/^{85}Kr$ ratios on linear and logarithmic scales based on a piston flow model. Activities are corrected to 31-Dec-1984	159

XXI

Figure 4.3.4: Calculated ^{3}H/^{85}Kr ratio as a function of the mean residence time of an exponential age distribution of water components. Calculation is done for three sampling dates 160

Figure 4.3.5: Calculated ^{85}Kr activity of water *versus* mean residence time based on the exponential model (LUDIN, 1989) 163

Figure 4.3.6: Calculated ^{3}H activity of water *versus* mean residence time based on an exponential model (LUDIN, 1989) 163

Figure 4.4.1: Corrected ^{39}Ar and uncorrected ^{14}C activities of samples from sediments. Measured δ^{13}C values are in parentheses 169

Figure 4.4.2: Corrected ^{39}Ar activity and measured ^{4}He concentrations of samples from sediments 169

Figure 5.1.1: Change of ^{13}C of total dissolved carbonate with dissolution of mineral carbonate 193

Figure 5.1.2: Change of ^{14}C content of total dissolved carbonate with dissolution of mineral carbonate 194

Figure 5.1.3: Change of apparent age calculated from ^{14}C content of total dissolved carbonate with dissolution of mineral carbonate 195

Figure 5.1.4: ^{13}C values plotted against ^{14}C contents of total dissolved carbonate in waters dissolving and reprecipitating mineral carbonate ... 196

Figure 5.1.5: δ^{13}C values plotted against ^{14}C contents as pmc and as apparent ages in waters dissolving and reprecipitating mineral carbonate .. 197

Figure 5.1.6: Cumulative probability of occurrence of δ^{13}C values 200

Figure 5.1.7: Histogram showing the distribution of modelled ^{14}C values 201

Figure 5.3.1: Oxygen isotopic composition of calcite at various depths in the Böttstein borehole 227

Figure 5.3.2: Carbon isotopic composition of calcite at various depths in the Böttstein borehole 227

Figure 5.3.3: Oxygen isotopic composition of calcite at various depths in the Schafisheim borehole 230

Figure 5.3.4: Oxygen isotopic composition of calcite at various depths in the Kaisten borehole 233

Figure 5.3.5:	Carbon isotopic composition of calcite at various depths in the Kaisten borehole	233
Figure 5.3.6:	Oxygen isotopic composition of calcite at various depths in the Leuggern borehole	236
Figure 5.3.7:	Carbon isotopic composition of calcite at various depths in the Leuggern borehole	236
Figure 6.1.1:	Schematic presentation of the various channels for subsurface production of isotopes in rock	239
Figure 6.2.1:	^{36}Cl concentrations *versus* chloride content all samples from Table 6.2.1	256
Figure 6.2.2:	Measured ^{36}Cl/Cl ratio in groundwater *versus* calculated equilibrium ^{36}Cl/Cl ratios in rock	257
Figure 6.2.3:	Measured ^{36}Cl/Cl ratios in water from the Muschelkalk *versus* calculated ^{36}Cl/Cl ratios in rock	260
Figure 6.2.4:	Measured ^{36}Cl/Cl ratios in water from the Buntsandstein *versus* calculated ^{36}Cl/Cl ratios in rock	260
Figure 6.2.5:	Measured ^{36}Cl/Cl ratios in water from the Permian *versus* calculated ^{36}Cl/Cl ratios in rock	262
Figure 6.2.6:	Measured ^{36}Cl/Cl ratios in water from the Crystalline *versus* calculated ^{36}Cl/Cl ratios in rock	262
Figure 6.3.1:	Measured ^{39}Ar concentrations in groundwater *versus* calculated equilibrium concentrations in rock.	269
Figure 6.3.2:	Measured ^{39}Ar concentrations in groundwater *versus* calculated equilibrium concentrations in rock for the Crystalline	269
Figure 6.3.3:	Measured ^{39}Ar concentrations in groundwater *versus* calculated equilibrium concentrations in rock for the Muschelkalk	271
Figure 6.3.4:	Measured ^{39}Ar concentrations in groundwater *versus* calculated equilibrium concentrations in rock for Buntsandstein and Permian samples	271
Figure 6.4.1:	^{3}He/^{4}He ratios and ^{4}He concentrations in Nagra samples. Samples plotted below the line lack ^{3}He data	278

XXIII

Figure 6.4.2:   Measured ^{3}He/^{4}He ratios in water (symbols) compared with the possible range of ^{3}He/^{4}He ratios (boxes) produced in the various formations 279

Figure 6.4.3:   Helium accumulation in a closed system calculated using average *in situ* production rates (Table 6.1.2) and the porosities indicated 280

Figure 6.4.4:   Helium concentrations *versus* chloride concentrations for the Crystalline, Buntsandstein and Permian aquifers 284

Figure 6.4.5:   Helium concentrations *versus* chloride for samples from all units . 286

Figure 6.5.1: ^{40}Ar/^{36}Ar ratios measured in large-volume gas samples *versus* those in water samples 289

Figure 6.5.2:   Columnar sections of deep boreholes with ^{40}Ar/^{36}Ar ratios of dissolved argon 290

Figure 6.5.3: ^{40}Ar/^{36}Ar ratios in water *versus* potassium concentrations in surrounding rock 293

Figure 6.5.4: ^{40}Ar/^{36}Ar ratios in water *versus* dissolved potassium concentrations 293

Figure 6.5.5: ^{40}Ar/^{36}Ar ratios in water *versus* dissolved helium concentrations .. 295

Figure 6.5.6:   Ratios of theoretical production rates of ^{40}Ar and ^{4}He with measured ^{40}Ar/^{36}Ar data and helium concentrations 295

Figure 7.1.1:   Comparison of δ^{18}O results on sulphate samples prepared by Fresenius with those prepared at Orsay 300

Figure 7.1.2: δ^{34}S *versus* δ^{18}O of sulphate dissolved in waters from Tertiary and Jurassic aquifers 302

Figure 7.1.3: δ^{34}S *versus* concentration of sulphate dissolved in waters from Tertiary and Jurassic aquifers 302

Figure 7.1.4: δ^{34}S *versus* δ^{18}O of sulphate minerals and of sulphate dissolved in waters from the Keuper 306

Figure 7.1.5: δ^{34}S *versus* concentration of sulphate dissolved in waters from the Keuper 306

Figure 7.1.6: δ^{34}S *versus* δ^{18}O of sulphate minerals and of sulphate dissolved in waters from the Muschelkalk 308

Figure 7.1.7:	$\delta^{34}S$ *versus* concentration of sulphate dissolved in waters from the Muschelkalk	308
Figure 7.1.8:	$\delta^{34}S$ *versus* $\delta^{18}O$ of sulphate minerals and of sulphate dissolved in waters from the Buntsandstein and Permian	310
Figure 7.1.9:	$\delta^{34}S$ *versus* concentration of sulphate dissolved in waters from the Buntsandstein and Permian	310
Figure 7.1.10:	$\delta^{34}S$ *versus* $\delta^{18}O$ of sulphate minerals and of sulphate dissolved in waters from the Crystalline	314
Figure 7.1.11:	$\delta^{34}S$ *versus* concentration of sulphate dissolved in waters from the Crystalline and low sulphate water from the Buntsandstein and Rotliegend	314
Figure 7.1.12:	$\delta^{18}O$ *versus* concentration of sulphate dissolved in waters from samples from the Crystalline, Buntsandstein and Rotliegend waters	315
Figure 7.2.1:	Plot showing $^{87}Sr/^{86}Sr$ ratio *versus* Sr contents of groundwaters	326
Figure 7.2.2:	Isochron diagram for values from crystalline basement rocks at Böttstein and Leuggern, and from illite vein fill (Leuggern, 1648.83 m)	329
Figure 7.3.1:	The three naturally occurring radioactive decay series	337
Figures 7.3.2 to 7.3.5:	Graphs of uranium content and $^{234}U/^{238}U$ activity ratio of groundwater samples from Nagra deep boreholes	344
Figure 7.3.6:	Location of uranium samples and of transition zone (shaded) between high-uranium, oxidising waters and low-uranium, reducing waters in the Buntsandstein and Crystalline	349
Figure 8.4.1:	Graph of the logarithms of bromide concentrations against those of chloride concentrations in waters from the Buntsandstein, Permian and Crystalline	389
Figure 8.4.2:	$\delta^{34}S$ and $\delta^{18}O$ values of sulphate dissolved in waters from the Buntsandstein, Permian and Crystalline and from minerals from the Permian of the Weiach borehole	389

1. INTRODUCTION

This volume is a report on the isotopic investigations of groundwater in northern Switzerland and adjacent regions carried out since 1981 by Nagra, the Swiss National Cooperative for the Storage of Radioactive Waste. As described in the preface, this study was undertaken to support a programme assessing potential sites for nuclear waste repositories. It is one of the most comprehensive isotope hydrology studies ever undertaken. It includes measurements on a large number of stable and radioisotopes and noble gases, supported by complete water chemical analyses and many rock and mineral analyses. A synthesis and interpretation of the data, along with the data themselves, are given here. The interpretation is an example of the present "state of the art" of isotope hydrology, and the data set will be useful to test new hypotheses and interpretive techniques in isotope hydrology.

1.1 Overview of This Report

The report was prepared by a Working Group comprising Nagra staff, contractors, and consultants. All members of this Group are listed as authors of the full report. Individual sections of Chapters 1 through 7 were prepared by various members of the Working Group, in some cases with assistance from colleagues who were not full members. Each chapter received scrutiny from all members of the Group. The names of the responsible authors are given at the beginning of each section. Sections without authors' names were prepared by the senior author, who also had editorial responsibility for the entire report. Chapter 8, the synthesis, was prepared by the entire group. Finally, the entire report received outside review from A. H. Bath, Fluid Processes Unit, British Geological Survey.

This introductory chapter includes a summary of the scope of the data presented, as well as a brief review of the conventions, units and standards used in reporting isotope data. It also includes overviews of the geology, stratigraphy, hydrogeology and hydrochemistry of northern Switzerland and adjacent regions as a framework within which to present the isotope data.

Chapter 2 assesses the quality of samples on which isotope analyses were performed and discusses the reliability and accuracy of the isotope data themselves. This discussion is based on comparison among analyses made on duplicate samples.

Chapters 3 through 7 present the results of various isotope analytical measurements. Each begins with reviews or brief presentations of the principles underlying the interpretation of the isotope results given. Each also gives conclusions of hydrochemical or hydrogeological significance which can be drawn from the data, as well as pointing out questions which are raised by the data.

Chapter 8 is a synthesis of the conclusions which can be drawn from consideration of the several isotopes in Chapters 3 through 7. The summary focuses on the information isotopes provide on patterns of groundwater flow and on the chemical evolution of groundwaters in northern Switzerland.

Chapter 9 is a summary of the entire report and Chapter 10 provides references to the literature cited.

1.2 Acknowledgments

The authors wish to acknowledge the contributions of a number of individuals to the preparation of this report. For scientific contributions, we thank Andreas Scholtis, Petra Blaser, and Pierre Wexsteen. The report has been much improved by the thoughtful and detailed comments of our colleague Adrian Bath. Finally, the senior author wishes to particularly acknowledge the contribution of Debra Murillo who prepared the various drafts and the final version of this report.

The work reported here was designed by Nagra and their geology consultants and carried out under the project management of Marc Thury (1980-1982), Klaus Gronemeier (1982-1983), Werner Kanz (1983-1986) and Andreas Gautschi (1986-1990). The authors and all others associated with this programme express their appreciation for the continuing scientific interest and administrative support of Nagra management.

1.3 Scope of Data Presented

The concentrations or ratios of a number of stable and radioactive isotopes were measured as part of the Nagra programme. In addition, the concentration of noble gases dissolved in many samples were determined. Table 1.3.1 presents a list of the isotope and gas analyses performed, the substances on which they were made, and the source of the analyses.

Samples of groundwater were taken from a number of locations in northern Switzerland and adjacent regions as part of the Nagra regional programme, as described by SCHMASSMANN AND OTHERS (1984), and groundwater and rock samples were taken as part of the Nagra deep drilling programme. The geology of the Nagra boreholes is described in reports by PETERS AND OTHERS (1986,1988a, and 1988b) on the Böttstein, Kaisten, and Leuggern boreholes, respectively, and by MATTER AND OTHERS (1987,1988a, and 1988b) on the Riniken, Weiach and Schafisheim boreholes, respectively. A description of the collection of water samples from the boreholes and the results of the chemical analyses made on them are given by WITTWER (1986). PEARSON AND OTHERS (1989) adjusted the raw analytical data for the effects of sampling and gives the probable chemistry of the groundwater *in situ*. The investigations carried out in connection with the drilling of the Böttstein borehole were summarized by NAGRA (1985). The chemistry and isotopic properties of water samples from Böttstein are discussed by PEARSON (1985) and BALDERER (1985b). All chemical and isotopic data collected in both the regional and borehole programmes, as well as those published by others on groundwater in northern Switzerland and adjacent regions, are assembled in a Nagra internal report (NAGRA, 1989).

Table 1.3.1: List of analyses made as part of the Nagra programme showing substances analysed, and the source of the analyses.

Determination	Substance	Analysed by
	Stable Isotope Ratios	
$^{2}H/^{1}H$	Water	GSF, Institut für Hydrologie, Neuherberg: W. Rauert and W. Stichler
$^{4}He/^{3}He$	Dissolved helium	University of Cambridge: Prof. K. O'Nions
$^{13}C/^{12}C$	Dissolved carbonate	AERE, Harwell: Dr. R. Otlet; LLC, Bern
	Carbonate minerals	University of Paris South, Orsay: Prof. J. Ch. Fontes
$^{18}O/^{16}O$	Water	GSF, Institut für Hydrologie, Neuherberg: W. Rauert and W. Stichler; LLC, Bern: Dr. U. Siegenthaler
	Carbonate minerals	University of Paris South, Orsay: Prof. J. Ch. Fontes
	Sulphate minerals	ditto
	Dissolved sulphate	ditto, On samples prepared at Orsay and by Institut Fresenius
$^{34}S/^{32}S$	Dissolved sulphate	ditto
		University of Göttingen: Prof. H. Nielsen
	Sulphate minerals	University of Paris South, Orsay: Prof. J. Ch. Fontes
	Dissolved sulphide	ditto
	Sulphide minerals	ditto
$^{40}Ar/^{36}Ar$	Dissolved argon	University of Bath: Prof. J. Andrews
	Radioisotopes	
^{3}H	Water	GSF, Institut für Hydrologie, Neuherberg: W. Rauert and W. Stichler: LLC, Bern
^{14}C	Dissolved carbonate	AERE, Harwell: Dr. R. Otlet; LLC, Bern
		ETH, Zürich-Hönggerberg, on samples prepared at LLC, Bern
^{36}Cl	Dissolved chloride	University of Rochester: Dr. D. Elmore, on samples prepared at University of Paris South, Orsay
	Chloride leached from rock	ETH, Zürich-Hönggerberg, on samples prepared by University of Bern, Abt.f. Isotopengeologie des Min. Petr. Institutes
^{39}Ar	Dissolved argon	LLC, Bern: Prof. Dr. H. H. Loosli
^{85}Kr	Dissolved krypton	ditto
$^{87}Sr/^{86}Sr$	Dissolved strontium	University of Bern, Abt.f. Isotopengeologie des Min. Petr. Institutes
	Minerals	ditto
Uranium and Thorium	Dissolved and Minerals	AERE, Harwell: Dr. M. Ivanovich; PSI, Würenlingen
	Dissolved Gases	
Helium	Dissolved	PSI, Würenlingen: O. Antonsen
		University of Bath: Prof. J. Andrews
		University of Cambridge: Prof. K. O'Nions
		Weizmann Institute, Rehovot: Prof. E. Mazor
Argon	Dissolved	PSI, Würenlingen: O. Antonsen
		University of Bath: Prof. J. Andrews
		Weizmann Institute, Rehovot: Prof. E. Mazor
Neon	Dissolved	University of Bath: Prof. J. Andrews
Krypton		Weizmann Institute, Rehovot: Prof. E. Mazor
Xenon		

The tables in this report give not only isotope and noble gas data on Nagra regional and borehole samples, but also isotope data published by others on groundwaters from Switzerland and adjacent regions.

In this report, sample locations are referred to by name, number or, generally, by both. Numbers are in the form XXX/YYY where XXX refers to the sampling location and YYY, to the sequential number of the sample taken at that location. Location numbers 301 through 307 are the deep boreholes. Sample numbers below 100 were taken as part of the Nagra programme. Samples numbered above 100 were reported by others.

Appendix table A.1, at the end of this report, is a list of the locations sampled. It is arranged in numerical order and includes the full name and community of the sampling point together with the short name used in this report. The geologic formation yielding the sample also appears, along with the Swiss national coordinates of the point, the date or range of dates on which samples were taken, and a reference to the source of data on samples not collected as part of the Nagra programme.

Plate I, also at the end of the report, is a map of the region studied showing the locations of the points sampled.

1.4 Reporting of Isotopic Data

Measurements of the stable isotopic compositions of natural substances are conventionally reported in the δ-notation:

$$\delta_X(\text{per mil}) = \left(\frac{R_X}{R_{\text{standard}}} - 1\right) \cdot 1000 \qquad (1.3.1)$$

where R_X is the isotope ratio ($^2H/^1H$, $^{13}C/^{12}C$ or $^{34}S/^{32}S$ for example) in the substance X, R_{standard} is the isotope ratio in a standard substance, and δ is expressed in parts per thousand (per mil or $^o/oo$).

Differences between the stable isotopic compositions of two substances A and B are given by the fractionation factor:

$$\alpha(A - B) = \frac{R_A}{R_B} \qquad (1.3.2)$$

δ-values for two substances <u>expressed relative to the same standard</u> are related to the fractionation factor, α, by the expression:

$$\alpha_{(A-B)} = \frac{1000 + \delta_A \text{ per mil}}{1000 + \delta_B \text{ per mil}} \qquad (1.3.3)$$

The term $1000 \ln \alpha$ is often used to express fractionation factors. The numerical value of α is close to one. Therefore:

$$1000 \ln \alpha_{(A-B)} \approx \delta_A(^o/oo) - \delta_B(^o/oo) \qquad (1.3.4)$$

The error in this approximation is less than 0.05 per mil when $\delta_A - \delta_B$ is less than ten per mil (FRIEDMAN AND O'NEIL, 1977, Table 1).

Isotope fractionation can also be expressed using the ϵ-notation (MOOK, 1983, 1986; FONTES, 1983), where ϵ is:

$$\epsilon (A\text{-}B) (^o/oo) = (\alpha_{(A-B)} - 1) \cdot 10^{-3} \qquad (1.3.5)$$

Again, because α is close to one,

$$\epsilon (A\text{-}B) (^o/oo) \approx 1000 \ln \alpha_{(A-B)}$$

$$\epsilon (A\text{-}B) (^o/oo) \approx \delta_A(^o/oo) - \delta_B(^o/oo)$$

Stable isotope measurements are reported relative to the following standards:

δ^2H: A water representative of ocean water and known as SMOW (Standard Mean Ocean Water)

$\delta^{13}C$: Calcium carbonate from a cretaceous Belemnite known as PDB

$\delta^{18}O$: The compositions of solid carbonates are usually referred to the carbonate PDB standard as $\delta^{18}O_{PDB}$ values. Water oxygen isotopic compositions are usually reported relative to the water SMOW standard as $\delta^{18}O_{SMOW}$ values. $\delta^{18}O$ values on the two scales are related by the equation (FRIEDMAN AND O'NEIL, 1977, MOOK, 1983):

$$\delta^{18}O_{SMOW} = 1.03086 \, \delta^{18}O_{PDB} + 30.86 \qquad (1.3.6)$$

$\delta^{34}S$: Sulphur from the mineral troilite, FeS, found in the Canyon Diablo meteorite, and indicated CD.

A more complete discussion of isotopes of particular interest in isotope hydrogeology is given in the Introduction to FRITZ AND FONTES (1980).

1.5 Regional Geology and Stratigraphy

A. Matter and Tj. Peters

1.5.1 Introduction

The main structural elements of northern Switzerland are the crystalline basement uplift exposed in the Black Forest (FRG), the Tabular Jura and the Folded Jura, and the northern part of the Molasse Basin (Figure 1.5.1 and Plate II). In the Tabular Jura, a thin sequence of Mesozoic-Tertiary sediments rests relatively undisturbed on the basement, whereas in the Folded Jura detachment of the cover took place along evaporitic units of the Muschelkalk (Figure 1.5.3).

The crystalline basement surface and Mesozoic cover dip southward beneath the Molasse Basin but are also exposed in the Alps (Aar Massif, see Figure 1.5.2). In the last 50 years, several ENE-trending Late Paleozoic troughs have been discovered in the subsurface of the Alpine foreland (Figure 1.5.3). One of these is the North-Swiss Permo-Carboniferous Trough, located just to the north of the Folded Jura. Nagra's geophysical surveys and drilling activities found this trough to be several kilometres deep but only 10 to 12 km wide, crossing the study area from WSW to ENE (SPRECHER AND MÜLLER, 1986; MATTER, 1987). The bounding faults of the trough were reactivated in the Palaeogene, and strongly influenced deformation of the Folded Jura, conditioning location of thrusts and folds developed during the Neogene compressional phase (LAUBSCHER, 1986,1987).

The study area is characterized by a fault pattern consisting of reactivated ENE faults and NNE trending ("Rhenish") faults related to the subsidence of the Rhine Graben in Eocene-Oligocene times. These faults affect both the basement and the Mesozoic cover. The continuity of aquifers may be affected by faulting; however, the faults may possibly represent major groundwater flow paths.

A schematic stratigraphic section emphasizing units of hydrogeological significance is given in Figure 1.5.4. Plate III shows the stratigraphy of the Nagra deep boreholes.

1.5.2 Crystalline Basement

Drilling for thermal waters, coal, oil, and, in particular, the Nagra deep drilling programme have shown that the crystalline basement of northern Switzerland is very similar to the exposed rocks of the neighbouring Black Forest. Petrographically, the basement consists of high grade metamorphic gneisses with intrusions of Variscan granites and syenites. Most of the gneisses have sedimentary (Precambrian?) protoliths; rare, coarse-grained orthogneisses represent older granites. Sillimanite-biotite gneisses, sillimanite-cordierite gneisses, cordierite-bearing tonalitic leucosomes, biotite-plagioclase gneisses with minor intercalations of banded amphibolites, and calc-silicate layers all date from Caledonian upper amphibolite facies metamorphism. The Variscan granites include: biotite-granites with large K-feldspar crystals (similar to the Albtal Granites); cordierite

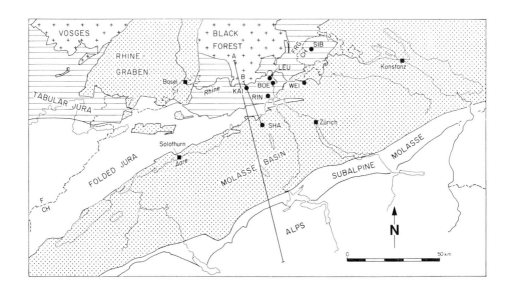

Figure 1.5.1: Map showing major structural units of northern Switzerland and adjacent areas and location of Nagra boreholes. A: Cross-section Figure 1.5.2, B: Cross-section Figure 1.5.3.

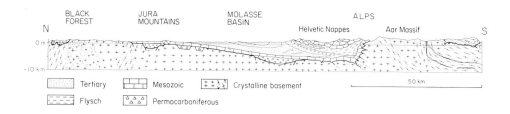

Figure 1.5.2: Geological cross section through central and northern Switzerland (from THURY AND DIEBOLD, 1987).

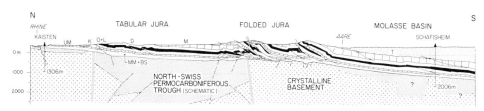

T= Tertiary, M= Malm, D= Dogger, O+L= Opalinus-Ton and Liassic, K= Keuper, UM= Upper Muschelkalk, MM-BS= Middle Muschelkalk to Buntsandstein.

Figure 1.5.3: Geological cross section showing Permo-Carboniferous Troughs, Tabular and Folded Jura, and the edge of the Molasse Basin (from DIEBOLD, 1986).

and andalusite-bearing two-mica granites (similar to the Säckingen Granite) and cordierite granites rich in gneissic xenoliths. A basic suite of biotite-rich syenite, monzonite and diorite was encountered in the Schafisheim borehole (PETERS, 1987).

The crystalline basement underwent several stages of deformation (MEYER, 1987) and hydrothermal alteration (PETERS, 1987). Deformation before granite intrusion resulted in the formation of ductile shear zones. Cataclastic deformation zones were formed during several stages of the history of the basement:

- As a result of granite intrusion in the gneisses;

- During vertical movements associated with Permo-Carboniferous basin subsidence and basement uplift;

- During the Early Permian as a result of extensional tectonics;

- During the uplift of the Black Forest and formation of the Rhine Graben in the Eocene-Oligocene and folding of the Jura and uplift of the Black Forest in the Late Tertiary.

The intensity of cataclastic deformation varies considerably from very low, in the drillholes in the Black Forest (< 3 per cent), to low, in the granites of Leuggern and Böttstein (5 per cent), and high, in the granite at Schafisheim (90 per cent) and in the gneisses (10 to 20 per cent). The cataclastic deformation zones seldom reach 1 m in width, and show sharp margins in the granites and aplites, but gradational margins with undeformed gneissic and syenitic host rocks. The abundance of fractures varies from four per metre in most drill cores to 40 per metre in Schafisheim. Only a few of the fractures (1 to 4 per cent) are open.

Hydrothermal alteration took place:

- During cooling after the peak of metamorphism and subsequent to the granite intrusions;

- During and after the deformation phases, when convective fluids circulated mainly along the cataclastic zones. Parts of the basement were severely altered with replacement by clay minerals leading to the development of diffuse microporosity.

Fluid inclusion studies (MULLIS, 1987) on calcite/quartz veins indicate that during the whole Mesozoic and the Early Tertiary extremely saline $NaCl(CaCl_2)$ hot solutions (up to 150°C) were present in the basement down to depths of more than 1500 m. Only after the Jura Folding in the Late Tertiary did the salinity of circulating solutions diminish. Deep groundwater flow paths occur mainly along open fractures, irregular solution cavities, cataclastic deformation zones and quartz-rich hydrothermal veins. Open fractures are present throughout the granites but in the mica-rich gneisses and syenites they are restricted to the quartz-rich, competent rocks such as aplites and pegmatites.

Calcite, quartz, barite, fluorite, and hematite are among the minerals growing into open fractures. They show little or no corroded surfaces and might be in equilibrium with modern deep groundwaters. Hematite is mainly restricted to the upper few hundred metres of the basement but is observed in some fractures at much deeper levels.

1.5.3 Sedimentary Cover

The sedimentary sequence ranges in age from Late Carboniferous (Stephanian) to Quaternary. The Permo-Carboniferous consists of a clastic succession up to several kilometres thick which was laid down in a series of ENE-trending troughs. These Late Paleozoic sediments are unconformably overlain by approximately 800 m of Mesozoic strata, mainly carbonates and shales, around 400 m of Tertiary Molasse, and a veneer of Quaternary moraine and associated fluvio-glacial sediments of highly variable thickness. The thickness given for the various stratigraphic units in this section refer to an area between the Rhine River and a line linking Konstanz, Zurich, Solothurn and Basel. Note that these thicknesses will not apply to the Rhine Graben.

The Upper Carboniferous was found only in the Weiach borehole, where it measures 572 m in thickness (Plate III). It is characterized by fluvial fining-upward cycles with breccias, dark grey cross-bedded sandstones, siltstones and bituminous mudstones. Numerous thin coal seams, and one seam approximately 6 m in thickness, were also found, as well as several thin tuffaceous ash beds (MATTER AND OTHERS, 1988a; MATTER, 1987).

A 135 m thick Lacustrine Series (Lower Rotliegend) of Early Permian (Autunian) age, comprising conglomerates, sandstones, siltstones and bituminous shales, rests unconformably upon the Carboniferous fluvial units at Weiach. The Lacustrine Series is overlain by the Lower Fanglomerate which is almost 200 m thick. The overlying Playa Mudstones are widespread in the North-Swiss Permo-Carboniferous Trough, but the overlying Upper Fanglomerate was encountered only at the northern and southern flank of the trough (MATTER, 1987). This sequence documents a change from a phase of tectonic quiescence when the hinterland was peneplained and the graben filled with mudstones, to a phase of tectonic rejuvenation and renewed deposition of coarse clastics of the Upper Rotliegend. No marine sediments are present. Therefore, it cannot be excluded that the uppermost part of the Rotliegend redbeds form a continental facies equivalent to the Zechstein, and may actually be of Late Permian age. In this report Rotliegend and Permian are used interchangeably.

The Buntsandstein rests unconformably upon crystalline basement or on Permian sediments. It was not involved in the Jura folding. The Buntsandstein decreases in thickness from around 100 m near Basel to zero in the southeastern part of the study area. It comprises a series of red and white sandstones and conglomerates with variegated jasper horizons of pedogenic origin. Total porosity values range up to 20 per cent, but as a result of different degrees of cementation and dissolution, the exact values vary widely from layer to layer at each locality. However, in the Siblingen well (307), the Buntsandstein is represented by a sequence of well-cemented and tight white sandstones. In this well, groundwater movement in the Buntsandstein occurs only along fractures, whereas in the other wells, flow also takes place through the pores.

The Lower Muschelkalk measures 10 to 50 m and consists, from base to top, of marine dolomites, a sequence of calcareous mudstones, and bituminous dolomitic mudstones. The Middle Muschelkalk, or "Anhydrite Group" (50 to 160 m), is dominated by evaporitic facies, with anhydrite at depth and gypsum at or near the surface. A rock salt layer of variable thickness (0 to 100 m) also occurs within this sequence. Near the edge of the salt basin, as in the Weiach borehole, the rock salt is absent due to leaching. The evaporitic sequence is capped by about 10 m of porous dolomite.

The Upper Muschelkalk (50 to 80 m) consists of marine limestones which are overlain by a bituminous dolomite (Trigonodus-Dolomit). The latter increases in thickness eastwards from about 20 m near Basel to 38 m in Weiach. The Upper Muschelkalk limestones show a corresponding decrease in thickness, reflecting dolomitization to progressively deeper levels. In the Nagra boreholes the dolomite is highly porous (up to 25 per cent open porosity) with drusy cavities up to 5 cm occurring in addition to the intercrystalline microporosity. As a result of telogenetic dissolution of anhydrite and carbonate, the porosity is higher in the Böttstein and Leuggern wells than in Weiach and Schafisheim, where the Trigonodus Dolomit is more deeply buried. Dissolution vugs are also frequently present in the underlying less porous limestones. The pores and druses are frequently lined, mainly with calcite. Large calcite and anhydrite crystals were also found in open fractures of the Upper Muschelkalk limestones in the Schafisheim borehole. The highly porous, fractured and karstified Upper Muschelkalk carbonates are an important regional aquifer.

The Keuper sequence (100 to 190 m in thickness) is dominated by evaporites, variegated mudstones, dark grey mudstones and sandstones. The Lettenkohle of the Lower Keuper (4 to 8 m) consists of beige porous dolomite beds separated by the thinner beds of the greenish to dark grey Esteria Shales. The facies of the Lettenkohle may be grouped with those of the Upper Muschelkalk, and also provide part of the Muschelkalk aquifer. The porosity of the Lettenkohle dolomites stems from leaching of gypsum or anhydrite nodules and from dissolution of bioclasts.

The Middle Keuper is mostly represented by the thick Gipskeuper (80 to 130 m), a heterogeneous formation comprising anhydrite or gypsum and dark grey shale and dolomitic marls. The Gipskeuper, especially its clayey upper part, is strongly veined with horizontal late satin spar (fibrous gypsum) veins and subvertical anhydritized early gypsum veins (DRONKERT AND OTHERS, 1989). The Gipskeuper is overlain by the Schilfsandstein, which represents a fluvial channel facies, and by the associated overbank deposits of the Untere Bunte Mergel, which mainly consist of variegated dolomitic marls. The total thickness varies from 10 to 30 m. A further sequence of variegated marls (Obere bunte Mergel) is separated from their lower marls by a .3 to 9 m thick dolomite (Gansinger Dolomit). The Upper Keuper (Rhaetian) is missing in that part of the study area east of a line between Rheinfelden and Olten due to either non-deposition as a result of marked Upper Keuper regression or to erosion during the subsequent Liassic transgression. The Upper Keuper is characterized by rapid lateral facies changes and variation in thickness from 0 to 10 m. Dark grey shales, fine grained quartz sandstones and bonebeds are the dominant lithologies.

The Liassic (15 to 50 m) consists of thick argillaceous intervals including the Posidonia Shales, with thin intercalated nodular and well-bedded limestones, and a few ferruginous oolites, which represent condensed horizons. Although generally well cemented, the Lower Liassic Arietenkalk is porous in some areas due to the occurrence of fractures and the presence of secondary porosity.

The Dogger section consists of several shallowing-upward regressive cycles which together form a shallowing-upward megacycle beginning with the Opalinus-Ton and ending with shallow water carbonates (BLÄSI, 1987). The Aalenian Opalinus-Ton (70 to 120 m thickness) is present over the entire study area. It consists of a monotonous sequence of dark grey, silty, micaceous clays, and is overlain by a complex Lower Bajocian succession of sandy limestones, shales and oolitic ironstone. The overlying Middle to Upper sequence is developed in two facies in the study area: the Celtic Platform facies in the northwest, where there are 50 to 80 m thick accumulations of oolites (Hauptrogenstein); and the Swabian Basin facies to the northeast, where mudrocks (*e.g.,* Parkinsoni Schichten, 10 to 60 m) dominate with only a few oolitic ironstone marker beds. The change in facies occurs just west of the lower reaches of the Arve river. Fracture porosity is developed in the competent strata of the Hauptrogenstein. This unit forms a regional aquifer. The Bathonian comprises up to *c.* 30 m of crinoidal limestones, marls and ferruginous oolites. During the Lower Callovian, crinoidal limestones (Dalle nacrée) with a thickness of *c.* 10 m were deposited in the western part of the study area. In the east, the Lower Callovian is represented by a condensed facies. This facies consisting of several dm- to 1 m thick oolitic ironstones and ferruginous shales extended over the entire study area in the Middle to Upper Callovian.

A lateral facies change occurs in the Oxfordian similar to that observed in the Bajocian. However, the edge of the platform had now retreated some 20 km to the northwest. Shallow water carbonates, including oolites and biohermal limestones, were deposited on the platform to the northwest while basinal argillaceous limestones and calcareous shales (Effinger Schichten) ranging from *c.* 30 to 215 m developed in the southeast.

The Kimmeridgian is represented by a platform facies in the northwest with massive and bedded micrites and a basinal facies consisting of marls, bedded limestones and glauconitic limestones in the southeast. This facies change is accompanied by a marked increase in thickness from 130 m to about 300 m in the southeastern part of the study area. The Kimmeridgian is absent in the northwestern part of the study area due to pre-Eocene karst erosion. Continental Eocene deposits comprise pisolitic iron ores, quartz sands and fire clays. These sediments partly fill karstic cavities and pockets. Karstic fissures filled with Eocene sediments are found in the Kimmeridgian limestones often several hundred m beneath the karst surface.

The Oligocene to Miocene Molasse rests upon the Jurassic carbonates, overlying a major unconformity below which part of the Upper Jurassic, the Cretaceous to Palaeocene, and locally the Eocene, are absent. Three of the Molasse units (Lower Freshwater Molasse, Upper Marine Molasse and Upper Freshwater Molasse) are present in the southern portion of the study area (*i.e.,* the northern margin of the Molasse Basin). As a result of the progressive northward progradation of clastic sedimentation onto the Jura Platform, successively younger Molasse Formations pinch-out in the study area. In addition, the thickness of each formation decreases rapidly across the area towards the northern basin margin.

The Lower Marine Molasse is absent in northern Switzerland due to non-deposition, except in the southern Rhine Graben. In the study area, the Lower Freshwater Molasse (0 to 1000 m) and the Upper Freshwater Molasse (0 to 500 m) consist mainly of fluvial sand bodies interbedded with thick overbank marls. Conglomerate horizons are present in the Upper Marine Molasse and the Upper Freshwater Molasse. The Upper Marine Molasse occurs between the two Freshwater Molasse units and is from 0 to 400 m thick in the study area. The Upper Marine Molasse consists of a fairly monotonous sequence of mainly medium-grained glauconitic and bioclastic sandstones. These sandstones are moderately cemented and fairly porous, and form an important aquifer.

The present landscape was, to a large extent, formed after post-molassic uplift and is the product of the dissolution of limestone and of the polyphase Pleistocene glaciation. Glaciers and rivers eroded deep valleys and transported debris, depositing glacial and fluvial sediments in the study area. Quaternary sediments include a veneer of lodgement till and drift as well as gravel terraces and fluvio-glacial and lacustrine valley fills up to several hundred metres thick. These sediments provide important local aquifers which are commonly exploited for water supply.

1.6 Hydrogeology and Hydrochemistry

M. Kullin and H. Schmassmann

1.6.1 General Situation

Figure 1.5.4 is a schematic stratigraphic and hydrogeological profile including the aquifers of northern Switzerland. These aquifers belong to three main stratigraphic groups, namely the Quaternary deposits, the Tertiary to Permo-Carboniferous bedrock sediments, and the crystalline basement. A full understanding of the complexity of the flow systems can only be achieved by understanding each single aquifer group and its interaction with the others.

a) The Quaternary consists mostly of unconsolidated sediments, is present over much of northern Switzerland and contains groundwaters. Several of man's activities, including pumping for water supplies, have modified the original flow systems of these groundwaters.

b) The bedrock Tertiary to Permo-Carboniferous sediments are generally less permeable than the Quaternary sediments, but contain several aquifers. These sediments also feed some thermal and mineral springs. The main aquifer groups are:

- Tertiary-Malm group with important aquifers in the sandstones of the Upper Marine Molasse and in the limestones of the Malm;

- Upper Muschelkalk (limestones and dolomites);

- Lower Triassic-Permian group with aquifers consisting of clastic sediments.

These main aquifer groups are generally separated by hydraulic barriers consisting of clays, marls, anhydrites, and even rock salt. Some spatially limited aquifers within these regional hydraulic barriers are present, *e.g.*, the Hauptrogenstein limestone in the western facies of the Middle Dogger and in some limestones, dolomites and sandstones of the Lias and the upper part of the Keuper.

c) The first few hundred metres of the crystalline basement exhibit a high hydraulic conductivity similar to that of some of the sedimentary aquifers. The groundwaters of the crystalline circulate mainly along tectonic fracture zones. In parts of northern Switzerland they are connected with the Lower Triassic-Permian aquifer group. In other parts of the region they are separated from the overlying sedimentary aquifers by a hydraulic barrier of sediments of the Permo-Carboniferous Trough.

Stratigraphy		Lithology	Thickness (m)	Lithological description	Aquitard / Aquifer	Hydrogeological Characterisation
QUATERNARY				Moraine, fluvio-glacial gravels and sands, lacustrine clays		Aquifer important for water supply, locally with low permeability beds
TERTIARY	Upper Freshwater Molasse (OSM)		0-500	Channel sandstones, marls and conglomerates		Various water conducting sandstone channels and layers
	Upper Marine Molasse (OMM)		0-400	Sandstones, glauconitic and bioclastic		Regional aquifer
	Lower Freshwater Molasse (USM)		0-1000	Channel sandstones and variegated marls		Various water conducting sandstone channels and layers
	Eocene					
MALM	Kimmeridgian		0-300	Micritic limestones, massive to well-bedded		Regional aquifer
	Oxfordian	Effinger Schichten	80-200	Coralline limestone, oolites / Alternation of argillaceous limestones and calcareous shales		Effinger Schichten: low permeability / Local aquifer in the western Jura ("Rauracian")
DOGGER	Bath. - Callov.		0-55	Bioclastic limestones, marls, Fe-oolites		Parkinsoni-Schichten: low permeability / Hr: local aquifer
	Bajocian	Parkinsoni-Schichten / Hr	Hr:50-80 / P:10-60	Oolites (Hr), Mudrocks (P)		
			15-65	Sandy, bioclastic limestones, shales, Fe-oolite		Low to very low permeability
	Aalenian	Opalinus-Ton	70-120	Monotonous sequence of dark grey silty, micaceous clays		
LIASSIC			15-50	Bioclastic limestones, sandy shales		Low permeability rocks with local aquifers
KEUPER		Gd / Sh	15-50	Variegated marls, dolomite variegated sandstone		
		Gipskeuper	80-130	Alternation of shales, nodular and bedded gypsum/ anhydrite Satin spar veins		Very low permeability
		Lk	4-8			
MUSCHELKALK	Upper	TD	50-80	Dolomite, porous / Limestones, bedded		Regional aquifer
	Middle		50-160	Dolomite, laminated / Alternation of shales, bedded and massive anhydrite / Rock salt		Very low permeability
	Lower		10-50	Silty mudstones, shales		
BUNTSANDSTEIN			0-100	Sandstone, porous to well-cemented		Regional aquifer
PERMO-CARBONIFEROUS / CRYSTALLINE BASEMENT				Permian (Rotliegendes): Red siltstones, sandstones and breccias / Carboniferous: Sandstones, siltstones, bituminous shales, breccias, coal seams / Basement: Gneisses with Variscan granite and syenite intrusions		Permo-Carboniferous: water conducting detrital layers / Crystalline basement: water conducting faults and fracture zones

Figure 1.5.4: Schematic stratigraphic section showing hydrogeological characteristics. (Hr = Hauptrogenstein, Gd = Gansinger Dolomit, Sh = Schilfsandstein, Lk = Lettenkohle, TD = Trigonodus-Dolomit).

The dominantly horizontal bedding of the sediment cover generally retards downflow of higher groundwater to depth and *vice versa*. Therefore, the deep groundwaters of the different aquifer groups exhibit individual characters. However, tectonic events may have produced hydraulic connections between individual aquifers.

Hydraulic conductivity data for the various formations are given in NAGRA (1985, Chapter 3.3.6.2) and in NAGRA (1988).

1.6.2   Data Base

This report is focused on deep groundwaters which are characterized by tritium activities generally below about 1 tritium unit (TU). Some deep groundwaters may have up to 20 TU, and mixtures between deep and shallow groundwaters may exceed 20 TU in a few cases. In this report the term "deep groundwater" refers not to the sampling depth but to the origin of the water. A number of ascending groundwaters were sampled near the surface.

Tables 1.6.1 through 1.6.5 list all deep and mixed groundwaters with isotope data from northern Switzerland and adjacent areas. The samples are grouped according to aquifers. In these tables the characteristic hydrochemical features, including total mineralization, chloride concentration and water type, are presented. Total mineralization is given in grams per litre (g/l) as well as in milliequivalents per litre (meq/l). The former quantity is the sum of the analysed dissolved constituents, and is roughly equivalent to the residues on evaporation used in subsequent chapters. The latter quantity is the arithmetic mean value of the measured cation and anion concentrations.

The description of the water type includes those ions which occur in concentrations higher than ten meq-per cent. The ions are listed in order of decreasing concentrations. Those with concentrations higher than 50 meq-per cent are underlined and those below 20 meq-per cent are given in parenthesis (JÄCKLI, 1970, p. 415; SCHMASSMANN AND OTHERS, 1984, p. 78). Analyses which are not from the Nagra programme are designated with an asterisk (*).

In the following sections the individual aquifers will be discussed. A location map of the individual sampling points is given in Plate I.

1.6.3   Groundwaters in Tertiary and Upper Malm Aquifers

In the Molasse Basin, hydrochemical and preliminary isotope results allowed a distinction of three main water types (Table 1.6.1, Figure 1.6.1) which are positioned one upon another (SCHMASSMANN AND OTHERS, 1984, Chapter 6.1; NAGRA, 1988, p. 104).

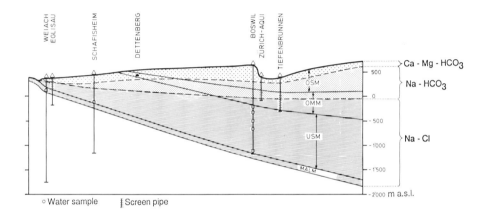

Figure 1.6.1: Distribution of water types in the Molasse Basin in northern Switzerland (from NAGRA, 1988, p. 104).

Table 1.6.1: Chemistry of deep groundwaters from the Malm and Tertiary aquifers.

Location Name	Number	Aquifer	Depth Range (m)	No. of Samples	Cl⁻ meq/l	Total mineralization meq/l	Total mineralization g/l	Water Type
Sauldorf	502*	Tertiary (OMM)	36 - 54	1	0.1	5.2	0.4	Ca-Mg-$\underline{HCO_3}$
Beuren F	149	Malm	30 - 60	2	0.3±0.0	6.7±0.0	0.5±0.0	Ca-Mg-$\underline{HCO_3}$
Dettenberg	170	Tertiary (OMM)	Tunnel Spring		<0.1	7.2±0.0	0.6±0.0	Ca-Mg-Na-$\underline{HCO_3}$-SO_4
Birnau	508*	Tertiary (OMM)	25 -150	1	0.3	4.4	0.4	Na-$\underline{HCO_3}$
Ravensburg	509*	Tertiary (OMM)	376.0-532.5	1	0.0	7.5	0.6	Na-$\underline{HCO_3}$
Konstanz	98	Tertiary (OMM)	500 -625	1	0.3	8.4	0.6	Na-$\underline{HCO_3}$-(SO_4)
Mainau	128	Tertiary (OMM)		1	0.1	12.7	1.0	Na-$\underline{HCO_3}$-SO_4
Zürich Aqui	1	Tertiary (OMM)		1	4.5	14.7	1.0	Na-$\underline{HCO_3}$-Cl-SO_4
Lottstetten	127	Malm	280 -500	6	1.6±0.0	12.5±0.0	1.0±0.0	Na-$\underline{HCO_3}$-(SO_4)-(Cl)
Singen	97	Malm	317 -387	1	3.2	19.9	1.5	Na-SO_4-$\underline{HCO_3}$-(Cl)
Zürich								
Tiefenbrunnen	2	Tertiary (OMM)	417 -685	2	32.9±0.2	56.5±0.1	3.6±0.0	Na-Cl-SO_4
Eglisau 1	141	Tert.(USM)-Malm	330 -726	1	23.4	36.7	2.4	Na-Cl-SO_4-(HCO_3)
Eglisau 2	142	Tertiary (USM)	10 -261.5	6	37.0±1.4	45.2±0.7	2.8±0.0	Na-Cl
Eglisau 4	144	Tertiary (USM)	60 -185	6	48.9±4.7	57.0±3.3	3.6±0.3	Na-Cl
Eglisau 3	143	Tertiary (USM)	58 -207.2	6	56.7±1.6	68.0±1.6	4.1±0.1	Na-Cl
Schafisheim	304/1,2	Tertiary (USM)	553 -563	2	146.9±0.4	149.4±0.1	8.8±0.1	Na-Cl
Weiach	302/5,6,8	Malm	242.9-267.0	3	107.1±0.8	115.5±1.6	2.3±0.0	Na-Cl
Rhine Graben								
Oberbergen	96	Tert. Volcanite	42 - 67	2	0.1±0.0	5.2±0.0	0.5±0.0	Ca-Mg-$\underline{HCO_3}$-(SO_4)
Schönenbuch	5	Tertiary (USM)		2	0.1±0.0	7.6±0.0	0.6±0.0	Mg-Ca-$\underline{HCO_3}$

Groundwaters from depths generally of less than 25 m and never more than 100 m are of the calcium-magnesium-bicarbonate type, and are chemically clearly distinct from the underlying water types. Commonly, the tritium contents indicate residence times of less than 35 years for these shallow groundwaters (SCHMASSMANN AND OTHERS, 1984, Chapter 6.1; NAGRA, 1988, p. 105).

In the Molasse Basin, sodium bicarbonate deep groundwaters are widely distributed at depths from 25 to 625 m. Their sodium source is extensive ion exchange rather than Na/Cl dissolution as demonstrated by Na/Cl ratios greater than one. These waters have tritium contents below the detection limit, increased helium concentrations, and indications from stable isotopes and noble gases contents of water infiltration during a colder geological period (SCHMASSMANN AND OTHERS, 1984, p. 120; BERTLEFF, 1986, p. 148; NAGRA, 1988, p. 107).

Sodium chloride groundwaters occur below the sodium bicarbonate waters, or where these are lacking at the northern border of the Molasse Basin (Figure 1.6.1), directly below calcium-magnesium-bicarbonate waters, *e.g.*, Eglisau. As revealed by several boreholes in the central and northeastern Swiss Molasse Basin, the upper limit of the sodium chloride water forms a slope inclined to the south-southeast (NAGRA, 1988, Figure 4.24).

The sodium chloride waters in the Molasse Basin contain parts that are older than the sodium carbonate waters. Particular chemical features can be used to distinguish Molasse waters with high chloride concentrations from those in other aquifers in northern Switzerland. For example, Molasse waters contain an increased iodine content which goes up to 8 to 10 mg/l in Weiach (302/5, 6, 8) and Schafisheim (304/1, 2) or higher bromide/-chloride and lower lithium/sodium ratios. (SCHMASSMANN AND OTHERS, 1984, p. 78, 127, 128, Beil. 12; NAGRA, 1985, p. 3.36). The sodium chloride content of the water in the Tertiary and Malm aquifers is attributed to connate marine waters which were later mixed with infiltrated meteoric waters.

Deep groundwaters of a different chemical type are known from the Tertiary of the southern Upper Rhine Graben. These include the alkaline-earth bicarbonate water with a rare magnesium dominance as encountered in Schönenbuch (5).

The waters of the Tertiary volcanics of the Kaiserstuhl as sampled at Oberbergen (96) constitute another unusual type.

1.6.4   Groundwaters in the Dogger, Lias and Keuper Aquifers

A number of analyses from deep groundwaters of the Hauptrogenstein aquifer in the Dogger are available from the Nagra regional programme (Table 1.6.2), but all samples are from the Upper Rhine Graben and its marginal fault blocks, and therefore can not directly be applied to other regions.

In the marginal fault blocks of the Rhine Graben, subthermal alkaline-earth bicarbonate groundwaters of low mineralization were sampled at Munzingen (109), Riedlingen Subtherme (106) and Müllheim (108). In the Rhine Graben system, the waters of Bad

Table 1.6.2: Chemistry of deep groundwaters from the Dogger, Lias, and Keuper aquifers.

Location Name	Number	Aquifer	Depth Range (m)	No. of Samples	Cl^- meq/l	Total mineralization meq/l	Total mineralization g/l	Water Type
Weissenstein-tunnel 1.480	56	Lias	Tunnel Spring	1	0.8	16.5	1.3	$\underline{Na}$-Ca-(Mg)-$\underline{HCO_3}$-SO_4
Beznau 104.5-109.5 m	171*	Keuper	105- 110	1	78.1	223.0	14.7	$\underline{Na}$-(Ca)-$\underline{SO_4}$-Cl
Riniken	303/1	Keuper	501- 530	1	78.3	234.2	15.5	$\underline{Na}$-(Ca)-$\underline{SO_4}$-Cl
Rhine Graben								
Munzingen	109	Dogger	Spring	1	0.4	5.0	0.4	$\underline{Ca}$-Mg-$\underline{HCO_3}$
Riedlingen Subtherme	106	Dogger	Spring	1	0.1	5.3	0.4	$\underline{Ca}$-Mg-$\underline{HCO_3}$
Müllheim	108	Dogger	Spring	1	1.2	9.4	0.7	$\underline{Ca}$-(Mg)-(Na)-$\underline{HCO_3}$-(Cl)-(SO_4)
Neuwiller	77	Dogger	974-1020	1	2.3	11.1	0.9	$\underline{Na}$-(Ca)-$\underline{HCO_3}$-Cl-(SO_4)
Bellingen 3	83	Dogger	539- 648	1	46.1	63.7	4.0	$\underline{Na}$-Ca-(Mg)-$\underline{Cl}$-HCO_3
Bellingen 1	81	Dogger	537- 643	1	55.3	73.0	4.6	$\underline{Na}$-Ca-(Mg)-$\underline{Cl}$-HCO_3
Steinenstadt Thermal	84	Dogger	390- 473	2	51.5±0.1	73.8±0.1	4.7±0.0	$\underline{Na}$-Ca-(Mg)-$\underline{Cl}$-(HCO_3)

Bellingen (81 and 83) and Steinenstadt (84) constitute a special case because their chemical properties are influenced by nearby Tertiary evaporites and by carbon dioxide ascending along tectonic faults (NAGRA, 1988, p. 117).

The borehole of Neuwiller (77) encountered a sodium bicarbonate water in a fault zone in the Hauptrogenstein in the southern Upper Rhine Graben. This water can be related hydrochemically to the waters of the same type in the Tertiary and Malm aquifers of the Molasse Basin.

The high Na/Cl ratio of deep groundwaters from the Lias and Keuper suggest that ion exchange processes have occurred. This, and the high total mineralization of the Beznau (171) and Riniken (303/1) samples, is consistent with the occurrence of these waters in limited aquifers enclosed by thick aquicludes. Waters with total mineralization of up to 77 g/l and chloride concentrations up to 1280 meq/l were reported from oil boreholes drilled in the Molasse Basin. It appears very unlikely that such highly saline waters are part of an active flow system (NAGRA, 1988, p. 118).

Except for restricted local aquifers, the formations between the Malm limestones and the Muschelkalk are considered to form a single hydraulic barrier. The Opalinus-clay as well as the Gipskeuper are formations with very low permeabilities. No indications could be found that waters from such deeper aquifers as the Muschelkalk have ever migrated to the Malm or Tertiary.

1.6.5 Groundwaters in the Upper Muschelkalk Aquifer

The limestones and dolomites of the upper Muschelkalk constitute the most important deep aquifer in northern Switzerland and adjacent areas. It has, therefore, been characterized by an extensive set of analyses. Table 1.6.3 lists the samples from this aquifer and illustrates the different prevailing water types.

In the Folded Jura, as well as in the western Tabular Jura, alkaline-earth bicarbonate sulphate and alkaline-earth sulphate bicarbonate deep groundwaters with low chloride concentrations were sampled at Lostorf 3 (123), Densbüren (14), Frenkendorf (9), and Magden (271 to 274). In this region, sodium chloride deep groundwaters and mixtures with them were also observed locally. The Pratteln (8) sample is the only one of this type discussed in this report.

In the eastern Tabular Jura, sodium alkaline-earth sulphate chloride to sodium chloride sulphate deep groundwaters were found in boreholes at Böttstein (301/2), Beznau (15) and Riniken (303/2). These waters have chloride concentrations from 35 to 170 meq/l and higher helium concentrations than waters from the western Tabular Jura (SCHMASSMANN AND OTHERS, 1984, p. 233, 319).

Along the northern border of the Molasse Basin, discharge of upper Muschelkalk waters with typically increased hydrogen sulphide concentrations (0.35 to 200 mg H_2S/l) are indicated by the thermal springs of Baden (201 to 221), Schinznach Bad (119, 120) and Lostorf 1 (121). Comparison of the ascending sodium chloride water of Lostorf 1 with

Table 1.6.3: Chemistry of deep groundwaters from the Upper Muschelkalk aquifer.

Name Location	Number	Aquifer	Depth Range (m)	No. of Samples	Cl$^-$ meq/l	Total mineralization meq/l	Total mineralization g/l	Water Type
Densbüren Felsbohrung	14	Muschelkalk	77 - 125	2	0.1±0.0	10.8±0.7	0.8±0.1	Ca-Mg-$\underline{SO_4}$-HCO$_3$
Lostorf 3	123	Muschelkalk	530 - 582.8	6	0.1±0.0	12.0±0.1	0.9±0.0	Ca-Mg-$\underline{SO_4}$-HCO$_3$
Frenkendorf	9	Muschelkalk	232 - 308	1	0.2	10.1	0.8	Ca-Mg-$\underline{HCO_3}$-SO$_4$
Magden Stockacher	274	Muschelkalk	200.2 - 272.8	1	0.1	10.4	0.8	Ca-Mg-$\underline{HCO_3}$-SO$_4$
Magden Falke	272	Muschelkalk	163 - 235	2	0.2±0.0	11.2±0.2	0.8±0.0	Ca-Mg-$\underline{SO_4}$-HCO$_3$
Magden Eich	273	Muschelkalk	191 - 248	2	0.3±0.0	13.6±0.1	1.0±0.0	Ca-Mg-$\underline{SO_4}$-HCO$_3$
Magden Weiere	271	Muschelkalk	208.7 - 282.0	3	1.6±0.1	20.1±0.6	1.4±0.1	Ca-Mg-Na-$\underline{SO_4}$-HCO$_3$
Pratteln	8	Muschelkalk	55.5 - 124.0	1	2.8	32.4	2.2	Ca-Mg-$\underline{SO_4}$-(HCO$_3$)
Leuggern	306/1	Muschelkalk	53.5 - 96.4	1	1.0	16.6	1.2	Ca-Mg-$\underline{SO_4}$-(HCO$_3$)
Böttstein	301/2	Muschelkalk	123.2 - 202.5	1	36.7	97.8	6.4	Na-Ca-(Mg)-$\underline{SO_4}$-Cl
Beznau	15	Muschelkalk	203 - 321.8	3	40.1±0.4	101.3±0.4	6.6±0.0	Na-Ca-(Mg)-$\underline{SO_4}$-Cl
Riniken	303/2	Muschelkalk	617.3 - 696.0	1	169.5	236.8	14.5	Na-(Ca)-$\underline{Cl}$-SO$_4$
Lostorf 1	121	Muschelkalk	12 - 22	1	13.7	24.7	1.6	Na-Ca-(Mg)-$\underline{Cl}$-SO$_4$-(HCO$_3$)
Lostorf 4	124	Muschelkalk	254 - 280	6	1.0±0.0	36.7±0.3	2.5±0.0	Ca-Mg-$\underline{SO_4}$-(HCO$_3$)
Schinznach Bad alt	119	Muschelkalk	9	6	9.7±0.6	29.4±1.2	2.0±0.1	Ca-Na-(Mg)-$\underline{SO_4}$-Cl-(HCO$_3$)
Schinznach Bad S2	120	Muschelkalk	65 - 90	6	13.6±0.4	41.8±1.0	2.8±0.1	Ca-Na-(Mg)-$\underline{SO_4}$-Cl-(HCO$_3$)
Weiach	302/10b	Muschelkalk	822.0 - 896.1	1	1.5	48.0	3.3	Ca-Mg-(Na)-$\underline{SO_4}$
Baden Verenahof	219	Muschelkalk	Spring	6	30.9±0.4	67.5±0.8	4.5±0.1	Na-Ca-(Mg)-$\underline{Cl}$-SO$_4$-(HCO$_3$)
Ennetbaden Schwanen	203	Muschelkalk	Spring	2	28.7±0.6	67.7±0.1	4.5±0.0	Na-Ca-(Mg)-$\underline{SO_4}$-Cl-(HCO$_3$)
Baden Heisse Steinq.	221	Muschelkalk	Spring	1	30.3	68.4	4.5	Na-Ca-(Mg)-$\underline{SO_4}$-Cl-(HCO$_3$)
Baden Limmatq.	210	Muschelkalk	Spring	5	31.7±0.5	69.1±0.9	4.6±0.1	Na-Ca-(Mg)-$\underline{Cl}$-SO$_4$-(HCO$_3$)
Baden Gr. Heisse St.	204	Muschelkalk	Spring	5	30.5±0.6	69.3±1.0	4.6±0.1	Na-Ca-(Mg)-$\underline{Cl}$-SO$_4$-(HCO$_3$)
Ennetbaden Allgemeine	201	Muschelkalk	Spring	5	31.0±0.4	70.2±0.7	4.6±0.0	Na-Ca-(Mg)-$\underline{Cl}$-SO$_4$-(HCO$_3$)
Schafisheim	304/3	Muschelkalk	1227.8-1293.0	1	147.4	149.3	15.3	Na-(Ca)-$\underline{Cl}$-SO$_4$
Rhine Graben								
Badenweiler 1	86	Muschelkalk	Gallery	2	0.24±0.0	4.6±0.0	0.4±0.0	Ca-Na-(Mg)-$\underline{HCO_3}$-SO$_4$
Zähringen	92	Muschelkalk	408 - 537	1	0.4	8.9	0.7	Ca-Mg-$\underline{SO_4}$-HCO$_3$
Liel Neuer Brunnen	129	Muschelkalk-Buntsandstein	- 740	1	1.5	20.9	1.5	Ca-Na-(Mg)-$\underline{SO_4}$-(HCO$_3$)
Krozingen 3	90	Muschelkalk	553 - 610	2	3.8±0.2	61.4±0.7	4.4±0.1	Ca-Na-(Mg)-$\underline{SO_4}$-HCO$_3$
Freiburg 2	94	Muschelkalk	722.9 - 857.9	1	3.0	65.5	4.7	Ca-Na-(Mg)-$\underline{SO_4}$-HCO$_3$

the calcium sulphate water of the nearby borehole of Lostorf 4 (124) demonstrates the strongly varying NaCl concentration within short distances in the same aquifer (NAGRA, 1988, p. 120).

The chemical and isotopic features of the samples from Schinznach Bad (119, 120) suggest mixing between recent and deep groundwaters.

The chemical compositions of the thermal waters of Baden can also be explained only as a mixture of different water types. These include a typical nonsaline Muschelkalk water like that of Lostorf 3 (123) or 4 (124), a more saline groundwater probably ascending from deeper formations, and subordinate amounts of a recent groundwater (SCHMASSMANN AND OTHERS, 1984, p. 238 to 240, 316; NAGRA, 1988, p. 120). Further east, the water of the Nagra Weiach borehole (302/10b) belongs to the same water type as Lostorf 4, but has undergone increased sodium-calcium exchange. Because at least an admixture of the Lostorf 4-Weiach water type is presumed in the waters of Schinznach Bad and Baden, a wide distribution of this water type along the northern border of the Molasse Basin is assumed. This alkaline-earth sulphate water type of comparatively low mineralization suggests a significantly high flow rate in the aquifer.

The deep upper Muschelkalk beneath the Molasse Basin yielded similar waters from both the Schafisheim (304/3 and 304/14) and the Riniken (303/2) boreholes. The sodium chloride water of Schafisheim is characterized by high methane (143 mg/l), ethane (14 mg/l), and hydrogen sulphide (780 mg/l) concentrations thought to originate from hydrocarbons of pre-Tertiary sediments beneath the Molasse Basin. Chloride concentrations from 860 to 1700 meq/l and total dissolved solids from 54 to 115 g/l, higher even than in Schafisheim, were found in Muschelkalk waters of oil boreholes from depths between 1500 to 2200 m (NAGRA, 1988, p. 120-121). These brines found in the central part of the Molasse Basin are presumably quasi-stagnant and not part of an active flow system. No isotope data are available from these oil test boreholes.

The samples from Zähringen (92) and Liel Neuer Brunnen (129) from the upper Muschelkalk of the Upper Rhine Graben are similar to types found in the Jura. In contrast, the waters of Krozingen 3 (90) and Freiburg 2 (94) exhibit high total dissolved carbonate concentrations originating from magmatic gases ascending along faults (SAUER, 1971, p. 92). A special case is the water of very low mineralization from the thermal spring of Badenweiler 1 (86) which is found in a totally silicified zone of upper Muschelkalk, the so called "Quarzriff", but ascends from the crystalline basement along the main fault of the Upper Rhine Graben. The same water type is obtained from deeper sedimentary formations by boreholes at the same location, *e.g.*, Badenweiler 3 (88).

1.6.6 Groundwaters of Lower Triassic and Permian Aquifers

The deep groundwaters of Lower Triassic and Permian aquifers are given in Table 1.6.4.

Lower Triassic (Buntsandstein) and Permian waters show typical features not observed in other waters. For instance, their fluoride to alkaline-earth ratios ($< 2.5 \cdot 10^{-3}$) and

Table 1.6.4: Chemistry of deep groundwaters from the Lower Triassic and Permian aquifers.

Location Name	Number	Aquifer	Depth Range (m)	No. of Samples	Cl⁻ meq/l	Total mineralization meq/l	Total mineralization g/l	Water Type
Grenzach 1	101	L.Muschelkalk-Buntsandstein	31	1	32.0	101.0	6.8	Na-Ca-$\underline{SO_4}$-Cl-(HCO_3)
Kaiseraugst	3	Buntsandstein-Permian	199.5- 213.5	2	72.4±0.1	193.8±0.8	12.9±0.1	$\underline{Na}$-(Ca)-$\underline{SO_4}$-Cl
Leuggern	306/2	Buntsandstein	208.2- 227.5	1	13.7	28.5	1.9	$\underline{Na}$-Cl-SO_4-(HCO_3)
Kaisten	305/2,3	Permian	276.0- 292.5	2	2.9±0.0	22.2±0.2	1.7±0.0	$\underline{Na}$-SO_4-HCO_3-(Cl)
Kaisten	305/1	Buntsandstein	97.0- 129.9	1	57.1	106.7	6.8	Na-Ca-$\underline{Cl}$-SO_4
Böttstein	301/5(6)	Buntsandstein-Crystalline	305.2- 327.6	1	19.6	34.3	2.3	$\underline{Na}$-Cl-SO_4-(HCO_3)
Riniken	303/3,4	Buntsandstein-Permian	793.0- 820.2	2	113.7±1.2	169.9±2.5	10.4±0.1	$\underline{Na}$-(Ca)-$\underline{Cl}$-SO_4
Riniken	303/5b	Permian	958.4- 972.5	1	256.1	295.7	17.9	$\underline{Na}$-$\underline{Cl}$-(SO_4)
Riniken	303/6(9)	Permian	977.0-1099.95	1	341.2	387.4	23.4	$\underline{Na}$-$\underline{Cl}$
Riniken	303/20	Permian	1355.4-1369.0	1	508.8	559.1	33.1	$\underline{Na}$-$\underline{Cl}$
Weiach	302/12(13)	Buntsandstein	981.0- 989.6	1	83.2	222.4	14.8	$\underline{Na}$-(Ca)-$\underline{SO_4}$-Cl
Weiach	302/19(20)	Permian	1109.2-1123.8	1	520.1	618.7	37.4	$\underline{Na}$-$\underline{Cl}$-(SO_4)
Weiach	302/17,18	Permian	1401.1-1415.7	2	1685.3±7.1	1699.5±6.5	98.2±0.4	Na-Ca-$\underline{Cl}$
Schafisheim	304/4,6,7	Buntsandstein-Crystalline	1476.0-1500.4	3	184.6±6.0	253.5±6.6	15.7±0.5	Na-(Ca)-$\underline{Cl}$-SO_4
Rhine Graben								
Badenweiler 3	88	Buntsandstein-Permian	293.0- 505.1	2	0.3±0.0	5.3±0.1	0.4±0.0	Ca-Na-(Mg)-$\underline{HCO_3}$-SO_4

25

fluoride to strontium equivalent ratios (≤ 0.5) are generally smaller than in the waters from the upper Muschelkalk and the crystalline. All Lower Triassic and Permian waters are characterized by high helium contents. Especially high helium concentrations ranging from 320 to 800 µg/kg were analysed in deep groundwaters of the Nagra boreholes at Weiach, Riniken and Schafisheim (SCHMASSMANN AND OTHERS, 1984, p. 262-263; NAGRA, 1988, p. 121).

Increased N_2/Ar ratios suggest a biochemical source, probably in Permo-Carboniferous bituminous shales and coals, in addition to an atmospheric nitrogen source (SCHMASSMANN AND OTHERS, 1984, p. 263, 264, 269; NAGRA, 1988, p. 121). In the Nagra boreholes Leuggern, Böttstein and Schafisheim where Permian sediments are lacking, the waters in the Buntsandstein always exhibit higher mineralization than was observed in the directly underlying crystalline aquifer.

In the Nagra boreholes at Riniken and Weiach, Buntsandstein and Permian sediments comprise a group of aquifers in which groundwater mineralization increases with depth. The Permian groundwaters of the Riniken and Weiach boreholes show the highest ion concentrations in the area. In one sample, the equivalent ion concentration is 2.7 times as high as in sea water. This suggests waters from continental salt lakes which would be consistent with Permian climatic conditions. Such connate waters could be mixing with waters taking part in the present active circulation system, but obviously have not been displaced completely so far. Another possibility would be that an infiltrated water has acquired its salinity by dissolution of traces of halite and sulphates precipitated within pore spaces by preexisting brines. In any case, this suggests that very little water circulation is taking place in this part of the aquifer (NAGRA, 1985, p. 3-41; NAGRA, 1988, p. 122).

1.6.7 Groundwaters in the Crystalline Basement

The water types from the crystalline basement are compiled in Table 1.6.5.

In the southern Black Forest, waters of low mineralization circulate along tectonic fracture zones as demonstrated by the subthermal spring of Bürchau (110) (NAGRA, 1985, p. 3-42). In the eastern Tabular Jura, north of the Permo-Carboniferous Trough, the upper part of the crystalline basement contains chemically similar water with comparatively low total mineralization of the sodium sulphate bicarbonate chloride type (Zurzach, 131 and 132, Leuggern, 306, Böttstein, 301, and Kaisten, 305). This chemical character is largely determined by cation exchange processes (SCHMASSMANN, 1987, p. 576). These waters also have high helium concentrations (NAGRA, 1985, p. 3-42).

Hydrodynamic modelling of the regional flow system was done assuming that recharge takes place in the Black Forest. Hydrochemical investigations are consistent with that assumption and show that groundwater flows in a northwestern direction underneath the eastern Tabular Jura and discharges ultimately in the Rhine River upstream of Säckingen (NAGRA, 1985, p. 3-42; SCHMASSMANN, 1987, p. 576; KANZ, 1987, p. 277).

A more highly mineralized water in the upper part of the crystalline basement at Leuggern (916.2 to 929.7 m) is a sodium sulphate type with a high calcium concentration and is considered to be an isolated groundwater similar to the water of Waldkirch (95) (KANZ, 1987, p. 274).

A highly mineralized sodium chloride water was found in the lower part of the crystalline basement in Böttstein (1321 to 1331 m) below the typical waters of low mineralization. The data presented in Table 1.6.5 are from a groundwater sample diluted with drilling fluid (deionized water), but PEARSON (1985, p. 104) estimated the original composition of this water from a series of samples and calculated a salinity of over 13 g/l. This water is quite distinct from that in the upper crystalline basement. It may have its origin in, or be influenced by, water of the Permo-Carboniferous Trough. Similar waters were found in Permo-Carboniferous sediments of the Weiach and Riniken boreholes. In the Leuggern borehole, at a greater distance from the Permo-Carboniferous sedimentary trough than the Böttstein borehole, a sodium chloride water of lower mineralization was encountered at 1427.4 to 1439.4 m as an intercalation in the depth range of upper crystalline type groundwaters.

The warm waters of Bad Säckingen (125, 126 and 161) probably also migrate from the sedimentary aquifers of the Permo-Carboniferous Trough into the crystalline basement where--after mixing with shallow young groundwaters--they ascend along tectonic faults (SCHMASSMANN, 1987, p. 576).

In the western Tabular Jura, close to the Upper Rhine Graben, a water with increased carbon dioxide and alkaline-earth concentrations was found in the Rheinfelden-Engerfeld borehole (SCHMASSMANN AND OTHERS, 1984, p. 307). This water has the highest carbon dioxide concentration reported from all deep groundwaters of northern Switzerland. This indicates gas and water flow paths from great depth (SCHMASSMANN, 1987, p. 573). The high N_2/Ar ratio of this water in addition to other hydrochemical features indicates a similarity to groundwaters of the Lower Triassic and Permian aquifers.

The groundwater obtained from the crystalline basement south of Permo-Carboniferous Trough was from Schafisheim. It is a saline water type with unusually high gas concentrations (NAGRA, 1985, p. 3-43; KANZ, 1987, p. 276; PEKDEGER AND BALDERER, 1987).

Table 1.6.5: Chemistry of deep groundwaters from the Crystalline aquifer. (Page 1 of 2)

Name	Location	Number	Aquifer	Depth Range (m)	No. of Samples	Cl⁻ meq/l	Total mineralization meq/l	Total mineralization g/l	Water Type
Bürchau		110	Crystalline	Spring	2	0.6±0.0	1.6±0.1	0.2±0.0	$\underline{Ca}$-Na-$\underline{HCO_3}$-(SO_4)
Zurzach 1		131	Crystalline	402.6- 428.5	6	3.8±0.0	14.1±0.0	1.0±0.0	$\underline{Na}$-$\underline{SO_4}$-HCO_3-Cl
Zurzach 2		132	Crystalline	439 - 469	6	3.7±0.0	14.1±0.1	1.0±0.0	$\underline{Na}$-$\underline{SO_4}$-HCO_3-Cl
Säckingen Margaretq.		126	Crystalline	18.9- 154.0	11	12.0±0.5	16.2±0.4	1.1±0.0	$\underline{Na}$-Ca-$\underline{Cl}$-(HCO_3)
Säckingen Badq.		125	Crystalline	81.9- 201.3	11	48.3±1.3	55.9±1.4	3.5±0.1	$\underline{Na}$-(Ca)-$\underline{Cl}$
Säckingen Stammelhof		161	Crystalline	380 - 505	1	102.7	119.6	7.4	$\underline{Na}$-(Ca)-$\underline{Cl}$
Rheinfelden		159/1-3	Permian-	336 - 344					
Waldkirch		95	Crystalline	404 - 600	3	17.8±0.3	62.6±0.5	4.6±0.0	$\underline{Na}$-Ca-HCO_3-SO_4-Cl
			Crystalline	- 660	2	4.3±0.1	79.1±0.4	5.6±0.0	$\underline{Na}$-Ca-$\underline{SO_4}$
Kaisten		305/4,5	Crystalline	299.3- 321.5	2	2.24±0.0	21.3±0.2	1.6±0.0	$\underline{Na}$-$\underline{SO_4}$-HCO_3-(Cl)
Kaisten		305/6,7	Crystalline	475.5- 489.8	2	1.7±0.0	20.8±0.1	1.6±0.6	$\underline{Na}$-$\underline{SO_4}$-HCO_3
Kaisten		305/9, 10,11							
Kaisten		305/12,13	Crystalline	816.0- 822.9	3	1.9±0.1	18.9±0.1	1.5±0.0	$\underline{Na}$-(Ca)-$\underline{SO_4}$-HCO_3-(Cl)
Kaisten		305/14,15	Crystalline	1021.0-1040.9	2	1.8±0.0	19.1±0.3	1.5±0.0	$\underline{Na}$-(Ca)-$\underline{SO_4}$-HCO_3
Kaisten		305/16,17	Crystalline	1140.8-1165.8	2	2.8±1.0	19.2±1.3	1.4±0.1	$\underline{Na}$-$\underline{SO_4}$-HCO_3-(Cl)
Leuggern		306/4	Crystalline	1238.0-1305.8	2	2.1±0.3	19.2±0.0	1.5±0.0	$\underline{Na}$-$\underline{SO_4}$-HCO_3-(Cl)
Leuggern		306/5,6	Crystalline	235.1- 267.5	1	5.0	17.6	1.3	$\underline{Na}$-$\underline{SO_4}$-Cl-HCO_3
Leuggern		306/7,8	Crystalline	440.4- 448.1	2	3.7±0.0	15.7±0.2	1.1±0.0	$\underline{Na}$-$\underline{SO_4}$-HCO_3
Leuggern		306/9,10	Crystalline	507.4- 568.6	2	3.4±0.0	14.9±0.2	1.1±0.0	$\underline{Na}$-$\underline{SO_4}$-Cl-HCO_3
Leuggern		306/17, 18,19	Crystalline	702.0- 709.5	2	3.1±0.0	16.1±0.0	1.2±0.0	$\underline{Na}$-$\underline{SO_4}$-HCO_3-(Cl)
Leuggern		306/11b, 12,13	Crystalline	834.5- 859.5	3	3.4±0.1	15.5±0.2	1.1±0.0	$\underline{Na}$-$\underline{SO_4}$-HCO_3-Cl
Leuggern		306/20, 21	Crystalline	916.2- 929.7	3	5.3±0.1	64.9±1.3	4.5±0.1	$\underline{Na}$-Ca-$\underline{SO_4}$
Leuggern		306/24, 25,26	Crystalline	1176.2-1227.2	4	3.6±0.1	13.7±0.3	1.0±0.0	$\underline{Na}$-$\underline{SO_4}$-HCO_3-Cl
Leuggern		306/16	Crystalline	1427.4-1439.4	3	14.9±1.7	24.3±1.5	1.6±0.1	$\underline{Na}$-Cl-$\underline{SO_4}$-HCO_3
Leuggern		306/23	Crystalline	1637.4-1649.3	1	3.5	14.0	1.1	$\underline{Na}$-$\underline{SO_4}$-HCO_3-Cl
Leuggern			Crystalline	1642.2-1688.9	1	3.5	13.9	1.0	$\underline{Na}$-$\underline{SO_4}$-HCO_3-Cl
Böttstein		301/8c,9 301/12b,	Crystalline	393.9- 405.1	2	3.4±0.0	16.6±0.2	1.2±0.0	$\underline{Na}$-$\underline{SO_4}$-HCO_3-Cl
Böttstein		13,16,17	Crystalline	608.0- 628.8	4	3.6±0.13	17.4±0.1	1.3±0.0	$\underline{Na}$-$\underline{SO_4}$-HCO_3-Cl
Böttstein		301/18	Crystalline	782.0- 802.8	1	4.0	18.1	1.3	$\underline{Na}$-$\underline{SO_4}$-HCO_3-Cl

Table 1.6.5: Chemistry of deep groundwaters from the Crystalline aquifer. (Page 2 of 2)

Location Name	Number	Aquifer	Depth Range (m)	No. of Samples	Cl^- meq/l	Total mineralization meq/l	Total mineralization g/l	Water Type
Böttstein	301/19,20, 21,22	Crystalline	1321.0-1331.4	4	95.4±6.5	111.5±7.5	6.7±0.5	Na-(Ca)-Cl-(SO$_4$)
Weiach	302/14,16	Crystalline	2211.6-2273.5	2	103.0±7.6	114.0±7.6	6.9±0.4	Na-Cl
Schafisheim	304/8,9	Crystalline	1564.5-1577.7	2	68.3±0.3	124.9±0.4	8.3±0.1	Na-(Ca)-Cl-SO$_4$-(HCO$_3$)
Schafisheim	304/10,11	Crystalline	1883.5-1892.3	2	100.3±2.1	131.8±2.0	8.5±0.1	Na-Cl-(SO$_4$)-(HCO$_3$)

2. LIMITATIONS OF SAMPLING AND ANALYTICAL PROCEDURES

F. J. Pearson, Jr., H. H. Loosli, and W. Balderer

Before using hydrochemical and isotopic data to support geochemical and hydrogeologic interpretations, it is necessary to assess whether the samples analysed were contaminated, and to evaluate the extent to which any such contamination could have influenced the analytical results. It is also important to examine all replicate analytical data and to select for interpretation those values that best represent the properties of the groundwater.

The quality of the chemical samples of groundwaters taken under the Nagra regional and borehole programmes are described using quality block designations. Block 1 samples have negligible possibility or extent of contamination. Block 2 samples are contaminated to some extent with nonformation fluid, but the amount of contamination and the composition of the contaminating fluid is known. Thus, the analyses can be corrected for the contamination if necessary for the intended use of the analytical data. Block 3 samples are so heavily contaminated that they can neither be used without correction nor corrected.

Geologisches Institut Dr. Schmassmann AG reviewed the quality of all Nagra samples and assigned the block designations which accompanied the analytical results given by NAGRA (1989). The review of the regional samples is based on the hydrogeologic setting of each of the sampling points as described by ISENSCHMID (1985/1986). In addition, WITTWER (1986) evaluated the contamination by drilling fluid of samples from Nagra test boreholes using the concentrations of drilling fluid tracers, the water chemistry itself, and ^{3}H.

The quality of the isotope samples is not necessarily the same as that of the chemical samples because they were collected at slightly different times and by different procedures. Generally, if the quality of the chemical samples is poor, the isotope samples will also be poor. However, as the discussion below will demonstrate, poor isotope results can also accompany good chemical samples. An assessment of the quality of the isotope samples forms part of this chapter. Contamination of isotope samples taken from the Böttstein borehole has been discussed in an earlier report (BALDERER, 1985b).

Virtually all of the samples discussed from the regional programme are of Block 1 chemical quality. Table 2.1.1 includes an indication of the chemical quality blocks of samples from the borehole programme.

Analyses for a number of isotopes and the noble gases were duplicated by several laboratories and these results are also discussed in this chapter. Two techniques were used for the analyses of ^{14}C with apparently systematically different results. Dissolved uranium concentrations measured by three laboratories are concordant except at lower concentrations.

Argon analyses were made by two laboratories and helium analyses by three (for some sampling points, four) laboratories. Differences among the results indicate the difficulty of collecting, transporting and storing representative samples of dissolved gases, particularly helium. When results effected by sampling difficulties were eliminated, there was good agreement among the dissolved gas analyses of the several laboratories.

2.1   Contamination of Samples

Groundwater samples may contain greater or lesser amounts of young water. The terms "young" and "very young" water will be used in the discussions which follow. Young water originated at the earth's surface since the beginning of the nuclear era. Very young water had a surface origin within two years of the date of sampling. ^{3}H and ^{85}Kr are useful for assessing the amount of young and very young waters present in groundwater samples and the residence times of young waters as described in Section 4.3.

Virtually all ^{3}H now present in the atmosphere and in surface water was produced by thermonuclear devices, the testing of which began in the early 1950s. Tests introducing significant amounts of ^{3}H to the atmosphere ended in the mid-1960s. Since then, the concentration of ^{3}H in precipitation and surface waters has been decreasing. Virtually all ^{85}Kr in the environment is a product of nuclear reactors. Its concentration in the atmosphere has been increasing since the early 1950s. The concentrations of ^{3}H and ^{85}Kr since 1950 are shown in Figures 4.3.1 and 4.3.2.

The presence of young water in a groundwater sample may be a result of natural hydrologic processes or of contamination. Components of young groundwater naturally present in some of the Nagra samples are discussed in Section 4.3. Young waters present in samples as a result of contamination by drilling fluid are discussed here.

Groundwater samples can be contaminated with fluids and other substances used for borehole drilling and construction. This type of contamination can influence the concentration and isotopic composition of dissolved solid constituents and gases, and the isotopic composition of the water itself. The quality of borehole samples has been assessed by WITTWER (1986). As pointed out above, the chemical analyses of all samples discussed from the regional programme are considered to be Block 1. Only results of analyses on Block 1 and 2 samples appear in the Nagra data report (NAGRA, 1989).

Except as noted below, the results of the isotopic analyses of Block 1 samples, like those of their chemical analyses, are generally useful for interpretation without further evaluation. Enough is known about Block 2 samples that their isotopic and chemical data are generally suitable for careful further use, either directly or after correction for contamination. Data for Block 3 samples, however, do not represent groundwaters and so are not used, except to support the correction of Block 2 samples.

The concentrations of some isotopes and gases can be changed during collection and transport, even when water uncontaminated by drilling fluid is being sampled. The large volume samples for ^{39}Ar and ^{85}Kr are susceptible to contamination by air during collection. The samples for the noble gases, especially helium, are difficult to store and ship without leakage. Thus, the concentrations of these substances in all samples must be examined for evidence of air contamination or leakage.

2.1.1 Large-Volume Gas Samples for ^{39}Ar and ^{85}Kr

Analyses of ^{39}Ar and (or) ^{85}Kr were made on samples collected from 19 zones in the test boreholes, and from 12 other boreholes as part of the regional programme. These 31 samples have been examined to estimate the extent to which they may have been contaminated by air, and the extent to which a young component, such as drilling fluid or young groundwater, may have been present in the sample.

For analyses of ^{39}Ar and ^{85}Kr, dissolved gas must be extracted from at least 10,000 litres of water. This is done in the field using a vacuum extraction apparatus, as described by BALDERER (1985b, Section 4.2.1). There is always the potential for atmospheric leakage into an apparatus of this sort, so the results of analyses of the large-volume gas samples were examined for possible air contamination using their oxygen and ^{85}Kr contents. The results of the ^{39}Ar and ^{85}Kr analyses are given in Table 2.1.1 at the end of this chapter. This table includes additional data used to assess the contamination of these samples. The hydrologic significance of the ^{39}Ar and ^{85}Kr results is discussed in Chapter 4.

The concentrations of gases dissolved in the water samples were analysed, as were the compositions of the large-volume gas samples. The proportion of oxygen in the gas sample is compared with the proportion of oxygen in the total gas dissolved in the water. A higher proportion of oxygen in the gas sample than in the water is an indication of air contamination of the gas sample.

The oxygen concentrations of the gas samples and of the total gas dissolved in equivalent water samples are given in Table 2.1.1. Both sets of values are from analyses made at the Paul Scherrer Institut (PSI). The per cent of air contamination of the gas samples calculated from the oxygen results is also given. The calculation assumed that the difference between the oxygen contents of the gas and equivalent water samples resulted from the addition of air containing 20.9 volume per cent oxygen. Of the 31 locations sampled, air contamination was below one per cent for 21 samples, between one and ten per cent for six samples, and greater than ten per cent for four samples.

The presence of ^{85}Kr could result from contamination of the large-volume gas sample, from drilling fluid contamination of the water sample, or from a young component of the groundwater itself. The activity of ^{85}Kr from contamination by air or drilling fluid during the period when these samples were collected was taken as 40 dpm/ml. The ^{85}Kr activity of groundwater will range from 0 to 40 dpm/ml depending on its age (see Section 4.3). Two interpretations of the ^{85}Kr content of these samples are possible. First, if a gas sample is relatively uncontaminated by air, as shown by its oxygen content, its ^{85}Kr content will measure the proportion of drilling fluid contamination or young water in the sample and should agree with the ^{3}H estimate. Second, if a gas sample is strongly contaminated by air, ^{85}Kr will be present but will not necessarily indicate drilling fluid contamination or the presence of a component of young groundwater.

The ^{85}Kr content can be used to estimate an upper limit for air contamination of the large-volume gas samples by assuming that the water samples contain no ^{85}Kr and that all ^{85}Kr measured is from contaminating air. The calculation, which is based on a mass balance equation for ^{85}Kr in the sample, is made using:

$$\text{Air Contamination (\%)} = \frac{^{85}A_{sm} \cdot \frac{Kr}{Ar} \cdot Ar\,(\%)}{^{85}A_{air} \cdot P_{Kr}} \qquad (2.1.1)$$

In equation 2.1.1, $^{85}A_{sm}$ and $^{85}A_{air}$ are the ^{85}Kr activities measured in the gas sample and in air, respectively, in disintegrations per minute per cm^3 Kr (dpm/cc Kr). The activity of air was taken to be 40 dpm/cc Kr. P_{Kr} is the concentration of Kr in air, $1.14 \cdot 10^{-6}$ cc/cc. The absolute Kr concentrations of the gas samples were not measured, but the Kr/Ar ratio of many of the samples were analysed at the University of Bath, and Ar concentrations at PSI. No Kr/Ar analyses were made on some of the gas samples. For these, the Kr/Ar ratio of the gas dissolved in equivalent water samples was used. The Kr/Ar ratios in the dissolved gases tend to be larger, by up to 40 per cent, than those of the gas samples, and this bias would translate directly to the values of air contamination calculated using them.

Table 2.1.1 gives values for gas sample contamination by air calculated from ^{85}Kr measurements, together with the additional analytical data required to use equation 2.1.1. Because the calculation assumes that all ^{85}Kr measured is from air contamination, all results are given as less than (<) values.

Alternately, ^{85}Kr measurements can be used to determine whether a water sample contains young water. To do this requires the assumption that all of the ^{85}Kr analysed was present in the water sample, and that none was introduced by contamination during collection of the large-volume gas sample. This assumption is the opposite of the one made above to use ^{85}Kr values to calculate air contamination of the gas samples.

Table 2.1.1 includes values for contamination by very young fluids calculated assuming that all analysed ^{85}Kr can be attributed to water containing ^{85}Kr with an activity of 40 dpm/cc Kr. If the contaminating water was drilling fluid of that activity, the amount of contamination calculated would be a maximum value. If the young water was groundwater with a lower ^{85}Kr activity, the sample could contain more young water than the calculated amount. If some of the ^{85}Kr measured was from air contamination of the gas sample, the water sample could contain less young water than calculated. Thus, the per cent of young water calculated from the ^{85}Kr measurements and given in Table 2.1.1 can be interpreted only in conjunction with other indicators of gas sample contamination and young water admixtures.

To use the ^{3}H content of the water samples to estimate their young water content requires information about the ^{3}H concentrations of young waters. A number of analyses of the ^{3}H content of the drilling fluid were made, with results ranging from 75 to 120 TU and a mean of about 90 TU. The tritium content of young groundwaters can be more or less than this value, as discussed in Section 4.3 of this report.

The ^{3}H contents of water samples equivalent to those from which the large gas samples were extracted are given in Table 2.1.1. The table also includes a column giving the per cent contamination with young water calculated using a ^{3}H content of 90 TU for the young water. This is the mean measured value for drilling fluid.

The table also gives values for the ratio of ^{3}H to ^{85}Kr for the samples. This ratio is useful for distinguishing whether ^{85}Kr in a sample results from gas sample contamination or from the presence of young water. As described in Section 4.3 of this report, the ^{3}H/^{85}Kr ratio of waters exposed to the atmosphere is > 1. A contaminated gas sample would have additional ^{85}Kr but no additional ^{3}H and its ^{3}H/^{85}Kr ratio would tend to be lower than one.

Table 2.1.1 includes comments on the extent of contamination of the samples from each zone. Although the calculated contamination for some samples is one per cent or below, Section 4.3 states that the smallest amount of young fluid that can be detected by this method is two or three per cent. Thus, two per cent is the minimum value given for the young water component. Several of the samples in the table have high concentrations of ^{85}Kr, oxygen concentrations higher than in the gas dissolved in equivalent water samples, and ^{3}H/^{85}Kr ratios below one. These samples, from the Muschelkalk at Frenkendorf (9/1), the Buntsandstein and Permian at Weiach (302/12 and 302/19) and the Buntsandstein of Leuggern (306/2) contain considerable contaminating air. The oxygen content of one sample from the crystalline at Böttstein (301/12b) suggests it may be contaminated by as much as four per cent air. Its relatively high ^{85}Kr content and low ^{3}H/^{85}Kr ratio support this interpretation.

Two samples, from the Muschelkalk at Magden Weiere (271/103) and the crystalline at Kaisten (305/4), have very high oxygen contents, but have low ^{85}Kr concentrations consistent with their low ^{3}H contents. It is possible that the aliquot of the gas sample which was taken for oxygen analysis was contaminated.

The O_2 contents of the remaining samples indicate air contamination of the extracted gas of less than two per cent. There is generally good agreement between the ^{3}H and the ^{85}Kr contents of these samples as well. For most, both ^{3}H and ^{85}Kr are low, indicating that the samples contain less than two per cent young water either from drilling fluid contamination or as a natural component of the groundwater. Both the ^{3}H and ^{85}Kr contents of the sample from the Riniken Permian (303/6) indicate that it contains two to three per cent of young water. This is consistent with WITTWER's (1986, Section 5.3.5) conclusion that the chemical sample contained about five per cent drilling fluid. The ^{3}H content of sample 301/23 from the Böttstein crystalline also suggests that it contains up to four per cent of young water, although the agreement between the ^{85}Kr and ^{3}H contents of this sample is less good. These samples are discussed further in Section 4.3.

High values of ^{3}H and ^{85}Kr suggest that there is a significant component of natural young water in the samples from the Muschelkalk at Leuggern (306/1), Pratteln (8/2), Densbüren (14/3), and Magden Stockacher (274/102). The sample from Säckingen Badquelle

(125/1) also contains high concentrations of ^3H and ^{85}Kr indicating a major component of young water. These samples are also discussed further in Section 4.3.

Finally, the ^{85}Kr of the sample from Lostorf 3 (123/110) suggests the presence of some younger water, although this sample contains no detectable ^3H. Later samples from the same source, however, contain up to 7.8 TU, and do suggest a significant natural young water component.

2.1.2 Samples for Carbon Isotopes

Contaminating ^{14}C could be added to a sample of water from a regional flow system by mixing with modern ^{14}C-bearing water, or during the sample collection process. The ^{14}C-containing water could be drilling fluid mixed with formation water during borehole construction, or young water naturally present in the environment mixed with the sample during pumping for collection or by natural processes.

The ^{14}C content of many of the borehole and regional samples approaches one per cent of modern ^{14}C (pmc). This is approximately the detection limit of the analytical methods used. Contamination of such samples by as little as a few pmc could lead to serious errors in interpretation. Therefore, the possibility of such contamination must be explored for all samples.

^{14}C samples were collected by direct precipitation, as $BaCO_3$, of the total carbonate dissolved in several tens of litres of water. M. Wolf and P. Fritz have noted (personal communication to F. J. P., 13-October-1989), that NaOH used in some $BaCO_3$ collection procedures virtually always contains some ^{14}C. Although this source usually contributes < 1 pmc to the ^{14}C content of the sample itself, contamination of up to 2 or 3 pmc is also possible when sampling groundwaters with low dissolved carbon concentration.

The procedure used to collect the Nagra samples used solutions of $Ba(OH)_2$ to precipitate $BaCO_3$. $Ba(OH)_2$ should contain little or no ^{14}C-bearing carbonate impurities, but no measurements were made to test for $Ba(OH)_2$ contamination.

A few samples were taken by vacuum extraction of CO_2 from small-volume water samples taken downhole under pressure. Results of the two extraction methods are compared in Section 2.2. This method should introduce no contamination into the ^{14}C sample collected.

Samples collected during or just after the drilling of a borehole may contain a proportion of the fluid used for drilling, or ^{14}C-bearing carbonate introduced by other drilling practices. Thus, it is necessary to evaluate the influence of drilling fluid contamination on the ^{14}C content of samples from the Nagra test borehole programme. Samples representing regional flow systems collected from old boreholes and springs are more likely to contain ^{14}C from mixing with natural modern waters than from residual drilling fluid. ^{14}C contamination of samples taken as part of the regional programme are therefore examined separately from those of the test borehole programme.

Section 5.2 discusses the significance of the ^{14}C contents of samples in terms of the ages and residence times of the groundwater they represent.

2.1.2.1 Samples from Test Borehole Programme

Samples from the test boreholes may be contaminated with ^{14}C from the fluid used to drill the boreholes. Drilling fluid can contain two types of ^{14}C-bearing carbon. The first type, dissolved carbonate, may contain ^{14}C from exchange with the atmosphere. The second type, modern organic carbon, may be present in substances such as starch, cellulose or other fillers added to the fluid to limit its loss to formations being drilled. Methods of recognizing whether contamination from either source is present and of estimating its extent are discussed separately.

The amount of contamination from ^{14}C-bearing drilling fluid depends on the amount of drilling fluid in the sample, the relative dissolved carbonate contents of the groundwater and the drilling fluid, and the ^{14}C content of the carbonate dissolved in the drilling fluid. BALDERER (1985b, Section 4.4) describes the calculations in detail. The increase in the ^{14}C content of a sample from drilling fluid contamination, $\Delta^{14}C_{sm}$, can be expressed:

$$\Delta^{14}C_{sm} = {}^{14}C_{df} \cdot \frac{C_{df}}{C_{sm}} \cdot P_{df} \qquad (2.2.1)$$

where ^{14}C and C are the ^{14}C and total carbonate contents and the subscripts sm and df refer to the sample and drilling fluid, respectively. P_{df} is the proportion of drilling fluid in the sample.

To calculate the maximum ^{14}C contamination from drilling fluid requires estimates of the amount of drilling fluid in the sample, P_{df}, the total dissolved carbonate in the drilling fluid, C_{df}, and the ^{14}C content of that dissolved carbonate, $^{14}C_{df}$. Estimates of drilling fluid contamination of the chemical samples from the boreholes are available from WITTWER (1986). Information about the total carbonate content of drilling fluid is less easily available but rough estimates, or at least ranges, of probable concentrations can be developed from measurements of total alkalinity (m-value) made as part of the drilling fluid monitoring programme (HAUG, 1985). Generally, the pH values of the drilling fluids were high enough that virtually all dissolved carbonate was present as bicarbonate and carbonate. Thus, it is assumed that the total carbonate dissolved in drilling fluid is not more than the measured drilling fluid alkalinity.

There are no measurements of the ^{14}C content of drilling fluid. Maximum concentrations, which would give rise to maximum contamination, would occur if the drilling fluid were in isotopic equilibrium with atmospheric CO_2. The ^{14}C content of atmospheric CO_2 in Switzerland from 1983 to 1986 was between 120 and 125 pmc. Isotopic fractionation between dissolved carbonate and CO_2 gas would increase the ^{14}C content of the dissolved carbonate by several per cent. Thus, an estimate of 125 pmc can be made for the maximum ^{14}C content of drilling fluid.

The second source of ^{14}C contamination is from carbon introduced into the borehole in drilling fluid additives. This ^{14}C would originally be present as organic carbon and, in this form, should not influence the ^{14}C content of the sample, which represents only dissolved carbonate. If this carbon were oxidized to carbonate in the borehole or in the formation, it could add ^{14}C to the dissolved carbonate.

LAMBERT (1987) sampled a number of wells from a gypsum-bearing dolomite in New Mexico. Modern organic carbon, such as cotton seed hulls and nut shells, had been added to these wells during drilling to stop the loss of drilling fluid. LAMBERT (1987) reported a strong correlation between the ^{14}C content and the alkalinity of samples taken up to several years after drilling these wells. He attributed this to the oxidation of organic carbon from the drilling fluid additives.

Nagra samples come from many aquifers and from wells much more widely separated than those sampled by LAMBERT (1987). Thus, his technique of plotting alkalinity against ^{14}C to recognize contamination is not applicable to the Nagra samples. Because this type of contamination is possible wherever drilling fluid additives containing modern ^{14}C have been used, records of drilling fluid additives for each interval sampled should be examined, if available. It is difficult to be certain that any interval exposed to additives bearing modern carbon has not been contaminated.

The carbon isotopic composition of samples from the test boreholes is given in Table 2.1.2 at the end of this chapter. The table also shows the quality block and the ^3H, dissolved organic carbon, and total dissolved carbonate contents of each sample. The total dissolved carbonate values are those calculated for the samples at equilibrium with calcite, as given by PEARSON AND OTHERS (1989).

One column of the table contains estimates of the maximum ^{14}C concentrations of the samples which could result from contamination with drilling fluid. These were calculated using equation 2.2.1. The alkalinity values for the drilling fluids required for this calculation were estimated from analyses of drilling fluids. For fluids containing mud, which were used for drilling the sedimentary sections, a value of 300 mg/l was chosen for C_{df}. Mixtures of deionized water and groundwater from the formations being drilled were used for the crystalline sections. A value of 200 mg/l was used for the alkalinity of these fluids. Both alkalinity values are probably higher than the actual concentrations. A value of 125 pmc, which was used for the ^{14}C content of the drilling fluid, $^{14}C_{df}$, is also likely to be a maximum value. Choosing maximum values for these parameters leads to a maximum estimate of possible contamination.

The proportion of drilling fluid in each sample, P_{df}, was taken from one of two sources. The amount of very young fluid present in some samples could be established using ^3H and ^{85}Kr as described in Sections 2.1.1 and 4.2. Very young fluid includes drilling fluid, so the amounts of very young fluid given in Table 2.1.1 are included in Table 2.1.2, and when available, were used for P_{df}.

The ^{3}H contents alone were used to estimate P_{df} of the remaining samples. The drilling fluid was taken to have a ^{3}H content of 90 TU, as it was for the calculations made for Table 2.1.1. The selection of 90 TU for this calculation does not necessarily lead to an estimate of the maximum drilling fluid contamination. The drilling fluids for some zones may have contained much less ^{3}H, in which case the actual proportion of drilling fluid in the sample could be larger than the amount calculated using 90 TU.

The proportion of drilling fluid in some samples from the deep boreholes could also be estimated from the drilling-fluid tracer data reported by WITTWER (1986). These estimates are in general agreement with those based on the isotope data alone.

Interpretation of the data on individual samples is given in the column of comments in Table 2.1.2. Certain general conclusions about contamination of the borehole ^{14}C samples can also be drawn from the table and are given here.

There is less than two per cent of drilling fluid or other very young fluid in most of the borehole samples. The maximum ^{14}C contamination of these samples from carbonate dissolved in the drilling fluid would be about 1 to 2 pmc. There are very few samples with analysed ^{14}C contents of less than 1 to 2 pmc, even though the hydrologic setting of many of the groundwaters sampled is such that they should contain no measurable ^{14}C. It appears that contamination from drilling fluid at the level of 1 to 2 pmc is common to many of the samples. This is also about the maximum contamination to be expected from ^{14}C present in the reagents used for sample collection, as described in the previous section.

Several samples with ^{14}C concentrations significantly above 2 pmc have low ^{3}H contents and should be representative of regional groundwaters. Such samples include those from the crystalline above 800 m at Böttstein (301/8 to 18) and from the 444.2 zone (306/5) and the 538.0 m zones (306/7) at Leuggern. The latter two samples, however, are close to 2 pmc and might also be treated as limiting values.

A few samples with high ^{14}C values also have high enough ^{3}H contents to account for virtually all of the ^{14}C measured as drilling fluid contamination. These include the 1326.2 m sample from Böttstein (301/21), the 2218.1 m sample from Weiach (302/16), and the 847.0 m sample from Leuggern (306/17). It is important to note that all but two of the samples discussed in this and the preceding paragraphs have dissolved organic carbon (DOC) contents of < 1 mg/l.

Inspection of Table 2.1.2 reveals an association between high DOC values and high ^{14}C contents. Samples from Weiach and from the Permian at Riniken, for example, are distinguished by DOC contents which are high relative to the DOC values of < 1 mg/l typical of the samples previously discussed. A number of zones from which these samples were taken, including the 859.1 m zone at Weiach (302/10b) and the 993.5 m zones at Riniken (303/6), as well as all zones below about 700 m at Leuggern (306/9 to 23), are noted by WITTWER (1986) as zones of drilling fluid loss. If the high DOC values of these samples can be attributed to the imperfect removal of organic drilling fluid additive from the zones being tested, the high ^{14}C contents of the samples could be explained as a product of oxidation of these additives in the formation. Unfortunately, the DOC values cannot be used for any quantitative correction for possible contamination by the oxidation of drilling fluid additives. They are useful, however, to indicate which samples could have been influenced by this type of contamination, and so have ^{14}C results which should be used only as limiting values, if at all.

The final group of samples are from hydrologic settings which could well yield ^{14}C-bearing water. This group includes the relatively shallow samples from the Buntsandstein/weathered crystalline at Böttstein (301/5 and 25b), and the samples from the 515.7 m zone at Riniken (303/1) and the 558.0 m zone at Schafisheim (304/2). These samples are discussed further in Section 5.2 of this report.

The ^{3}H content of the sample from the Muschelkalk at Leuggern (306/1) suggests that its entire ^{14}C content could be a result of mixing with young water. This young water is not likely to be drilling fluid, however, because WITTWER (1986) includes the chemical sample from this zone in quality Block 1. This sample probably contains a significant component of natural young water, and so it is discussed in detail in Chapter 4.

2.1.2.2   Other Samples

Many ^{14}C samples are from the Nagra regional programme and from other studies in northern Switzerland and adjacent areas. Most were taken from springs or from boreholes which had been constructed some time before sample collection and so should have minimal residual contamination from drilling fluids or additives. However, they could well contain ^{14}C from young groundwater mixed with them as a result of pumping or of natural processes taking place near the sampling point. Because these samples are intended to support the interpretation of regional flow systems, they must be examined for evidence of local mixing which would make them non-representative of regional systems.

Table 2.1.3, at the end of this section, gives the carbon isotope results, ^{3}H, dissolved organic carbon, and total dissolved carbonate values for samples from the Nagra regional programme. It also includes values culled from the literature on samples taken from the same and additional collection points in and adjacent to northern Switzerland. The total dissolved carbonate values are calculated after adjusting the samples to calcite saturation, as were those given for the borehole samples in Table 2.1.2.

The table includes comments on many of the samples. Those without comments have ^{3}H contents below about 1 or 2 TU, the analytical detection limit. It is probable that the samples from these points contain less than one or two per cent young water, so that their ^{14}C contents are likely to be representative of intermediate and regional scale regional flow systems.

There is also ^{85}Kr evidence, shown in Table 2.1.1 and discussed in Section 4.2, that samples from several locations contain less than two or three per cent young water. These locations, which are also noted in Table 2.1.3, include:

1	Zürich Aqui
2	Zürich Tiefenbrunnen
9	Frenkendorf
15	Beznau
271	Magden Weiere
272	Magden Falke
273	Magden Eich

Samples from locations 1, 2 and 15 also contain no measurable ^{14}C.

Samples from many points have both measurable ^3H and measurable ^{14}C. The ^3H indicates that these samples contain at least some young water. It is important to explore how the ^{14}C contents of these samples are related to their young water content.

In Figure 2.1.1, the measured ^3H contents of the samples of Table 2.1.3 are plotted against their ^{14}C contents. Results from the Muschelkalk samples from the Weiach (302/10a and b) and Leuggern (306/1) boreholes are also included. Most points fall near or to the upper left of the dashed line on the figure. The samples comprising the line, in order of their source formations, are:

Quaternary (q):

    501   Reichenau

Middle Keuper (km):

    23   Eptingen
    171   Beznau 104.5 to 109.5 m

Middle Keuper (and Muschelkalk?) (km +m?):

    19   Meltingen 2

Upper Muschelkalk (mo):

    12   Kaisten Felsbohrung
    16   Windisch BT 2
    119   Schinznach Bad alt
    201   Ennetbaden Allgemeine
    203   Ennetbaden Schwanen
    206   Baden Limmatquelle
    219   Baden Verenahof
    306/1   Leuggern 53.5 to 96.4 m.

Crystalline (KRI):

    125   Säckingen Badquelle

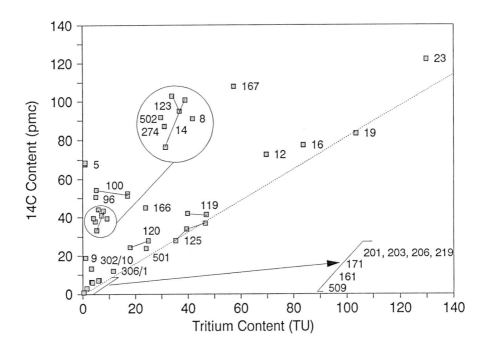

Figure 2.1.1: Graph of ^{3}H against ^{14}C of samples from northern Switzerland and adjacent areas.

The ^{14}C and ^{3}H contents of the samples from Kaisten Felsbohrung, Windisch BT 2, and Meltingen 2 (12, 16 and 19) are in the ranges found in modern water. These samples can be taken as virtually 100 per cent young water. The sample from Eptingen (23), although it too has high ^{3}H and ^{14}C values, is excluded from this discussion. It has a relatively positive δ^{13}C value, -3.6 per mil, suggesting that isotope exchange effects, associated perhaps with the evolution of CO_2 gas, may have influenced its carbonate chemistry, as discussed in Section 5.2.2.1.

The other samples along the line have ^{3}H/^{14}C ratios closely similar to those of the three modern samples. This indicates that the other samples also have a young water component, although their lower ^{3}H and ^{14}C values show that they contain lower proportions of young water. Estimates of these proportions are given in the comments in Table 2.1.3 and range from approximately 60 per cent, for Schinznach Bad alt (119) to less than five per cent for the samples from Beznau 104.5 to 109.5 m (171).

Samples from several points with measurable ^{3}H contents plot above and to the left of the band of samples. Waters with compositions in this range could be mixtures of young waters with older, lower ^{3}H waters. They could also be unmixed young waters recharged at a time when the ^{3}H/^{14}C ratio in surface water differed from that within the samples comprising the band. The presence of young water in these samples is noted in the comments to Table 2.1.3.

The final group of samples are those which have ^{3}H contents below detection (generally < 2 TU). As mentioned, these samples contain so little young water that their ^{14}C contents can be taken to represent those of water from regional flow systems.

SCHMASSMANN AND OTHERS (1984) discuss the chemistry, isotopic composition, and hydrologic setting of Nagra samples from northern Switzerland and adjacent regions. Their discussion include estimates of groundwater ages and describe the mixed character of many of these samples. Their conclusions are shown in Table 2.1.3 and are consistent with those drawn from the ^{3}H data alone.

2.2 Comparison of Carbon Isotope Results From Conventional and AMS Measurements

Table 2.1.2 indicates the procedure by which the ^{14}C measurements were made on samples from the test boreholes. Most ^{13}C analyses were made by conventional mass spectrometric analyses of CO_2 evolved from $BaCO_3$ precipitated from a large volume of sample water. Most ^{14}C measurements were made by counting ^{14}C decay in material also prepared from the $BaCO_3$. A few samples were analysed for both ^{13}C and ^{14}C by accelerator mass spectroscopy (AMS). The samples for AMS measurement were CO_2 extracted from small-volume water samples. Some of these yielded enough CO_2 that their δ^{13}C values could be measured by conventional mass spectroscopy. These are designated by (MS) in Table 2.1.2.

Inspection of Table 2.1.2 reveals consistent differences between the conventional and AMS results on duplicate samples. The most striking difference is that the δ^{13}C values of the small-volume AMS samples are significantly more negative than those measured on conventional samples of the same or hydrogeologically similar waters. Two hypotheses can be advanced to account for this difference.

First, a number of AMS analyses were made on samples with relatively high DOC values which may have been contaminated by carbonate from the oxidation of organic additives to the drilling fluid. The contaminating carbonate in such a case would be both depleted in ^{13}C and would be ^{14}C-bearing. Samples 302/16, 302/19, 304/6, 306/20 and 306/26 have relatively high DOC and ^{14}C contents and relatively negative δ^{13}C values. These samples could be accounted for by this hypothesis.

The second hypothesis has to do with the manner in which the samples are prepared. The $BaCO_3$ precipitate should include all dissolved carbonate species. Carbon samples prepared quantitatively from it will represent the isotopic composition of the total dissolved carbonate of the original water. The preparation of the small-volume gas samples, on the other hand, includes at least one step in which isotope fractionation is possible. CO_2 extracted from water under other-than-acid conditions will be depleted in the heavy isotopes relative to the total dissolved carbonate. The amount of fractionation will depend on the water chemistry, the temperature and the details of the extraction procedure, and will be of the order of seven to eight per mil for ^{13}C and 1.5 per cent for ^{14}C.

There are two pairs of samples in which the ^{13}C values of the small-volume samples are about seven to eight per mil more negative than those measured conventionally, and with ^{14}C values by the AMS method about 1.5 per cent lower than those measured by counting. These are samples 305/1 and 305/2.

The differences in the ^{14}C contents of the pair of analyses from sample 302/19 is 1.4 per cent, consistent with the fractionation hypothesis. However, the difference in $\delta^{13}C$ values for this pair is 28 per mil, which is too high to be explained by the fractionation hypothesis. However, this sample has a high DOC content. Oxidation of this carbon could have influenced the two types of analyses differently.

Only AMS analyses were made on samples from 301/21 and 302/18. The results of 301/21 can be attributed to the large quantity of drilling fluid in the sample. The results for the other sample appears reasonable. Because 302/18 has a relatively high DOC content, it should be used with caution.

The small-volume member of the pair of results on the samples from 306/1 is inconsistent with the conventional results by either hypothesis. This sample cannot presently be explained.

2.3 Evaluation of Replicate Analytical Data

Replicate chemical and isotopic data were collected on many Nagra water samples. Analyses were made on samples taken from the same source under different conditions. These include surface and downhole pressure vessel samples from the same zones of the deep boreholes, or samples taken at different times from the same well or spring. There are also analyses for the same constituent or property on replicate samples made by different laboratories, in many cases using different sampling and measurement techniques. When replicate data exist, it is important to choose for interpretation those which best represent the character of the groundwater under formation conditions.

Replicate analyses of dissolved solid species and major gases of samples from the Nagra test boreholes are given by WITTWER (1986). PEARSON AND OTHERS (1989) derive groundwater chemical compositions from them. This section describes the replicate data on dissolved gases and isotopes from both the deep borehole and the regional programmes, and the rationale followed in judging which values are most representative of groundwaters.

The sampling and analytical procedures used by the various laboratories are important in evaluating the replicate data. Information on these procedures is given in reports by

BALDERER (1985b), KUSSMAUL AND ANTONSEN (1985), and SCHMASSMANN AND OTHERS (1984).

2.3.1 Dissolved Gases

The dissolved gases comprise one of the largest groups of substances for which replicate analyses are available. Gases of relevance to this report include dissolved oxygen and the noble gases helium, neon, argon, krypton and xenon.

2.3.1.1 Oxygen

Analyses of oxygen dissolved in samples from both the deep boreholes and regional sampling points were made in the field by Fresenius. Samples of water collected to minimize gas loss were taken for laboratory analysis by PSI (KUSSMAUL AND ANTONSEN, 1985). Samples from the test boreholes were also taken downhole, under pressure, for analysis at PSI as described by HAUG (1985). Finally, aliquots of the large-volume gas samples extracted in the field for ^{39}Ar and ^{85}Kr analysis, were also analysed at PSI.

The dissolved oxygen contents of the water and large-volume gas samples were used to evaluate air contamination of the large-volume samples as described in Section 2.1.1. This evaluation used PSI results to assure consistency of analytical techniques.

2.3.1.2 Noble Gases

Analyses of some or all of the noble gas were made in several laboratories on samples collected by various techniques. Samples specifically for noble gases were collected in copper tubing and analysed at the University of Bath by Professor J. Andrews. These were taken under pressure at the surface from the regional sampling points and from some intervals of the test boreholes. Other test borehole samples were taken downhole. A few samples taken early in the regional programme were analysed at the Weizmann Institute by Dr. E. Mazor.

Table 2.3.1 compares the results of the analyses performed at the Weizmann Institute with those made at the University of Bath. There are only three locations from which samples for noble gases were analysed by both laboratories. As the table shows, the results for all gases except helium, while not identical, agree reasonably well.

The results of the Bath analyses for neon, argon, krypton and xenon were used to calculate infiltration temperatures. As described in Section 3.3.2, inconsistencies among the temperatures calculated for a given sample using the several noble gases can be an indication of difficulties during sample collection, storage or analysis.

Figure 2.3.1 compares the results of the Bath argon analyses with those made at PSI. Two sets of data are distinguished in the figure. The solid circles are the Bath results on

Table 2.3.1: Noble gas analyses on replicate samples made at Weizmann Institute, the University of Bath, and PSI.

Location /Sample Number	Name	Helium 10^{-8} cc/cc Bath	Weizmann	PSI	Neon 10^{-7} cc/cc Bath	Weizmann	Argon 10^{-4} cc/cc Bath	Weizmann	PSI	Krypton 10^{-8} cc/cc Bath	Weizmann	Xenon 10^{-8} cc/cc Bath	Weizmann
1/1 (OMM)	Zürich Aqui	468 501	920 740 750	1200	3.2 4.4	7.8 3.4	5.7 10.4	4.7	4.9	27.6	16.9 2.9	1.6	1.6
123/1 (mo)	Lostorf 3		12 16	n.n.		3.6 4.2		3.4 3.5	4.5		13.3 12.1		1.4 1.2
124/1 (mo)	Lostorf 4		66 30 36	n.n.		3.3 2.7 2.1		3.5 3.0	3.9		13.7 11.1		1.6 1.2
125/1 (KRI)	Säckingen Badquelle	16686	10500 13600	6000 12000	3.3	2.0 3.4	5.2	2.7 3.4	2.9	13.5	10.4 12.3	2.0	1.3 1.3
132/1 (KRI)	Zurzach 2	10000	11000	18000		3.1 3.2		3.7 3.6	4.9		14.0 11.3		1.6 1.2
201/1 (mo)	Ennetbaden Allgemeine	1052	10600 9720	18000	3.5	0.9 0.8	2.7	2.2 1.9	2.3	6.5	9.7 8.1	1.2	1.6 1.0
219/1 (mo)	Baden Verenahof		9100 7400	18000 12000		0.8 0.6		2.1 1.8	2.4 2.2		7.9 6.9	1.1	0.9

samples in which noble gas analyses lead to consistent recharge temperatures. The open circles are results for samples which lead to inconsistent recharge temperatures, and which are, therefore, considered unacceptable analyses. While there is good agreement between the PSI results and the acceptable Bath analyses, many of the unacceptable Bath results are considerably lower than the corresponding PSI result.

The techniques used to collect samples for gas analyses are described by BALDERER (1985b) and HAUG (1985). Samples analysed at Bath consisted of about 5 cm^3 of water taken under pressure in copper tubing which was clamped and sealed before pressure release. The PSI samples were of 0.5 l or more water, and were taken under pressure both at depth and at the surface. After collection, the PSI samples were taken directly to the laboratory and were analysed within a few days. The Bath samples, on the other hand, were shipped to the laboratory in groups, and the time between sampling and analysis often reached several months. If some of the Bath samples lost gas between sampling and analysis, it would account for the low argon contents of many of the unacceptable samples relative to the PSI analyses, as shown in Figure 2.3.1.

Two samples are not included in Figure 2.3.2, numbers 306/4, from the Leuggern, and 304/10 from the Schafisheim borehole. One of the samples from 306/4 analysed at Bath contained 3.7 mg Ar/kg compared with 0.9 mg/kg in the other sample analysed at Bath, and 0.8 mg/kg analysed at PSI. The PSI sample of 304/10 had 6.2 mg Ar/kg, while three samples measured at Bath had concentrations ranging from 0.6 to 0.9 mg/kg. The high

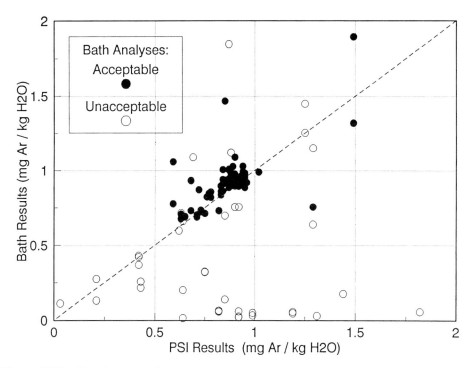

Figure 2.3.1: Graph comparing argon analyses made at the University of Bath with those made at PSI.

values in both cases may well have resulted from the entrainment of gas bubbles during the collection procedure.

Helium analyses were made by Fresenius, PSI, at Bath, and at Weizmann. Helium concentrations were also measured at the University of Cambridge on samples which were collected in copper tubes, like the Bath samples, and analysed for their ^{3}He/^{4}He ratios. The Weizmann helium results are included in Table 2.3.1. Results of replicate analyses made at Bath, Cambridge and PSI are given in Table 2.3.2, at the end of this chapter. The Bath results for samples with noble gas analyses unacceptable for recharge temperature calculations are indicated in this table.

Figure 2.3.2 shows the Bath and Cambridge results plotted against the PSI results. The Bath results accepted for noble gas calculations are differentiated from the unacceptable values as they were in Figure 2.3.1.

Most of the unacceptable Bath analyses, and some of the Cambridge results as well, have helium contents far smaller than those reported by PSI. This is consistent with the possibility of gas loss from the Bath samples as proposed to account for the argon results discussed above. The Cambridge samples were collected, stored and shipped using the same equipment and procedures as the Bath samples. Thus, gas loss could also have occurred from some of the Cambridge samples.

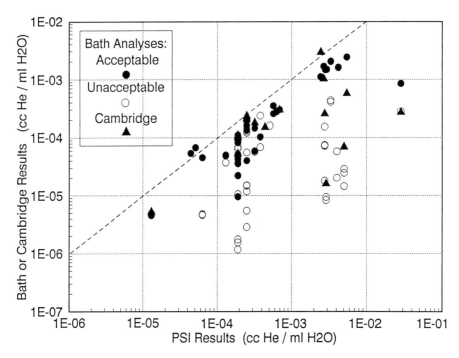

Figure 2.3.2: Graph comparing helium analyses made at the University of Bath with those made at the University of Cambridge and at PSI.

The acceptable Bath samples generally agree with the PSI results, although, as Figure 2.3.2 shows, there remains a tendency for the Bath results to be lower than the PSI values by a factor of about two. The reason for this difference is not known. However, because the helium concentrations have a range of nearly three orders of magnitude, uncertainties of a factor of two in the helium concentrations do not significantly affect their interpretation.

The sample with the highest helium content measured by PSI ($3 \cdot 10^{-2}$ cc He/ml H_2O) is 304/10. PSI also reported an exceptionally high Ar content for this sample, although the Bath results are acceptable for noble gas temperature calculations. The high helium content of this sample supports the explanation proposed for its high argon value--that a bubble of free gas was entrained in this sample during its collection.

2.3.2 Uranium, Thorium and Daughter Elements

Ratios among various uranium, thorium, and radium isotopes, as well as uranium and radium concentrations, were measured by the Atomic Energy Research Establishment, Harwell. Uranium concentrations and $^{234}U/^{238}U$ ratios were also measured by PSI, and uranium and radium concentrations by Fresenius. The Fresenius and PSI results on samples from the deep boreholes are given by WITTWER (1986). The detection limit of the Fresenius method for uranium analysis is from 0.1 to 0.5 μg/l, considerably larger than those of the other laboratories. Thus, the Fresenius uranium results can be compared with those of the other laboratories only at higher concentrations.

Figure 2.3.3 compares the uranium analyses of PSI, Fresenius and Harwell. As this graph shows, there is generally good agreement among all three sets of data. The uranium results are discussed in Section 7.3.2, along with the significance of analytical uncertainty.

There are major differences between the $^{234}U/^{238}U$ ratios reported by PSI, which are given by WITTWER (1986), and those from Harwell. In general, differences between replicate analyses by PSI are significantly larger than those between analyses by Harwell. Thus, only the Harwell $^{234}U/^{238}U$ values are used in this report.

Radium concentrations measured by Fresenius and by Harwell are given in Table 2.3.3, and displayed in Figure 2.3.4. As the figure shows, there is only qualitative agreement between the two sets of results. It may be that analyses of substances present in such low concentrations should not be expected to agree within factors smaller than two to five.

Radon (^{222}Rn) activities were measured by both Fresenius and PSI. The measurement techniques and comparisons between the two sets of values are discussed in Section 7.3.2.

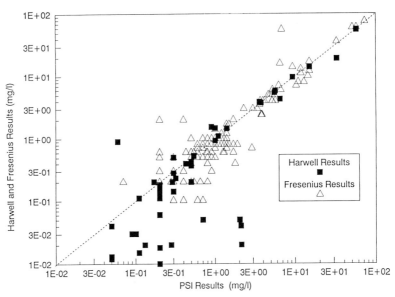

Figure 2.3.3: Graph comparing uranium concentrations measured at Fresenius, PSI and AERE, Harwell.

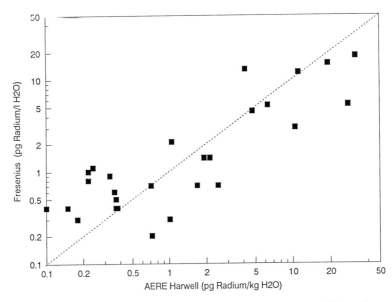

Figure 2.3.4: Graph comparing radium concentrations measured at Fresenius and AERE, Harwell.

Table 2.1.1: Evaluation of contamination of large-volume gas samples. (Page 1 of 5)

Sample Location Number	WATER SAMPLES				LARGE GAS SAMPLES						GAS SAMPLE CONTAMINATION (Percent) Calculated from		VERY YOUNG FLUID CONTAMINATION (Percent) Calculated from		Comments
	Quality Block	O2 (vol% Total Gas)	Kr/Ar 10-3 cc/cc	3H (TU)	O2 (vol%)	Ar (vol%)	Kr/Ar 10-3 cc/cc	39Ar (%mdn)	85Kr (dpm/cc)	3H/85Kr (TU/(dpm/cc))	O2	85Kr	85Kr	3H	
1/1	1	tr.	0.208 0.265	< 0.6	<0.03	1.86 ± 0.04	- -	< 4.7	0.089	< 6.7	< 0.1	< 0.8 < 1.0	0.2	0.7	Uncontaminated gas samples; < 2% young water component.
2/3	1	0.02	0.211	< 0.9	0.05 ± 0.01	1.87 ± 0.04	0.209	8 ± 5	< 0.17	< 5.3	0.1	< 1.5	< 0.4	1.0	< 1% gas sample contamination; < 2% young water component.
8/2	1	0.55	0.219	9.2 ± 0.8	0.80	1.82	0.156	255 ± 11 249 ± 11	3.42 ± 0.22	2.7 ± 0.2	1.2	< 21.3	8.6	10.2	< 2% gas sample contamination; significant natural young water.
9/2	1	tr.	0.219	< 1.1	12.1	1.20	0.155	32.8 ± 3.8	6.87 ± 0.27	< 0.2	57.9	< 28.0	17.2	1.2	Considerable gas sample contamination; 3H suggests < 2% young water component.
14/3	1	0.01	0.204	7.8 ± 0.9	tr.	1.71	0.153	63.6 ± 3.5	0.79 ± 0.03	9.9 ± 0.7	0.0	< 4.5	2.0	8.7	Uncontaminated gas sample; Significant young water component.
15/1	1	0.04	0.365	< 0.6	<0.03	1.06 ± 0.02	- -	< 4.6	0.09 ± 0.01	< 6.7	0.0	< 0.8	0.2	0.7	Uncontaminated gas sample; < 2% young water component.
123/110				1.7 ± 2.0	tr.	- -	- -	52.5 ± 4.2	0.3 to 0.9	< 5.7	0.0	- -	0.8	1.9	Uncontaminated gas sample; 85Kr suggests < 5% young water; 3H of later samples ranges up to 7.8 TU.

Table 2.1.1: Evaluation of contamination of large-volume gas samples. (Page 2 of 5)

Sample Location Number	WATER SAMPLES			LARGE GAS SAMPLES						3H/85Kr (TU/(dpm/cc))	GAS SAMPLE CONTAMINATION (Percent) Calculated from		VERY YOUNG FLUID CONTAMINATION (Percent) Calculated from		Comments
	O2 Qual-(vol% ity Total Block Gas)	Kr/Ar 10-3 cc/cc	3H (TU)	O2 (vol%)	Ar (vol%)	Kr/Ar 10-3 cc/cc	39Ar (%mdn)	85Kr (dpm/cc)			O2	85Kr	85Kr	3H	
125/1	1.50	0.261	46.5 ± 2.0	1.48 ± 0.06	0.73 ± 0.02	--	224 ± 7	10.20 ± 0.30		4.6 ± 0.2	0.0	< 42.7	25.5	51.7	Uncontaminated gas sample; significant young water component.
271/103	--	0.224	< 0.8	15.07	1.17	--	17.3 ± 3.2	0.27 ± 0.11		< 3.0	72.1	< 1.6	0.7	0.9	Sample for O2 may be contaminated; 85Kr and 3H reflect < 2% young water.
271/104	--	--	< 1.1	0.01	1.55	--	11.3 ± 2.1	0.15 ± 0.10		< 7.3	0.0	--	0.4	1.2	Uncontaminated gas sample; 85Kr and 3H reflect < 2% young water.
272/102	--	0.230	< 1.6	--	--	--	16.3 ± 2.8	--		--	--	--	--	1.8	
272/103	--	--	< 1.0	0.02	1.62	--	--	0.10 ± 0.01		< 9.8	0.1	--	0.3	1.1	< 1% gas sample contamination; 85Kr and 3H reflect < 2% young water.
273/102	--	--	< 1.9	--	--	--	23.0 ± 2.4	0.072 ± 0.006		< 26.4	--	--	0.2	2.1	
273/103	--	--	< 0.9	0.01	1.43	--	23.7 ± 2.4	0.101 ± 0.008		< 8.9	0.0	--	0.3	1.0	< 1% gas sample contamination; 85Kr and 3H reflect < 2% young water.
274/102			4.7 ± 0.7	0.02		--	41.9 ± 3.1	0.609 ± 0.037		7.7 ± 0.7	0.1	< 0.0	1.5	5.2	Uncontaminated gas sample; significant young water component.

Table 2.1.1: Evaluation of contamination of large-volume gas samples. (Page 3 of 5)

Sample Location Number	WATER SAMPLES			LARGE GAS SAMPLES						GAS SAMPLE CONTAMINATION (Percent) Calculated from		VERY YOUNG FLUID CONTAMINATION (Percent) Calculated from		Comments
	Qual-ity Block	O2 (vol% Total Gas)	Kr/Ar 10-3 cc/cc	3H (TU)	O2 (vol%)	Ar (vol%)	Kr/Ar 10-3 cc/cc	39Ar (%mdn)	85Kr (dpm/cc)	3H/85Kr (TU/(dpm/cc))	O2	85Kr	85Kr	3H	
301/2	1	0.09	0.227 ± 0.227	< 1.0	0.13 ± 0.02	1.03 ± 0.02	0.195	< 6.0	0.15 ± 0.05	< 6.7	0.2	< 0.7	0.4	1.1	< 1% gas sample contamination; < 2% young water component.
301/5	2	0.11 0.11	- -	< 0.7	0.56 ± 0.10 0.26 ± 0.10	1.74 ± 0.03 1.73 ± 0.03	0.201	148 ± 5 149 ± 4	0.33 ± 0.04	< 2.1	2.2 0.7	< 2.5 < 2.5	0.8	0.8	c. 1-3% gas sample contamination; < 2% young water component.
301/12b	1	0.02 0.02 0.02	0.211 ± 0.212	1.0 0.7	0.59 ± 0.1 0.60 ± 0.1 0.83 ± 0.1	1.74 ± 0.03 1.63 ± 0.03 1.62 ± 0.03	0.224	450 ± 10	1.75 ± 0.09	0.6 ± 0.2	2.7 2.8 3.9	< 15.0 < 14.0 < 13.9	4.4	1.1	Some contamination of gas sample; < 2% young water component.
301/18	1	tr.	0.231 ± 0.234	0.9 0.9	tr.	1.65	0.177	523 ± 11	- -	- -	0.0	- -	- -	1.0	Uncontaminated gas sample; < 2% young water component.
301/23	2	- -	0.224 ± 0.224	3.2 0.7	0.10	1.65	- -	510 ± 12	0.28 ± 0.02	11.4 ± 1.5	0.5	< 2.3 < 2.3	0.7	3.6	< 1% gas sample contamination; < 4% young water component.
302/10 a and b	1	0.04	0.231 ± 0.230	3.2 0.7	tr.	1.58 1.47	0.168	16.5 ± 3.6 15.5 ± 1.6	0.65 ± 0.04	4.9	0.0	< 3.8	1.6	3.6	Uncontaminated gas sample; 3H suggests 3-5% young water component.
302/12	1	tr.	0.217 ± 0.224	< 0.8	2.82	0.93	0.165	44.8 ± 3.8	6.17 ± 0.26	< 0.1	13.5	< 20.8	15.4	0.9	Considerable gas sample contamination; < 2% young water component.

Table 2.1.1: Evaluation of contamination of large-volume gas samples. (Page 4 of 5)

Sample Loca-tion Number	WATER SAMPLES				LARGE GAS SAMPLES						GAS SAMPLE CONTAMINA-TION (Percent) Calculated from		VERY YOUNG FLUID CONTAMI-NATION (Percent) calculated from		Comments
	Qual-ity Block	O2 (vol% Total Gas)	Kr/Ar 10-3 cc/cc	3H (TU)	O2 (vol%)	Ar (vol%)	Kr/Ar 10-3 cc/cc	39Ar (%mdn)	85Kr (dpm/cc)	3H/85Kr (TU/(dpm/cc))	O2	85Kr	85Kr	3H	
302/19	2	0.01	0.224 ± 0.221	2.2 ± 0.6	0.98	0.47	– –	114 ± 6	20.5 ± 3.1	0.1 ± 0.0	4.6	< 47.2 46.6	51.3	2.4	Considerable gas sample contamina-tion; 3H suggests < 3% young water component.
303/2	1	tr.	0.219 < 0.242	1.1	tr.	1.63	– –	< 7.0	– –	– –	0.0	– –	– –	1.2	Uncontaminated gas sample; < 2% young water component.
303/3	1	0.01	0.194 ± 0.188	0.7 ± 0.7	tr.	0.38	0.213	106.5 ± 6.6	0.16 ± 0.01	4.4 ± 2.2	0.0	< 0.3	0.4	0.8	Uncontaminated gas sample; < 2% young water component.
303/6	2	n.d.	0.187 ± 0.236 0.239	2.6 ± 1.0	0.01	0.34	– –	270 ± 15	0.96 ± 0.03	2.7 ± 0.5	0.0	< 1.3 < 1.7 < 1.7	2.4	2.9	Negligible gas sample contamina-tion; c. 2-3% young water compo-nent; Chemistry supports c. 5% drill fluid.
304/14	2	– –	0.304 ± 0.294	1.4 ± 0.7	0.00	0.18	– –	14.2 ± 4.3	– –	– –	0.0	– –	– –	1.6	Uncontaminated gas sample; < 2% young water from 3H and chemistry.
305/1	1	0.04	0.188 ± 0.206	0.4 ± 0.2	0.06	0.97	– –	57.7 ± 5.2 57.0 ± 4.6	0.74 ± 0.06	0.5 ± 0.1	0.1	< 3.0 < 3.3	1.9	0.4	< 1% gas sample contamination; < 2% young water component.
305/2	1	tr.	0.211 ± 0.209	0.7 ± 0.3	tr.	1.40	– –	192 ± 6	0.35 ± 0.02	2.0 ± 0.4	0.0	< 2.3 < 2.2	0.9	0.8	Uncontaminated gas sample; c. 2% young water compo-nent.

Table 2.1.1: Evaluation of contamination of large-volume gas samples. (Page 5 of 5)

Sample Location Number	WATER SAMPLES			LARGE GAS SAMPLES						GAS SAMPLE CONTAMINATION (Percent) Calculated from		VERY YOUNG FLUID CONTAMINATION (Percent) Calculated from		Comments
	Qual-ity Total Block Gas)	O2 (vol%	Kr/Ar 10-3 cc/cc	3H (TU)	O2 (vol%)	Ar (vol%)	Kr/Ar 10-3 cc/cc	39Ar (%mdn)	85Kr (dpm/cc)	3H/85Kr (TU/(dpm/cc)	O2	85Kr	85Kr	3H	
305/4	1	tr.	0.199 0.206 ±	1.1 0.7	9.50	1.24	--	183 ± 8	0.50 ± 0.01	2.2 ± 0.7	45.5	< 2.7 < 2.8	1.3	1.2	Sample for O2 may be contaminated; < 2% young water component.
306/1	1	0.09 0.09	0.228 0.222 ±	11.6 1.1 11.6 1.1	0.28 0.20	1.84 1.85 1.94	--	53.0 ± 3.7	3.42 ± 0.13 3.27 ± 0.14	3.4 ± 0.2 3.5 ± 0.2	0.9 0.5	< 31.5 < 30.7 < 30.1	8.6 8.2	12.9	< 1% gas sample contamination; significant young component; Chemistry suggests natural young water.
306/2	2	0.02	0.210 0.211 0.193	< 0.8	1.70	1.61	0.164	173 ± 7	5.44 ± 0.39	< 0.1	8.0	< 31.5	13.6	0.9	Considerable gas sample contamination; 3H suggests < 1% young water component.
306/9	1	0.03	0.239 0.243	< 1.3	tr.	1.48	--	168 ± 5	0.22 ± 0.02	5.9	0.0	< 1.7 < 1.7	0.6	1.4	Uncontaminated gas sample; < 2% young water.
306/16	1	0.01	0.218 0.181 ±	0.7 0.7	0.24	1.53	0.153	586 ± 10	0.73 ± 0.06	1.0 ± 0.5	1.1	< 3.7	1.8	0.8 0.8	< 2% gas sample contamination; < 2% young water.

Table 2.1.2: Estimates of 14C contamination of samples from Nagra deep boreholes. (Page 1 of 4)

Sample Location Number	Source Formation	Mean Depth (m)	Quality Block (GSF) (Wittwer, 1986)	3H (TU)	Dissolved Organic Carbon (mg/l)	Total Dissolved Carbonate (mmol) (NTB 86-19)	Young Fluid Contents (per cent)	14C Contamination from Fluid (pmc) (See Text)	Delta 13C (per mil)	14C (pmc)	Type of Analysis	Comments
BÖTTSTEIN			Estimated Drill Fluid Alkalinity 300									
301/2	mo	162.9	1	< 1.0	3.5	8.71	< 2	< 1.4	-4.8	1.49 ± 0.12	Count	Probably all 14C from drilling fluid.
301/25b	s-KRI	312.5	2	0.2 ± 0.2	2.90	6.16		< 0.2	-9.3	14.03 ± 0.13	Count	Negligible contamination from drilling fluid.
301/5	s-KRI	316.6	2	< 0.7	11.10	6.74	< 1	< 0.9	-6.3	12.13 ± 0.25	Count	Negligible contamination from drilling fluid; Relatively high DOC.
			Estimated Drill Fluid Alkalinity 200									
301/8c	KRI	399.5	1	< 1.0	0.90	6.14		< 1.1	-9.3	8.25 ± 0.13	Count	Negligible contamination from drilling fluid.
301/16	KRI	618.4	1	1.0 ± 0.7	0.74	6.18		< 1.1	-9.0	7.89 ± 0.09	Count	Negligible contamination from drilling fluid.
301/12b	KRI	621.3	1	1.0 ± 0.7	0.82	6.45		< 1.1	-8.8	< 1.30	Count	Contamination below detection.
301/23	KRI	649.0	2	3.2 ± 0.7	- -	6.40	< 4	< 2.6	-9.1	13.80 ± 0.13	Count	Negligible contamination from drilling fluid.
301/18	KRI	792.4	1	0.9 ± 0.9	0.53	6.34		< 1.0	-9.1	7.64 ± 0.07	Count	Negligible contamination from drilling fluid.
301/21	KRI	1326.2	2	57 ± 5	0.20	1.94			-6.3 ± 2.9	154.08 ± 2.16	AMS	All 14C from drilling fluid.
WEIACH			Estimated Drill Fluid Alkalinity 300									
302/6	joki	255.0	2	< 5.7	18.00	8.79		< 4.4	-4.2 (MS)	12.47 ± 0.25	AMS	Some 14C probable from drilling fluid and (or) from additives.
302/10a,b	mo	859.1	1	3.2 ± 0.7	34.00	7.22	< 5	< 4.3	-4.6 ± 0.5	13.28 ± 0.14	Count	Probable natural young groundwater.
302/12	s	985.3	1	< 0.8	3.90	12.99	< 2	< 0.9	-8.3	3.63 ± 0.15	Count	Negligible contamination from drilling fluid; Low DOC for this borehole.
302/19	r	1116.5	2	2.2 ± 0.6	74.00	13.00	< 3	< 1.4	-25.4 (MS) 2.6	12.60 ± 0.30 13.96 ± 0.50	AMS Count	Negligible contamination from drilling fluid; Possible contamination from additives.

Table 2.1.2: Estimates of 14C contamination of samples from Nagra deep boreholes. (Page 2 of 4)

Sample Location Number	Source Formation	Mean Depth (m)	Quality (GSF) Block (Wittwer, 1986)	3H (TU)	Dissolved Organic Carbon (mg/l)	Total Dissolved Carbonate (mmol) (NTB 86-19)	Young Fluid Contents (per cent)	14C Contamination from Fluid (pmc) (See Text)	Delta 13C (per mil)	14C (pmc)	Type of Analysis	Comments
	WEIACH (continued)		Estimated Drill Fluid Alkalinity					300				
302/18	r	1408.3	2	6.4 ± 0.9	21.50	3.88		11.3	-4.4 ± 4.9	< 0.40	AMS	Contamination below detection.
302/16	KRI	2218.1	2	2.7 ± 0.7	7.80	1.60		11.5	-26.7 ± 2.0	10.10 ± 0.40	AMS	Some 14C probable from drilling fluid.
	RINIKEN		Estimated Drill Fluid Alkalinity					300				
303/1	km	515.7	1	< 1.2	1.4	3.87	< 2	2.1	-8.2	3.35 ± 0.30	Count	Significant contamination from drilling fluid; Low DOC.
303/2	mo	656.7	1	< 1.1	1.0	6.65	< 2	1.8	-4.8	0.90 ± 0.20	Count	Probably all 14C from drilling fluid; Low DOC.
303/3	s-r	806.6	1	0.7 ± 0.7	0.68	25.18	< 2	0.5	-9.2	0.55 ± 0.20	Count	Probably all 14C from drilling fluid; Low DOC.
303/5b,c	r	965.5	2	< 1.1	9.2	10.70		0.7	-7.6	10.85 ± 0.20	Count	Negligible contamination from drilling fluid; Possible contamination from additives.
303/6	r	993.5	2	2.6 ± 1.0	144	11.50	< 5	2.7	-14.6	40.81 ± 0.60	Count	Negligible contamination from drilling fluid; Possible contamination from additives.
	SCHAFISHEIM		Estimated Drill Fluid Alkalinity					300				
304/1,2	USM	558.0	1	< 1.2	2.0	2.31		3.5	-8.5	40.46 ± 0.70	Count	Moderate contamination from drilling fluid; Low DOC.
304/3	mo	1260.4	1	< 0.9	4.0	12.50		0.5		4.41 ± 1.20	Count	Negligible contamination from drilling fluid; Possible contamination from additives.
304/5,6	s-KRI	1488.2	2	1.8 ± 0.7	95.0	20.86		0.6	-21.2 ± 2.0	1.40 ± 0.40	AMS	Significant contamination from drilling fluid; Probable contamination from additives.
304/10	KRI	1887.9	2	1.5 ± 0.9	0.3	26.01		0.4	-9.2	< 0.70	Count	Contamination below detection.

Table 2.1.2: Estimates of 14C contamination of samples from Nagra deep boreholes. (Page 3 of 4)

Sample Loca- tion Number	Source For- ma- tion	Mean Depth (m)	Qual- ity Block (GSF) (Wittwer, 1986)	3H (TU)	Dis- solved Organ- ic Carbon (mg/l)	Total Dis- solved Carbor- ate (mmol) (NTB 86-19)	Young Fluid Con- tents (per cent)	14C Con- tami- nation from Fluid (pmc) (See Text)	Delta 13C (per mil)	14C (pmc)	Type of Anal- ysis	Comments
KAISTEN			Estimated Drill Fluid Alkalinity 300									
305/1	s	113.5	1	0.4 ± 0.2	0.34	7.52	< 2	< 1.6	-17.2 (MS) -9.8	< 0.50 1.64 ± 0.30	AMS Count	Contamination below detection. Probably all 14C from drilling fluid; Low DOC.
305/2	r	284.3	1	0.7 ± 0.3	0.69	6.83		< 0.7	-17.8 (MS)	0.80 ± 0.20	AMS	Possible contamination from drilling fluid.
			Estimated Drill Fluid Alkalinity 200									
305/4	KRI	310.4	1	1.1 ± 0.7	0.33	7.13	< 2	< 1.1	-9.8	0.98 ± 0.20	Count	Probably all 14C from drilling fluid; Low DOC.
305/6	KRI	482.6	1	1.3 ± 0.7	0.55	6.49		< 0.9	-10.4	0.86 ± 0.20	Count	Probably all 14C from drilling fluid; Low DOC.
305/9	KRI	819.4	1	0.9 ± 0.7	0.58	6.70		< 0.6	-9.9	0.63 ± 0.20	Count	Probably all 14C from drilling fluid; Low DOC.
305/12	KRI	1031.0	1	0.8	< 0.10	7.06		< 0.5	-9.2	0.84 ± 0.20	Count	Significant contamination from drilling fluid; Low DOC.
305/16	KRI	1271.9	1	1.2 ± 0.7	0.95	6.57		< 0.8	-10.1	1.56 ± 0.30	Count	Significant contamination from drilling fluid; Relatively high DOC for borehole.
LEUGGERN			Estimated Drill Fluid Alkalinity 300									
306/1	mo	74.9	1	11.6 ± 1.1	0.30	4.98	13	16.0	-15.8 (MS) -6.4	12.00 ± 0.30 11.23 ± 0.30	AMS Count	Probable natural young water.
306/2	s	217.9	2	< 0.8	0.91	5.02	< 2	2.4	-9.3	1.56 ± 0.20	Count	Probably all 14C from drilling fluid; Moderate DOC for borehole.
			Estimated Drill Fluid Alkalinity 200									
306/4	KRI	251.2	1	0.8 ± 0.7	0.22	4.83		< 0.8	-10.1	1.36 ± 0.30	Count	Significant contamination from drilling fluid; Low DOC.

Table 2.1.2: Estimates of 14C contamination of samples from Nagra deep boreholes. (Page 4 of 4)

Sample Location Number	Source Formation	Mean Depth (m)	Quality Block (GSF) (Wittwer, 1986)	3H (TU)	Dissolved Organic Carbon (mg/l)	Total Dissolved Carbonate (mmol) (NTB 86-19)	Young Fluid Content (per cent)	14C Contamination from Fluid (pmc) (See Text)	Delta 13C (per mil)	14C (pmc)	Type of Analysis	Comments
	LEUGGERN (continued)											
306/5	KRI	444.2	1	< 1.2	0.70	4.01	Estimated Drill Fluid Alkalinity 200	< 1.4	-10.1	3.19 ± 0.20	Count	significant contamination from drilling fluid; Moderate DOC for borehole.
306/7	KRI	538.0	1	< 1.0	0.28	3.20		< 1.4	-10.2	2.41 ± 0.20	Count	significant contamination from drilling fluid; Low DOC.
306/9	KRI	705.7	1	< 1.3	0.30	3.63	< 2	< 2.3	-10.2	1.40 ± 0.30	Count	Probably all 14C from drilling fluid; Low DOC.
306/17	KRI	847.0	2	3.5 ± 0.7	2.20	3.91		< 4.1	-9.8	2.18 ± 0.20	Count	Probably all 14C from drilling fluid; High DOC for borehole.
306/20	KRI	1203.2	2	1.3 ± 0.7	0.93	3.93		< 1.5	-26 ± 5	4.50 ± 0.30	AMS	Moderate contamination from drilling fluid; High DOC for borehole.
			2	1.4 ± 0.9	4.00	3.29		< 1.9	-23 ± 5	11.70 ± 0.30	AMS	Moderate contamination from drilling fluid; Probable contamination from additives.
306/16	KRI	1643.4	1	0.7 ± 0.7	0.34	4.59	< 2	< 1.8	-9.3	1.19 ± 0.20	Count	Probably all 14C from drilling fluid; High DOC for borehole.

Table 2.1.3: Young water contents of regional programme samples. (Page 1 of 3)

Sample Location Number	Source Formation	Tritium (TU)	Dissolved Organic Carbo (mg/l)	Total Dissolved Carbonate (mmol)	Delta 13C (per mil)	14C (pmc)	Comments
1/1	tOMM	< 0.7	0.50	6.80	-3.0	< 1.3	< 2% young water; Sect. 4.2 and 14C.
2/1	tOMM	< 0.7	0.25	4.91	-3.1	< 2.2	< 2% young water; Sect. 4.2 and 14C.
3/1	s-r	< 0.7	0.30	17.14	-8.3	2.85 ± 0.28	
3/3	s-r	< 0.7	1.00	17.98	-9.2	2.55 ± 0.30	
5/1	tUSM	0.9 ± 0.7	0.40	8.71	-14.2	67.7 ± 0.9 / 68.5 ± 1.1	
8/2	mo	9.2 ± 0.8	1.00	4.46	-9.5	39.4 ± 0.4	c. 15% young water; Sect. 4.2.
9/2	mo	< 1.1	0.42	6.30	-9.8	18.97 ± 0.3	< 2% young water; Sect. 4.2.
12/1	mo	69.8 ± 4.8	0.40	8.10	-14.0	75.6 ± 0.6	c. 100% young water; Sect. 2.1.1.
14/1	mo	5.2 ± 0.6	0.70	6.60	-8.9	33.3 ± 0.4	7-40% young water; Sect. 4.2.
14/3	mo	7.8 ± 0.9	0.39	6.62	-9.6	43.4 ± 0.4	
15/1	mo	< 1.0	0.40	7.69	-5.4	< 0.4	< 2% young water; Sect. 4.2 and 14C.
16/1	mo	83.7 ± 5.6	0.6	8.18	-13.4	77.3 ± 0.6	c. 100% young water; Sect. 2.1.1; Young, mixed water: NTB 84-21.
19/1	km (+?m)	103.5 ± 6.8	1.40	9.79	-15.4	83.2 ± 0.8	c. 100% young water; Sect. 2.1.1; < 10 years: NTB 84-21.
23/1	km	129.8 ± 8.5	3.30	2.87	-3.6	122.0 ± 0.9 / 121.9 ± 1.0	c. 100% young water; Sect. 2.1.1; < 10 years: NTB 84-21.
77/1	jmHR	< 0.7	0.31	7.70	-4.4	1.3 ± 0.2 / 1.2 ± 0.3	
95/2	KRI	< 0.9	0.46	6.76	-8.0	0.69 ± 0.2	
96/2	tvu	4.9 ± 0.7	0.47	4.70	-10.5	50.67 ± 0.5	c. 7% young water: 3H.
97/1	jo	< 0.7	0.40	6.98	-5.2	< 2.2 / < 2.2	
97/103	jo					10.7 ± 0.4	14C grossly inconsistent with other samples from location 97.
97/105	jo	0.6		5.54	-6.1	0.5	

Table 2.1.3: Young water contents of regional programme samples. (Page 2 of 3)

Sample Location Number	Source Formation	Tritium (TU)	Dissolved Organic Carbo (mg/l)	Total Dissolved Carbonate (mmol)	Delta 13C (per mil)	14C (pmc)	Comments
98/1	tOMM	< 0.7	0.40	6.65	-4.3	< 1.3 < 1.3	
98/102	tOMM	0.5		6.89	-4.0	0.6	
100/1	jo (-q)	17.0 ± 1.3 17.0		6.62	-11.2	52.4 ± 0.5 51.2 ± 0.5	c. 20% young water; 3H: (Chemistry and 3H from 149/1).
100/101	jo (-q)	5.1 ± 1.1				54.3 ± 1.0	c. 7% young water; 3H.
119/118	mo	46.9 ± 5.1		7.07	-12.0	41.4 ± 0.5	c. 60% young water: Sect. 2.1.1; More young water than 120.
119/119	mo	39.8 ± 4.5		6.73	-10.4	42.0	
120/1	mo			5.45	-9.3	27.1 ± 0.4	25-35% young water: 3H; 30-50% young water: NTB 84-21.
120/118	mo	18.0 ± 2.1		7.50	-10.4	24.5 ± 0.3	
120/119	mo	24.9 ± 4.2		8.50	-9.4	28.0	
123/110	mo				-9.4	44.0 ± 1.2	< 5% young water; Sect. 4.2; 5-10% young water: 3H.
123/111	mo	5.8 ± 0.7		5.21	-9.6	44.1 ± 0.5	
123/113	mo	7.1 ± 3.3		5.29	-9.3	41.0	
124/109	mo				-6.8	14.8 ± 0.8	
124/110	mo	< 1.0		5.26	-6.8	15.5 ± 0.2	
124/111	mo	< 0.8		5.27	-9.3	15.0	
125/1	KRI	46.5 ± 2.0	0.20		-13.0	36.9 ± 0.4	c. 50% young water; Sect. 4.2; c. 60% young water; Sect. 2.1.1.
125/108	KRI	39.4 ± 4.2		6.08	-14.2	34.0 ± 0.4	
125/109	KRI	35.3 ± 4.3		6.22	-12.4	28.0	
127/1	jo	< 0.7	0.70	9.18	-3.8	2.1 ± 0.4 2.0 ± 0.3	
127/104	jo	0.8 ± 0.7		9.01	-3.9	3.5	
128/1	tOMM	< 0.9	0.10	8.96	-5.5	< 2.2 < 2.2	
128/101	tOMM	0.3		8.93	-5.7	0.5	
131/105	KRI				-9.6	4.62 ± 0.16	
131/115	KRI	< 2.0			-9.5	6.78 ± 0.64	
132/103	KRI				-9.4	4.67 ± 0.16	
132/131	KRI	< 1.0		4.48	-9.4	3.50	
141/1	tUSM	< 0.7	0.40	4.75	-8.7	11.1 ± 2.1	
159/4	KRI	< 0.9			-7.1	2.2	
161/1	KRI	1.5 ± 0.7	0.37	12.32	-10.9	2.9 ± 0.2	2-5% young water; 3H and 14C.
166/1		23.8 ± 1.8	0.85	2.33	-13.2	45.0 ± 0.5	30-50% young water; 3H and 14C.
167/1		57.3 ± 4.0	0.76	1.82	-21.2	107.8 ± 1.1	80-100% young water; 3H and 14C.

Table 2.1.3: Young water contents of regional programme samples. (Page 3 of 3)

Sample Location Number	Source Formation	Tritium (TU)	Dissolved Organic Carbo (mg/l)	Total Dissolved Carbonate (mmol)	Delta 13C (per mil)	14C (pmc)	Comments
171/101	km3b	5.6 ± 3.9		4.17	-6.4	< 4.7	< 10% young water; 3H and 14C.
172/101	mo3			7.84	-6.4	0.7 ± 0.3 0.7 ± 0.3	
173/101	mo3			7.80	-6.9	1.1 ± 0.2 1.1 ± 0.2	
174/101	mo1			8.45	-6.4	1.7 ± 0.3 1.0 ± 0.3	
175/101	mm4			8.52	-6.7	1.0 ± 0.2 0.5 ± 0.3	
201/1 201/3	mo mo	3.3 ± 0.8 3.4 ± 0.6	0.62	15.31	-7.0	6.3 ± 0.5	5-10% young water; 3H and 14C; 3-5% young water: NTB 84-21.
203/3 203/112	mo mo	6.7 ± 0.9 6.2 ± 0.6	0.85	14.27	-8.5	7.2 ± 0.3	5-10% young water; 3H and 14C; 3-5% young water: NTB 84-21.
206/102	mo	3.6 ± 0.5		15.31	-9.9	6.0	5-10% young water; 3H and 14C.
219/112	mo	5.8 ± 0.7		15.38	-8.0	7.0	5-10% young water; 3H and 14C; 3-5% young water: NTB 84-21.
271/101 271/103 271/104	mo mo mo	< 0.8 < 1.1 < 1.1		5.92 5.98 5.76	-8.1 -6.8 -8.2	4.0 ± 0.3 8.5 ± 0.3 10.3 ± 0.3	< 2% young water; Sect. 4.2.
272/102 272/103	mo mo	< 1.6 < 1		6.22 6.30	-9.6 -9.3	28.1 ± 0.4 5.5 ± 0.2	< 2% young water; Sect. 4.2.
273/102 273/103	mo mo	< 1.4 < 0.9		6.30 6.06	-10.5 -9.5	19.9 16.7 ± 0.6	
274/102	mo	4.7 ± 0.7		6.22	-10.6	37.8 ± 0.7	< 15% young water; Sect. 4.2.
302/10a	mo	3.2 ± 0.7	34.0	7.22	-4.6	13.3 ± 0.1	< 5% young water; Sect. 4.2.
306/1	mo	11.6 ± 1.1	0.3	4.98	-15.8 -6.4	12.0 ± 0.3 11.2 ± 0.3	10-15% young water; Sect. 4.2.
501/102	q	24.0		5.05	-15.2	24.0	20-30% young water; Sect. 2.1.1.
502/101	OMM	4.0 ± 0.6		5.33	-10.8	39.6 ± 2.4	< 5% young water: 3H.
508/101	OMM	0.5		3.36	-6.4	1.9	
509/101	OMM	0.4 ± 0.6		7.08	-4.7	0.8 ± 0.8	

Table 2.3.2: Dissolved helium concentrations measured on replicate samples at the Universities of Cambridge and Bath and at PSI.

Location/Sample	Helium Concentrations (cc STP/cc H_2O) Cambridge	Bath	PSI	Location/Sample	Helium Concentrations (cc STP/cc H_2O) Cambridge	Bath	PSI
1/1		4.68E-06	1.30E-05	304/5		1.45E-03	2.84E-03
		5.01E-06 +		304/6		5.85E-05 +	4.04E-03
2/3		5.42E-05	4.40E-05			2.03E-05 +	
3/1		1.60E-04 +	5.10E-04	304/7		1.48E-03	2.97E-03
3/3		2.79E-04	6.30E-04	304/8	5.73E-04	2.02E-03	5.49E-03
		2.72E-04				2.74E-03	
15/1		4.03E-05	2.50E-04	304/10	2.68E-04	2.80E-04 +	2.84E-02
95/2		3.51E-04	5.70E-04			7.81E-04	
		3.47E-04				9.10E-04	
161/1		2.71E-04	5.70E-04	305/1	2.97E-04	2.96E-04	6.90E-04
		2.39E-04				3.02E-04	
201/1		1.05E-05 +	1.90E-04	305/2		9.15E-05	1.90E-04
233/1		4.56E-06	1.30E-05			9.29E-05	
234/1		4.95E-05	1.30E-04	305/3		1.01E-04 +	1.90E-04
301/1		1.53E-04 +	2.50E-04	305/4	5.34E-05	1.10E-04	1.90E-04
		1.71E-04 +				1.10E-04	
301/2	2.00E-04	2.12E-04	2.50E-04	305/6	1.10E-04	1.00E-04	1.90E-04
		2.17E-04				1.01E-04	
301/6	2.15E-04	1.33E-04	2.50E-04	305/9	1.09E-04	1.02E-04	1.90E-04
301/8c		1.90E-04	2.50E-04			9.94E-05	
		2.24E-04				9.77E-05	
301/9		1.61E-04	2.50E-04	305/12		5.14E-05	1.90E-04
301/12b	2.14E-04	2.00E-04	2.50E-04			5.22E-05	
		2.11E-04				5.29E-05	
301/13	2.33E-04	1.96E-04	2.50E-04	305/14		1.57E-06 +	1.90E-04
301/16		1.41E-04	2.50E-04	305/15		4.58E-05	6.30E-05
		1.42E-04				1.02E-05	
301/17		1.41E-04	2.50E-04	305/16		5.06E-05	1.30E-04
301/18	1.80E-04	1.47E-04	3.20E-04			3.53E-05	
		1.44E-04				2.39E-05	
301/25b		1.01E-04	1.90E-04	305/17		9.55E-06	1.90E-04
		1.07E-04		306/1		4.65E-06	6.30E-06
302/6	1.50E-04		4.40E-04			4.53E-06	
302/10b	2.50E-06	1.68E-06		306/2		8.82E-05	1.90E-04
		1.72E-06				8.65E-05	
302/12	1.02E-03	1.81E-03	2.71E-03			8.28E-05	
		1.51E-03		306/3		3.74E-05 +	1.90E-04
302/14		7.30E-05 +	2.78E-03	306/4		2.22E-05	1.90E-04
		1.80E-05 +				4.82E-05 +	
302/16	1.60E-05	8.32E-06 +	2.90E-03	306/5		4.23E-05	1.90E-04
		9.41E-06 +				4.57E-05 +	
302/18	6.98E-05	1.45E-05 +	5.05E-03	306/6		3.59E-05	1.90E-04
		2.48E-05 +		306/7	1.01E-04	4.20E-05	1.90E-04
		2.91E-05 +				6.23E-05	
302/19	2.55E-04	7.54E-05 +	2.78E-03	306/8		6.94E-05 +	1.90E-04
		1.53E-04 +		306/9		8.90E-05	1.90E-04
302/20		1.59E-03				7.67E-05	
		1.59E-03 +	4.23E-03	306/10		8.30E-05	1.90E-04
303/1		1.18E-05 +	2.50E-04			6.46E-05 +	
		1.22E-04 +		306/11b		5.52E-06 +	2.50E-04
		1.27E-04 +				5.72E-06	
303/2		6.79E-05	5.10E-05	306/12		1.50E-05 +	2.50E-04
		3.97E-05				1.48E-05 +	
303/3	2.88E-03	1.10E-03	2.46E-03	306/13		5.62E-05 +	2.50E-04
		1.11E-03		306/16		1.09E-04	1.90E-04
303/6		2.01E-03	3.35E-03			1.12E-04	
		4.36E-04 +		306/17		5.96E-05	3.20E-04
		4.10E-04 +				5.76E-05 +	
304/1	5.24E-06		1.30E-05	306/20		1.17E-06 +	1.90E-04
304/2		4.66E-06 +	6.30E-05			1.76E-06 +	
		4.80E-06 +		306/21		3.76E-05 +	1.30E-04
304/3		2.37E-04 +	3.80E-04	306/23		5.17E-05	1.90E-04
	1.03E-04			306/26		2.88E-06 +	2.50E-04
		6.92E-05 +					

\+ samples unacceptable for recharge temperature calculation

Table 2.3.3: Radium concentrations measured on replicate samples at Fresenius and by AERE, Harwell.

Location /Sample	Harwell dpm/kg ^{226}Ra	^{228}Ra	Harwell pg/kg ^{226}Ra	^{228}Ra	pg/kg Ra tot	Fresenius pg/l Ra tot
301/5	1.55		0.71	0.00	0.71	0.7
302/19	4.64	17.10	2.11	0.03	2.14	1.4
303/6	41.70	157.10	19.00	0.26	19.26	15.0
304/1	24.50		11.16	0.00	11.16	12.0
304/3	70.00		31.89	0.00	31.89	18.0
304/14	60.80	3.20	27.70	0.01	27.71	5.3
304/8	8.99	37.10	4.10	0.06	4.16	13.0
304/10	2.14	18.50	0.97	0.03	1.00	0.3
305/1	4.21		1.92	0.00	1.92	1.4
305/2	0.72	0.50	0.33	0.00	0.33	0.9
305/4	0.78	1.21	0.36	0.00	0.36	0.6
304/6	0.84	1.31	0.38	0.00	0.38	0.4
305/9	23.00	1.70	10.48	0.00	10.48	3.0
305/12	2.31	0.68	1.05	0.00	1.05	2.1
305/14	13.90		6.33	0.00	6.33	5.2
305/16	5.49		2.50	0.00	2.50	0.7
(305/16)	3.70		1.69	0.00	1.69	0.7
306/1	10.40		4.74	0.00	4.74	4.5
306/2	1.57		0.72	0.00	0.72	0.2
306/4	0.80	0.70	0.36	0.00	0.37	0.4
306/5	0.21		0.10	0.00	0.10	0.4
306/7	0.33	0.58	0.15	0.00	0.15	0.4
306/9	0.40		0.18	0.00	0.18	0.3
306/17	0.48		0.22	0.00	0.22	0.8
306/18	0.48		0.22	0.00	0.22	1.0
306/20	0.14		0.06	0.00	0.06	0.7
306/26	0.52		0.24	0.00	0.24	1.1
306/16	0.80	0.34	0.36	0.00	0.37	0.5

3. INFILTRATION CONDITIONS

Many groundwater samples from the hydrochemical programme in northern Switzerland were analysed for their δ^2H and $\delta^{18}O$ values and for their ^3H (tritium) contents. The concentrations of the noble gases helium, neon, argon, krypton and xenon dissolved in a number of them were also measured. The stable isotope ratios and the concentrations of the noble gases are influenced by the temperature and pressure (altitude) prevailing during recharge. The results of these analyses are given in this chapter and are interpreted in terms of the conditions under which the groundwaters infiltrated.

To interpret groundwater isotope data requires knowledge of the ^2H and ^{18}O content of recharge. The composition of modern recharge is therefore described before discussing the Nagra groundwaters.

The stable hydrogen and oxygen isotopic compositions are given as δ^2H and $\delta^{18}O$ values, respectively, following the conventional notation (Chapter 1; INTERNATIONAL ATOMIC ENERGY AGENCY, 1983a). Both are given relative to the SMOW standard. The principles governing the interpretation of the isotopic composition of groundwater are described by INTERNATIONAL ATOMIC ENERGY AGENCY (1983a), and references given therein. A summary discussion of the Nagra ^2H and ^{18}O results was given by BALDERER (1985a) and BALDERER AND OTHERS (1987a).

3.1   Isotopic Composition of Modern Recharge

M. Kullin and H. Schmassmann

The environmental stable isotopes (^{18}O, ^2H) in modern precipitation, and waters from springs, rivers and boreholes were correlated to altitude (z) for various regions of Switzerland and adjacent areas (Figure 3.1.1). All data are from Nagra projects or the publications listed in Table 3.1.1 at the end of Section 3.1.

It must be emphasized that the interpretations of this section should be considered only a first attempt at compiling and formally interpreting the stable isotope data available. The authors are aware of the heterogeneity of the data, which represent waters of different types including groundwater, surface water, and precipitation. Data were also collected at different times and for various periods, and are scarce in certain areas. Furthermore, a number of processes may influence the isotopic composition during infiltration. Such processes were not a subject of this study, but they could add uncertainties of an unknown extent. For a comprehensive understanding of all these problems, further investigation would have to be carried out. The principal aim of this section was to derive stable isotope-temperature relationships of recent waters for comparison with the infiltration temperatures obtained by the use of noble gases. This could be reasonably well done only for the area of northern Switzerland and the Mittelland.

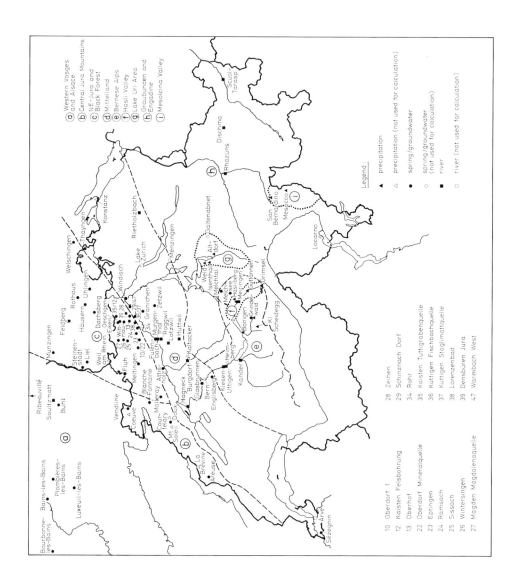

Figure 3.1.1: Location map of stable isotope data (δ^2H or δ^{18}O) of modern recharge in Switzerland and adjacent regions.

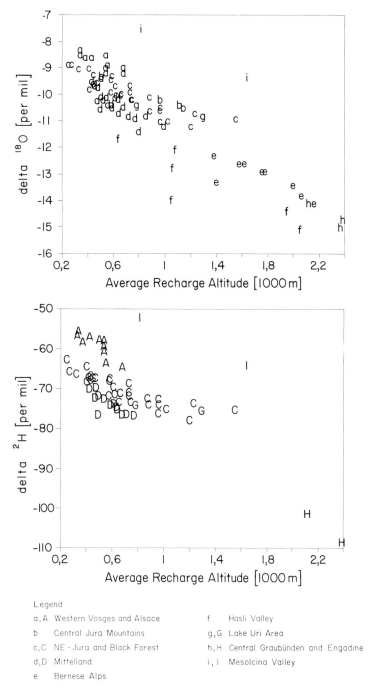

Figure 3.1.2: Graph of stable isotope values against recharge altitudes, grouped according to climatic regions.

3.1.1 Stable Isotope-Altitude Relationships

Regression analyses were performed to derive relationships between the mean isotopic composition of modern waters and average recharge altitude for various regions. The only data used for the calculations were either from precipitation or on water from springs, rivers and boreholes with average recharge altitudes which could be reliably estimated. All data available were plotted on stable isotope-altitude diagrams (Figure 3.1.2). Several populations were evident and led to the delineation of the different areas discussed here. These areas also coincide in most cases with the climatic regions presented by MÜLLER (1980, Anhang I) (Figure 3.1.3).

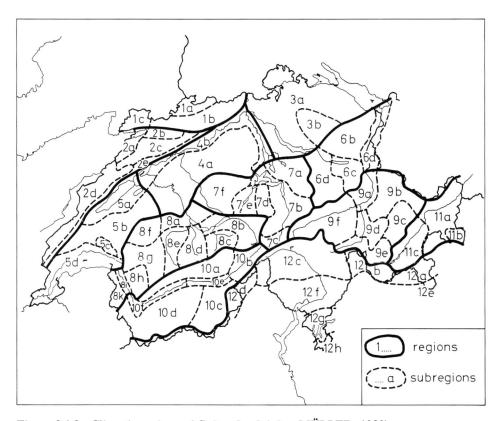

Figure 3.1.3: Climatic regions of Switzerland (after MÜLLER, 1980).

Stable isotope-average recharge altitude relationships are based either on spring, river and borehole samples, or on precipitation data. There are slight differences between the two data sets probably caused by uneven seasonal infiltration processes. The compositions of most springs, especially those of the Western Vosges and Alsace area, the Northeastern Jura and Black Forest area, and the Mittelland are based only on single measurements. For these data to be useful for the purposes of this section, it has been assumed that the waters of these springs have such long residence times that their measured stable isotope values represent mean long-term values.

In the following sections, the data of each area will be discussed briefly. The stable isotope-average recharge altitude equations developed are given in Table 3.1.2. Figure 3.1.4 displays stable isotope-average recharge altitude diagrams with the data points, calculated regression line(s), and 95 per cent confidence intervals for each area. Bold data points were used for the calculation of the regression line, open ones are given for additional information. In the following discussion, the areas are lettered a through i; lower case refers to $\delta^{18}O$-, and upper case to δ^2H-relationships.

a) <u>Western Vosges and Alsace</u>

   Stable isotope data of springs and rivers with corresponding estimated average recharge altitudes for the western Vosges and Alsace region (Figures 3.1.4a and 3.1.4A) were published by GARCIA (1986, p. 25).

b) <u>Central Jura Mountains</u>

   The regression line was calculated from data of four springs ($\delta^{18}O$ values only) read from Figure 3 in SIEGENTHALER AND OTHERS (1983, p. 478). Precipitation data of four additional stations were also available and, except for the summit station on Mt. Soleil (Figure 3.1.4b), adequately fit the regression line calculated. BLAVOUX AND OTHERS (1979, p. 299) give an equation for $\delta^{18}O$ as a function of altitude (z, metres) for an extended area comprising the French as well as the western and central Swiss Jura. Their equation is

   $$\delta^{18}O = -0.0018 \cdot z - 8.62$$

   and is based on values measured during low water in September, which correspond to long-term groundwater values. It is consistent with the result obtained here for the central Jura Mountains.

c) <u>Northeastern (NE) Jura and Black Forest</u>

   This area includes the Jura mountains and their northern foreland but only the southeastern slope of the Black Forest. The stable isotope-average recharge altitude relationships are based on data from the Nagra and the Swiss National Energy Research Foundation (NEFF) programmes (Figures 3.1.4c and 3.1.4C) published in SCHMASSMANN AND OTHERS (1984, p. 73, Beilage 4) and DUBOIS AND FLÜCK (1984, p. 26), respectively. For these data, the spring altitudes are known. Statistical average recharge altitudes were estimated by multiplying the spring altitudes by an empirical factor 1.2 and are therefore presented in brackets in Table 3.1.1. This factor, which is applicable to this particular area of northern Switzerland and the southeastern slope of the Black Forest, seems to be statistically correct, but does not allow a calculation of the average recharge altitude of a spring in each single case.

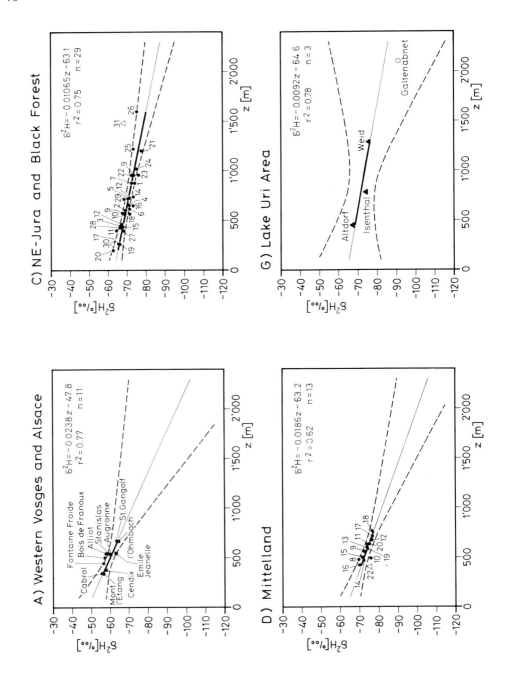

Figure 3.1.4: Relationships between stable isotope data (δ^2H or δ^{18}O) of modern waters and mean recharge altitude for areas of Switzerland and adjacent regions. (Page 1 of 4)

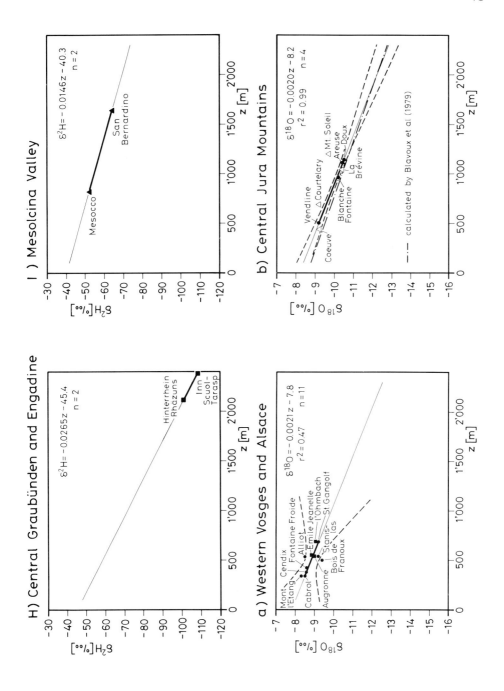

Figure 3.1.4: Relationships between stable isotope data (δ^2H or $\delta^{18}O$) of modern waters and mean recharge altitude for areas of Switzerland and adjacent regions. (Page 2 of 4)

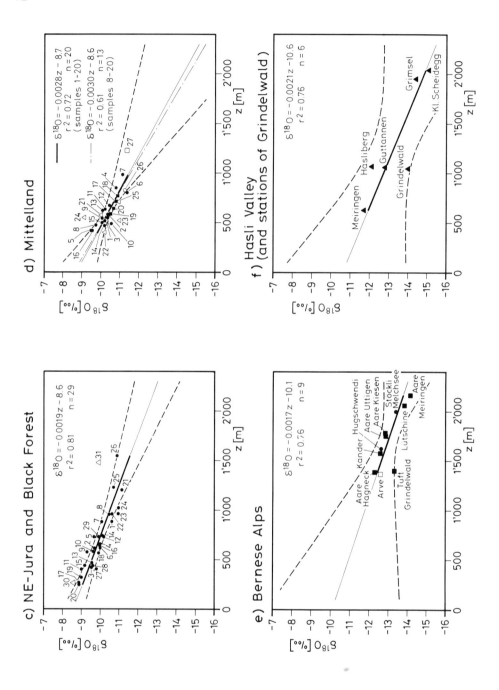

Figure 3.1.4: Relationships between stable isotope data (δ^2H or $\delta^{18}O$) of modern waters and mean recharge altitude for areas of Switzerland and adjacent regions. (Page 3 of 4)

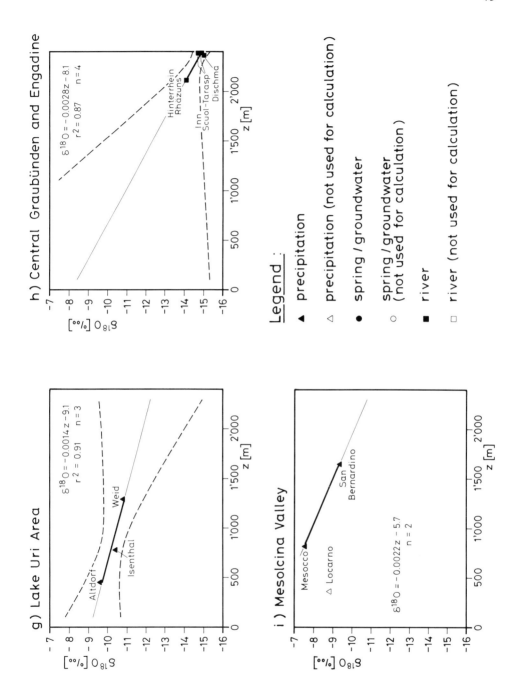

Figure 3.1.4: Relationships between stable isotope data (δ^2H or δ^{18}O) of modern waters and mean recharge altitude for areas of Switzerland and adjacent regions. (Page 4 of 4)

Additional precipitation data from two stations, Weil am Rhein and Feldberg, were available covering a period of several years. The station at Weil am Rhein fits the calculated regression line, but the summit station on Feldberg is characterized by more positive values. The quality of the data from the Feldberg station is doubtful, however (W. Stichler, Institut für Hydrologie, GSF, oral communication, September 1988).

d) <u>Mittelland</u>

This area also includes data from the southern slope of the Jura Mountains. The calculated regression lines for this area (Figures 3.1.4d and 3.1.4D) are based on data both from the Nagra regional programme and from the literature. The statistical average recharge altitudes were calculated for the springs of the Nagra regional programme using the empirical factor 1.2 as described in Section (c). For comparison, a second regression line was computed for the $\delta^{18}O$-z relationship based only on samples for which both $\delta^{18}O$ and δ^2H data were available (Figure 3.1.4d, dashed line; Table 3.1.2). Of the 20 samples used for the calculation, only three samples (Sézegnin, Thayngen, Rietholzbach) are located outside of the major climatic region of the central Mittelland (Figure 3.1.3).

The data point of Lake Zürich is above the calculated regression line. This can be explained by the fact that a considerable part of the catchment area is influenced by the Föhn as is the Lake Uri Area (cf. Section 3.1.1g).

e) <u>Bernese Alps</u>

The data set for this area comprises $\delta^{18}O$ values and average estimated altitudes of the recharge area of six rivers and three springs from the northern slope of the Bernese Alps. As is evident from Figure 3.1.4e, the value for the Arve River fits the regression line very well. The recharge area of the Arve river is west of this area but is part of the same major climatic region (MÜLLER, 1980).

f) <u>Hasli Valley (and stations of Grindelwald)</u>

Precipitation data from six stations in this area were used for the calculation (Figure 3.1.4f). Data on the three stations that are part of the network of the Swiss Meteorological Service, Meiringen, Guttannen and Grimsel, and on the Hasliberg station have been published along with their mean annual temperatures.

g) <u>Lake Uri area</u>

As part of the Nagra research project at Oberbauenstock in the Lake Uri area of central Switzerland, precipitation was measured at stations at Altdorf, Isenthal and Weid. Monthly precipitation was sampled and every other sample was analysed for δ^2H and $\delta^{18}O$ for a period of one year. The weighted annual means of $\delta^{18}O$ and δ^2H were used for the calculations (Figures 3.1.4g and 3.1.4G).

A slight uncertainty results from the fact that the February 1988 sample from Weid covers a period of only eight days, and has a δ^2H value which seems too positive compared with the Altdorf and the Isenthal samples taken in the same month. The weighted mean value for Weid is expected to be more negative, resulting in a slightly more negative gradient for the stable isotope-altitude relationship.

The δ^2H value of a spring in the Muota valley (Galtenäbnet) is consistent with the calculated regression line (Figure 3.1.4G).

h) <u>Central Graubünden and Engadine</u>

Four $\delta^{18}O$ values from the Hinterrhein, Inn and Dischma River were available from literature, but only two δ^2H values were measured. Because of the scarcity of data, the narrow altitude range and large areas drained by each river, and the fact that the rivers are in different climatic regions (Figure 3.1.1 and 3.1.3), these stable isotope-altitude relationships should be considered only gross approximations. The uncertainty is indicated by the wide 95 per cent confidence interval (Figures 3.1.4h and 3.1.4H).

i) <u>Mesolcina Valley</u>

As part of the Nagra research project at Piz Pian Grand in the Mesolcina Valley on the southern slope of the Alps, precipitation was measured at stations in San Bernardino and in Mesocco, and samples were taken for stable isotope analyses. Samples from every second month from July 1987 to January 1988 were analysed for δ^2H and $\delta^{18}O$, and the mean amount-weighted values were used for the calculation (Figures 3.1.4i and 3.1.4I). Because the data do not cover a period of even one year, the results must be considered only as a rough indication. The precipitation data from Locarno, which is a station of the network of the Swiss Meteorological Service on the southern slope of the Alps, but situated in another climatic and topographic region, do not fit the regression line calculated for the Mesolcina Valley.

Figures 3.1.5 and 3.1.6 show the stable isotope-average recharge altitude relationships of the areas under study. All lines are characterized by similar slopes but have different δ-intercepts.

When considering only the regression lines <u>based on groundwater and river data</u>, the following sequence from more positive to more negative δ-values can be recognized, reflecting zones with increasing continentality:

1) Western Vosges and Alsace (lines a, A);
2) Central Jura Mountains and NE-Jura and Black Forest area (lines b, B and c, C);
3) Mittelland, Bernese Alps and central Graubünden and Engadine (lines d, D; e, E; and h, H).

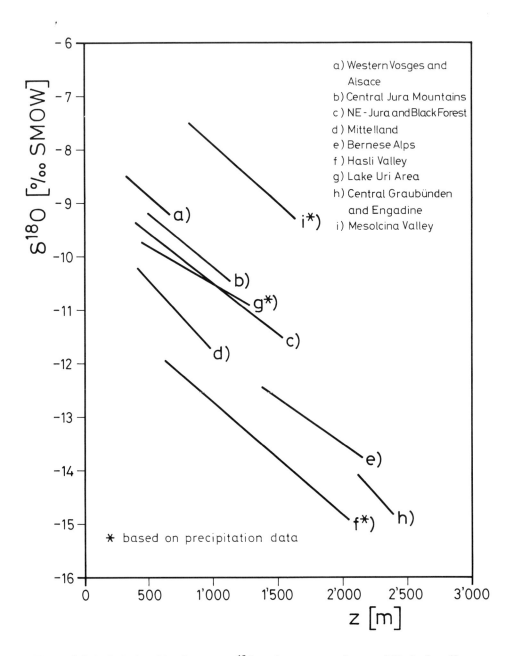

Figure 3.1.5: Relationships between $\delta^{18}O$ and average recharge altitude for all areas.

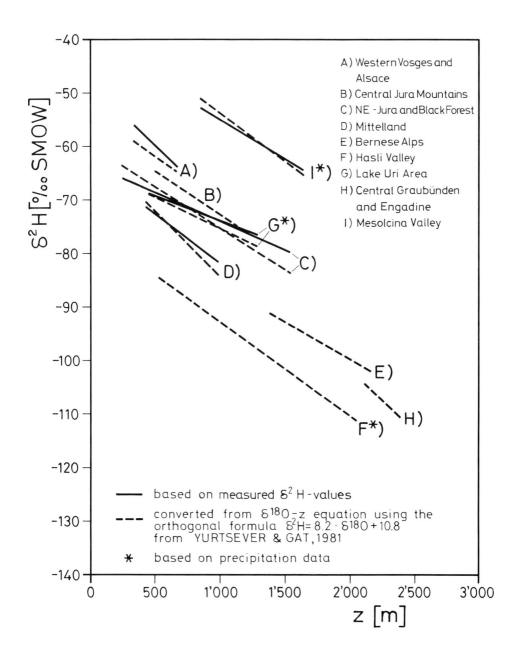

Figure 3.1.6: Relationships between δ^2H and average recharge altitude for areas with δ^2H values and from conversion of δ^{18}O relationships.

The data of these three continental zones belong to statistically different populations. The similarity of the lines of the Central Jura Mountains (b, B) and the NE-Jura and the Black Forest area (c, C) is not surprising because these two areas are adjacent. On the other hand, it was unexpected to find the Mittelland (d, D), the Bernese Alps (e, E) and the central Graubünden and Engadine (h, H) in the same continental zone. These three areas were not included in a single group because they represent different altitude ranges.

All stable isotope-altitude relationships based on precipitation data, from the Lake Uri Area (g, G), the Mesolcina Valley (i, I) and the Hasli Valley (f, F), do not fit well into this pattern as demonstrated below:

- Regression lines based on groundwater and river data should be compared carefully, if at all, with regression lines based on precipitation data. This is illustrated by the two areas of the Hasli Valley and the Bernese Alps. Because the Hasli Valley (f, F) is part of the area of the Bernese Alps (e, E), one would expect very similar regression lines of these two areas, but the line of the Hasli Valley (based on precipitation data) is shifted to more negative δ-values by about 1.2 per mil for $\delta^{18}O$;

- The Mesolcina Valley (i, I), situated at the southern slope of the Alps, exhibits the most positive stable isotope values observed in this study. This shift to the more positive δ-values can be explained by the proximity and strong influence of the Mediterranean;

- The Lake Uri area (g, G) is also shifted to more positive stable isotope values than expected. Based on the location of this area, one would expect a regression line similar to that of the Bernese Alps (e, E). This shift to more positive δ-values can be explained by a strong influence of the Föhn, especially in the Uri and also in the Muota valley (SCHREIBER AND OTHERS, 1977; Beilage 1:500'000). This is also demonstrated by the northwards directed bulge of isotherms in this area (SCHÜEPP, 1981, Tafel 11).

In spite of the climatic explanations for the more positive δ-values of the regression lines of the Mesolcina Valley and the Lake Uri Area, it must be emphasized that the difference of stable isotope values between groundwater samples and river water and precipitation samples is still a problem which needs further investigation. Therefore, comparison between groundwater data and river water and precipitation data should be done very carefully, and in a qualitative sense.

3.1.2 $\delta^{18}O$-δ^2H Relationship

Regression of the $\delta^{18}O$ and δ^2H values from 64 samples leads to the equation:

$$\delta^2H = 7.55 \cdot \delta^{18}O + 4.8$$

This line, its 95 per cent confidence band, and the data points on which it was based are shown in Figure 3.1.7.

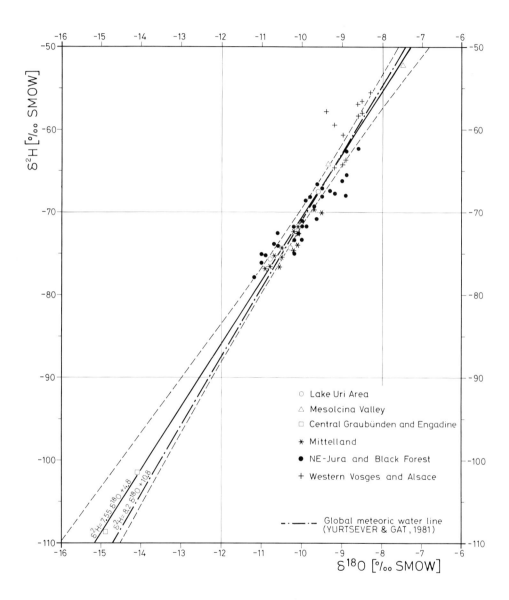

Figure 3.1.7: Relationships between $\delta^{18}O$ and δ^2H for modern waters in Switzerland and adjacent areas.

The global orthogonal regression line for precipitation published by YURTSEVER AND GAT (1981, p. 123) is:

$$\delta^2H = 8.2\ \delta^{18}O + 10.8$$

This regression line is also shown in Figure 3.1.7. It is within the 95 per cent confidence interval of the line calculated here for Switzerland and adjacent areas and almost coincides at the centre of the data population. The result obtained confirms that the global orthogonal regression line from YURTSEVER AND GAT (1981) also describes the relationship between $\delta^{18}O$ and δ^2H in the area under investigation.

Table 3.1.2 summarizes the equations relating $\delta^{18}O$, δ^2H, and z for the different areas. The equations in the first two columns of the table given are all based on data. The third column gives δ^2H-z equations calculated from the $\delta^{18}O$-z equations of the first column and the global orthogonal $\delta^{18}O$-^2H equation of YURTSEVER AND GAT (1981). This was done for two reasons: 1) to provide ^2H-z equations where no δ^2H data were available, and 2) to compare the converted formula with the formula based on data.

As Figure 3.1.6 shows, the calculated and converted lines match reasonably well in the data range under consideration.

3.1.3 Local Altitude-Temperature Equations

Infiltration temperatures calculated from dissolved noble gases are discussed in Section 3.3 of this chapter. To compare the noble gas and stable isotope results, the local stable isotope-altitude relationships were converted to stable isotope-temperature relationships using altitude-temperature equations.

The average decrease of air temperature with increasing altitude exhibits both regional and seasonal variations. In the winter, the gradients are slightly lower than in summer because of frequent temperature inversion during the cold season. SCHÜEPP (1981, Tafel 11) and KIRCHHOFER (1982, Tafel 6.1) present a mean annual temperature gradient of 0.5°C per 100 m for Switzerland. For northern Switzerland, a mean annual air temperature of 10.6°C for a reduced altitude of sea level can be found from the temperature-altitude relationship presented by BIDER (1978, p. 137). This conclusion is supported by the 8.5°C isotherm (reduced to 500 m) which runs across the southern part of northern Switzerland (SCHÜEPP, 1981, Tafel 11). Based on these references, the following T-z conversion equation was used for the area of northern Switzerland:

$$T = -0.005 \cdot z + 10.6\ [°C]$$

No consideration of uncertainties is given in the literature for the data used.

Table 3.1.2: Stable isotope-altitude equations.

Area	$\delta^{18}O$-z [+]	δ^2H-z [+]	δ^2H-z [*]
a) Western Vosges and Alsace	$\delta^{18}O = (-0.0021\pm0.0007)z-7.8\pm0.3$ n = 11 $r^2 = 0.47$	$\delta^2H = (-0.0238\pm0.0043)z-47.8\pm1.6$ n = 11 $r^2 = 0.77$	$\delta^2H = -0.0172z-53.2$
b) Central Jura Mountains	$\delta^{18}O = (-0.0020\pm0.0001)z-8.2\pm0.1$ n = 4 $r^2 = 0.99$		$\delta^2H = -0.0164z-56.4$
c) NE-Jura and Black Forest	$\delta^{18}O = (-0.0019\pm0.0002)z-8.6\pm0.3$ n = 29 $r^2 = 0.81$	$\delta^2H = (-0.0107\pm0.0018)z-63.1\pm1.9$ n = 29 $r^2 = 0.75$	$\delta^2H = -0.0156z-59.7$
d) Mittelland	$\delta^{18}O = (-0.0028\pm0.0004)z-8.7\pm0.3$ n = 20 $r^2 = 0.72$ $\delta^{18}O = (-0.0030\pm0.0007)z-8.6\pm0.3$ n = 13 $r^2 = 0.61$	$\delta^2H = (-0.0186\pm0.0043)z-63.2\pm1.6$ n = 13 $r^2 = 0.61$	$\delta^2H = -0.0230z-60.5$ $\delta^2H = -0.0246z-59.7$
e) Bernese Alps	$\delta^{18}O = (-0.0017\pm0.0005)z-10.1\pm0.4$ n = 9 $r^2 = 0.64$		$\delta^2H = -0.0139z-72.0$
f) Hasli Valley	$\delta^{18}O = (-0.0021\pm0.0006)z-10.6\pm0.7$ n = 6 $r^2 = 0.76$		$\delta^2H = -0.0172z-76.1$
g) Lake Uri Area	$\delta^{18}O = (-0.0014\pm0.0004)z-9.1\pm0.3$ n = 3 $r^2 = 0.91$	$\delta^2H = (-0.0092\pm0.0049)z-64.7\pm3.0$ n = 3 $r^2 = 0.78$	$\delta^2H = -0.0115z-63.8$
h) Central Graubünden and Engadine	$\delta^{18}O = (-0.0028\pm0.0007)z-8.1\pm0.2$ n = 4 $r^2 = 0.87$	$\delta^2H = -0.0265z-45.4$ n = 2	$\delta^2H = -0.0230z-55.6$
i) Mesolcina Valley	$\delta^{18}O = -0.0022z-5.7$ n = 2	$\delta^2H = -0.0146z-40.3$ n = 2	$\delta^2H = -0.0180z-35.9$

[+] Formula calculated from data
[*] Formula converted from $\delta^{18}O$-z relationship with $\delta^2H = 8.2 \cdot \delta^{18}O + 10.8$ (YURTSEVER AND GAT, 1981)

3.1.4 Stable Isotope-Temperature Relationships

By simple conversion of the stable isotope-altitude equations using the regional temperature-altitude equations, it is possible to calculate stable isotope-temperature equations for each area. This was done for two areas of particular interest for the Nagra programme, namely the NE-Jura and Black Forest area (c, C) and the Mittelland area (d, D). The correlation coefficients for the stable isotope-altitude relationships for these areas are from 0.78 to 0.9. The following stable isotope-temperature relationships were calculated using the T-z equation presented in Section 3.1.3:

NE-Jura and Black Forest

$$\delta^{18}O = (0.38 \pm 0.02) \cdot T - 12.6 \pm 0.5 \quad c)$$
$$\delta^{2}H = (2.13 \pm 0.24) \cdot T - 85.7 \pm 4.2 \quad C)$$

Mittelland

$$\delta^{18}O = (0.56 \pm 0.08) \cdot T - 14.6 \pm 0.3 \quad d)$$
$$\delta^{2}H = (3.72 \pm 0.86) \cdot T - 102.7 \pm 1.6 \quad D)$$

As demonstrated by YURTSEVER AND GAT (1981, p. 117) the $\delta^{18}O$-T relationship probably is characterized by a curved rather than by a linear regression line for a temperature range from -4°C to +28°C. However, for the temperature range under consideration, a linear regression line is a good estimate. Figure 3.1.8 displays the calculated lines showing the variation of stable isotopes with mean annual air temperatures with their 95 per cent confidence intervals.

Figure 3.1.8 (top) also exhibits the global $\delta^{18}O$-T regression line presented by YURTSEVER AND GAT (1981, p. 116) which is based on 91 samples from a worldwide net of stations. For Figure 3.1.8 (bottom) the global line was converted with the orthogonal $\delta^{18}O$-$\delta^{2}H$ equation from YURTSEVER AND GAT (1981) used above. Both plots in Figure 3.1.8 show that stable isotope-temperature lines calculated for northern Switzerland are subparallel to the global line, but shifted to more negative stable isotope values dependent on the continental position of the area under investigation.

Theoretically, one would expect lines parallel to the global line, but with more negative stable isotope values, dependent on the continentality of the area under question. Considering the actual data range (bold lines) in Figure 3.1.8, the Mittelland (d, D) and the NE-Jura and Black Forest (c, C) lines overlap at higher temperatures (at about > 8°C). The different slopes of the two lines could be within the statistical error or could be real. If the difference is real, it would reflect an overlap in the lower altitude (higher temperature) region of Schaffhausen and the Bodensee (Lake Constance). To verify this point, a larger data set is needed for this particular area.

It is not possible to present a single stable isotope-temperature (altitude) relationship that is applicable for the whole of Switzerland. Rather, one would expect a set of lines parallel to the global line presented by YURTSEVER AND GAT (1981, p. 116), but with different δ-intercepts dependent on continentality as well as on local climatic conditions of the particular area. Figure 3.1.8 exhibits such subparallel stable isotope-temperature lines for northern Switzerland (c, C and d, D), which are shifted to more negative δ-values than the global line, as was expected from their continental position.

The line of SIEGENTHALER AND OESCHGER (1980), which can be considered as a first approach for a line of Switzerland, is shifted to more negative δ-values than the lines c and d in Figure 3.1.8. This is not at all surprising because this line is based on five stations (Bern, Meiringen, Guttanen, Grimsel and Locarno) all situated--except Bern--further south of the area of northern Switzerland. The fact that some of the precipitation stations even belong to different continental zones may explain the steeper slope of this line compared to the global line. However, the line from SIEGENTHALER AND OESCHGER (1980) should be compared to the lines of northern Switzerland only with the understanding that this line is based on precipitation data.

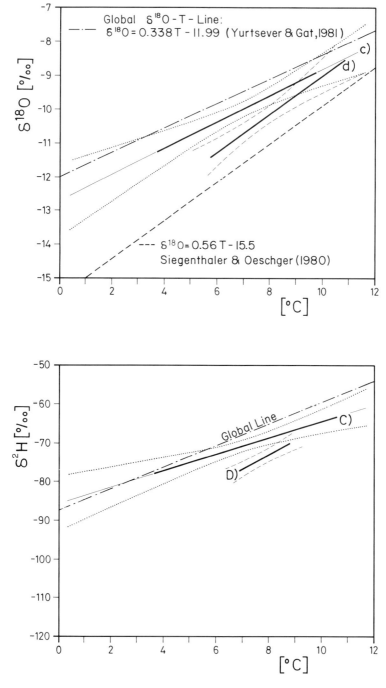

Figure 3.1.8: Relationships between stable isotopes and mean annual air temperature relationships for northern Switzerland.

Table 3.1.1: Sources of data on isotopic composition of modern recharge. (Page 1 of 5)

Location	Date or time range	Sam-ples	Val-ue(s) pub-lished	Used for Calc	Type	Mean annu-al temp.(°C)	Alti-tude of sprg. (m)	Average re-charge alti-tude m a s l	Delta 18O (per mil)	Delta 2H (per mil)	Reference
a) Western Vosges and Alsace											
Stanislas (Plombières)	9.6.83	1	Yes	+	S			537	-9.20	-59.5	Garcia (1986)
Alliot (Plombières)	20.7.77	1	Yes	+	S			537	-8.50	-58.0	Garcia (1986)
Augronne (Plombières)	15.6.77	2	Yes	+	R			537	-9.00	-60.7	Garcia (1986)
Cendix (Bains)	15.6.77	1	Yes	+	S			375	-8.60	-58.3	Garcia (1986)
Bois de Franoux (Bains)	16.12.83	1	Yes	+	S			500	-9.40	-57.8	Garcia (1986)
Fontaine Froide (Luxeuil)	7.6.83	1	Yes	+	S			425	-8.60	-57.0	Garcia (1986)
Mont l'Etang (Bour-bonnes)	16.6.77	1	Yes	+	S			337	-8.50	-56.6	Garcia (1986)
Cabrol (Bourbonnes)	9.6.83	1	Yes	+	S			337	-8.30	-55.6	Garcia (1986)
L'Ohmbach (Soutzmatt)	14.6.83	1	Yes	+	R			675	-9.20	-64.6	Garcia (1986)
St. Gangolf (Buhl)	14.6.83	1	Yes	+	S			675	-9.00	-64.3	Garcia (1986)
Emile Jeanelle (Ribeauvillé)	15.6.83	1	Yes	+	R			550	-8.90	-63.7	Garcia (1986)
b) Central Jura Mountains											
Vendline	1.67-7.68		Yes	+	S			500	-9.30		Siegenthaler et al. (1970)
Blanche Fontaine	≥ 1 year		+)	+	S			960	-10.20		Siegenthaler et al. (1983)
Areuse	≥ 1 year		Yes	+	S			1110	-10.40		Siegenthaler et al. (1983)
Doux	≥ 1 year		+)	+	S			1140	-10.50		Siegenthaler et al. (1983)
Coeuve	≥ 1 year		Yes	–	P	9.6		430	-9.20		Siegenthaler et al. (1970)
La Brévine	10.77-9.79		Yes	–	P			1042	-10.40		Müller (1980), SMA
Courtelary			+)	–	P			692	-9.20		Oeschger & Siegenthaler (1972) SMA
Mt. Soleil			+)	–	P	5.8		1183	-9.70		Oeschger & Siegenthaler (1972) SMA

Table 3.1.1: Sources of data on isotopic composition of modern recharge. (Page 2 of 5)

Location	Date or time range	Samples	Value(s) published	Used for CalcType	Mean annual temp. (°C)	Altitude of sprg. (m)	Average recharge altitude (m a s l)	Delta 18O (per mil)	Delta 2H (per mil)	Reference
c) NE-Jura and Black Forest										
1 Malleray (7)	9.9.81	1	Yes	+	S	735	(882)	-10.61	-74.1	Schmassmann et al. (1984)
2 Oberdorf (10)	24.11.81	1	Yes	+	S	506	(607)	-9.67	-69.4	Schmassmann et al. (1984)
3 Kaisten Felsbohrung (12)	15.7.81	1	Yes	+	S	353	(423)	-9.48	-67.2	Schmassmann et al. (1984)
4 Oberhof (13)	10.9.81	1	Yes	+	S	550	(660)	-9.96	-71.2	Schmassmann et al. (1984)
5 Meltingen (19)	8.9.81	1	Yes	+	S	605	(726)	-9.65	-70.9	Schmassmann et al. (1984)
6 Oberdorf (22)	16.7.81	1	Yes	+	S	515	(618)	-10.01	-71.5	Schmassmann et al. (1984)
7 Eptingen (23)	16.7.81	1	Yes	+	S	730	(876)	-10.09	-72.5	Schmassmann et al. (1984)
8 Ramsach (24)	10.9.81	1	Yes	+	S	800	(960)	-10.46	-73.9	Schmassmann et al. (1984)
9 Sissach (25)	17.7.81	1	Yes	+	S	485	(582)	-9.29	-67.5	Schmassmann et al. (1984)
10 Wintersingen (26)	17.7.81	1	Yes	+	S	480	(576)	-9.43	-68.2	Schmassmann et al. (1984)
11 Magden Magdalenaq. (27)	25.11.81	1	Yes	+	S	335	(402)	-9.01	-64.4	Schmassmann et al. (1984)
12 Zeihen (28)	15.7.81	1	Yes	+	S	610	(732)	-10.18	-71.9	Schmassmann et al. (1984)
13 Flüh (31)	8.9.81	1	Yes	+	S	388	(465)	-9.55	-67.3	Schmassmann et al. (1984)
14 Rohr (34)	13.7.81	1	Yes	+	S	620	(744)	-10.17	-73.4	Schmassmann et al. (1984)
15 Kaisten Tuttigraben (35)	15.7.81	1	Yes	+	S	365	(438)	-9.23	-67.8	Schmassmann et al. (1984)
16 Densbüren (39)	14.7.81	1	Yes	+	S	545	(654)	-10.01	-73.4	Schmassmann et al. (1984)
17 Steinenstadt Mark (85)	18.8.81	1	Yes	+	S	225	(270)	-8.88	-65.6	Schmassmann et al. (1984)
18 Welschingen (99)	19.7.81	1	Yes	+	S	478	(574)	-9.89	-71.8	Schmassmann et al. (1984)
19 Liel Schlossbrunnen (105)	18.8.81	1	Yes	+	S	270	(324)	-9.02	-66.2	Schmassmann et al. (1984)
20 Munzingen (109)	27.11.81	1	Yes	+	S	207	(248)	-8.88	-62.7	Schmassmann et al. (1984)
21 Rothaus (150)	3.3.82	1	Yes	+	S	1000	(1200)	-11.20	-77.9	Schmassmann et al. (1984)
22 Dachsberg Hierbach (194)	10.82-6.83	2	Yes	+	S	800	(960)	-10.60	-72.6	Dubois & Flück (1984)
23 Uehlingen Stollenmund (192)	10.82-6.83	3	Yes	+	S	800	(960)	-11.00	-76.1	Dubois & Flück (1984)
24 Uehlingen Giessbach (193)	10.82-6.83	2	Yes	+	S	850	(1020)	-11.00	-75.1	Dubois & Flück (1984)
25 Häusern Sägtobel (191)	10.82-6.83	2	Yes	+	S	1025	(1230)	-10.70	-73.6	Dubois & Flück (1984)
26 Feldberg	6.83	1	Yes	+	S	1290	(1548)	-10.90	-75.3	Dubois & Flück (1984)
27 Oeschgen Volliweid	10.82-6.83	3	Yes	+	S	340	(408)	-9.80	-68.2	Dubois & Flück (1984)
28 Eiken Brieglibrunnen	10.82-6.83	3	Yes	+	S	365	(438)	-9.60	-66.7	Dubois & Flück (1984)

Table 3.1.1: Sources of data on isotopic composition of modern recharge. (Page 3 of 5)

	Location	Date or time range	Sam- ples	Val- ue(s) pub- lished	Used for Calc	Type	Mean annu- al temp.(°C)	Alti- tude of sprg. (m)	Aver- age re- charge alti- tude m a s l	Delta 18O (per mil)	Delta 2H (per mil)	Reference
29	Wittnau Wasserfallen	10.82-6.83	3	Yes	+	S		605	(726)	-9.90	-68.6	Dubois & Flück (1984)
30	Weil am Rhein	1983-86		Yes	–	P			270	-8.56	-62.4	Jahresb. d. Inst. f. Hydrolo- gie, GSF
31	Feldberg	1982-85		Yes	–	P			1493	-9.89	-68.1	Jahresb. d. Inst. f. Hydrolo- gie, GSF

d) Mittelland

	Location	Date or time range	Sam- ples	Val- ue(s) pub- lished	Used for Calc	Type	Mean annu- al temp.(°C)	Alti- tude of sprg. (m)	Aver- age re- charge alti- tude m a s l	Delta 18O (per mil)	Delta 2H (per mil)	Reference
1	Thayngen	≥1 year		+)	+	S			515	-10.20		Siegenthaler et al. (1983)
2	Zetzwil	≥1 year		+)	+	S			580	-10.40		Siegenthaler et al. (1983)
3	Glasbrunnen Bern	≥1 year		+)	+	S			550	-10.40		Siegenthaler et al. (1983)
4	Englisberg	≥1 year		+)	+	S			850	-10.80		Siegenthaler et al. (1983)
5	Sézegnin	≥1 year		Yes	+	S			420	-9.50		Siegenthaler et al. (1983)
6	Rietholzbach	≥1 year		+)	+	S			795	-11.40		Siegenthaler et al. (1983)
7	Emme Burgdorf	≥1 year		+)	+	R			990	-11.20		Siegenthaler et al. (1983)
8	Schinznach Dorf (29)	23.7.81	1	Yes	+	S		395	(474)	-9.73	-69.7	Schmassmann et al. (1984)
9	Attisholz (32)	9.9.81	1	Yes	+	S		448	(538)	-10.12	-72.6	Schmassmann et al. (1984)
10	Fulenbach (33)	14.7.81	1	Yes	+	S		410	(492)	-10.55	-76.6	Schmassmann et al. (1984)
11	Küttigen Fischb. (36)	10.9.81	1	Yes	+	S		535	(632)	-10.18	-74.7	Schmassmann et al. (1984)
12	Küttigen Stäglim.(37)	14.7.81	1	Yes	+	S		560	(672)	-10.46	-76.5	Schmassmann et al. (1984)
13	Lorenzenbad (38)	11.9.81	1	Yes	+	S		510	(612)	-10.12	-74.0	Schmassmann et al. (1984)
14	Warmbach (47)	23.7.81	1	Yes	+	S		390	(468)	-10.22	-72.3	Schmassmann et al. (1984)
15	Gränichen (130)	26.11.81	1	Yes	+	S		415	(498)	-10.11	-71.8	Schmassmann et al. (1984)
16	Windisch (16)	23.7.81	1	Yes	+	S		355	(426)	-9.52	-70.1	Schmassmann et al. (1984)
17	Langeten Lotzwil	3.78-6.79		Yes	+	R			713	-10.80	-76.5	Moser et al. (1981), *)
18	Langeten Huttwil	3.78-6.79		Yes	+	R			766	-10.90	-76.8	Moser et al. (1981), *)
19	Rot Roggwil	3.78-6.79		Yes	+	R			586	-10.50	-74.2	Moser et al. (1981) *)
20	Murg Murgenthal	3.78-6.79		Yes	+	R			637	-10.70	-75.3	Moser et al. (1981), *)
21	Bern	1971-78	8	Yes	–	P	8.8		572	-10.00		Siegenthaler & Oeschger (1980)
22	Konstanz	1981-86		Yes	–	P	8.4		403	-10.18	-75.1	Jahresb. d. Inst. f. Hydrolo- gie, GSF
23	Burgdorf			+)	–	P	9.2		525	-11.00		Oeschger & Siegenthaler (1972) SMA
24	Heimberg			+)	–	P	8.2		550	-9.10		Oeschger & Siegenthaler (1972) SMA
25	Kaltacker			+)	–	P			710	-10.90		Oeschger & Siegenthaler (1972)
26	Menzingen ZG	2.1988	2		–	S			825	-11.30		SMA
27	Lake Zürich	---		Yes	–	R	4.7		1220	-11.40		Lister (1989)

Table 3.1.1: Sources of data on isotopic composition of modern recharge. (Page 4 of 5)

Location	Date or time range	Samples	Values published	Used for Calc	Type	Mean annual temp. (°C)	Altitude of sprg. (m)	Average recharge altitude m a s l	Delta 18O (per mil)	Delta 2H (per mil)	Reference
e) Northern slope of Bernese Alps:											
Aare Hagneck	≥ 1 year		+)	+	R			1380	-12.30		Siegenthaler et al. (1983)
Aare Meiringen	≥ 1 year		+)	+	R			2160	-14.10		Siegenthaler & Schotterer (1977)
Aare Kiesen	≥ 1 year		Yes	+	R			1780	-12.90		Siegenthaler et al. (1983)
Aare Uttigen	≥ 1 year		Yes	+	R			1760	-12.90		Siegenthaler & Schotterer (1977)
Kander	≥ 1 year		+)	+	R			1580	-12.60		Siegenthaler et al. (1983)
Lütschine Bönigen	≥ 1 year		+)	+	R			2060	-13.60		Siegenthaler et al. (1983)
Stockli Melchsee	≥ 1 year		Yes	+	S			2000	-13.40		Siegenthaler et al. (1983)
Tuft Grindelwald	≥ 1 year		+)	+	S			1400	-13.30		Schotterer et al. (1982)
Hugschwendi Melchtal	≥ 1 year		+)	+	S			1620	-12.60		Siegenthaler et al. (1983)
Arve	≥ 1 year		+)	-	R			1370	-12.60		Amberger et al. (1981), *
f) Hasli Valley (and stations of Grindelwald)											
Grimsel	1971-78	8	Yes	-	P	1.2		1950	-14.40		Siegenthaler & Oeschger (1980)
Guttannen	1971-78	8	Yes	-	P	6.3		1055	-12.80		Siegenthaler & Oeschger (1980)
Meiringen	1971-78	8	Yes	-	P	7.9		632	-11.70		Siegenthaler & Oeschger (1980)
Grindelwald	10.79-9.81	2	Yes	-	P			1050	-14.00		Schotterer et al.(1982)
Kl. Scheidegg	10.79-9.81	2	Yes	-	P			2050	-15.10		Schotterer et al.(1982)
Hasliberg			+)	+	P	6.3		1075	-12.10		Oeschger & Siegenthaler (1972)
g) Lake Uri Area											
Altdorf	1987	6	Yes	+	P			449	-9.59	-67.4	Scholtis (1988a)
Isenthal	1987	6	Yes	+	P			778	-10.39	-74.3	Scholtis (1988a)
Weid	1987	6	Yes	+	P			1290	-10.80	-75.7	Scholtis (1988a)
Galtenäbnet	8./9./78	4	Yes	-	S		2000	2100		-91.5	Moser et al.(1981)

Table 3.1.1: Sources of data on isotopic composition of modern recharge. (Page 5 of 5)

Location	Date or time range	Samples	Value(s) published	Used for Calc	Type	Mean annual temp. (°C)	Altitude of sprg. (m)	Average recharge altitude m a s l	Delta 18O (per mil)	Delta 2H (per mil)	Reference
h) Central Graubünden und Engadine											
Dischma	≥ 1 year		+)	+	R			2372	-15.00		Siegenthaler et al. (1983), *)
Inn (Scuol-Tarasp)	1965-1976	6	Yes	+	R			2390	-14.70		Siegenthaler (1980)
Inn (Scuol-Tarasp)	1984-1985	3	Yes	+	R			2390	-14.90	-108.7	Wexsteen (1987)
Hinterrhein Rhäzüns	11.84-6.85	3	Yes	+	R			2118	-14.10	-101.5	Wexsteen (1987)
i) Mesolcina Valley											
Mesocco	7.87-1.88	4	Yes	+	P			815	-7.53	-52.2	Scholtis (1988b)
San Bernardino	7.87-1.88	4	Yes	+	P			1639	-9.36	-64.2	Scholtis (1988b)
Locarno	1973-78	5	Yes	-	P	11.		366	-8.80		Siegenthaler & Oeschger (1980)

P = precipitation
R = river
S = spring/groundwater
+) = read out of graphic
*) = average recharge altitude from "Hydrologisches Jahrbuch der Schweiz"
SMA Schweizerische Meteorologische Anstalt station altitude for type P waters

3.2 Isotopic Composition of Groundwater

W. Balderer, F. J. Pearson, Jr., W. Rauert, and W. Stichler

Samples taken as part of the Nagra programme were analysed for ^{2}H, ^{18}O and ^{3}H by the Institut für Hydrologie, GSF München, Neuherberg. Some samples were also analysed for ^{3}H and ^{18}O by the Low Level Counting Laboratory (LLC), Physikalisches Institut der Universität Bern. Additional published data on waters from Switzerland and adjacent regions were used to augment and assist with the interpretation of the Nagra results. δ^2H, δ^{18}O and ^{3}H results obtained by GSF on samples from the Nagra deep boreholes are given by WITTWER (1986). Results from many samples from the regional programme appear in SCHMASSMANN AND OTHERS (1984) and are summarized by BALDERER AND OTHERS (1987a).

The results of all analyses are given in Tables 3.2.1 through 3.2.5. The total dissolved solids contents of many of the samples, represented by their residues-on-evaporation at 110°C and their chloride contents, are also included. No errors were reported with the tritium results from the Weizmann Institute. The 2σ uncertainty in these values is probably between 10 and 20 per cent.

A few samples from the Nagra test boreholes given in Table 3.2.5 required correction for drilling fluid contamination. The values given in the table are corrected values and are considered to represent the composition of the groundwater. The correction procedure is discussed in Section 3.2.5.

3.2.1 Samples from Quaternary, Tertiary and Malm Aquifers

Analyses of waters from this group of formations are given in Table 3.2.1 at the end of Section 3.2. Most of the samples are from the jurassic Malm (jo) and from the Tertiary Upper Marine Molasse (tOMM) and Lower Freshwater Molasse (tUSM). Samples taken from Quaternary sediments, other units of the Molasse, and other Tertiary formations are also included. In addition to results from the Nagra programme, the table contains a number of samples from the Molasse and Malm from southern Germany as given by BERTLEFF (1986).

The samples are plotted in Figure 3.2.1. The line on this figure corresponds to the global meteoric water line of YURTSEVER AND GAT (1981) discussed above in Section 3.1.2.

Most of the samples fall within the range of values typical of recent water in northern Switzerland. These samples, which are not numbered in Figure 3.2.1, are principally superficial Ca-HCO$_3$ waters as described by SCHMASSMANN AND OTHERS (1984, Sections 6.1.1 and 6.2.1) from the Molasse and the Malm. Samples from other Tertiary and Jurassic formations are also included as described in Section 6.1.2 and 6.2.3 of SCHMASSMANN AND OTHERS (1984). One sample from the quaternary at Bözberg 39.498 (112) in Switzerland is also in this group. All but of one of the samples containing more than 20 tritium units are in this group.

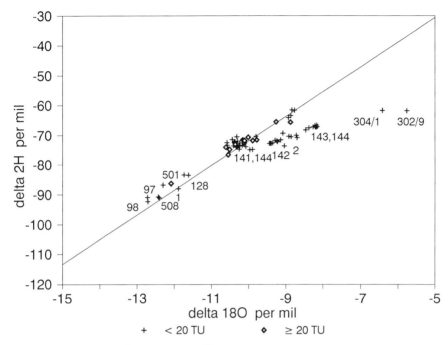

Figure 3.2.1: Graph of δ^2H against $\delta^{18}O$ for groundwaters from the Quaternary, Tertiary, and Malm.

As Figure 3.2.1 shows, there is a group of samples falling along the meteoric water line, but depleted in ^2H and ^{18}O relative to the more numerous group just discussed. These include the deep Na-HCO$_3$ waters of SCHMASSMANN AND OTHERS (1984, Section 6.1.3) from Zürich Aqui (1), Singen (97), Konstanz (98), and Mainau (128). Samples from Reichenau (501) and Birnau (508), described by BERTLEFF (1986), are also of this type. Waters from Lottstetten (127) are chemically in the group of deep Na-HCO$_3$ waters described by SCHMASSMANN AND OTHERS (1984), but their δ^2H and $\delta^{18}O$ values do not differentiate them from the group of superficial Ca-HCO$_3$ waters.

Two hypotheses can be advanced to account for the light isotopic composition of these samples. Konstanz (98), Mainau (128), Reichenau (501) and Birnau (508) are close to the Bodensee, while Zürich Aqui (1) is adjacent to Zürichsee. The average isotopic composition of the Bodensee at Konstanz during 1979 and 1980 was: δ^2H = -88.8 per mil and $\delta^{18}O$ = -12.35 per mil. The composition of the Rhine River at Steigen, which represents the outflow from the Bodensee, averaged δ^2H = -87.8 and $\delta^{18}O$ = -12.16 per mil during the same period. The lakes could be the sources of at least part of the water sampled at these five locations.

The second hypothesis is that the depleted waters were recharged during a colder climatic period such as existed during parts of the Pleistocene, for example. All samples but that from Reichenau (501) have ^3H contents below detection, and adjusted ^{14}C ages of greater than 13 thousand years (ka). As discussed in Section 3.3, the noble gas contents of many of these samples are consistent with recharge temperatures below those of the present. These points suggest that the depletion of ^{2}H and ^{18}O in these samples is best explained by this second hypothesis.

The ^{3}H content of the Reichenau sample, on the other hand, suggests that it contains about 30 per cent young water. Young water so depleted in ^{2}H and ^{18}O can only have been derived from lake water.

Samples shown on Figure 3.2.1 and not yet discussed are from Eglisau (141, 142, 143 and 144), Zürich Tiefenbrunnen (2), and the Schafisheim (304/1) and Weiach (302/6 and 302/9) test boreholes. All are grouped as deep Na-Cl waters by SCHMASSMANN AND OTHERS (1984, Section 6.1.4). The Eglisau samples are from mineral water boreholes in the Lower Freshwater Molasse. Samples 302/6 and 302/9 are from the Malm Massenkalk in the Weiach test borehole. SCHMASSMANN AND OTHERS (1984, Section 6.1.4) show that the chemistry and the isotopic composition of Eglisau samples 1, 2, and 4 (141, 142 and 144) can be represented by mixing of waters of the types found in Eglisau (143) and Mainau (128).

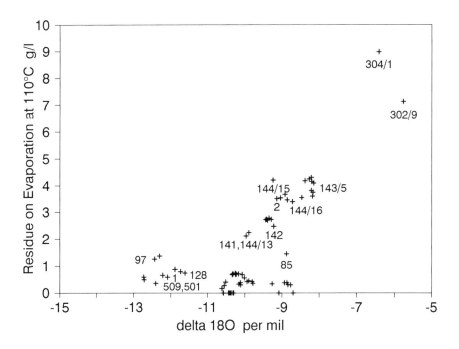

Figure 3.2.2: δ^{18}O values of waters from the Tertiary and Malm *versus* their residues on evaporation.

Figure 3.2.2 is a graph of the δ^{18}O values of the Tertiary and Malm waters against their residues on evaporation at 110°C. As this figure shows, the relationship between δ^{18}O and salinity extends from dilute isotopically-light waters, such as Mainau (128), through the Eglisau samples to the Weiach Massenkalk sample (302/6). Thus, the mixing trend noted by SCHMASSMANN AND OTHERS (1984) with Eglisau (143) as an end member may be part of a larger trend with an end member closer to the Weiach samples (302/6 and 302/9). The Tiefenbrunnen sample (2) also fits this model. The sample from the Lower Freshwater Molasse at Schafisheim (304/1) does not fit this model. Perhaps this is because it was taken so far from the region where the other samples were collected.

This model may well be oversimplified, however, because the Schafisheim and Weiach (304/1, 302/6 and 302/9) samples have ^{14}C levels suggesting both could be young water (Table 5.2.1). Also, the δ^2H content of the Schafisheim sample does not correlate as well as does its $\delta^{18}O$ with a mixing line defined by the Eglisau samples.

3.2.2 Samples from the Dogger and Lias

Analyses of a few waters from these formations were made with the results given in Table 3.2.2 at the end of Section 3.2. Most samples are superficial Ca-HCO$_3$ waters from the Hauptrogenstein of the Dogger as discussed by SCHMASSMANN AND OTHERS (1984, Section 6.2.2), or are treated as special cases of waters from the Jurassic by SCHMASSMANN AND OTHERS (1984, Section 6.2). As shown in Figure 3.2.3, the δ^2H and $\delta^{18}O$ values of these samples are within the range of modern recharge as shown in Figure 3.1.5. They are not numbered in Figure 3.2.3.

A group of samples plot below the range of recent recharge. Samples from the thermal springs at Bellingen (81 and 83), and from Steinenstadt Thermal (84) are deep Na-Cl waters. Neuwiller (77) is a deep Na-HCO$_3$ water. All are from the Upper Rhine Graben and are discussed by SCHMASSMANN AND OTHERS (1984, Section 6.2.4 and 6.1.3). All samples except Weissenstein 1.480 (56) and the four deep samples from the Upper Rhine Graben contain tritium and, therefore, include a young component.

3.2.3 Samples from the Keuper

The results of analyses on samples from the Keuper are given in Table 3.2.3 at the end of Section 3.1. All samples except Lostorf Gipsquelle (21) and Bözberg 39.502 (113) are discussed by SCHMASSMANN AND OTHERS (1984, Section 6.3). Lostorf Gipsquelle is a subthermal gypsum spring. Location 113 is within the Bözberg railway tunnel near Schinznach Dorf.

The δ^2H and $\delta^{18}O$ values of these samples are shown in Figure 3.2.4. All samples except 303/1 contain tritium and a young component.

Sample 303/1 is from the Riniken test borehole. It is considerably enriched in both ^2H and ^{18}O not only with respect to the other Keuper samples, but also with respect to most of the samples analysed as part of this programme. This water cannot have been recharged during a climatic regime like that of the present.

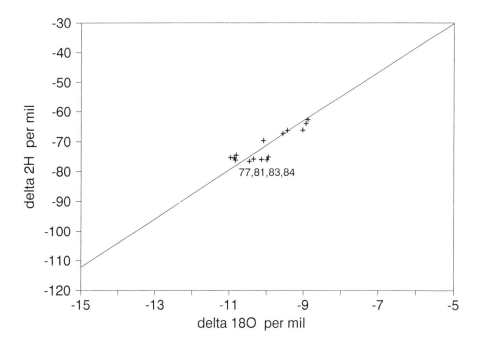

Figure 3.2.3: Graph of δ^2H against δ^{18}O for groundwaters from the Dogger and Lias.

The sample has a total dissolved solids content of nearly 16 grams per litre and an adjusted ^{14}C age of > 17 ka (Table 5.2.1). Several hypotheses can be advanced to account for this. Based on mass balances using ^{2}H, ^{18}O, and total dissolved solids, it could represent a mixture of 35 to 45 per cent normal sea water with water of the type found in the other Keuper samples. However, the chloride content of this sample is only about 15 per cent of that of sea water, which would seem to rule out a sea-water mixing hypothesis. The ^{18}O content of groundwaters can increase above that in recharge as a result of water-rock reactions. Such reactions may also give rise to high total dissolved solid contents, but require considerable time. The enriched ^{2}H in this sample argues against this hypothesis for its origin. One other sample of high salinity from the Keuper was collected from the Beznau borehole (171/101). No δ^2H value is available for this sample but as Table 3.2.3 shows, its ^{18}O content, -8.77 per mil, is closer to that of the bulk of the Keuper sample than to that of 301/1. Sample 171/101 contains some ^{3}H, however. So the formation water it represents may be of higher salinity and ^{18}O content. With the present data, neither 303/1 nor 171/101 can be unequivocally interpreted.

3.2.4 Samples from the Muschelkalk

Results of analyses of many groundwaters from this unit for δ^2H, δ^{18}O, and ^{3}H are given in Table 3.2.4 at the end of Section 3.1.

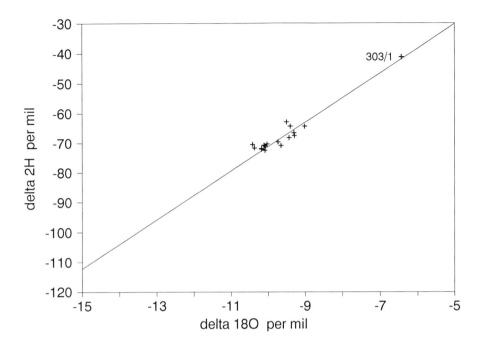

Figure 3.2.4: Graph of δ^2H against δ^{18}O for groundwaters from the Keuper.

Figure 3.2.5 shows the δ^2H and δ^{18}O values of young Muschelkalk waters, those with ^{3}H contents > 20 TU. These samples all cluster around the meteoric water line, except that from Windisch Reussuferquelle (48). There is no obvious reason why the ^{2}H content of this sample should be so much higher than those of the other Muschelkalk samples. Most of the samples shown in Figure 3.2.5 are classified as superficial waters of the upper and middle Muschelkalk and as mixed superficial waters and deeper waters of the upper Muschelkalk by SCHMASSMANN AND OTHERS (1984, Sections 6.4.1, 6.4.2 and 6.4.3).

Figure 3.2.6 shows the δ^2H and δ^{18}O values of samples with < 20 tritium units. A number of these results fall within the range of young Muschelkalk waters shown in Figure 3.2.5. A number of other samples, however, can be distinguished from the young waters based on their δ^2H and δ^{18}O values. These samples deserve further discussion.

A large group of samples from locations 201 through 221 have ^{2}H contents similar to the majority of Muschelkalk samples but are enriched in ^{18}O relative both to other Muschelkalk samples and to normal meteoric water. These samples are from the springs at Baden and Ennetbaden. These springs emerge at temperatures of 46° to 48°C and, as Table 3.2.4 shows, have closely similar dissolved solids contents, and ^{3}H concentrations from 3 to 6 TU for samples taken from 1982 to 1984.

The Baden springs are classified chemically as deep Na-Ca-Cl-SO$_4$ waters of the upper Muschelkalk by SCHMASSMANN AND OTHERS (1984, Section 6.4.5). These authors

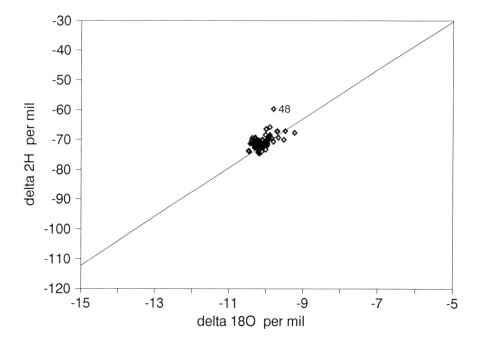

Figure 3.2.5: Graph of δ^2H against $\delta^{18}O$ for groundwaters from the Muschelkalk with ^3H contents greater than 20 TU.

show (Figure 31) that the tritium contents of the various springs correlate with certain of their chemical properties. This supports the hypothesis that the waters sampled represent deep, warm Muschelkalk water mixed with a young superficial component. Based on their ^{14}C and tritium contents, these springs are estimated above (Table 2.1.3) to contain from 5 to 10 per cent young water.

SCHMASSMANN AND OTHERS (1984) suggest that the deep component of the Baden spring waters could itself be a mixture between a deep water of the type found at Böttstein (301/2) and Beznau (15) and a water enriched in both ^2H and ^{18}O. Such enrichment could have been produced by evaporation and the high-^2H and ^{18}O water would also be expected to be highly mineralized.

Isotopically enriched waters with relatively high salinities are found in the Lower Freshwater Molasse (USM) at Schafisheim (304/1) and in the Malm at Weiach (302/6 and 302/9), as well as in the Buntsandstein, Rotliegend, and (or) crystalline at Riniken (303/3b) and Schafisheim (304/5, 304/8, 304/10 and 304/13). If a water with an isotopic composition similar to these seven waters was mixed in about a one-to-one ratio with relatively depleted waters, such as are found in the Muschelkalk at Böttstein (301/2) and Beznau (15), the resulting water would have the ^2H and ^{18}O contents of the Baden springs. The chemistry of the waters from the Baden and Ennetbaden springs argue against their production by such a mixing process. The total mineralization of the spring

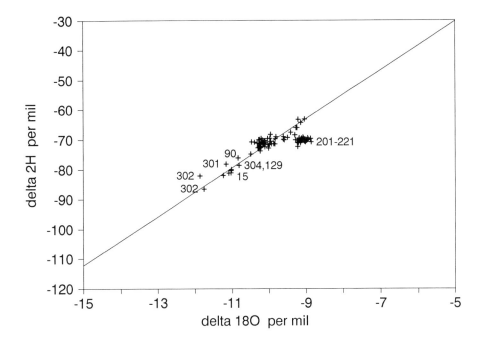

Figure 3.2.6: Graph of δ^2H against $\delta^{18}O$ for groundwaters from the Muschelkalk with ^3H contents less than 20 TU.

waters (c. 4.7 g/l) and their chloride contents (c. 1.1 g/l) are lower than those of both proposed end-members. The total dissolved solids and chloride contents of the Böttstein (301/2) and Beznau (15) samples are about 6.5 and 1.4 g/l, respectively. The dissolved solids in the seven isotopically enriched samples range from 7 to 16 g/l, and their chloride contents from 2.4 to 6.4 g/l.

Groundwater can be enriched in ^{18}O by isotope exchange with aquifer rock given a sufficiently long residence time and (or) a relatively high temperature (SAVIN, 1980). The enrichment in ^{18}O observed in the samples from the Baden springs can be attributed to such exchange.

As Figure 3.2.6 shows, a number of samples are depleted in both ^2H and ^{18}O relative to the younger Muschelkalk waters. These samples are from the Böttstein (301), Weiach (302), and Schafisheim (304) test boreholes, and from the Beznau borehole (15), and are grouped chemically with the Baden spring samples by SCHMASSMANN AND OTHERS (1984, Section 6.4.5). Riniken sample 303/2 is also in this chemical group, although its δ^2H and $\delta^{18}O$ values plot within the group of samples isotopically equivalent to young water.

The Muschelkalk aquifer from which this group of samples comes is a dolomitic limestone with disseminated gypsum or anhydrite. This mineral assemblage will exert a very strong effect on the chemistry of groundwater, possibly overwhelming chemical evidence of the origin of the water itself. The differences in isotopic composition between these isotopically light samples and the Baden and Ennetbaden samples (201 through 221) may be evidence of differences of their origin and evolution in spite of their similar chemistries.

The most depleted Muschelkalk samples are from the Weiach test borehole (302). Few other samples in this study are more depleted in ^{2}H and ^{18}O than Weiach water. Those which are more depleted include waters from crystalline rock in the Alps (231, 232, 233 and 234; Table 3.2.5 and Figure 3.2.7) and samples from the Molasse and Malm at Zürich Aqui (1), Singen (97), Konstanz (98), Reichenau (501) and Birnau (508). As discussed in Section 3.1.2, all but the Reichenau sample probably represent Pleistocene recharge. The Reichenau sample, however, has a significant young component, the isotopic composition of which corresponds to water from the Bodensee.

Water from the Weiach Muschelkalk contains measurable tritium (Table 3.2.4) and has ^{14}C and ^{39}Ar contents corresponding to a water age of modern to not more than 1 ka (Tables 5.2.1 and 4.2). This borehole is close to the Rhine, the isotopic composition of which, like that of the Bodensee, resembles alpine precipitation rather than local precipitation. The probable explanation for the depleted δ^2H and δ^{18}O values in the Weiach Muschelkalk is that water in it is derived, in least in part, from the Rhine River.

Samples 15, 301, and 304 are from the Beznau, Böttstein and Schafisheim boreholes. Adjusted ^{14}C ages can be calculated only for the Beznau and Böttstein samples. They are > 29 ka and > 17 ka, respectively (Tables 5.2.2 and 5.2.1), which would be consistent with recharge during a cooler period.

The two remaining isotopically light samples in Figure 3.2.6 are from Krozingen 3 (90) and from Liel Schlossbrunnen (105). These waters most probably originate from recharge in areas of higher altitude in the adjacent Black Forest Massif. They could result from infiltration of river water along tectonic fault zones within the eastern border of the Rhine Graben. They could also be a result of direct internal connection between waters of the Black Forest crystalline and the downfaulted Muschelkalk within the Rhine Graben itself (SCHMASSMANN AND OTHERS, 1984, Section 6.4.6.5).

3.2.5 Samples from the Buntsandstein, Permian and Crystalline

Data on samples from these water-bearing units are given in Table 3.2.5 and displayed in Figure 3.2.7. This figure differs from the previous δ^2H versus δ^{18}O graphs by covering a wider range of compositions and by including two meteoric water lines. The solid line is the global meteoric water line of YURTSEVER AND GAT (1981) shown in previous figures. The dashed line is the regression line for modern waters from Switzerland and adjacent areas developed in Section 3.1.2. As the figure shows, most samples have δ^{18}O values between -9 and -11 per mil and fit either line equally well. Samples 231 through 234 are from tunnels in the Alps and are depleted in heavy isotopes, consistent with their recharge from precipitation at high altitudes. These samples correspond better with the Swiss line than with the global line.

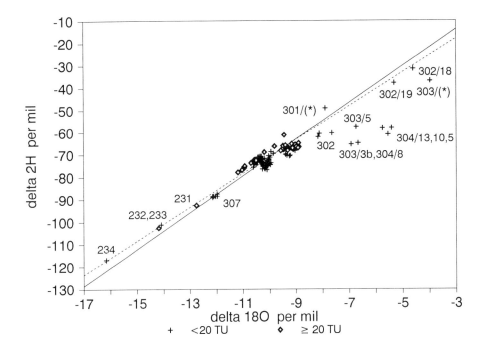

Figure 3.2.7: Graph of δ^2H against $\delta^{18}O$ for groundwaters from the Buntsandstein, Permian and Crystalline.

Most samples group around a $\delta^{18}O$ value of c. -10 per mil. A second group apparent in Figure 3.2.7, with $\delta^{18}O$ values around -9 per mil, comprises almost entirely samples from one location, Säckingen Badquelle (125). As discussed in Section 4.2, samples from this location are mixtures of about 50 per cent tritium-free, deep crystalline water and 50 per cent tritium-bearing young water.

Waters from the Buntsandstein and Permian are discussed in Section 6.5 of SCHMASSMANN AND OTHERS (1984). The samples from Kaiseraugst (3) and Waldkirch (95) of the regional programme are included along with samples from Weiach (302/12), Riniken (303/3, 5 and 6), and Kaisten (305/1) of the Nagra test boreholes.

Although the Waldkirch borehole (95) is completed in gneiss, its water chemistry is more like Buntsandstein or Permian water than water from crystalline rock elsewhere. SCHMASSMANN AND OTHERS (1984) include it with Buntsandstein and Permian waters in their Section 6.5. As discussed below, however, the Waldkirch samples also resemble water from a zone of low yield at 923.0 m in the Leuggern test borehole (306/11b, 12, 13).

The Badenweiler 3 borehole (88) is completed in the Buntsandstein and Permian but yields water chemically equivalent to that from the other Badenweiler boreholes, which are completed in the crystalline. Thus, it is treated as a crystalline sample by

SCHMASSMANN AND OTHERS (1984) and discussed in their Section 6.6.1. Results from some of the borehole samples were not available until after the report of SCHMASSMANN AND OTHERS (1984) was completed. These include Buntsandstein and Permian samples from Böttstein (301/5, 6), Weiach (302/18, 19), Riniken (303/20), and Leuggern (306/2). Samples from the Permian at Kaisten (305/2, 3) are chemically similar to water from the underlying crystalline and are described by SCHMASSMANN AND OTHERS (1984) in their Section 6.6.2.

Four groups of crystalline waters are discussed by SCHMASSMANN AND OTHERS (1984). Superficial and deep Ca-Na-HCO_3 and Ca-Na-HCO_3-SO_4 waters are described in their Section 6.6.1. This group includes samples from Bürchau (110) and Rothaus (150), as well as the samples from Badenweiler 3 (88) mentioned above. These samples together with those from locations 162 through 194 in Table 3.2.5 are generally ^{3}H-bearing and represent young waters in the crystalline of the Black Forest.

SCHMASSMANN AND OTHERS (1984) describe deep Na-SO_4-HCO_3-Cl waters in their Section 6.6.2. This includes samples from thermal water boreholes 1 and 2 at Zurzach (131, 132), from the upper granite at Böttstein, (301/8 to 18), and from the Permian and gneiss at Kaisten, (305/2 to 6). The remaining samples from Kaisten, (305/9 to 16) and from the crystalline at Leuggern (306/4 to 9, 17, 18, 20) are similar to this group chemically.

The ^{2}H and ^{18}O contents of all samples discussed to this point are within the main group in Figure 3.2.7. They have $\delta^{18}O$ values between c. -10 and -10.5 per mil and are isotopically similar to meteoric water. The remaining samples from Leuggern also fall in this range of isotopic compositions, as do the samples from the Rheinfelden borehole (159/2 to 4), which are described as Na-Ca-HCO_3-SO_4-Cl waters by SCHMASSMANN AND OTHERS (1984, Section 6.6.4).

As Figure 3.2.7 shows, there are a number of samples which are enriched both in ^{2}H and ^{18}O or in ^{18}O alone relative to the bulk of the samples. Samples with < 20 TU are shown on an expanded scale in Figure 3.2.8.

As discussed by WITTWER (1986) and in Chapter 2 of this text, some samples from the test boreholes are contaminated, primarily from mixing between drilling fluid and groundwater. Such contamination can influence the stable isotopic composition of the sample as well as its ^{3}H and chemical composition. Six particularly contaminated samples were corrected for their drilling fluid content so that the chemical and isotopic data given in Table 3.2.5 and shown in Figures 3.2.7 through 3.2.10 would represent groundwater values. The corrected values for two samples, the deep granite at Böttstein (301/(*); see Table 3.2.5) and the 965 m Permian at Riniken (303/5b), were taken from the references given in Table 3.2.5. The correction to 303/5b changes only its chemistry. Corrections for the remaining samples are shown in Table 3.2.6 and were made as follows:

> WITTWER (1986), using drilling fluid tracer measurements, chemistry, borehole history, and whatever data were available has estimated drilling-fluid contamination. For the samples of interest here, the drilling fluid was deionized water. In the absence of measurements of the δ^2H and $\delta^{18}O$ content of the drilling fluid

itself at the time of drilling, the isotopic composition of the contaminant has been taken as approximately that of the Rhine River.

Table 3.2.6: Corrections to chemistry and isotopic composition of contaminated samples.

	Residue at 110°C (mg/l)	Chloride (mg/l)	δ^2H (per mil)	$\delta^{18}O$ (per mil)
WEIACH 2267.0 m Sample 302/14			Reference: 5.2.5	
as Analysed	7649	3924	-62.7	-8.06
Corrected for 9% Deionised Water	8405	4312	-60.2	-7.65
RINIKEN 1361.5 m Sample 303/20			Reference: 5.3.7	
as Analysed	33180	18040	-48.6	-5.88
Corrected for 23% Deionised Water	43091	23429	-36.8	-3.98
SCHAFISHEIM 1488.2 m Sample 304/5			Reference: 5.4.3	
Analysed, 304/6	15993	6405		
Analysed, 304/5			-61.6	-6.24
Corrected for 12% Deionised Water			-58.0	-5.42
SCHAFISHEIM 1980.4 m Sample 304/13			Reference: 5.4.6	
as Analysed	6800	2549	-62.6	-6.73
Corrected for 15% Deionised Water	8000	2999	-58.1	-5.76
Deionised Water:	0	0	-88	-12.25

References are to sections in WITTWER (1986)

Table 3.2.6 shows the residue on evaporation, chloride content, and δ^2H and $\delta^{18}O$ values for the samples, both as analysed and as corrected for the indicated amount of drilling fluid contamination. The amounts of contamination are as given by WITTWER (1986) in the sections cited in the tables. The properties assumed for the drilling fluid are also given. The corrected values appear in Table 3.2.5 and in the figures.

Groundwaters with < 20 TU are shown in Figure 3.2.8, a conventional δ^2H-$\delta^{18}O$ plot, Figure 3.2.9, a plot of δ^2H against residue on evaporation, and in Figure 1.1.18, a plot of $\delta^{18}O$ against residue on evaporation.

BALDERER AND OTHERS (1987b) interpret the ^{18}O, ^2H, and total dissolved solids contents of samples from the Buntsandstein, Permian, and crystalline in terms of mixing and exchange processes. Minerals and rock contain much less hydrogen than oxygen. Therefore, changes in the hydrogen isotopic composition of water by exchange with minerals are much less profound than changes in water oxygen. Mixing effects uncomplicated by exchange can best be seen using the δ^2H values as shown in Figure 3.2.9.

Two bands are drawn on Figure 3.2.9. One encompasses samples from the Permian and Buntsandstein. The δ^2H values and residues on evaporation of these samples could be

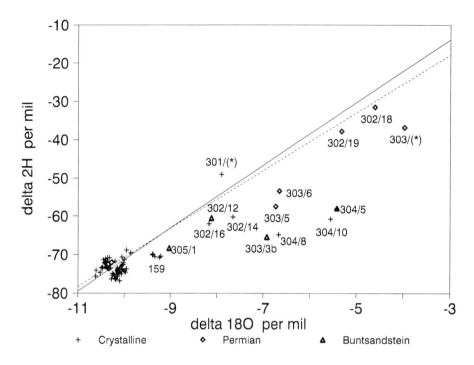

Figure 3.2.8: Graph of δ^2H against $\delta^{18}O$ for relatively enriched groundwaters from the Buntsandstein, Permian and Crystalline.

developed by mixing waters like those from 1116 m at Weiach (302/19), or 1362 m at Riniken (303/(*)) with water of low residue and δ^2H values like those in most of the samples from the crystalline and in young waters. Examination of the chemistry of the samples in this band shows that differences in their sodium and chloride contents account for virtually all of the differences in their residues on evaporation.

Two locations yield waters outside this band. The 1408 m zone at Weiach (302/18) yields water with higher ^2H and ^{18}O contents than the 1116 m zone (302/19), and with much higher residues on evaporation. Water from the Buntsandstein and Permian at Kaiseraugst (3) is virtually chemically identical with the 985 m sample from the Buntsandstein at Weiach (302/12), but has considerably lower ^2H and ^{18}O contents.

The second band in Figure 3.2.9 includes samples from the deep crystalline. Both the chemical and isotopic values of many of these points result from the correction of contaminated samples (see Table 3.2.5 and 3.2.6), and so conclusions from them must be used with care. The values in this band could be developed by mixing water like that found at 1326 m at Böttstein (301/(*)) with waters like those found in the upper crystalline. These upper Crystalline waters appear on Figure 3.2.9 and 3.2.10 as the series of unlabelled points with low residues on evaporation and relatively negative δ^2H and $\delta^{18}O$ values.

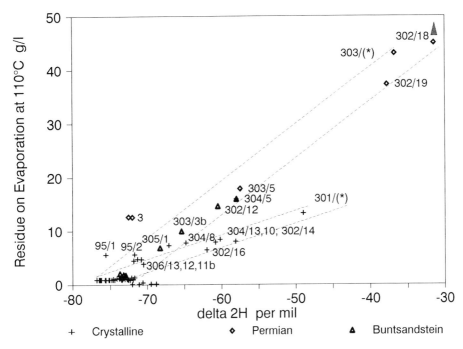

Figure 3.2.9: **Graph of δ^2H against residue on evaporation for relatively enriched waters from the Buntsandstein, Permian and Crystalline.**

Figure 3.2.10 shows the δ^{18}O values of these samples *versus* their residues on evaporation. The bands of possible mixing evident on the δ^2H figure (Figure 3.2.9) are not clear in this δ^{18}O plot. For example, the δ^{18}O values for the Buntsandstein samples from 1488 m at Schafisheim (304/5) and 807 m at Riniken (303/3b) as well as of the samples from the crystalline at Schafisheim (304/8, 10 and 13) are relatively more positive than their δ^2H values. On the other hand, the 1326 m sample from the deep crystalline at Böttstein (301/(*)) has a δ^{18}O value relatively more negative than its δ^2H value. The ^{18}O content of groundwater can be changed more readily than its ^{2}H content by isotope exchange with aquifer rock. For such exchange to be evident requires high temperatures and (or) very long times of contact between water and rock.

Finally, consideration must be given to the origin of the end-member waters of the two mixing groups suggested in Figure 3.2.9. The low-residue, low ^{18}O and ^{2}H waters are those from the crystalline at shallow depths, and are equivalent isotopically to young, high ^{3}H waters (Figure 3.2.7). The waters with the highest residues on evaporation and the highest ^{2}H and ^{18}O contents are from the Permian at Weiach (302/18, 19) and Riniken (303/(*)). These samples have higher residues on evaporation than normal sea water (c. 35 g/l), but are depleted in both ^{18}O and ^{2}H relative to sea waters. Waters of this chemical character can be formed by evaporative concentration of continental meteoric water, but this process normally leads to relatively higher enrichment in ^{18}O than in ^{2}H.

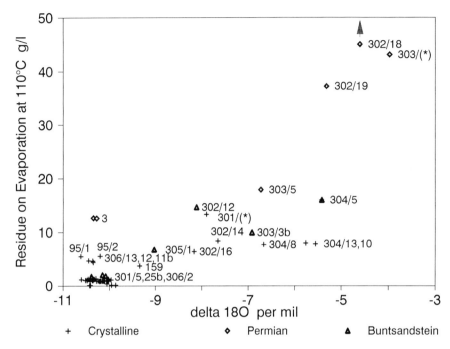

Figure 3.2.10: Graph of $\delta^{18}O$ against residue on evaporation for relatively enriched waters from the Buntsandstein, Permian and Crystalline.

The isotopic composition of the Riniken sample (303/(*)) is consistent with such an origin but those of the Weiach samples (302/18, 19) are not.

The two enriched Weiach samples may represent normal meteoric water recharged at a higher temperature or in a region closer to the ocean than prevails in northern Switzerland today. The high dissolved solids of these waters would then be a product of water-rock interaction. These waters could also be mixtures of highly evaporated sea water with normal meteoric waters similar to young waters found today in the upper crystalline of northern Switzerland.

The high ^2H water from the crystalline could represent recharge at a higher temperature or a location closer to the ocean than prevail in northern Switzerland today and give rise to water in the shallow crystalline. Under this hypothesis the dissolved solids would have resulted from water-rock reactions and the varying $\delta^{18}O$ values from varying amounts of isotope exchange. The Böttstein (301), Weiach (302), and Schafisheim (304) boreholes are so located that water from the Permo-Carboniferous Trough could also be a source of the deep crystalline water in them. The differences in the ^2H and ^{18}O contents of the high-residue Permian and crystalline waters could result from mixing with lower residue waters and subsequent water-rock exchange. However, as pointed out, the uncertainties in the isotopic and chemical compositions of many of these deep crystalline waters are such that an exhaustive interpretation of them is not warranted.

Table 3.2.1: Isotopic composition of groundwaters from the Quaternary, Tertiary and Malm. (Page 1 of 2)

Location Sample (Table A1)	Date Sampled	Source Formation	Residue at 110° (mg/l)	Chloride (mg/l)	Delta 2H (per mil)	Delta 18O (per mil)	Tritium (TU ± 2SD)	Laboratory
1/1	21JUL1981	tOMM	872	158.9	-88.0	-11.89	<0.7	GSF
1/1	21JUL1981	tOMM	872	158.9			<0.6	LLC Bern
2/1	21JUL1981	tOMM	3538	1170	-73.6	-9.04	<0.7	GSF
2/3	30OCT1982	tOMM	3510	1160	-71.6	-9.14	<0.9	GSF
5/1	08SEP1981	tUSM	381	2.8	-63.4	-8.87	0.9 ± 0.7	GSF
5/3	06FEB1984	tUSM	376	2.9	-64.1	-8.93	1 ± 0.6	GSF
7/1	09SEP1981	jo	168	1.5	-74.1	-10.61	93.3 ± 6.2	GSF
30/1	21JUL1981	jo	563	50.1	-70.7	-10.01	88.8 ± 5.9	GSF
32/2	09SEP1981	tUSM	383	8.8	-72.6	-10.12	68.1 ± 4.5	GSF
33/2	14JUL1981	tUSM	273	4.85	-76.6	-10.55	87 ± 5.7	GSF
42/1	21JUL1982	tOSM	324	2.71	-71.6	-10.16	86.8 ± 6	GSF
80/2	22JUL1982	jo	345	5.01	-65.5	-9.26	47.1 ± 3.2	GSF
85/1	18AUG1981	teo	1447	177	-65.6	-8.88	95.1 ± 6.2	GSF
96/1	27NOV1981	tvu	308	4.49	-61.5	-8.84	3.2 ± 0.8	GSF
96/2	11APR1984	tvu	300	2.78	-61.6	-8.77	4.9 ± 0.7	GSF
97/1	28JUL1981	jo	1254	114.3	-90.7	-12.43	<0.7	GSF
97/105	21MAR1983	jo	1367	114.16	-86.8	-12.30	<0.6	Heidelberg
98/1	29JUL1981	tOMM	495	12.3	-92.3	-12.71	<0.7	GSF
98/102	21MAR1983	tOMM	587.5	13.12	-90.9	-12.72	<0.52	Heidelberg
99/1	29JUL1981	jo(-q)	448	50.1	-71.8	-9.89	109.6 ± 7.3	GSF
112/101	06DEC1982	q		3.2	-72.5	-10.38	68.6 ± 4.6	GSF
112/102	24FEB1983	q		4.6	-73.1	-10.36	65.4 ± 5.8	GSF
127/1	24JUL1981	jo	713	53.6	-73.9	-10.32	<0.7	GSF
127/2	23NOV1981	jo	682	55	-73.9	-10.07	<0.7	GSF
127/3	04MAR1982	jo	708	54.3	-73.6	-10.17	<0.7	GSF
127/4	19AUG1982	jo	692	59.1	-73.6	-10.24	<0.7	GSF
127/5	19AUG1982	jo	692	59.1	-73.6	-10.24	<0.7	GSF
127/8	16MAY1983	jo	734	55.4	-73.6	-10.26	<0.8	GSF
127/11	24FEB1984	jo	683	54.2	-74.1	-10.34	<0.7	GSF
127/14	16NOV1984	jo	702	54.3	-73.4	-10.24	<0.7	GSF
127/16	12AUG1985	jo	705	53.4	-74.7	-10.25	<1.4	GSF
127/17	27MAY1986	jo	692	57	-72.6	-10.23	<1.0	GSF
127/102	08DEC1982	jo		55.8	-72.1	-10.36	1.6 ± 1.8	GSF
127/103	25FEB1983	jo		54.12	-70.5	-10.32	<0.4	GSF
127/104	26JUN1983	jo		53.1	-72.0	-10.30	0.8 ± 0.7	GSF
128/1	29JUL1981	tOMM	737	4.14	-83.5	-11.62	<0.9	GSF
128/101	21MAR1983	tOMM	794.6	4.96	-83.4	-11.74	<0.3	Heidelberg
130/2	26NOV1981	tOMM-tUS	304	9.7	-71.8	-10.11	68.1 ± 4.5	GSF
141/1	24JUL1981	tUSM	2236	828.1	-74.8	-9.89	<0.7	GSF
142/2	04MAR1982	tUSM	2728	1354	-71.7	-9.29	<1.0	GSF
142/4	19AUG1982	tUSM	2738	1354	-72.7	-9.41	1 ± 0.7	GSF
142/7	20MAY1983	tUSM	2468	1254	-72.2	-9.22	0.7 ± 0.7	GSF
142/10	23FEB1984	tUSM	2694	1242	-72.8	-9.39	1 ± 0.7	GSF
142/13	15NOV1984	tUSM	2716	1324	-72.7	-9.44	0.9 ± 0.7	GSF
142/15	14AUG1985	tUSM	2778	1336	-72.6	-9.35	1.4 ± 0.8	GSF
143/2	28OCT1981	tUSM	4098	2017	-67.1	-8.15	<0.7	GSF
143/5	19AUG1982	tUSM	4150	2081	-67.2	-8.20	<0.8	GSF
143/8	20MAY1983	tUSM	3600	1806	-67.3	-8.18	<0.7	GSF

Table 3.2.1: Isotopic composition of groundwaters from the Quaternary, Tertiary and Malm. (Page 2 of 2)

Location Sample (Table A1)	Date Sampled	Source Formation	Residue at 110° (mg/l)	Chloride (mg/l)	Delta 2H (per mil)	Delta 18O (per mil)	Tritium (TU ± 2SD)	Laboratory
143/11	24FEB1984	tUSM	4240	2136	-67.2	-8.27	<0.7	GSF
143/14	15NOV1984	tUSM	4182	2133	-67.5	-8.38	<0.9	GSF
143/16	14AUG1985	tUSM	3753	1895	-66.6	-8.17	<1.6	GSF
143/17	27MAY1986	tUSM	4300	2200	-66.7	-8.21	<1.3	GSF
144/2	28OCT1981	tUSM	3814	1966	-67.0	-8.21	<0.8	GSF
144/5	19AUG1982	tUSM	3458	1778	-70.5	-8.86	<0.7	GSF
144/8	20MAY1983	tUSM	3391	1678	-70.0	-8.72	<0.8	GSF
144/11	23FEB1984	tUSM	3667	1840	-70.4	-8.92	<0.7	GSF
144/13	16NOV1984	tUSM	2110	981	-74.9	-9.97	<0.7	GSF
144/15	14AUG1985	tUSM	4205	2156	-71.9	-9.24	<1.6	GSF
144/16	27MAY1986	tUSM	3550	1830	-68.2	-8.47	<0.7	GSF
149/1	28JUL1981	jo (-tUSM)	422	9.4	-70.4	-9.80	17 ± 1.3	GSF
149/2	09FEB1984	jo (-tUSM)	359	11.6	-71.6	-9.78	26.9 ± 1.9	GSF
160/1	09FEB1984	q	409	12.5	-74.8	-10.51	71.5 ± 5	GSF
170/2	02JUL1986	tOMM	360	<1	-71.6	-10.01	1.0 ± 0.7	GSF
207/2	30MAY1986	tOMM			-69.3	-9.08	3.8 ± 0.7	GSF
302/6	06MAR1983	joki	7129	3844			<5.7	GSF
302/9	05MAR1983	joki			-61.8	-5.76	<3.6	GSF
304/1	20DEC1983	USM	8985	5193	-61.7	-6.41	<1.2	GSF
501/102	21MAR1983	q	589.28	17.37	-86.3	-12.09	24	
502/101	12OCT1982	tOMM	420.28	2.84		-9.94	4 ± 0.6	GSF
508/101	22MAR1983	tOMM	354.95	12.05	-91.1	-12.40	<0.5	Heidelberg
509/101	07JUL1983	tOMM	651.1	0.35		-12.22	0.4 ± 0.6	GSF

Table 3.2.2: Isotopic composition of groundwaters from the Dogger and Lias.

Location Sample (Table A1)	Date Sampled	Source Formation	Residue at 110° (mg/l)	Chloride (mg/l)	Delta 2H (per mil)	Delta 18O (per mil)	Tritium (TU ± 2SD)	Laboratory
19/1	08SEP1981	km(+?m)	2339	3.4	-70.9	-9.65	103.5 ± 6.8	GSF
21/1	16AUG1985	k	1505	15.8	-70.8	-10.11	26.7 ± 1.9	GSF
21/2a	26MAY1986	k	1240	16	-71.1	-10.08	31.1 ± 2.2	GSF
23/1	16JUL1981	km	1453	5.06	-72.5	-10.09	129.8 ± 8.5	GSF
23/4	16AUG1985	km	1412	5.2	-70.4	-10.03	74 ± 4.9	GSF
25/1	17JUL1981	km	2585	7.72	-67.5	-9.29	75.4 ± 5.2	GSF
26/1	17JUL1981	km	2664	5.08	-68.2	-9.43	94.4 ± 6.2	GSF
27/1	25NOV1981	km	2571	6.72	-64.4	-9.01	57.1 ± 3.9	GSF
28/2	15JUL1981	km	2473	2.47	-71.9	-10.18	125.1 ± 8.4	GSF
28/3	10FEB1984	km	2374	3.1	-72.1	-10.18	91 ± 6.0	GSF
29/1	23JUL1981	km	2516	8.89	-69.7	-9.73	64.2 ± 4.3	GSF
29/105	15MAY1979	km		10.6	-64.4	-9.40	82	Weizmann
29/106	24JUL1979	km		9.04	-62.9	-9.50		Weizmann
29/107	31OCT1979	km		9.57	-66.6	-9.30	62	Weizmann
113/101	06DEC1982	km		2.7	-70.5	-10.42	89.4 ± 5.9	GSF
113/102	24FEB1983	km		3.1	-71.7	-10.37	86 ± 6.3	GSF
171/101	14JAN1980	km3b	14390	2770	--	-8.77	5.6 ± 3.9	LLC
303/1	25JUL1983	km	15704	2775	-41.2	-6.43	<1.2	GSF

Table 3.2.3: Isotopic composition of groundwaters from the Keuper.

Location Sample (Table A1)	Date Sampled	Source Formation	Residue at 110° (mg/l)	Chloride (mg/l)	Delta 2H (per mil)	Delta 18O (per mil)	Tritium (TU ± 2SD)	Laboratory
31/2	08SEP1981	jmHR	321	10	-67.3	-9.55	55.4 ± 4.0	GSF
37/2	14JUL1981	ju	478	3.1	-76.5	-10.46	102.5 ± 6.8	GSF
56/2	21JUL1982	ju	965	26.9	-75.3	-10.96	<1.2	GSF
68/2	20JUL1982	jmHR	503	24.6	-69.7	-10.08	66.3 ± 4.4	GSF
76/2	20JUL1982	jmHR	245	2.37	-75.4	-10.86	84.1 ± 5.5	GSF
76/102	11NOV1982	jmHR		2.36	-76.1	-10.83	89.4 ± 6.1	GSF
76/103	23FEB1983	jmHR		2.84	-74.5	-10.81	82.6 ± 6.3	GSF
77/1	07SEP1981	jmHR	615	81	-75.8	-10.35	<0.7	GSF
81/1	17AUG1981	jmHR	4242	1961	-75.1	-9.95	<0.7	GSF
83/1	17AUG1981	jmHR	3610	1636	-76.0	-9.99	<1.0	GSF
84/1	18AUG1981	jmHR	4187	1821	-75.9	-10.14	<0.7	GSF
105/1	18AUG1981	jmHR	704	55	-66.2	-9.02	78.5 ± 5.2	GSF
106/1	20AUG1981	jmHR	305	4.6	-66.3	-9.43	10.2 ± 0.9	GSF
108/1	19AUG1981	jmHR	502	43	-64.0	-8.93	18.1 ± 1.3	GSF
109/1	27NOV1981	jmHR	273	15.7	-62.7	-8.88	17.8 ± 1.5	GSF

Table 3.2.4: Isotopic composition of groundwaters from the Muschelkalk. (Page 1 of 5)

Location Sample (Table A1)	Date Sampled	Source Formation	Residue at 110° (mg/l)	Chloride (mg/l)	Delta 2H (per mil)	Delta 18O (per mil)	Tritium (TU ± 2SD)	Laboratory
8/2	07FEB1984	mo	2230	100.3	-66.7	-9.29	9.2 ± 0.8	GSF
9/1	26NOV1981	mo	647	9.6	-63.2	-9.04	2.6 ± 0.7	GSF
9/2	08FEB1984	mo	616	7.2	-64.4	-9.14	<1.1	GSF
10/1	24NOV1981	mo (+?km)	1830	7.91	-69.4	-9.67	70.1 ± 5	GSF
11/2	09FEB1984	mo	445	4.34	-70.2	-9.94	77.0 ± 5	GSF
12/1	15JUL1981	mo	1188	8.34	-67.2	-9.48	69.8 ± 4.8	GSF
13/1	10SEP1981	mo	1296	3.5	-71.2	-9.96	75.4 ± 5	GSF
14/1	14JUL1981	mo	732	4.16	-72.3	-10.23	5.2 ± 0.6	GSF
14/3	10FEB1984	mo	623	2.3	-73.8	-10.24	7.8 ± 0.9	GSF
15/1	22JUL1981	mo	6567	1398	-81.2	-11.09	<1.0	GSF
15/1	22JUL1981	mo	6567	1398			<0.6	LLC Bern
15/3	08FEB1984	mo	6708	1416	-81.1	-11.03	<0.7	GSF
15/4	13NOV1984	mo	6744	1446	-82.0	-11.24	<0.8	GSF
16/1	23JUL1981	mo	912	35.6	-70.1	-9.52	83.7 ± 5.6	GSF
16/101	12DEC1979	mo		31.7	-65.9	-9.90	90.0	Weizmann
17/101	05FEB1980	mo		108	-72.4	-10.20	76.0	Weizmann
22/1	16JUL1981	mm	1904	4.23	-71.5	-10.01	73.6 ± 5	GSF
24/1	10SEP1981	mm	1300	3.9	-73.9	-10.46	104.2 ± 7	GSF
34/2	13JUL1981	mo	1004	2.03	-73.4	-10.17	79.1 ± 5.4	GSF
35/2	15JUL1981	mo	951	5.23	-67.8	-9.23	63.9 ± 4.1	GSF
36/2	10SEP1981	mo	882	5.9	-74.7	-10.18	68.1 ± 4.5	GSF
38/2	11SEP1981	mo	583	5.1	-74.0	-10.12	66.5 ± 4.6	GSF
39/2	14JUL1981	mo	727	3.4	-73.4	-10.01	61.2 ± 4.2	GSF
47/1	23JUL1981	mo	720	4.55	-72.3	-10.22	71.0 ± 4.7	GSF
47/101	15MAY1979	mo		4.79	-69.2	-9.90	99.0	Weizmann
47/102	24JUL1979	mo		4.61	-70.0	-10.10		Weizmann
47/103	01NOV1979	mo		5.25	-66.5	-10.00	77.0	Weizmann
48/101	23JUL1979	mo		41.7	-59.7	-9.80	72.0	Weizmann
65/2	19JUL1982	mo	856	13.8	-71.4	-10.32	48.4 ± 3.3	GSF
65/101	11NOV1982	mo		8.54	-71.1	-10.30	48.7 ± 3.3	GSF
66/2	19JUL1982	mo	872	6.08	-71.9	-10.27	35.9 ± 2.7	GSF
66/103	11NOV1982	mo		5.89	-71.4	-10.43	37.6 ± 2.6	GSF
66/104	23FEB1983	mo		11.04	-70.8	-10.20	48.3 ± 3.9	GSF
70/2	20JUL1982	mo	776	6.06	-70.1	-9.98	66.5 ± 5	GSF
70/101	11NOV1982	mo		4.71	-71.6	-10.22	67.0 ± 4.5	GSF
70/102	23FEB1983	mo		4.4	-69.7	-9.94	43.7 ± 4.3	GSF
71/2	20JUL1982	mo	1016	13.9	-71.7	-9.96	47.0 ± 3.6	GSF
71/102	11NOV1982	mo		5.3	-74.4	-10.21	45.3 ± 3.3	GSF
71/103	23FEB1983	mo		5.23	-71.4	-10.38	49.9 ± 4	GSF
72/2	20JUL1982	mo	1839	5.72	-69.7	-9.84	21.4 ± 1.7	GSF
73/1	06JUN1984	mo	1688	7	-69.6		5.2 ± 0.7	GSF
74/2	20JUL1982	mo	659	7.69	-73.8	-10.48	60.8 ± 4.2	GSF
86/2	02MAR1982	mo	281	8.15	-69.1	-9.81	6.3 ± 0.7	GSF
90/2	02MAR1982	mo	4005	139.1	-76.1	-10.84	<0.9	GSF
92/1	01MAR1982	mo	569	14.3	-63.3	-9.21	4.9 ± 0.7	GSF

Table 3.2.4: Isotopic composition of groundwaters from the Muschelkalk. (Page 2 of 5)

Location Sample (Table A1)	Date Sampled	Source Formation	Residue at 110° (mg/l)	Chloride (mg/l)	Delta 2H (per mil)	Delta 18O (per mil)	Tritium (TU ± 2SD)	Laboratory
94/1	01MAR1982	mo-s	4340	106.6	-71.2	-10.32	<0.7	GSF
101/1	25NOV1981	mu-so	6569	1134	-72.5	-10.21	17.0 ± 1.3	GSF
114/101	06DEC1982	mo		1.2	-69.6	-10.32	62.4 ± 4.3	GSF
114/102	24FEB1983	mo		2.7	-69.5	-10.38	64.7 ± 4.9	GSF
115/101	06DEC1982	mo		2.9	-71.0	-10.36	88.6 ± 6.4	GSF
115/102	24FEB1983	mo		3.8	-70.4	-10.40	85.2 ± 6.2	GSF
119/2	30OCT1981	mo	1768	288.3	-72.3	-10.13	59.4 ± 4.6	GSF
119/5	17AUG1982	mo	1652	291.5	-72.0	-10.13	57.0 ± 3.8	GSF
119/8	18MAY1983	mo	1822	315	-72.7	-10.32	47.0 ± 3.1	GSF
119/11	01MAR1984	mo	2035	371	-72.5	-10.16	34.0 ± 2.5	GSF
119/14	14NOV1984	mo	2170	416	-71.3	-10.26	28.2 ± 2.1	GSF
119/17	27MAY1986	mo	1960	365	-71.6	-10.10	26.0 ± 1.9	GSF
119/106	16JUN1977	mo		408	-71.9	-9.96	48.0	Weizmann
119/109	07MAR1978	mo		398	-71.1	-10.18	62.0	Weizmann
119/113	15MAY1979	mo		398	-67.2	-9.70	51.0	Weizmann
119/114	24JUL1979	mo		430	-70.8	-9.80	52.0	Weizmann
119/115	31OCT1979	mo		457	-67.4	-9.70	43.0	Weizmann
119/116	08JAN1980	mo		451	-70.3	-10.20	36.0	Weizmann
119/118	22JUN1983	mo		345	-72.1	-10.19	46.9 ± 5.1	GSF
120/2	30OCT1981	mo	2513	433.3	-72.6	-10.10	24.3 ± 2	GSF
120/5	17AUG1982	mo	2698	499.3	-72.5	-10.24	28.4 ± 2.1	GSF
120/8	18MAY1983	mo	2796	529	-73.0	-10.25	20.5 ± 1.4	GSF
120/11	01MAR1984	mo	2850	500.8	-72.8	-10.30	19.4 ± 1.5	GSF
120/14	14NOV1984	mo	2702	496	-72.3	-10.31	23.9 ± 1.7	GSF
120/17	27MAY1986	mo	2750	510	-70.8	-10.18	16.5 ± 1.2	GSF
120/102	25FEB1980	mo		348	-70.3	-10.10	63.0	Weizmann
120/103	27FEB1980	mo		275	-72.4	-10.10	70.0	Weizmann
120/104	28FEB1980	mo		245	-69.3	-10.30	82.0	Weizmann
120/107	02MAR1980	mo		347	-70.9	-10.00	57.0	Weizmann
120/116	11NOV1982	mo		486	-71.9	-10.18	27.5 ± 3.5	GSF
120/117	23FEB1983	mo		500	-72.4	-10.11	20.0 ± 2.3	GSF
120/118	22JUN1983	mo		547	-72.7	-10.23	18.0 ± 2.1	GSF
121/1	13JUL1981	mo(-q)	1620	487.3	-74.2	-10.22	94.0 ± 6.3	GSF
121/103	26JUN1972	mo(-q)		850		-11.00	106.0	LLC Bern
121/104	16JUN1977	mo(-q)		1630	-72.0	-10.18	91.0	Weizmann
121/107	07MAR1978	mo(-q)		1090	-72.3	-10.30	119.0	Weizmann
122/101	19JUN1972	mo		48		-10.10	2.2 ± 0.8	LLC Bern
123/2	26OCT1981	mo	782	4.78	-70.9	-10.08	6.0 ± 0.7	GSF
123/5	16AUG1982	mo	763	2.68	-69.7	-10.04	6.4 ± 0.7	GSF

Table 3.2.4: Isotopic composition of groundwaters from the Muschelkalk. (Page 3 of 5)

Location Sample (Table A1)	Date Sampled	Source Formation	Residue at 110° (mg/l)	Chloride (mg/l)	Delta 2H (per mil)	Delta 18O (per mil)	Tritium (TU ± 2SD)	Laboratory
123/8	18MAY1983	mo	813	2.29	-72.4	-10.27	5.5 ± 0.6	GSF
123/11	20FEB1984	mo	783	1.97	-70.9	-10.15	7.0 ± 0.7	GSF
123/14	12NOV1984	mo	748	1.91	-70.1	-10.28	7.3 ± 0.7	GSF
123/16	15AUG1985	mo	771	2.5	-71.5	-10.03	6.9 ± 0.7	GSF
123/17	26MAY1986	mo	764	3.5	-70.0	-10.26	6.8 ± 0.7	GSF
123/101	26JUN1972	mo		11.7		-10.00	2.4 ± 0.8	LLC Bern
123/103	16JUN1977	mo		7.5	-70.8	-10.47	98.0 **	Weizmann
123/106	07MAR1978	mo		7.5	-71.3	-10.17		Weizmann
123/111	23FEB1982	mo		2.02	-71.4	-10.15	5.8 ± 0.7	GSF
123/112	11NOV1982	mo		2.65	-71.0	-10.18	6.9 ± 0.8	GSF
124/2	26OCT1981	mo	2358	37.8	-70.9	-10.15	<1.1	GSF
124/5	16AUG1982	mo	2571	37.4	-70.4	-10.13	<1.2	GSF
124/8	18MAY1983	mo	2640	35.3	-70.7	-10.21	0.4 ± 0.2	GSF
124/11	20FEB1984	mo	2587	34.8	-70.8	-10.13	<0.6	GSF
124/14	12NOV1984	mo	2624	35.3	-70.2	-10.21	<1.2	GSF
124/16	15AUG1985	mo	2610	35.5	-70.4	-10.11	<1.2	GSF
124/17	26MAY1986	mo	2700	38	-69.7	-10.21	<0.9	GSF
124/102	16JUN1977	mo		38.5	-70.9	-10.39		Weizmann
124/105	07MAR1978	mo		37.5	-71.6	-10.26		Weizmann
124/109	10OCT1979	mo				10.33	<0.5	LLC Bern
124/110	23JUN1983	mo		34.8	-70.2	-10.18	<1.0	GSF
129/1	18AUG1981	mo-s	1404	52.5	-80.3	-11.02	<1.0	GSF
139/101	15MAY1979	mo		4.26	-68.5	-9.90	115.0	Weizmann
139/102	24JUL1979	mo		4.43	-70.0	-10.10		Weizmann
139/103	01NOV1979	mo		4.93	-66.6	-10.00	83.0	Weizmann
148/101	11NOV1982	mo		5	-71.7	-10.34	65.5 ± 4.5	GSF
148/102	23FEB1983	mo		5.3	-68.5	-10.02	22.6 ± 2.5	GSF
172/101	04FEB1980	mo3	6660	1440		-11.18		LLC Bern
173/101	05MAR1980	mo3	6804	1500		-11.17		LLC Bern
174/101	28FEB1980	mo1	6615	1410		-11.60		LLC Bern
175/101	26FEB1980	mm4	6580	1300		-11.33		LLC Bern
196/101	12APR1983	mo-mm		56.6	-70.1	-9.59	8.1 ± 1	GSF
196/102	13APR1983	mo-mm		24.27	-69.9	-9.61	3.1 ± 0.8	GSF
196/103	14APR1983	mo-mm		12.78	-67.6	-9.41	1.2 ± 0.6	GSF
197/101	19APR1983	mo		3851	-71.5	-9.93	10.3 ± 1.1	GSF
197/102	25APR1983	mo		4725	-71.7	-9.99	2.2 ± 0.6	GSF
197/103	26APR1983	mo		3745	-72.9	-10.01	12.3 ± 1	GSF
198/101	04APR1972	mo		44.3		-10.00	<2.0	LLC Bern

Table 3.2.4: Isotopic composition of groundwaters from the Muschelkalk. (Page 4 of 5)

Location Sample (Table A1)	Date Sampled	Source Formation	Residue at 110° (mg/l)	Chloride (mg/l)	Delta 2H (per mil)	Delta 18O (per mil)	Tritium (TU ± 2SD)	Laboratory
201/1	08JUL1981	mo	4685	1113	-70.3	-8.99	3.3 ± 0.8	GSF
201/1	08JUL1981	mo	4685	1113			3.3 ± 0.8	LLC Bern
201/3	18AUG1982	mo	4491	1104	-70.4	-9.02	3.4 ± 0.6	GSF
201/6	17MAY1983	mo	4596	1091	-70.6	-9.03	3.6 ± 0.6	GSF
201/9	21FEB1984	mo	4336	1053	-70.8	-9.14	3.4 ± 0.7	GSF
201/12	13NOV1984	mo	4590	1117	-69.7	-9.19	3.5 ± 0.7	GSF
201/14	13AUG1985	mo	4568	1138	-70.3	-8.99	4.0 ± 1.1	GSF
201/15	29MAY1986	mo	4700	1140	-69.9	-9.14	3.8 ± 0.7	GSF
201/102	06DEC1982	mo		1131	-69.6	-9.12	3.4 ± 1	GSF
201/103	22FEB1983	mo		1018	-69.2	-9.08	3.7 ± 0.6	GSF
203/3	18AUG1982	mo	4401	1020	-70.0	-9.03	6.7 ± 0.9	GSF
203/103	16JUN1977	mo		1050	-70.8	-8.86	5.0	Weizmann
203/107	07MAR1978	mo		1110	-71.0	-9.20	3.0	Weizmann
203/111	08DEC1982	mo		1087	-69.5	-9.18	6.5 ± 1.3	GSF
203/112	23FEB1983	mo		1050	-68.4	9.01	6.2 ± 0.6	GSF
204/2	29OCT1981	mo	4369	1076	-69.7	-8.88	3.7 ± 0.7	GSF
204/5	17MAY1983	mo	4545	1084	-70.2	-9.11	3.8 ± 0.6	GSF
204/8	21FEB1984	mo	4243	1008	-70.4	-9.12	4.2 ± 0.7	GSF
204/11	13NOV1984	mo	4528	1095	-70.2	-9.10	5.4 ± 1	
204/13	13AUG1985	mo	4572	1131	-70.7	-9.01	4.0 ± 0.7	
204/14	29MAY1986	mo	4630	1130	-70.1	-9.28	3.6 ± 0.7	
206/102	23JUN1983	mo		1087	-70.0	-8.99	3.6 ± 0.5	
210/3	18AUG1982	mo	4304	1103	-70.3	-9.21	4.9 ± 0.8	
210/6	17MAY1983	mo	4534	1136	-70.4	-8.99	3.7 ± 0.6	
210/9	21FEB1984	mo	4264	1070	-71.1	-9.21	4.2 ± 0.7	
210/12	13NOV1984	mo	4518	1136	-70.5	-9.15	4.1 ± 0.7	
210/14	13AUG1985	mo	4563	1170	-70.8	-9.01	4.3 ± 0.8	
210/15	29MAY1986	mo	4600	1140	-70.9	-9.21	3.9 ± 0.7	
219/2	29OCT1981	mo	4306	1087	-69.9	-8.94	4.8 ± 0.7	
219/5	18AUG1982	mo	4167	1055	-70.4	-9.06	5.0 ± 0.7	GSF
219/8	17MAY1983	mo	4427	1109	-70.1	-8.96	4.3 ± 0.6	GSF
219/11	21FEB1984	mo	4179	1056	-71.0	-9.23	5.1 ± 0.7	GSF
219/14	13NOV1984	mo	4458	1118	-69.6	-9.07	4.6 ± 0.7	GSF
219/16	13AUG1985	mo	4463	1146	-69.5	-8.95	5.4 ± 0.7	GSF
219/17	29MAY1986	mo	4500	1140	-69.3	-9.08	3.8 ± 0.7	GSF
219/103	16JUN1977	mo		1065	-70.5	-9.03	12.0	Weizmann
219/107	07MAR1978	mo		1105	-71.1	-9.15	6.0	Weizmann
219/111	27MAR1983	mo		1062	-69.9	-9.09	4.2 ± 0.6	GSF
221/3	18AUG1982	mo	4376	1075	-70.4	-9.10	4.5 ± 0.7	GSF
270/101	09NOV1983	mo		35		-9.57	1.1 ± 2.3	GSF
270/101	09NOV1983	mo		35		-9.57	1.1 ± 2.3	LLC Bern
271/101	14SEP1983	mo	1475	67	-70.8	-10.14	<0.8	GSF
271/102	09MAY1985	mo	1764	57	-71.4	-9.87	<1.1	GSF
271/103	11JUN1985	mo		57	-74.8	-10.49	<0.8	GSF
271/104	28OCT1986	mo		50	-71.5	-9.84	<1.1	GSF
272/102	25NOV1985	mo		4.5	-66.1	-9.24	<1.6	GSF
272/103	28OCT1986	mo		6.4	-72.3	-9.22	<1.0	GSF

Table 3.2.4: Isotopic composition of groundwaters from the Muschelkalk. (Page 5 of 5)

Location Sample (Table A1)	Date Sampled	Source Formation	Residue at 110° (mg/l)	Chloride (mg/l)	Delta 2H (per mil)	Delta 18O (per mil)	Tritium (TU ± 2SD)	Laboratory
273/101	10FEB1986	mo		14	-70.6	-9.95	<1.3	GSF
273/102	16MAY1986	mo	902	12	-69.8	-9.83	<1.9	GSF
273/103	28OCT1986	mo	783	12	-69.3	-9.50	<0.9	GSF
274/102	28OCT1986	mo	601	4.3	-68.5	-9.30	4.3 ± 0.7	GSF
275/101	25FEB1987	mo		5.3	-66.1	-9.27	<0.7	GSF
276/101	11NOV1986	mo	7529	3990	-69.1	-9.61	2.3 ± 0.7	GSF
301/2	01NOV1982	mo	6480	1302	-78.2	-11.16	<1.0	GSF
301/2	01NOV1982	mo	6480	1302			2.0 ± 2.7	LLC Bern
302/10a	03APR1983	mo			-82.2	-11.87	3.2 ± 0.7	GSF
302/10b	04APR1983	mo	3409	53	-86.6	-11.76		GSF
303/2	17AUG1983	mo	14970	6008	-71.6	-10.14	<1.1	GSF
304/3	17FEB1984	mo	15914	6400	-80.0	-11.02	<0.9	GSF
304/14	21JAN1985	mo	18480	8090	-78.6	-10.81	1.4 ± 0.7	GSF
306/1	18JUL1984	mo	1089	34.3	-68.3	-9.96	11.6 ± 1.1	GSF

** May have been contaminated during sampling (VUATAZ, 1982, p. 41)

Table 3.2.5: Isotopic composition of groundwaters from the Buntsandstein, Permian and Crystalline. (Page 1 of 3)

Location Sample (Table A1)	Date Sampled	Source Formation	Residue at 110° (mg/l)	Chloride (mg/l)	Delta 2H (per mil)	Delta 18O (per mil)	Tritium (TU ± 2SD)	Laboratory
3/1	16JUL1981	s-r	12645	2568	-72.5	-10.35	<0.7	GSF
3/3	06FEB1984	s-r	12580	2563	-72.0	-10.28	<0.7	GSF
88/2	02MAR1982	s-r	328	9.53	-70.6	-10.05	<0.9	GSF
95/1	01MAR1982	KRI	5537	155.2	-75.6	-10.62	<0.9	GSF
95/2	11APR1984	KRI	5590	151.7	-71.7	-10.20	<0.9	GSF
110/2	20AUG1981	KRI	121	1.9	-69.5	-9.86	11.0 ± 1.0	GSF
110/4	20FEB1984	KRI	115	2.21	-68.8	-9.95	10.4 ± 1.0	GSF
125/1	22JUL1981	KRI	3052	1487	-66.6	-8.89	46.5 ± 2.0	GSF
125/2	28OCT1981	KRI	3083	1572	-66.6	-9.31	48.0 ± 3.6	GSF
125/3	05MAR1982	KRI	3194	1619	-67.1	-8.97	45.8 ± 3.3	GSF
125/5	20AUG1982	KRI	3498	1817	-66.4	-8.94	40.6 ± 2.9	GSF
125/8	16MAY1983	KRI	3247	1644	-67.8	-9.10	43.1 ± 2.9	GSF
125/11	22FEB1984	KRI	3466	1720	-67.2	-9.17	37.8 ± 2.8	GSF
125/14	14NOV1984	KRI	3820	1890	-65.2	-9.16	33.5 ± 2.4	GSF
125/16	12AUG1985	KRI	3940	1994	-67.3	-9.07	29.6 ± 3.0	GSF
125/17	27MAY1986	KRI	3400	1790	-67.1	-9.13	30.5 ± 2.1	GSF
125/106	08DEC1982	KRI		1617	-65.5	-9.03	41.5 ± 3.0	GSF
125/107	25FEB1983	KRI		1530	-64.9	-8.90	43.9 ± 3.6	GSF
125/108	27JUN1983	KRI		1613	-66.4	-8.99	39.4 ± 4.2	GSF
125/109	15DEC1983	KRI		1739			35.3 ± 4.3	GSF
126/2	27OCT1981	KRI	871	383	-67.0	-8.92	66.7 ± 4.7	GSF
126/3	04MAR1982	KRI	1050	472.3	-67.3	-9.30	56.9 ± 4.0	GSF
126/5	20AUG1982	KRI	999	453	-66.9	-9.44	59.1 ± 4.2	GSF
126/8	16MAY1983	KRI	1019	425.3	-67.7	-9.46	63.6 ± 4.2	GSF
126/11	22FEB1984	KRI	814	327.5	-67.7	-9.49	61.2 ± 4.1	GSF
126/14	14NOV1984	KRI	1122	482	-67.7	-9.52	47.0 ± 3.2	GSF
126/16	12AUG1985	KRI	1124	485	-68.1	-9.31	42.2 ± 3.0	GSF
126/17	26MAY1986	KRI	1150	463	-68.3	-9.58	36.6 ± 2.5	GSF
126/103	08DEC1982	KRI		318	-66.1	-9.50	67.0 ± 4.7	GSF
126/104	25FEB1983	KRI		440	-65.8	-9.40	56.4 ± 4.9	GSF
126/105	27JUN1983	KRI		456	-61.1	-9.45	49.9 ± 4.5	GSF
131/2	27OCT1981	KRI	861	134	-76.8	-10.10	<0.7	GSF
131/5	17AUG1982	KRI	877	138.1	-75.1	-10.06	<0.9	GSF
131/8	19MAY1983	KRI	881	135	-76.4	-10.28	0.7 ± 0.2	GSF
131/11	23FEB1984	KRI	881	133.5	-76.5	-10.18	<0.6	GSF
131/14	15NOV1984	KRI	880	135	-76.3	-10.20	<0.7	GSF
131/16	15AUG1985	KRI	881	134	-74.8	-10.11	<0.8	GSF
131/17	27MAY1986	KRI	880	135	-72.5	-10.02	<1.0	GSF
131/107	16JUN1977	KRI		135	-75.0	-10.18	2.0	Weizmann
131/111	07MAR1978	KRI		131	-75.4	-10.26		Weizmann
131/115	18JUL1979	KRI				-10.30	<0.5	LLC Bern
131/117	20MAY1980	KRI				-10.31	<1.0	LLC Bern
132/2	27OCT1981	KRI	851	133	-75.6	-10.21	<1.2	GSF
132/5	17AUG1982	KRI	897	134.4	-74.6	-10.12	<1.1	GSF
132/7	25FEB1983	KRI	880	138	-75.2	-10.22		GSF
132/8	19MAY1983	KRI	896	132	-75.6	-10.15	0.8 ± 0.2	GSF
132/11	23FEB1984	KRI	885	131	-75.7	-10.14	<0.6	GSF
132/14	15NOV1984	KRI	890	133	-76.2	-10.19	<0.8	GSF

Table 3.2.5: Isotopic composition of groundwaters from the Buntsandstein, Permian and Crystalline. (Page 2 of 3)

Location Sample (Table A1)	Date Sampled	Source Formation	Residue at 110° (mg/l)	Chloride (mg/l)	Delta 2H (per mil)	Delta 18O (per mil)	Tritium (TU ± 2SD)	Laboratory
132/16	15AUG1985	KRI	900	132	-74.4	-10.05	<0.7	GSF
132/17	27MAY1986	KRI	890	136	-73.4	-10.15	<0.9	GSF
132/103	01JAN1976	KRI				-10.30	<1.0	LLC Bern
132/105	16JUN1977	KRI		132	-75.1	-10.23		Weizmann
132/113	06DEC1977	KRI		133	-74.9	-10.26		Weizmann
132/130	11NOV1982	KRI		130	-73.9	-10.16	0.4 ± 0.6	GSF
132/131	24FEB1983	KRI		123	-73.7	-10.09	<1.0	GSF
133/1	11SEP1981	KRI			-76.2	-10.20	11.9 ± 3.0	GSF
150/1	03MAR1982	KRI	74	5.21	-77.9	-11.20	76.2 ± 5.0	GSF
159/101	11APR1983	KRI	3726	635	-70.5	-9.34	2.2 ± 3.0	GSF
159/102	20APR1983	KRI			-70.1	-9.39	1.6 ± 0.5	GSF
159/103	09MAY1983	KRI			-69.8	-9.38	<1.1	GSF
159/106	30NOV1983	KRI			-70.4	-9.21	<0.9	GSF
159/106	15DEC1983	KRI			-70.7	-9.24	<2.0	GSF (duplicate)
161/1	22FEB1984	KRI	7249	3640	-67.1		1.5 ± 0.7	GSF
162/1	19SEP1985	KRI	64	1	-72.0	-10.43	15.2 ± 1.3	GSF
163/1	19SEP1985	KRI	67	0.9	-71.1	-10.41	11.3 ± 1.1	GSF
164/1	19SEP1985	KRI		1.5	-70.8	-10.32	10.3 ± 1.1	GSF
165/1	19SEP1985	KRI	67	2.3	-72.3	-10.56	48.4 ± 3.4	GSF
166/1	18SEP1985	KRI	175	3.1	-66.3	-9.82	23.8 ± 1.8	GSF
167/1	18SEP1985	KRI	50	1	-69.0	-10.21	57.3 ± 4.0	GSF
191/101	27OCT1982	KRI		1.47	-73.8	-10.71	47.5 ± 4.3	GSF
191/102	27JUN1983	KRI		1.78	-73.4	-10.60		GSF
192/101	27OCT1982	KRI		11.07	-76.0	-10.97	66.6 ± 5.6	GSF
192/102	25FEB1983	KRI		16.94	-76.9	-11.05	59.6 ± 4.6	GSF
192/103	27JUN1983	KRI		11.05	-75.5	-10.97		GSF
193/101	27OCT1982	KRI		9.25	-75.1	-10.95	59.5 ± 4.9	GSF
194/101	27OCT1982	KRI		5.95	-72.5	-10.60	45.7	GSF
194/102	27JUN1983	KRI		4.42	-72.6	-10.53		GSF
231/1	11SEP1985	KRI	120	2.6	-92.6	-12.77	41.4 ± 2.8	GSF
232/1	11SEP1985	KRI	84	1.9	-102.6	-14.19	37.7 ± 2.7	GSF
233/1	12SEP1985	KRI	152	8.8	-101.2	-14.09	15.8 ± 1.4	GSF
234/1	12SEP1985	KRI	2465	246	-116.9	-16.16	2.7 ± 0.7	GSF
301/5	16NOV1982	s-KRI	2044	694.8	-73.7	-10.15	<0.7	GSF
301/8c	14DEC1982	KRI	1045	120.9	-72.7	-10.05	<1.0	GSF
301/12b	22JAN1983	KRI	1017	120.5	-72.7	-10.04	1.0 ±	GSF
301/16	09AUG1983	KRI	1088	124.4	-74.1	-10.01	1.0 ±	GSF
301/18	16AUG1983	KRI	1119	141.8	-73.7	-9.96	0.9 ± 0.9	GSF
301/23	20OCT1983	KRI			-74.6	-9.97	3.2 ± 0.7	GSF
301/24a	04NOV1983	KRI			-74.3	-9.99	1.2 ± 0.7	GSF
301/24b	05NOV1983	KRI			-73.7	-9.95	2.2 ± 0.7	GSF
301/25b	20JAN1984	s-KRI	1846	637.6	-73.2	-10.08	0.2 ± 0.2	GSF
301/(*)		KRI	13320		-49	-7.9		
Extrapolated from samples 301/19-22: PEARSON (1985, Section 7.1.3).								
302/12	19JUL1983	s	14648	2948	-60.5	-8.12	<0.8	GSF
302/14	04APR1984	KRI	8405	4312	-60.2	-7.65		GSF
Corrected for 9% deionised water contamination: See Table 3.1.8.								

Table 3.2.5: Isotopic composition of groundwaters from the Buntsandstein, Permian and Crystalline. (Page 3 of 3)

Location Sample (Table A1)	Date Sampled	Source Formation	Residue at 110° (mg/l)	Chloride (mg/l)	Delta 2H (per mil)	Delta 18O (per mil)	Tritium (TU ± 2SD)	Laboratory
302/16	27APR1984	KRI	6392	3382	-62.0	-8.17	2.7 ± 0.7	GSF
302/18	15JUN1984	r	118560	59500	-31.5	-4.61	6.4 ± 0.9	GSF
302/19	28JUN1984	r	37245	18440	-37.8	-5.32	2.2 ± 0.6	GSF
303/3	16SEP1983	s-r	9950	4072	-65.4	-6.92	0.7 ± 0.7	GSF
303/5	04OCT1983	r	17898	9081	-57.5	-6.73	<1.1	GSF
Chemistry corrected for 0.85% brine contamination: WITTWER (1986, Section 5.3.4).									
303/6	01NOV1983	r	19483	9900	-53.4	-6.65	2.6 ± 1.0	GSF
303/(*)		r	43091	23429	-36.8	-3.98		GSF
Sample 303/20 corrected for 23% deionised water contamination: See Table 3.1.6.									
304/5	01APR1984	s-KRI	15993	6405	-58.0	-5.42	1.8 ± 0.7	GSF
Corrected for 12% deionised water contamination: See Table 3.1.8.									
304/8	02MAY1984	KRI	7715	2430	-64.8	-6.66	2.5 ± 0.7	GSF
304/10	17JUN1984	KRI	7835	3630	-60.8	-5.55	1.5 ± 0.9	GSF
304/13	27AUG1983	KRI	8000	3000	-58.1	-5.76		GSF
Corrected for 15% deionised water contamination: See Table 3.1.8.									
305/1	22FEB1984	s	6764	2025	-68.3	-9.03	0.4 ± 0.2	GSF
305/2	01MAR1984	r	1409	103.5	-73.5	-10.36	0.7 ± 0.3	GSF
305/3	01MAR1984	r		104.9	-73.6	-10.35	0.7 ± 0.7	GSF
305/4	15MAR1984	KRI	1350	79.4	-73.5	-10.29	1.1 ± 0.7	GSF
305/6	03APR1984	KRI	1299	60.7	-72.7	-10.34	1.3 ± 0.7	GSF
305/9	02MAY1984	KRI	1217	62.4	-73.8	-10.32	0.9 ± 0.7	GSF
305/12	05JUN1984	KRI	1183	62.5	-74.1	-10.61	<0.8	GSF
305/14	13AUG1984	KRI	1303	135	-71.7	-10.40	<0.8	GSF
305/16	27AUG1984	KRI	1275	63.2	-73.2	-10.46	1.2 ± 0.7	GSF
306/2	08AUG1984	s	1750	486	-72.9	-10.39	<0.8	GSF
306/4	14AUG1984	KRI	1077	178	-73.3	-10.49	0.8 ± 0.7	GSF
306/5	14SEP1984	KRI	1030	131.5	-73.0	-10.35	<1.2	GSF
306/7	27SEP1984	KRI	1012	120	-71.9	-10.37	<1.0	GSF
306/9	17OCT1984	KRI	1051	110	-74.6	-10.52	<1.3	GSF
306/11b	28NOV1984	KRI	4420	180	-71.8	-10.35	8.1 ± 0.9	GSF
306/12	30NOV1984	KRI	4660	185	-70.8	-10.36	4.9 ± 0.7	GSF
306/13	07DEC1984	KRI	4710	194	-71.3	-10.45	2.8 ± 0.7	GSF
306/16	15FEB1985	KRI	916	125	-71.9	-10.25	0.7 ± 0.7	GSF
306/17	25MAR1985	KRI	982	120.2	-73.5	-10.48	3.5 ± 0.7	GSF
306/18	26MAR1985	KRI	995	123.1	-72.6	-10.36	2.2 ± 0.7	GSF
306/20	22APR1985	KRI	854	132	-72.3	-10.42	1.3 ± 0.7	GSF
306/25	11MAY1985	KRI	4	528	-71.2	-9.99	4.2 ± 0.7	GSF
306/26	14MAY1985	KRI	1306	422	-72.1	-10.02	1.4 ± 0.9	GSF
307/2b	17OCT1988	s	536	26	-86.3	-12.05	<0.8	GSF
307/3	09NOV1988	KRI		22	-86.9	-12.07		GSF
Corrected for 33% drilling fluid contamination: Blaser and others (in prep.).									
307/4b	26NOV1988	KRI	519	25	-84.7	-11.87		GSF
Corrected for 11% drilling fluid contamination: Blaser and others (in prep.).									
307/6b	21FEB1989	KRI	529	25	-86.5	-12.03	<1.2	GSF
307/7b	31MAR1989	KRI		27	-86.2	-11.78	0.9 ± 0.7	GSF

3.3 Noble Gases in Groundwater

D. Rauber, H. H. Loosli, H. Schmassmann, and J. N. Andrews

3.3.1 Introduction

The solubility of the noble gases helium, neon, argon, krypton and xenon is temperature-dependent, as shown in Figure 3.3.1 (IUPAC, 1979). To obtain the dissolved concentrations of these gases in equilibrium with the atmosphere, the values in the figure must be multiplied by the partial pressures of the gases. These partial pressures, in atmospheres are: $p_{He} = 5.24 \cdot 10^{-6}$; $p_{Ne} = 18.18 \cdot 10^{-6}$: $p_{Ar} = 0.00934$; $p_{Kr} = 1.14 \cdot 10^{-6}$; $p_{Xe} = 8.6 \cdot 10^{-8}$; $p_{N2} = 0.78$. Because of this temperature dependence, the measured absolute noble gas contents of a groundwater sample can be used to calculate the temperature prevailing during infiltration. Applications of this method have been described by ANDREWS AND LEE (1979) and by HERZBERG AND MAZOR (1979). While chemical reactions in the groundwater can be ruled out for these gases, the following physical processes must be taken into consideration:

- During infiltration, gas bubbles with atmospheric composition can penetrate into the water. They then dissolve in the water as a result of the increase in hydrostatic pressure during infiltration which, in turn, leads to the groundwater being oversaturated with gas at atmospheric pressure. This additional air component, called "excess air", is most evident in the case of neon. Groundwater samples often have neon contents which are greater than the range of saturation values shown in Figure 3.3.1. When calculating the recharge temperatures, this additional air content must be calculated and subtracted;

- Two noble gases, helium and argon, can be produced underground. ^{4}He is produced during the natural decay series of uranium and thorium. Because the amount of ^{4}He produced *in situ* very often considerably exceeds the amount of atmospheric He, the ^{4}He/^{3}He ratio cannot be used to assess atmospheric air. Thus, no recharge temperature can be calculated using the measured ^{4}He content of a water. ^{40}Ar is produced *in situ* by decay of ^{40}K so that the ^{40}Ar/^{36}Ar ratio in water samples may be higher than the atmospheric ratio of 295.5. However, the measured argon concentration can be corrected for *in situ* production using the measured ^{40}Ar/^{36}Ar ratio, and a recharge temperature can then be calculated from the corrected Ar concentration;

- Heating the water can lead to degassing if the hydrostatic pressure is not sufficiently high. If this process occurs relatively slowly, a correction can be undertaken, assuming that the gas loss occurred at equilibrium with the solution;

- Degassing frequently occurs during sampling but not in the aquifer itself. For example, with high concentrations of CO_2, H_2S or CH_4, the risk of degassing is particularly great because high pressure is needed during sampling to prevent loss of any gas. No correction is possible unless the degassing occurs in thermodynamic equilibrium.

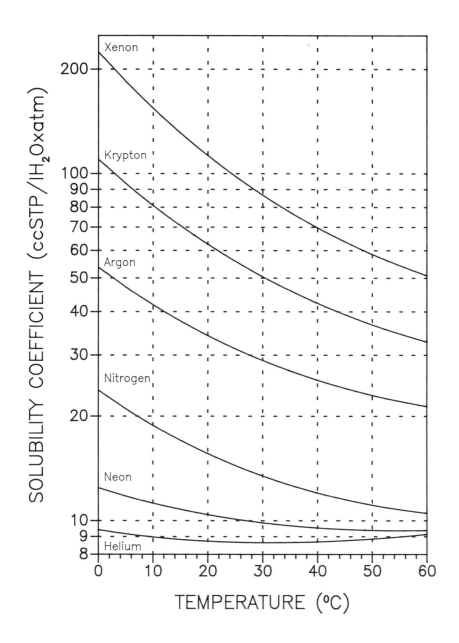

Figure 3.3.1: Solubilities of the noble gases He, Ne, Ar, Kr, Xe and of N_2 in ccstp/$1H_2O$ · atm (IUPAC, 1979).

Statements about recharge temperature contain an element of uncertainty due to the corrections which must be made. The corrections themselves, however, may give information on such underground processes as heating of the water with consequent degassing, accumulation of nuclides produced underground as an indication of age, *etc.* Comparison between calculated noble gas recharge temperatures and information from the $\delta^{18}O$ and $\delta^{2}H$ results is the central issue of this section, so possible information in the corrections themselves has not been explored.

The different sampling techniques have already been described by BALDERER (1985b) and HAUG (1985). The best results were obtained for samples from freely flowing sources (pumped or artesian). Sampling with evacuated cylinders (bailer samples) proved completely unsuitable for the calculation of recharge temperatures because the samples lost up to 90 per cent of the dissolved gas on inflow. On the other hand, the GTC--method (developed by Geological Testing Consultants, now Intera Inc.) has proved itself in practice; this method also uses steel cylinders but these are flooded with nitrogen which restricts degassing.

3.3.2 Calculation of Recharge Temperature Corrections

The quantities of noble gas dissolved are a function of the temperature of the water and of the partial pressure of the gas itself. This allows the temperature which prevailed during the solution process, *i.e.,* during infiltration, to be calculated from the measured concentration of each noble gas in a water sample. For calculations of degassing processes, the influence of salinity on gas solubility must also be included.

The solubilities used were calculated with the approximation formula (as a function of temperature and salinity) given in IUPAC (1979) by Weiss and others. For He, Ne, Ar, Kr and N_2, the coefficients could be taken directly from the approximation formula of Weiss and others, but those for Xe had to be calculated from the coefficients of a similar IUPAC (1979) formula with the solubilities in molar fractions. The coefficients which take account of salinity were calculated using data from KESTER (1975) (see RAUBER, 1987).

The recharge temperatures calculated from the measured concentrations of each of the noble gases generally do not correlate well with one another. For helium, it is usually impossible to determine any temperature at all. Because gas concentrations can be altered by various processes during infiltration, in the aquifer and during sampling, a number of corrections are necessary. These are as follows:

a) <u>Excess air (EA)</u>

For all samples, it is necessary to correct for any dissolved gas of atmospheric composition which is additional to the equilibrium concentration. This excess air is probably caused by changes in the level of the groundwater table in the unsaturated zone of the aquifer. Oversaturation is clearest in the case of Ne because the Ne/Ar ratio in air-saturated water is lower than in air, while, for example, the Kr/Ar ratio in water is higher than in the atmosphere. Therefore, air contamina-

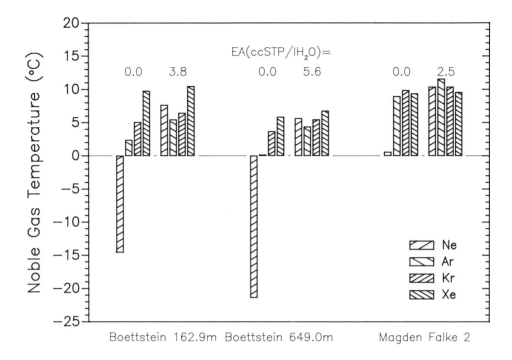

Figure 3.3.2: The effect of excess air on the calculated recharge temperatures of three samples before (left) and after (right) correction.

tion affects the Ne concentration in water more than it does the Ar or Kr concentrations. Figure 3.3.2 shows the uncorrected (left) and corrected (right) temperatures calculated from the Ne, Ar, Kr and Xe contents of three samples. The volume of excess air (EA) in ccSTP/lH$_2$O is given for each data set. For each sample, the corrected noble gas temperatures appear to agree better with one another, and the average recharge temperatures based on the corrected data are higher. Note that the Ne contents of these samples are $2.25 \cdot 10^{-4}$ and $3.16 \cdot 10^{-4}$ ccSTP/lH$_2$O (see Table 3.3.2) while the Ne content of air-saturated cold water can be at most about $2.3 \cdot 10^{-4}$ ccSTP/lH$_2$O.

The correction for oversaturation is made using a programme which calculates the recharge temperature from the measured concentrations by varying the value for excess air so as to minimize the sum of the quadratic deviations between the temperatures calculated for the individual gases.

The measured N$_2$/Ar volume ratios can also be used to correct for oversaturation with air. The N$_2$/Ar ratio of air-saturated water is 38, that of air itself is 85, that of water with dissolved excess air may be up to 50, and values measured in groundwaters range from 43 to 216. However, calculations to date indicate that using N$_2$/Ar ratios gives a significantly larger quantity of excess air than is the case with the noble gases. It follows from this that additional processes must be occurring which introduce mainly N_2 into solution (*e.g.*, in the uppermost soil layers and (or) underground). For this reason, the N_2/Ar ratios have not been used to determine oversaturation in this study.

b) <u>Underground production of noble gases</u>

Helium and argon can be produced underground in quantities which rule out calculation of recharge temperatures from helium concentrations and necessitate a correction for argon concentration.

Helium is produced as α-particles in the natural decay series of uranium and thorium. The amount produced, in ccSTP per cm³ of rock per year, can be calculated from the uranium and thorium contents of the rock in ppm and the density ρ in g/cm³ using the following equation (see Chapter 6.1):

$$^4He = \rho \cdot \{1.17 \cdot 10^{-13} [U] + 2.85 \cdot 10^{-14} [Th]\} \qquad (3.3.1)$$

Depending on the sample, the proportion of helium produced underground exceeds the quantity of gas dissolved in equilibrium with the atmosphere by a factor of 10 to 5000 and, in extreme cases, by up to 40,000. Thus, a correction for radiogenic helium and calculation of an He recharge temperature is not possible. The radiogenic helium concentrations can, however, give qualitative indications of water residence times in the aquifer, as discussed in Section 6.4 of this report.

Argon can also be produced underground as discussed in Chapter 6. ^{40}Ar is produced by electron capture during decay of ^{40}K. The production rate, in ccSTP/cc rock and year equals:

$$^{40}Ar = \rho \cdot 3.87 \cdot 10^{-14} [K] \qquad (3.3.2)$$

ρ is the density of the rock and [K] is the potassium content of the rock in weight percent. The production rate of ^{36}Ar from the β-decay of ^{36}Cl is significantly lower than that of ^{40}Ar and can generally be ignored.

Radiogenic Ar can be present in groundwaters at concentrations up to 20 per cent of those of Ar in air-saturated water. Atmospheric argon originally dissolved in water has a $^{40}Ar/^{36}Ar$ ratio of 295.5. The amount of argon produced underground can be calculated from the increased $^{40}Ar/^{36}Ar$ ratio and the argon concentration originally present on infiltration can then be reconstructed. An increase of one per cent in the $^{40}Ar/^{36}Ar$ ratio has the effect of altering the recharge temperature of argon by around 0.5°C. The effect of this correction on the mean recharge temperature calculated from all four noble gases is very slight. However, the temperatures agree better with one another, which results in a smaller scatter range for the calculated recharge temperatures. Examples of the effects of this correction on three samples are given in Figure 3.3.3. The four uncorrected noble gas temperatures are given on the left and those corrected with the measured $^{40}Ar/^{36}Ar$ ratio are given on the right. For the first two examples, the effect of the correction is to bring the argon temperature to values which correlate with those determined from other gases; in the third case, even with the correction, there is poor agreement among the four gas temperatures.

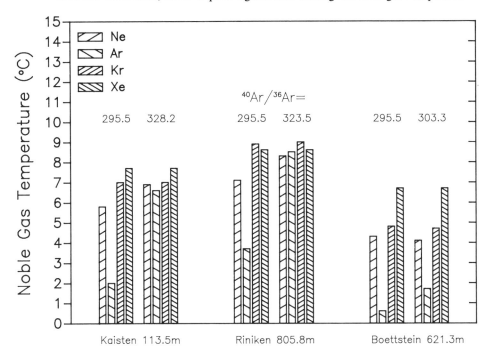

Figure 3.3.3: **Effect of $^{40}Ar/^{36}Ar$ correction on the calculated recharge temperatures.**

c) <u>Pressure</u>

A basic uncertainty in calculating recharge temperatures is the elevation of the infiltration area. This is because the dissolved gas quantities are functions of gas partial pressures and, therefore, of atmospheric pressure. At higher elevations the atmospheric pressure and the amount of dissolved gas are lower. Ignoring this elevation effect would result in the calculated recharge temperatures being too high. Since the infiltration elevations for the present samples are unknown, they have to be estimated. Initially, a pressure of 950 mbar was used, which is about the mean pressure at an elevation of 500 m (see Table 3.3.1). Consequently, the stable isotopes cannot be used to make corrections since the connection between their concentrations and the recharge temperatures still has to be investigated. Estimating the infiltration elevation from the hydrostatic pressure measured in the undisturbed aquifer is also unsuitable because it is determined by many factors in addition to the recharge altitude. Therefore, the estimates of the infiltration altitude are somewhat rough. Figure 3.3.4 gives some examples of the effect of the elevation correction on three samples. The four noble gas temperatures for a normal pressure of 1013 mbars are given on the left and those for adjusted elevation are given on the right. For the first two waters from the Alps, it can be seen that the excessively high recharge temperatures are reduced to reasonable values.

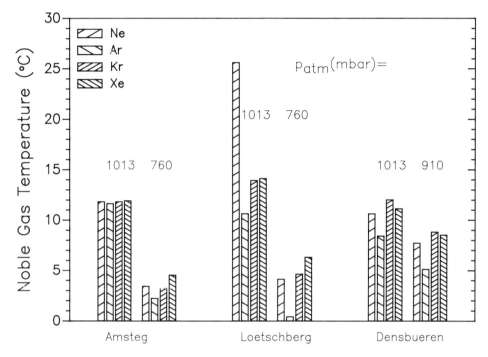

Figure 3.3.4: Effect of the elevation (and pressure) correction on the calculated recharge temperatures.

d) <u>Degassing processes</u>

For some samples, a recharge temperature can only be calculated if it is assumed that the samples have been degassed.

Such samples often have noble gas contents which are significantly lower than the ranges given in Figure 3.3.1. A correction is possible only if the degassing process has occurred in thermodynamic equilibrium. If so, the gas bubble being lost would have had a composition defined by the measured noble gas concentrations of the water. Since the partial pressure in the gas phase for a given concentration in the water is a function of temperature and salinity, these two parameters must be known. The correction calculation uses the sampling temperature and the measured salinity, but these do not necessarily have to correspond with the conditions prevailing during degassing. For the correction, the volume of the bubble produced is varied, the gas quantities resulting from this are added to the measured quantities, and the recharge temperature is calculated. As for the case of varying oversaturation, the value for the gas volume determined by this procedure is that which gives the best agreement between the recharge temperatures for the four gases.

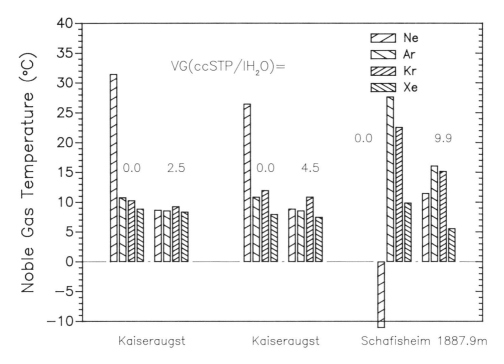

Figure 3.3.5: **Effect of a correction for gas losses of the water sample assuming equilibrium degassing.**

For most waters where the noble gas measurements indicate degassing processes, such a correction is not feasible. This is the case if the calculation for the different gases gives widely differing temperatures, thus raising the error in the mean value, or gives negative temperatures. The reason for these discrepancies is that degassing processes may have occurred during sampling, often with bailer samples. In this case, the gas bubble produced had no time to reach equilibrium with the water, so that the gases were fractionated. Figure 3.3.5 gives some examples of these corrections to three samples. The uncorrected (left) and corrected (right) noble gas temperatures are given together with the degassed volume VG (in ccSTP/lH_2O) assumed for the optimization process. It can be seen that the correlation between the noble gas temperatures improves and that those for neon, which is apparently affected most by the degassing process, can be corrected successfully.

e)   Combination of the different corrections

When calculating recharge temperatures from measured noble gas concentrations, it is possible that most, or even all, of the above corrections will have to be applied. The corrections are applied as follows:

- If the $^{40}Ar/^{36}Ar$ ratio is available, the Ar concentration is corrected according to the following equation:

$$Ar_{corrected} = \frac{(^{40}Ar/^{36}Ar)_{Air}}{(^{40}Ar/^{36}Ar)_{measured}} \cdot Ar_{measured} \qquad (3.3.3)$$

- Because the elevation of the infiltration area, and thus the atmospheric pressure which prevailed during infiltration of the water, is unknown, a value must be assumed for the recharge elevation. The corresponding pressure is specified at the beginning of the calculations and is taken into account when optimizing agreement among the four noble gas temperatures;

- The amount of excess air and the value for the degassed volume are determined by a double iteration procedure, which leads to values for which the deviations between the noble gas temperatures are minimum. Only positive values are permitted for excess air but the values for degassed volume can be negative. Negative degassing volumes can be interpreted as capture of gas bubbles which originate from a previous degassing process.

There are uncertainties involved in calculating the recharge temperatures, so criteria were developed to decide whether calculated results should be accepted or rejected. Difficulties can arise in interpreting the results because the underground processes are more complex than assumed in the correction models, and because misinterpretation can occur in the sampling and, possibly, in the analyses. An indication of underground processes exists if the samples have a gas content lower than would correspond to air-saturation as a result of subsurface outgassing at a high temperature stage of the water evolution, gas stripping by the outgassing of large amounts of CO_2 or CH_4 or other processes. Another indication would be poor agreement between the amount of excess air from the noble gases and from the N_2/Ar ratio. Problems in sampling and analysis can be identified if different laboratories give differing results, or if results from duplicate samples deviate widely from one another. Rapid degassing during sampling or inclusion of gas bubbles lead to irregular variations in the results which can be corrected only to a limited extent.

Nevertheless, about three-quarters of all results give useable recharge temperatures once the corrections above have been carried out. Criteria for deciding on the acceptability of the results were:

- The scatter of the calculated gas temperatures as measured by the standard deviation of their mean. If this is around 5°C or more, not much credibility can be attached to the results. Therefore, Table 3.3.1 gives only those results which have a standard deviation of 5°C or less;

- Agreement between the results from duplicate samples or from repeated samples from the same aquifer, evaluated using the statistical error. In the table of results (Table 3.3.2), reference is made to sample horizons from which all results were not used. It can be seen that a significant majority of the samples gave results which correlated well and that only a few had to be ruled out for this reason;

- The assumptions of the correction models. For example, duplicate samples should show comparable quantities of excess air, but any bubbles captured during sampling, which appear as negative degassed volume, must also be considered. Table 3.3.1 gives the amounts of excess air (EA) and degassed volume (DV) determined *via* the optimization process. It can be seen that relatively large degassed volumes also lead to large errors in the mean temperature values and the reliability of the results appears to be low. Nevertheless, most of the excess air corrections furnish useable results and, for some samples, assumptions of degassing (or bubble capture) lead to improved results.

To summarize, it can be said that groups of results which correspond with one another can be seen as reliable (*e.g.,* many crystalline waters from Böttstein, Kaisten and Leuggern or most of the Muschelkalk waters). Results from individual horizons are credible if they were determined on duplicate samples, while single results should be interpreted with some caution.

It should also be stressed that many parameters which are as yet unknown or cannot be modelled affect the gas contents of infiltrating waters (*e.g.,* the ratio of summer-to-winter precipitation, differing δ^2H-δ^{18}O relationships depending on region). These factors lead to real variations in the results and increase the uncertainty involved in interpreting the modelled temperatures.

3.3.3 Correlation Between Recharge Temperature and δ^2H and δ^{18}O Values

Because the noble gas contents and the isotope ratios of the stable nuclides of H and O are both functions of temperature and, consequently, of the altitude of the infiltration area, relationships exist between these parameters which will be identified repeatedly during measurements of groundwater samples. Relationships between the stable isotopes δ^2H and δ^{18}O of recent waters and temperature for northern Switzerland were calculated and summarized in Section 3.1 on the basis of data already published or newly acquired by the Nagra programme.

The very similar gradients for the isotopes and the temperature in the areas investigated ("NE-Jura and Black Forest" and "Mittelland") are noticeable; these vary only slightly from the global gradients of 2.77 per mil δ^2H per °C calculated by YURTSEVER AND GAT (1981) (converted from the δ^{18}O-T relationship). However, the location of the lines (axis intercept) can alter depending on continental or orographic location of the area, as is demonstrated in Section 3.1.

It is possible to derive information on the elevation of the infiltration area from stable isotope values only if one knows the temperature and altitude relationships of the δ^2H or $\delta^{18}O$ values for the infiltration area. The relationships between δ^2H and recharge temperatures (R_T) of recent waters discussed in Section 3.1 (Figure 3.1.8) are used for the following discussion of the noble gas recharge temperatures. Figure 3.1.6 gives the relationships between δ^2H and altitude for the various areas of Switzerland. It must be stressed that the regression lines in Section 3.1 should not be considered as final because some are supported by few data and others are based on estimates of the mean altitude of the infiltration area. Also, all are based on an equation of unknown precision relating altitude and mean temperature.

It should also be noted that the recharge temperatures derived from noble gas concentrations correspond to mean soil temperatures and not to air temperatures. Provisional analysis by Institut Schmassmann of data published in Annalen der Schweizerische Meteorologischen Anstalt suggest that the mean annual soil temperature could be around two degrees higher than the air temperature.

In an ideal situation, it would be expected that the measured δ^2H and $\delta^{18}O$ values and the recharge temperatures calculated from noble gas concentrations would correspond to the recent δ^2H- and $\delta^{18}O$-temperature relationships of modern groundwaters for the area. However, various processes can alter both the stable isotope contents and the noble gas concentrations in the groundwater. The following paragraphs describe eight processes which result in shifts on the δ^2H- and $\delta^{18}O$-temperature diagrams; these processes are shown schematically in Figure 3.3.6.

a) A groundwater sample can be a mixture of different components which infiltrated into the soil under different temperature conditions. Within the scope of measuring accuracy, a linear relationship can be assumed between noble gas concentrations and recharge temperature, so that the mixed values lie on a line. In the case of such mixing processes, the measured points remain on the line for the appropriate regional relationship;

b) If there is an exchange between the stable isotopes in the water and the aquifer rock, as observed in some geothermal systems, the $\delta^{18}O$ values increase with time. Exchange of deuterium is less likely, which is the reason for preferring this isotope when discussing the results;

c) Evaporation prior to infiltration can enrich the heavy isotopes in the water leading to higher $\delta^{18}O$ and δ^2H values for the same recharge temperatures. This process is insignificant in northern Switzerland because at most only locally increased evaporation will occur, and this will be balanced out on a regional scale;

d) Old waters may exhibit low recharge temperatures and more negative δ^2H values. A reason why this might not be observed could be that during periods

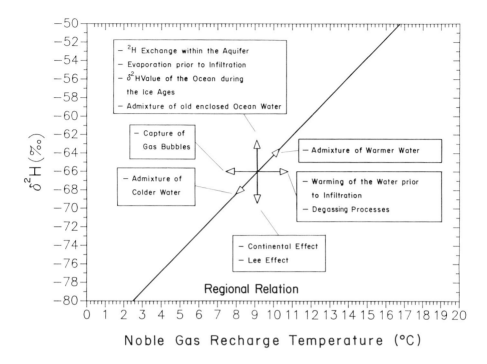

Figure 3.3.6: Schematic representation of processes of potential influence on the location of points in the δ^2H/recharge temperature diagram.

of glaciation, increased quantities of water were stored on the continents in the form of ice, which resulted in the ocean having increased values for δ^{18}O and δ^2H (around 1 to 1.5 per mil for δ^{18}O and 8 to 12 per mil for δ^2H). If the circulation conditions in the atmosphere and the temperature differences between the place of origin and formation of the precipitation do not differ significantly from present-day conditions, higher values must be expected for the same precipitation temperatures from that time. In the δ^2H-temperature diagram, such samples lie above the lines for present-day precipitation;

e) Gas loss and capture in a sample have no effect on the stable isotopes, but cause a change in the calculated recharge temperatures if not corrected as described above. In the δ^2H-temperature diagram, this causes a shift parallel to the temperature axis, towards lower temperatures in the case of gas capture and towards higher temperatures in the case of gas loss. Gas losses may accompany difficulties in collecting samples which contain high concentrations of other gases such as CO_2 and CH_4, and can cause a shift of several degrees in the calculated temperatures. Gas losses prior to infiltration cannot be ruled out if the water warms up significantly at this stage. Re-equilibrium as well as excess air entrainment are conceivable in the case of waters which are transported from higher catchment areas to lower regions by rapidly flowing streams and rivers and which infiltrate only after warming up. Subsurface gas loss is also possible if the

f) groundwater has been subjected to geothermal heating. Cooling prior to infiltration has the same effect as the capture of gas bubbles;

f) As air masses move further from the coast, the heavy isotope component of the precipitation decreases (*e.g.,* ROZANSKI AND OTHERS, 1982). This phenomenon is known as the continental effect and, for the same temperatures, can lead to varying $\delta^{18}O$ and $\delta^{2}H$ values for different regions. Such waters are distinguished by their location in $\delta^{18}O$- and $\delta^{2}H$-temperature diagrams. The lee-side effect, which causes the precipitation falling on the wind-shadow side of a mountain to be depleted in heavy isotopes, can also lead to lower $\delta^{18}O$ and $\delta^{2}H$ values than would be expected. These effects are discussed in more detail in Section 3.1 for Switzerland and the surrounding regions;

g) If one groundwater is recharged primarily by summer precipitation and another by winter precipitation, the calculated recharge temperatures (and also the $\delta^{18}O$ and $\delta^{2}H$) will show the difference. A correction is not possible here but, when interpreting the recharge temperatures, this effect should not be ignored as a cause of variations;

h) Underground admixing of old, previously trapped ocean water leads to an increase in the $\delta^{18}O$ and $\delta^{2}H$ values.

3.3.4 Results

Corrected recharge temperatures are acceptable for around 130 out of 169 individual samples analysed. For the remainder of the samples, degassing and other processes have rendered estimation of a recharge temperature impossible. As a rule, two samples were collected at each depth level and the noble gas contents were measured. Table 3.3.2, at the end of Section 3.3 gives a summary of the results of all analyses, including N_2/Ar and $^{40}Ar/^{36}Ar$ ratios. Where several samples were taken at one particular depth but not all were used for calculating the recharge temperatures, those used are indicated * and those not used **. In the case of the $^{40}Ar/^{36}Ar$ ratios, a distinction is drawn between measurements on water samples enclosed in Cu-tubes and gas extracted for ^{39}Ar. The meaning of these two $^{40}Ar/^{36}Ar$ ratios are discussed in Section 6.5.2.

Table 3.3.1 gives all calculated recharge temperatures arranged by borehole and depth. "$R_{T,M}$" is the mean value for multiple samples from the same water and "R_T used" is the mean value shown in the figures. The errors given for the R_T values correspond to the standard deviation of the mean of the temperatures calculated from each of the gases. The error given for the mean temperature, $R_{T,M}$, is the larger of the errors determined by two procedures. One is based on the errors of the individual results on the repeated samples, and the other is based on the variability of the individual R_T values. The use of the larger of the two errors is an attempt to allow for uncertainties due to the optimization in the correction procedure and due to replicate sample variation. The numbering of the samples was selected arbitrarily for the following discussion, and does not correspond to the system used elsewhere in this report, although the latter is also given. The amounts of excess air (EA) and the degassed volume (DV) determined *via* the optimization procedure are also given. A negative degassed volume signifies the supposed capture of a gas bubble. Note that recharge temperatures calculated for replicate multiple samples from the same source agree in most cases within the given error range, with the notable exception of those from Schafisheim 1571 m. The EA and DV values determined by the optimization procedure also generally agree well for replicate samples.

The recharge temperatures were calculated for the atmospheric pressure given, which is a very rough estimate. Table 3.3.1 also gives the measured values for $\delta^{18}O$, $\delta^{2}H$, ^{14}C and $\delta^{13}C$ for the purposes of the following discussion.

Table 3.3.1 also included temperatures for several non-Nagra samples from southern Germany. The values for Freiburg im Br. are given by RUDOLPH AND OTHERS (1983) and those for Reichenau (501), Birnau (508), Mainau (128) and Singen (97) were calculated from the raw data of BERTLEFF (1986). The optimization procedure already described was applied to these four samples, assuming 960 mbars for the pressure and 302 for the $^{40}Ar/^{36}Ar$ ratio. The data given by Bertleff for Konstanz (98) gave negative recharge temperatures and were therefore omitted from the table.

Many of the samples from the crystalline of the Böttstein (301), Kaisten (305) and Leuggern (306) boreholes have similar noble gas temperatures with a 1-σ range from 2° to 5°C. Apparently we are dealing with a water homogenous with respect to recharge temperatures and $\delta^{18}O$ and $\delta^{2}H$. For this reason, later figures often give only one measured value for these crystalline waters.

3.3.5 Discussion

Figure 3.3.7 shows the measured $\delta^{2}H$ values plotted against the mean values for the calculated noble gas temperatures. It is a summary figure only, and suggests that separate discussions are required on waters of the different groups. The figure also shows the $\delta^{2}H$-temperature relationships for recent recharge in the "NE-Jura and Black Forest" and "Mittelland" areas discussed in Section 3.1, which cover the area studied. Black Forest in this connection refers exclusively to the SE slope of the Black Forest. The $\delta^{2}H$-temperature lines are based on the stable isotope values dominantly from springs with ^{3}H contents above 20 TU and from rivers.

Figures 3.3.8 a to d show the results presented in Figure 3.3.7 by groups including the sample identification numbers and error bars. The division into groups is based on hydrogeological criteria and will be discussed. For comparison purposes, these four detailed figures contain at least one of the two $\delta^{2}H$-temperature relationships for modern ground and surface water in northern Switzerland.

The results are considered in five groups based on the formation from which the sample originated. In a few cases, the water originated from a neighbouring formation. Thus, for example, the sample from the Permian of Kaisten (305/2) is assigned to the crystalline water group.

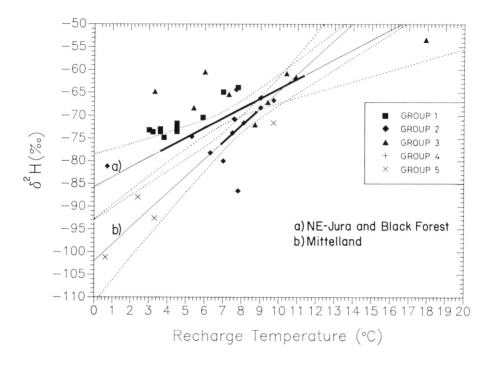

Figure 3.3.7: Summary of measured δ^2H values *versus* noble gas recharge temperatures for all samples. See text for description of Groups.

The groups are as follows:

Group 1

Crystalline   Böttstein (301/8 to 23), Kaisten (305/4, 6, 8, 9, 12, 15, 16), Leuggern, (306/4, 5, 6, 7, 9, 17, 16, 23), Waldkirch (95/2), Rheinfelden (159/4) and Freiburg im Br. (RUDOLPH AND OTHERS, 1983, samples 40, 41 and 42)

Buntsandstein-Crystalline   Böttstein (301/3, 6, 7, 25a and b) and Leuggern (306/2)

Permian   Kaisten (305/2)

Group 2

All Muschelkalk samples

Group 3

Crystalline	Schafisheim (304/10) and Bad Säckingen (161/1)
Buntsandstein	Kaisten (305/1) and Weiach (302/12)
Buntsandstein-Permian	Riniken (303/3) and Kaiseraugst (3/3)
Buntsandstein-Crystalline	Schafisheim (304/5, 7)
Permian	Riniken (303/6)

Group 4

Tertiary	Konstanz (98), Mainau (128/2)
Tertiary	Reichenau (501/1), Birnau (508/2)
Tertiary	Tiefenbrunnen (2/3) and Aqui (1/2)
Malm	Singen (97/5)

Group 5

Crystalline (Alps)	Lötschberg (234/1), Amsteg (231/1) and Grimsel (233/1)

The following conclusions can be drawn from the results:

Group 1: Crystalline, Buntsandstein-crystalline and Permian waters
Figure 3.3.8a (Numbers are those from Table 3.3.1)

- The multiple samples from the crystalline of Böttstein (4), Kaisten (7) and Leuggern (11) all have closely similar recharge temperatures (2° to 5°C), stable isotopic compositions, salinities and $^{40}Ar/^{36}Ar$ ratios. However, these samples clearly have lower temperatures or higher δ^2H values than correspond to either of the two δ^2H-temperature relationships for modern water developed in Section 3.1 and shown in Figure 3.1.8b. This is supported by the low relative uncertainty of the calculated temperatures;

- The samples from the Permian of Kaisten 285.3 m (6) and the Buntsandstein-crystalline transition zones at Leuggern (9) and Böttstein (2) give practically the same values as the crystalline samples mentioned above; a largely crystalline origin can be assumed for these samples on the basis of other parameters, such as their general chemistry and sulphate isotopic composition;

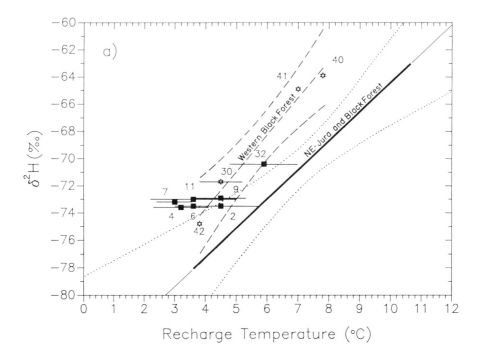

Figure 3.3.8a: Samples of Group 1, numbered according to Table 3.3.1. Black Forest results (stars) are from RUDOLPH AND OTHERS (1983).

- One sample from the Freiburg im Br. region (42) has a low ^{14}C value and correlates well with the above samples. This would imply a place of origin for this sample with a similar δ^2H-temperature relationship. The sample from Waldkirch (30) in the southern Black Forest also lies close to this group in the δ^2H-temperature diagram, although the water shows a Permian influence (SCHMASSMANN AND OTHERS, 1984, p. 259-271);

- Crystalline waters with high ^{14}C contents (40 and 41) have recharge temperatures around 3°C higher and δ^2H values around nine per mil more positive than those for the main crystalline group;

- A regression line calculated for the crystalline waters from the Western Black Forest (40 to 42 and 30) is shown in Figure 3.3.8a as the dashed line "Western Black Forest" with its 95 per cent confidence interval. It has the following equation:

$$\delta^2H = 2.72 \cdot T - 84.5 \text{ (per mil), with } r^2 = 0.98$$

This regression line has a similar slope to the line "NE-Jura and Black Forest" (SE slope of the Black Forest) but is shifted to more positive δ^2H values.

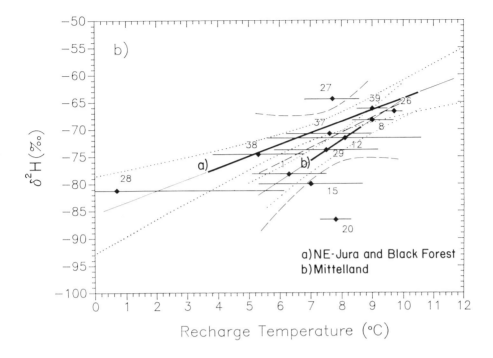

Figure 3.3.8b: Samples of Group 2, numbered according to Table 3.3.1.

This result fits very well with the sequence of parallel δ^2H-temperature relationships lying one above the another as a function of continental location described in Section 3.1. In Figure 3.1.8, the "Western Black Forest" line would then lie more or less in the position of the global precipitation lines of YURTSEVER AND GAT (1981). This corresponds to the picture of decreasing continental character from the southeast to the western Black Forest, which would be expected for a recent δ^2H-temperature relationship for the Black Forest;

- The large group of crystalline waters from the Nagra boreholes correlate best with the "Western Black Forest" δ^2H-temperature relationship. A prerequisite in this case is that the altitude corrections for the four samples used (40 to 42 and 30) and those for the crystalline waters, as carried out here and by RUDOLPH AND OTHERS (1983), should lead to values which are comparable. Therefore, the Black Forest must be considered as a potential catchment area for these waters;

- The position of the crystalline waters from the Nagra boreholes on the extension of the "Western Black Forest" line can be taken as an indication that these waters infiltrated during a climatic period colder than the present one. This requires that the relationship between stable isotopes and temperature in the Black Forest infiltration area during this colder period be similar to the relationship today. The suggestion of recharge during a colder period agrees with the measured ^{14}C values for most of the waters of this group, except for some of the Böttstein waters.

Group 2: Muschelkalk
    Figure 3.3.8b:

- These samples have a greater variation in noble gas temperatures and measured stable isotopes than those of Group 1, but are in a similar range. Even Beznau (28) lies within this range if its large error is taken into account. The δ^2H value for Weiach 859.1 m (20) is significantly more negative than other members of the group, however;

- These samples have similar deuterium contents but significantly higher noble gas temperatures than those of the first group. The calculated noble gas temperatures lie between 6° and 10°C;

- The dashed regression line with confidence interval shown in Figure 3.3.8b was calculated from all the Muschelkalk waters except Beznau (28) and Weiach (20). Both lines for recent waters from northern Switzerland, "NE-Jura and Black Forest" and "Mittelland" (Section 3.1.1), correspond within the 95 per cent confidence intervals with the line determined for the Muschelkalk waters. The regression line for the Muschelkalk waters is described by the following mathematical equation:

$$\delta^2H = 2.77 \cdot T - 92.8 \text{ (per mil), with } r^2 = 0.5$$

Therefore, the samples from the Muschelkalk agree more or less with the recent infiltration conditions in northern Switzerland;

- All Muschelkalk samples with significant ^{14}C contents (samples 20, 26, 27, 29, 39) have noble gas temperatures which correspond to the current mean annual temperatures for the infiltration altitude assumed. This supports the conclusion that the noble gas temperatures calculated on the basis of the correction models more or less accurately reflect the recharge temperatures of these Muschelkalk waters;

- As mentioned, the δ^2H value for Weiach 859.1 m (20) is more negative than the values for the other Muschelkalk waters, while its noble gas temperature agrees. For this sample, it was necessary to carry out a degassing correction as well as an altitude correction, so that the value for the calculated temperature has a greater element of uncertainty than the statistical error given in Figure 3.3.8b. If, for example, this sample contained water from the Rhine (see Section 3.2.4), the observed shift of the noble gas temperature towards higher temperatures while the δ^2H value remains the same could be explained by recharge in part by water from the Alps which has been warmed.

Group 3: Crystalline, Buntsandstein and Permian
    Figure 3.3.8c:

- The noble gas temperatures of some of the waters of this group are less reliable than those in Groups 1 and 2. This is probably because all the samples in this group have a large (sometimes extremely large) excess of N_2. This is shown by the measured N_2/Ar ratios which range from around 90 to 200. Sampling without gas loss is difficult

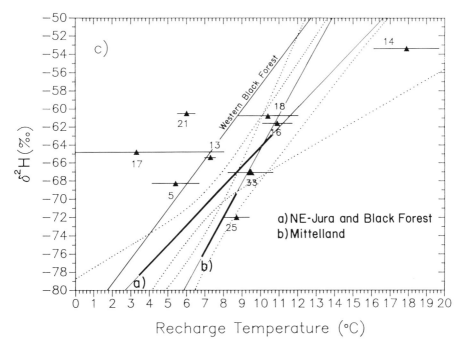

Figure 3.3.8c: Samples of Group 3, numbered according to Table 3.3.1.

under these conditions, and the corrections may only increase the uncertainty of the calculated temperatures. Some samples also have very high $^{40}Ar/^{36}Ar$ ratios. Even after extensive corrections, discrepancies still exist;

- The duplicate samples from Riniken at 805.8 m (13, Buntsandstein-Permian) give a useable result after taking into account a considerable quantity of excess air;

- The duplicate samples from Kaisten at 113.5 m (5, Buntsandstein) and at Weiach 985.3 m (21, Buntsandstein) also give useable results after a correction for excess air;

- The Buntsandstein waters (5, 13, 21; total of six samples) have positive δ^2H values or low recharge temperatures, relative to present δ^2H-temperature relationships in northern Switzerland (Figure 3.3.8c). The results from Kaisten 113.5 m (5) and Riniken (13) are in the δ^2H-temperature range of the granite waters of Group 1 (Figure 3.3.8a) and fit well with the δ^2H-temperature relationship "Western Black Forest." Therefore, the Black Forest is to be considered as a potential infiltration area for these two waters. This agrees with the fact that chemical analyses of these Buntsandstein waters indicate mixing with waters from the deeper lying Permian;

- The results for Riniken 993.5 m (14, Permian) should be interpreted with caution because only one of three samples gives a useable noble gas temperature (Table 3.3.2);

- The estimated noble gas temperature of 27.6 ± 6.2°C from Weiach 1116.5 m (22, Permian; not shown in Figure 3.3.8a) is unusually high (see Table 3.3.1) and is not altered significantly by large altitude and degassing corrections. It, too, must be used carefully, because only one of several samples from this interval yielded a useful temperature;

- The interpretation of the recharge temperatures of Permian waters and of Buntsandstein-Permian transition waters from Weiach (21) is still fraught with considerable difficulties. The noble gas temperature values for pure Permian waters (*e.g.,* 14, 22) have large uncertainties. However, it is noticeable that relative to present-day waters in northern Switzerland, the two Permian samples are clearly shifted towards higher temperatures and more positive δ^2H values. This could indicate infiltration during a warmer climatic period. Another less probable explanation of the higher δ^2H values would be deuterium exchange in the aquifer;

- The sample from Kaiseraugst (25) required a gas loss correction. Two out of three samples gave a useable result; these lie in the range of the present-day δ^2H-temperature relationships for the Mittelland;

- The water from the crystalline at Bad Säckingen Stammelhof (33) is located in the overlap zone of the two lines for northern Switzerland;

- The two samples from Schafisheim (16, from the Buntsandstein-crystalline transition zone at 1488.2 m and 18, from the crystalline at 1887.8 m) differ significantly from the samples from the crystalline north of the Permo-Carboniferous Trough (Group 1). Their noble gas temperatures and the values for the stable isotopes are shifted to positive values. They also have $^{40}Ar/^{36}Ar$ and N_2/Ar ratios more similar to those of Permian samples than to those of granite samples of Group 1. As Figure 1.4.3 shows, the crystalline at Schafisheim appears to be surrounded by Permian sediment. Thus, it is not surprising that water of the character of that found elsewhere in deep Permian sediments should also be present in the crystalline here. The results for these two Schafisheim samples (16 and 18) lie in the confidence interval of both the current Mittelland and NE-Jura and Black Forest δ^2H-temperature lines;

- As Table 3.3.1 shows, there is poor agreement among the four samples from the crystalline at 1571.1 m at Schafisheim (17), even after the various corrections. The 1-σ range of the mean recharge temperature includes temperatures below 0°C, and, although the sample appears in Figure 3.3.8c, it cannot be used.

Group 4: Tertiary and Malm
    Figure 3.3.8d:

- The calculated noble gas temperature for Tiefenbrunnen (24) has a large error;

- The values for the samples from Aqui (23), Reichenau (43), Birnau (44), Mainau (45) and Singen (46) all lie close to the present-day δ^2H-temperature lines for northern Switzerland, but have lower temperatures than found in modern waters (heavy solid line);

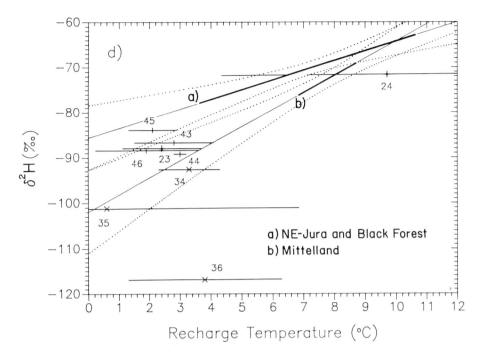

Figure 3.3.8d: Samples of Groups 4 and 5, numbered according to Table 3.3.1. All values are results from BERTLEFF (1986), with the exception of (23) and (24).

- All temperatures calculated using the noble gases, with the exception of Tiefenbrunnen (24), indicate infiltration during a colder period. This is consistent with the low measured ^{14}C values of all samples except Reichenau. Other ways of reaching such low δ^2H and recharge temperature values (*e.g.*, infiltration primarily of winter precipitation at the assumed infiltration altitudes) appear unlikely.

Group 5: Alpine Waters
    Figure 3.3.8d:

- There are only three samples with noble gas temperatures from the Alps and no reliable recent δ^2H-temperature relationships. Thus, little interpretation of the samples Amsteg (34), Grimsel (35) and Lötschberg (36) is possible;

- Compared with the Amsteg (34) and Grimsel (35) samples, the value for Lötschberg (36) is clearly shifted to low δ^2H values. However, based on the local orographic situation, it can be presumed that the precipitation fell in a high catchment area and reached lower regions via surface flow, only infiltrating into the soil after warming up.

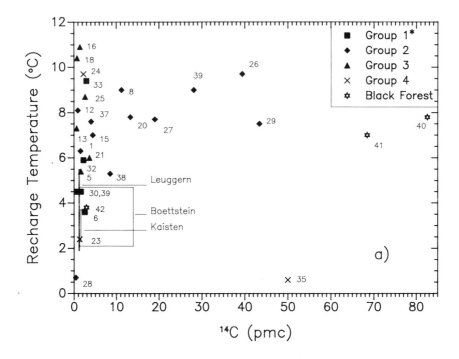

Figure 3.3.9: Recharge temperatures calculated from noble gas contents plotted against measured ^{14}C of samples from all groups.

3.3.6 Recharge Temperatures and ^{14}C Results

Figures 3.3.9 and 3.3.10 show temperatures calculated from noble gas concentrations plotted against measured ^{14}C values. For the multiple samples from the crystalline of Kaisten and Leuggern and from the crystalline and the Buntsandstein-crystalline transition zone in Böttstein, only the ranges are given in Figure 3.3.10. The values are taken from Table 3.3.1 and the numbering of the individual samples in the figures corresponds to that used in the table. The points for Riniken 993.5 m and Weiach 1116.5 m (14 and 22, both from the Permian) are omitted because the ^{14}C values are not reliable.

- The Black Forest waters with high ^{14}C values (40 and 41) clearly have higher noble gas temperatures than those with low ^{14}C values (42 and 30). The results for the three Black Forest waters 40 to 42 are taken from RUDOLPH AND OTHERS (1983);

- Both the ^{14}C values and noble gas temperatures of the samples from the crystalline of Kaisten and Leuggern lie close together and are comparable with the samples from Freiburg im Br. (42) and Waldkirch (30). This confirms the conclusion drawn earlier that the crystalline waters with a low ^{14}C content infiltrated under colder climatic conditions than the present-day Black Forest waters;

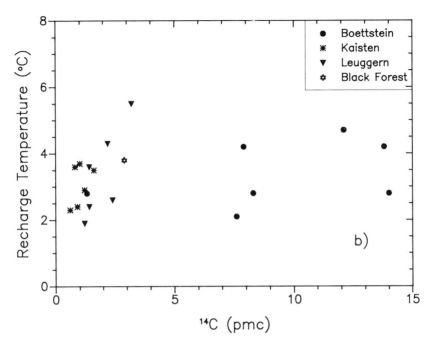

Figure 3.3.10: Calculated temperature plotted against measured ^{14}C of samples from the crystalline of Böttstein, Leuggern and the Black Forest.

- For Böttstein on the other hand, the noble gas temperatures lie close together but the ^{14}C values vary from 1.5 to 14 pmc, *i.e.*, the samples with ^{14}C contents over 5 pmc cannot be distinguished within the error in the recharge temperature from old crystalline waters. If the error is taken into account, the samples with increased ^{14}C values could lie on the line describing mixtures of young and old Black Forest waters;

- The weighted mean noble gas temperatures for the Muschelkalk samples with ^{14}C contents above 10 pmc, (8, 20, 26, 27, 29 and 39) is 9.0 ± 0.4°C, while the mean value for the samples with ^{14}C contents below 5 pmc (1, 12, 15, 28 and 37) is 6.9 ± 0.5°C. The difference between the two mean values is small, but is significant within 2 σ but not within 3 σ. The conclusion that Muschelkalk waters with low ^{14}C contents infiltrated under colder conditions than those with ^{14}C content above 10 pmc should be considered as probable, but by no means certain. In particular, the two groups do not have as large a difference in recharge temperatures as would be expected between waters which infiltrated under present-day conditions and ice-age conditions;

- All samples of Groups 3 and 4 (except Grimsel 35) have low ^{14}C values but their noble gas temperatures range from one or two over 10°C. The results of the 1488.2 m (16) and 1887.8 m (18) samples from Schafisheim and from Bad Säckingen (33) of around 9° to 11°C appear to be reliable. Because, to a certain extent, the samples of Groups 3 and 4 have nothing to do with each other, a mean value has little significance.

Nevertheless, it is possible to state that some waters (*e.g.*, 16, 18, 33) did not infiltrate during a cold period because, despite their low ^{14}C content, the recharge temperatures point rather to a warmer period. For these samples, high ^{40}Ar/^{36}Ar and ^{4}He values indicate a relatively high age. The measured point for Aqui (23) coincides with the Tertiary and Jurassic waters investigated by BERTLEFF (1986), for which, based on the low ^{14}C value, infiltration under cold climatic conditions appears likely;

- It is notable in Figure 3.3.9 that, for waters with ^{14}C contents below 5 pmc, the whole range of recharge temperatures from 3° to 11°C is found. These observed differences in the recharge temperatures for similar ^{14}C contents must be real if they are larger than the errors associated with the temperatures. This is supported by the fact that six crystalline waters from Leuggern, six from Kaisten and seven crystalline and Buntsandstein-crystalline waters which are part of Group 1 show very similar values for recharge temperature (Figure 3.3.10). For these samples at least, the optimization procedure yields reliable values;

- If the weighted mean values for noble gas temperatures are calculated for the samples from the crystalline at Kaisten and Leuggern, the figures obtained are 2.7 ± 0.2°C and 3.5 ± 0.5°C, respectively. These values differ significantly from those of younger waters, *e.g.*, Muschelkalk waters with ^{14}C contents > 10 pmc (9.0 ± 0.4°C). This temperature difference of around 6°C is consistent with values from the literature for the difference between present-day and ice-age climatic conditions as summarized in Figure 7 of NAGRA (1984, NTB 84-26).

3.3.7 Summary

For around 130 of the 169 individual measured samples, recharge temperatures could be derived from the noble gas contents after corrections (*e.g.*, for oversaturation, ^{40}Ar production, altitude and degassing) had been carried out. The majority of the unuseable results were those obtained with evacuated containers (bailer samples). These extraction methods have proved unsuitable for noble gas measurements since the gas loss of up to 90 per cent does not occur in thermodynamic equilibrium and cannot be corrected.

The corrections reduce the reliability of even the acceptable temperatures to a certain extent and temperatures on which the corrections have a large influence should be interpreted with caution. There are also uncertainties in the relationships adopted between recharge temperatures and the stable isotope contents of infiltrating water. Therefore, the conclusions drawn from the data differ in their reliability. Those which are least uncertain are discussed here.

The results from the regional programme are a useful supplement to the data from the deep-drilling programme in that multiple comparisons are made possible. The relationships between δ^2H and recharge temperature in particular can be examined more closely.

Despite these difficulties, several important conclusions can be considered relatively certain. These are summarized as follows:

Group 1: Crystalline, Buntsandstein-crystalline and Permian waters

- The multiple samples from the crystalline of Böttstein (4), Kaisten (7) and Leuggern (11) all have closely similar recharge temperatures (2° to 5°C) and stable isotopic compositions;

- The samples from the Permian of Kaisten at 285.3 m (6) and from the Buntsandstein-crystalline transition zones of Leuggern (9) and Böttstein (2) give practically the same values as the crystalline samples mentioned above, and based on other parameters, the water can be concluded as having primarily a crystalline origin;

- However, these samples clearly have lower temperatures, or more positive δ^2H values than indicated by the δ^2H-temperature relationships for samples of modern waters from northern Switzerland;

- The sample from the Freiburg im Br. region with a low ^{14}C value (42) and the sample from Waldkirch (30) in the southern Black Forest agree well with the above mentioned samples, which would indicate that the waters have a comparable region of origin. The waters from Freiburg im Br., which can be ranked as young on the basis of the ^{14}C results, have δ^2H values which are around nine per mil higher than those of the crystalline group mentioned above and their recharge temperature is some 3°C warmer;

- The results for the four samples from the Freiburg im Br. (40 to 42) and Waldkirch (30) describe a relationship between δ^2H and temperature. The crystalline waters from the Nagra boreholes fit better on this δ^2H-temperature relationship than on those plotted for the NE-Jura and Black Forest based on modern waters as described in Section 3.1. This would support the hypothesis that the Black Forest is the relevant infiltration area. The position of the Nagra crystalline waters in the δ^2H-temperature diagram can be taken as an indication that most of the Nagra crystalline waters investigated infiltrated during a colder climatic period than the present one. Except for some of the Böttstein waters, this statement is consistent with the measured ^{14}C values;

- If the weighted mean values for recharge temperature are calculated for the crystalline samples from Kaisten and Leuggern, the figures obtained are 2.7 ± 0.2°C and 3.5 ± 0.5°C respectively. These values differ clearly from those for younger waters, *e.g.*, Muschelkalk waters with ^{14}C contents > 10 pmc (9.0 ± 0.4°C). This temperature difference of around 6°C agrees well with literature values for the difference between present-day and ice-age conditions.

Group 2: Muschelkalk

- A regression line calculated from the δ^2H and noble gas temperature values for the Muschelkalk samples coincides with the δ^2H-temperature relationships for recent water from northern Switzerland. Compared with the crystalline waters, the Muschelkalk samples have significantly higher recharge temperatures (6° to 10°C) for similar δ^2H values;

- All Muschelkalk samples with ^{14}C values have above 10 pmc recharge temperatures which correspond to the present-day mean annual temperatures for the assumed infiltration altitudes. This supports the conclusion that the recharge temperatures calculated for these Muschelkalk waters on the basis of correction models are correct.

Group 3: Crystalline and Buntsandstein-Permian

- Six individual samples from the Buntsandstein and the Buntsandstein-Permian transition have more positive δ^2H values or lower recharge temperatures than predicted by the δ^2H-temperature relationships for recent water from northern Switzerland. The values from the Kaisten 113.5 m and Riniken 805.8 m samples are on the δ^2H-temperature relationship for the Black Forest. Therefore, the Black Forest is to be seen as a potential infiltration area for these two waters. The δ^2H value for Weiach is still higher;

- Interpretation of the recharge temperatures of Permian waters is difficult, but it is noticeable that these waters have relatively high temperatures compared to present waters in northern Switzerland;

- The two Schafisheim samples from the Buntsandstein-crystalline transition zone (1488.2 m) and from the crystalline (1887.8 m) differ significantly from samples from the crystalline north of the Permo-Carboniferous Trough. The Schafisheim samples have higher recharge temperatures than samples north of the trough. With respect to the ^{40}Ar/^{36}Ar and N$_2$/Ar ratio, they also show more similarity to the Permian samples than to the granite samples mentioned above;

- The results for the Schafisheim samples lie in the 95 per cent confidence interval of the δ^2H-temperature relationships of both the Mittelland and the NE Jura and Black Forest;

- The Kaiseraugst (Buntsandstein-Permian) and Bad Säckingen (crystalline) samples lie close to the current δ^2H-temperature relationships for northern Switzerland;

- It is possible to say that some waters (*e.g.,* Schafisheim 1488.2 m, Schafisheim 1887.8 m and Säckingen) did not infiltrate during the ice-age because, despite their low ^{14}C content, the recharge temperature points rather to a relatively warm period. Increased ^{40}Ar/^{36}Ar and ^{4}He values, nevertheless, point to a relatively great age for these samples.

Group 4: Tertiary-Malm

- The values for the Tertiary and Jurassic waters from Aqui, Reichenau, Birnau, Mainau and Singen all lie together in the range of current δ^2H-temperature relationships for northern Switzerland, but they have lower noble gas temperatures. The most likely explanation for this is the assumption that these waters infiltrated during a colder period than the present one.

General

- It is noticeable that, for waters with ^{14}C below 5 pmc, the whole range of recharge temperatures from 3° to 11°C is found. Some of the relatively large differences observed in the recharge temperatures (with similar ^{14}C contents) must be real.

Table 3.3.1: Summary of calculated recharge temperatures. (Page 1 of 5)

R_T calculated recharge temperatures for individual samples.

$R_{T,M}$ arithmetic mean of calculated recharge temperatures for individual parallel samples.

R_T used mean value of all samples collected at a specific depth or in a specific formation. These are the values presented in the figures and used for the discussion.

The atmospheric pressure used for the calculation (p), the values for excess air (EA) and degassed volume (DV) obtained via optimisation and the measured δ^2H, $\delta^{18}O$, ^{14}C and $\delta^{13}C$ values are also given. The footnotes at the end of the Table should be consulted for the depth intervals. The sample numbers are those used in the text and figures of Section 3.3.

Sample No.	NAGRA No.	Avg. Depth m	p mbar	EA ccSTP/l	DV cm³/l	R_T °C	$R_{T,M}$ °C	R_T used °C	δ^2H per mil	$\delta^{18}O$ per mil	^{14}C pmc	$\delta^{13}C$ per mil
301/2-1			950	4.2	0.0	6.6 ± 1.8						
301/2-2			950	4.2	0.0	6.0 ± 1.5	6.3 ± 1.2					
1 Böttstein		162.9 m Muschelkalk (mo)						6.3 ± 1.2	-78.2	-11.16	1.5	-4.8
301/25a-1		312.5	950	4.8	0.0	3.0 ± 1.0						
301/25a-2		312.5	950	5.9	0.0	3.1 + 0.8						
301/25b-1		312.5	950	5.0	0.0	2.9 ± 0.8						
301/25b-2		312.5	950	7.5	0.0	2.1 ± 0.6	2.8 ± 0.4					
301/7-1		315.0	950	6.8	0.0	4.2 ± 1.4						
301/7-2		315.0	950	3.4	0.0	5.4 ± 1.1	4.8 ± 0.9					
301/6-1		316.6	950	10.0	0.0	4.7 ± 0.7	4.7 ± 0.7					
301/3-1		321.2	950	5.3	0.0	5.5 ± 1.0	5.5 ± 1.0					
2 Böttstein Buntsandstein-Crystalline (s-KRI) 1)								4.5 ± 1.2	-73.5	-10.08		2)
301/8c-1		399.5	950	8.0	0.0	3.2 ± 2.2						
301/8c-2		399.5	950	9.9	0.0	2.6 ± 2.4						
301/9-1		399.5	950	4.1	0.0	2.6 ± 2.6	2.8 ± 1.4					
301/16-1		618.4	950	7.8	0.0	3.6 ± 0.2						
301/16-2		618.4	950	7.5	0.0	3.1 ± 0.2						
301/17-1		618.4	950	0.0	-7.0	5.9 ± 1.6	4.2 ± 1.5					
301/12b-1		621.3	950	5.0	0.0	2.9 ± 2.2						
301/12b-2		621.3	950	6.3	0.0	2.5 ± 2.0						
301/13-1		621.3	950	4.7	0.0	3.1 ± 2.1	2.8 ± 1.2					
301/23-1		649.0	950	6.0	0.0	4.5 ± 0.6						
301/23-2		649.0	950	6.1	0.0	3.9 ± 0.2	4.2 ± 0.4					
301/18-1		792.4	950	6.5	0.0	1.4 ± 0.8						
301/18-2		792.4	950	7.3	0.0	2.7 ± 0.6	2.1 ± 0.9					
4 Böttstein Crystalline (KRI)								3.2 ± 0.9	-73.6	-10.01		3)

Table 3.3.1: Summary of calculated recharge temperatures. (Page 2 of 5)

Sample No.	NAGRA No.	Avg. Depth m	p mbar	EA ccSTP/l	DY cm³/l	R_T °C	$R_{T_C}M$	R_T used °C	δ^2H - per mil -	$\delta^{18}O$	^{14}C pmc	$\delta^{13}C$ per mil
305/1-1		113.5	950	4.6	0.0	5.3 ± 2.5						
305/1-2		113.5	950	4.1	0.0	5.5 ± 0.6	5.4 ± 1.3					
5 Kaisten Buntsandstein (s)								5.4 ± 1.3	-68.3	-9.03	<0.5	-17.2
305/2-1		285.3	950	6.0	0.0	3.0 ± 1.0						
305/2-2		285.3	950	7.3	0.0	4.1 ± 0.7	3.6 ± 0.6					
6 Kaisten Permian								3.6 ± 0.6	-73.5	-10.36	1.7	-17.8
305/4-1		310.4	950	6.2	0.0	4.1 ± 2.3						
305/4-2		310.4	950	5.2	0.0	3.2 ± 2.0	3.7 ± 1.5					
305/6-1		482.7	950	6.5	0.0	2.7 ± 1.5						
305/6-2		482.7	950	5.9	0.0	2.1 ± 1.4	2.4 ± 1.0					
305/8-1		656.5	950	5.8	0.0	2.3 ± 1.0						
305/8-2		656.5	950	7.2	0.0	3.4 ± 0.9	2.9 ± 0.7					
305/9-1		819.5	950	7.6	0.0	2.5 ± 1.0						
305/9-2		819.5	950	6.8	0.0	2.1 ± 1.1						
305/9-3		819.5	950	6.9	0.0	2.3 ± 1.1	2.3 ± 0.6					
305/12-1		1031.0	950	6.7	0.0	4.4 ± 5.5						
305/12-2		1031.0	950	7.2	0.0	3.6 ± 3.2						
305/12-3		1031.0	950	6.6	0.0	2.9 ± 2.0	3.6 ± 2.2					
305/15-1		1153.3	950	6.4	0.0	2.4 ± 3.2						
305/15-2		1153.3	950	4.0	0.0	3.2 ± 1.2	2.8 ± 1.7					
305/16-1		1271.9	950	8.5	0.0	2.7 ± 2.4						
305/16-2		1271.9	950	3.6	0.0	2.7 ± 3.3						
305/16-3		1271.9	950	7.1	0.0	4.8 ± 1.5						
305/17-1		1271.9	950	7.8	0.0	3.6 ± 2.4	3.5 ± 1.2					
7 Kaisten Crystalline (KRI)								3.0 ± 0.6	-73.2	-10.39	4)	
306/1-1		75.0	950	3.2	0.0	9.0 ± 0.2						
306/1-2		75.0	950	3.2	0.0	9.0 ± 1.4	9.0 ± 0.7					
8 Leuggern Muschelkalk (mo)								9.0 ± 0.7	-68.3	-9.96	12.0	-15.8
306/2-1		217.9	950	5.4	0.0	3.8 ± 1.2						
306/2-2		217.9	950	5.3	0.0	3.8 ± 1.1						
306/2-3		217.9	950	5.5	0.0	5.9 ± 2.4	4.5 ± 1.2					
9 Leuggern Buntsandstein (s)								4.5 ± 1.2	-72.9	-10.39	1.6	-9.3
306/4-1		251.3	950	5.1	0.0	3.6 ± 2.3	3.6 ± 2.3					
306/5-1		444.3	950	5.1	0.0	4.9 ± 2.7						
306/6-1		444.3	950	12.0	0.0	6.1 ± 0.2	5.5 ± 1.4					
306/7-1		538.0	950	6.0	0.0	1.8 ± 0.7						
306/7-2		538.0	950	7.9	0.0	3.3 ± 1.4	2.6 ± 0.8					
306/9-1		705.8	950	5.5	0.0	2.3 ± 0.8						
306/9-2		705.8	950	5.6	0.0	2.4 ± 1.2	2.4 ± 0.7					
306/17-1		847.0	950	8.7	0.0	4.3 ± 0.7	4.3 ± 0.7					

Table 3.3.1: Summary of calculated recharge temperatures. (Page 3 of 5)

Sample No.	NAGRA No.	Avg. Depth m	p mbar	EA ccSTP/l	DY cm³/l	R_T °C	R_{T_CM} °C	R_T used °C	δ^2H - per mil -	$\delta^{18}O$	^{14}C pmc	$\delta^{13}C$ per mil
	306/16-1	1643.4	950	5.9	0.0	1.6 ± 1.1						
	306/16-2	1643.4	950	41.3	0.0	2.2 ± 2.5	1.9 ± 1.4					
	306/23-1	1665.6	950	0.0	-8.1	5.1 ± 0.9	5.1 ± 0.9					
11 Leuggern Crystalline (KRI)								3.6 ± 1.4	-73.0	-10.41	5)	
	303/2-1	656.7	950	25.0	0.0	7.3 ± 4.0						
	303/2-2	656.7	950	8.6	0.0	8.8 ± 3.2	8.1 ± 2.5					
12 Riniken Muschelkalk (mo)								8.1 ± 2.5	-71.6	-10.14	0.9	-4.8
	303/3-1	805.8	950	11.7	0.0	7.1 ± 0.2						
	303/3-2	805.8	950	14.0	0.0	7.4 ± 0.5	7.3 ± 0.3					
13 Riniken Buntsandstein-Permian (s-r)								7.3 ± 0.3	-65.4	-6.92	0.6	-9.2
	303/6-1	993.5	950	3.2	0.0	17.9 ± 1.8	17.9 ± 1.8					
14 Riniken Permian (r)								17.9 ± 1.8	-53.4	-6.65	40.8	-14.6
	304/3-2	1260.0	950	0.5	0.0	7.0 ± 1.7	7.0 ± 1.7					
15 Schafisheim Muschelkalk (mo)								7.0 ± 1.7	-80.0	-11.02	4.4	
	304/5-1	1488	950	6.8	0.0	11.8 ± 1.0						
	304/7-1	1488	950	6.7	0.0	10.0 ± 1.1	10.9 ± 1.3					
16 Schafisheim Buntsandstein-Crystalline (s-KRI)								10.9 ± 0.8	-61.6	-6.24	1.4	-21.2
	304/8-1	1571	950	1.6	-10.2	5.8 ± 0.2						
	304/8-2	1571	950	0.0	-11.0	-3.6 ± 2.3						
	304/9-1	1571	950	0.1	-4.5	4.2 ± 2.1						
	304/9-2	1571	950	0.0	-5.1	6.9 ± 2.2	3.3 ± 4.8					
17 Schafisheim Crystalline 1571m								3.3 ± 4.8	-64.8	-6.66		
	304/10-2	1888	950	3.3	0.0	9.9 ± 2.8						
	304/10-3	1888	950	6.4	0.0	10.9 ± 1.4	10.4 ± 1.6					
18 Schafisheim Crystalline 1888m								10.4 ± 1.6	-60.8	-5.55	0.7	-9.2
	302/10a-1	859	950	45.7	35.7	6.9 ± 0.4						
	302/10a-2	859	950	13.6	9.3	9.1 ± 0.1						
	302/10b-1	859	950	6.2	4.2	7.5 ± 0.4						
	302/10b-2	859	950	3.8	2.5	7.6 ± 0.4	7.8 ± 0.9					
20 Weiach 859.1m Muschelkalk (mo)								7.8 ± 0.5	-86.6	-11.76	13.3	-4.6
	302/12-1	985	950	8.0	0.0	6.2 ± 0.6						
	302/12-2	985	950	5.6	0.0	5.7 ± 0.4	6.0 ± 0.5					
21 Weiach 985.3m Buntsandstein (s)								6.0 ± 0.5	-60.5	-8.12	3.6	-8.3
	302/20-1	1116.5	950	0.0	-4.4	27.6 ± 6.2	27.6 ± 6.2					
22 Weiach 1116.5m Permian (r)								27.6 ± 6.2	-37.8	-5.32	12.6	-25.4

Table 3.3.1: Summary of calculated recharge temperatures. (Page 4 of 5)

Sample No.	NAGRA No.	Avg. Depth m	p mbar	EA ccSTP/l	DV cm³/l	R_T °C	$R_{T_c}M$	R_T used °C	δ^2H per mil	$\delta^{18}O$ per mil	^{14}C pmc	$\delta^{13}C$ per mil
	001/1-1		960	5.7	0.0	1.4 ± 2.8						
	001/2-1		960	5.6	0.0	2.6 ± 2.0						
	001/2-2		960	5.3	0.0	3.3 ± 2.1	2.4 ± 1.3					
23	Aqui Tertiary (t)							2.4 ± 1.3	-88.0	-11.89	<1.3	-3.0
	002/3-1		960	5.7	0.0	9.7 ± 5.4						
24	Tiefenbrunnen Tertiary (t)							9.7 ± 5.4	-71.6	-9.04	2.2	-3.1
	003/3-1		960	0.7	2.5	8.6 ± 0.4						
	003/3-2		960	2.8	4.5	8.9 ± 1.4	8.7 ± 0.7					
25	Kaiseraugst Buntsandstein-Permian (s-r)							8.7 ± 0.7	-72.0	-10.28	2.6	-9.2
	008/2-1		960	1.1	0.0	9.7 ± 0.3						
26	Pratteln Muschelkalk (mo)							9.7 ± 0.3	-66.7	-9.29	39.4	-9.5
	009/2-1		960	4.0	0.0	7.7 ± 0.9	7.7 ± 0.9					
27	Frenkendorf Muschelkalk (mo)							7.7 ± 0.9	-64.4	-9.14	19.0	-9.8
	015/1-1		960	0.0	1.4	0.7 ± 5.5	0.7 ± 5.5					
28	Beznau Muschelkalk (mo)							0.7 ± 5.5	-81.2	-11.09	<0.4	-5.4
	014/3-1		910	6.7	0.0	7.5 ± 1.7	7.5 ± 1.7					
29	Densbüren Muschelkalk (mo)							7.5 ± 1.7	-73.8	-10.24	43.4	-9.6
	095/2-1		980	2.9	0.0	4.6 ± 1.1						
	095/2-2		980	3.2	0.0	4.3 ± 0.8	4.5 ± 0.7					
30	Waldkirch Crystalline (KRI)							4.5 ± 0.7	-71.7	-10.20	0.7	-8.0
	159/4-3		970	7.8	0.0	5.9 ± 1.1	5.9 ± 1.1					
32	Rheinfelden Crystalline (KRI)							5.9 ± 1.1	-70.4	-9.21	<2.2	-7.1
	161/1-1		950	4.2	0.0	9.5 ± 1.2						
	161/1-2		950	2.1	0.0	9.2 ± 1.1	9.4 ± 1.2					
33	Bad Säckingen Crystalline (KRI)							9.4 ± 1.2	-67.1		2.9	-10.9
	231/1-1		760	3.8	0.0	3.3 ± 1.0	3.3 ± 1.0					
34	Amsteg Crystalline (KRI)							3.3 ± 1.0	-92.6	-12.77		
	233/1-2		760	4.1	0.0	0.6 ± 6.3	0.6 ± 6.3					
35	Grimsel Crystalline (KRI)							0.6 ± 6.3	-101.2	-14.09	≈50	
	234/1-1		760	1.0	0.0	3.8 ± 2.5	3.8 ± 2.5					
36	Lötschberg Crystalline (KRI)							3.8 ± 2.5	-116.9	-16.16		

Table 3.3.1: Summary of calculated recharge temperatures. (Page 5 of 5)

Sample No.	NAGRA No.	Avg. Depth m	p mbar	EA ccSTP/l	DV cm³/l	R_T °C	$R_{T_{C}M}$	R_T used °C	δ^2H - per mil -	$\delta^{18}O$	^{14}C pmc	$\delta^{13}C$ per mil
	271/101-1		960	5.4	0.0	7.1 ± 1.4						
	271/101-1		960	8.3	0.0	8.1 ± 2.4	7.6 ± 1.4					
37 Magden Weiere Muschelkalk (mo)								7.6 ± 1.4	-70.8	-10.14	4.0	-8.1
	271/103-1		960	3.3	0.0	5.6 ± 1.8						
	271/103-2		960	3.2	0.0	6.0 ± 1.3						
	271/103-3		960	3.7	0.0	4.3 ± 3.9	5.3 ± 1.5					
38 Magden Weiere Muschelkalk (mo)								5.3 ± 1.5	-74.6	-10.49	8.5	-6.8
	272/102-1		960	2.8	0.0	8.9 ± 0.8						
	272/102-2		960	1.5	0.0	9.1 ± 0.7	9.0 ± 0.5					
39 Magden Falke 2 Muschelkalk (mo)								9.0 ± 0.5	-66.2	-9.24	28.1	-9.6
40 Freiburg im Br. (KRI) 6)								7.8	-63.9		82.5	
41 Freiburg im Br. (KRI) 6)								7.0	-64.9		68.4	
42 Freiburg im Br. (KRI) 6)								3.8	-74.8		2.9	
	1(2)-8		960	10.7	0.0	2.8 ± 1.3						
43 Reichenau (Quaternary) 7)								2.8 ± 1.3	-86.7	-12.09		
	2-7		960	4.6	0.0	3.0 ± 0.2						
44 Birnau (Quaternary) 7)								3.0 ± 0.2	-89.2	-12.40		
	2-8		960	7.4	0.0	2.1 ± 0.8						
45 Mainau (OMM) 7)								2.1 ± 0.8	-83.9	-11.74		
	5-7		960	10.3	0.0	1.9 ± 1.7						
46 Singen (Malm) 7)								1.9 ± 1.7	-88.4	-12.30		

2) BOE 320m ^{14}C: 14.0 and 12.1 pmc $\delta^{13}C$: -9.3 and -6.3 per mil
3) BOE 400-792m ^{14}C: 1.3 to 13.8 pmc $\delta^{13}C$: -8.8 to -9.3 per mil
4) KAI 310-1271m ^{14}C: 1.0 to   1.6 pmc $\delta^{13}C$: -9.2 to -10.4 per mil
5) LEU 251-1666m ^{14}C: 1.2 to   3.2 pmc $\delta^{13}C$: -9.3 to -10.1 per mil
6) Data from RUDOLF AND OTHERS, 1983
7) Raw data from BERTLEFF, 1984

FORMATIONS   mo Muschelkalk   r   Rotliegend (Permian)
             s   Buntsandstein   tOMM Upper Marine Molasse
             KRI Crystalline

Table 3.3.2: Noble gas concentrations of samples from the deep boreholes and the regional programme. (Page 1 of 4)

$^{40}Ar/^{36}Ar$ values are given both for gases extracted directly from the water samples and for large samples extracted in the field for ^{39}Ar measurements.

** Unacceptable samples (for example, Bailer samples).
*   Designates one of a set of duplicate samples which can be used in spite of the fact that the other duplicate sample must be rejected.
AL Surface Discharge Sample
DST Drill Stem Test
GTC Geological Testing Consultants
RFT Repeat Formation Test
BAI Preussag Bailer
GWT Ground Water Temperature
Min Mineralisation

Sample/ NAGRA No.	Avg. Depth (m)	Geo- logy	Sampling Method	GWT (°C)	Min. (g/kg)	He $\times 10^{-8}$	Ne $\times 10^{-7}$	Ar $\times 10^{-4}$	Kr $\times 10^{-8}$	Xe $\times 10^{-8}$	N2/Ar	Argon 40/36 Water Sample	Argon 40/36 Gas Sample
BÖTTSTEIN		(BOE)											
301/2-1	162.9	mo	AL	19.7	6.38	21200	2.77	4.71	10.70	1.34	53.2		309.9 ± 1.5
301/2-2	162.9	mo	AL	19.7	6.38	21700	2.77	4.75	10.76	1.39	53.2		309.9 ± 1.5
301/1-1 **	162.9	mo	DST	19.7	6.34	15300	5.11	7.04	14.87	1.84			
301/1-2 **	162.9	mo	DST	19.7	6.34	17100	7.48	8.14	16.05	1.90			
301/25a-1	312.5	s-KRI	AL	26.0	0.00	8750	2.95	5.31	11.29	1.60		313.8 ± 2.7	
301/25a-2	312.5	s-KRI	AL	26.0	0.00	10100	3.14	5.38	11.26	1.63		313.8 ± 2.7	
301/25b-1	312.5	s-KRI	AL	26.0	0.00	10100	2.99	5.32	11.33	1.63	49.6	313.8 ± 2.7	
301/25b-2	312.5	s-KRI	AL	26.0	0.00	10700	3.46	5.63	11.82	1.72	49.6	313.8 ± 2.7	
301/7-1	315.0	s-KRI	RFT	26.0	0.00	9530	3.29	5.37	11.31	1.51	40.1		
301/7-2	315.0	s-KRI	RFT	26.0	0.00	8140	2.65	4.87	10.50	1.44	40.1		
301/6-1	316.6	s-KRI	DST	26.0	2.28	13300	3.86	5.56	11.39	1.56	51.6		300.9
301/3-1	321.2	s-KRI	DST	26.0	0.00	13600	2.99	5.06	10.56	1.47			
301/8c-1	399.5	KRI	AL	28.0	1.19	19000	3.53	5.49	11.41	1.55	51.3		
301/8c-2	399.5	KRI	AL	28.0	1.19	22400	3.88	5.77	11.80	1.59	51.3		
301/9-1	399.5	KRI	GTC	28.0	1.25	16100	2.83	5.28	11.16	1.53			
301/16-1	618.4	KRI	AL	36.0	1.32	14100	3.48	5.04	11.57	1.64	47.4		295.2 ± 4.9
301/16-2	618.4	KRI	AL	36.0	1.32	14200	3.44	5.10	11.53	1.70	47.4		295.2 ± 4.9
301/17-1	618.4	KRI	GTC	36.0	1.24	14100	5.72	5.11	11.49	1.71			
301/12b-1	621.3	KRI	AL	36.0	1.30	20000	2.98	5.35	11.30	1.53	52.6		303.3 ± 1.5
301/12b-2	621.3	KRI	AL	36.0	1.30	21100	3.22	5.49	11.62	1.57	52.6		303.3 ± 1.5
301/13-1	621.3	KRI	GTC	36.0	1.27	19600	2.92	5.28	11.21	1.52			
301/23-1	649.0	KRI	AL	36.0	0.00	19300	3.13	4.99	11.19	1.54	57.6	306.1 ± 1.8	
301/23-2	649.0	KRI	AL	36.0	0.00	18800	3.16	5.00	11.20	1.63	57.6	306.1 ± 1.8	
301/18-1	792.4	KRI	AL	42.0	1.33	14700	3.29	5.12	11.84	1.86	56.6		290.8 ± 1.9
301/18-2	792.4	KRI	AL	42.0	1.33	14400	3.41	5.02	11.77	1.75	56.6		290.8 ± 1.9
301/21-1 **	1326.2	KRI	BAI			14000	1.08	0.62	0.91	0.17			
301/21-2 **	1326.2	KRI	BAI			7980	0.40	0.38	0.66	0.14			

Table 3.3.2: Noble gas concentrations of samples from the deep boreholes and the regional programme. (Page 2 of 4)

Sample/ NAGRA No.	Avg. Depth (m)	Geo-logy	Sampling Method	GWT (°C)	Min. (g/kg)	He $\times 10^{-8}$	Ne $\times 10^{-7}$	Ar $\times 10^{-4}$	Kr $\times 10^{-8}$	Xe $\times 10^{-8}$	N2/Ar	Argon 40/36 Water Sample	Argon 40/36 Gas Sample
WEIACH		**(WEI)**											
302/3-1 **	550.0	jm	DST	35.0		148	7.56	6.33	10.79	1.23			
302/10a-1	859.1	mo	AL	50.0		57	2.41	3.52	8.12	1.17	50.1		294.4 ± 1.5
302/10a-2	859.1	mo	AL	50.0		82	2.39	3.76	8.65	1.25	50.1		294.4 ± 1.5
302/10b-1	859.1	mo	AL	50.0	3.31	168	2.24	3.87	9.35	1.33			
302/10b-2	859.1	mo	AL	50.0	3.31	172	2.17	3.89	9.43	1.34			
302/12-1	985.3	s	AL	55.0	14.84	181000	3.46	4.82	10.46	1.54	96.5		306.4 ± 0.7
302/12-2	985.3	s	AL	55.0	14.84	151000	3.04	4.69	10.50	1.50	96.5		306.4 ± 0.7
302/19-1 **	1116.5	r	AL	59.0	37.42	7540	0.43	1.44	3.22	0.76	141.0	395.4 ± 3.9	
302/19-2 **	1116.5	r	AL	59.0	37.42	15300	0.46	1.21	2.67	0.58	141.0	395.4 ± 3.9	
302/20-1 *	1116.5	r	GTC	59.0	31.93	159000	3.19	3.97	6.42	1.01			
302/20-2 **	1116.5	r	GTC	59.0	31.93	159000	3.50	4.02	6.39	0.36			
302/18-1 **	1408.4	r	BAI	70.0	97.76	1450	2.45	1.12	0.93	0.09	182.0	375.9 ±10.2	
302/18-2 **	1408.4	r	BAI	70.0	97.76	2910	0.21	0.08	0.08	0.02	182.0	375.9 ±10.2	
302/16-1 **	2219.1	KRI	BAI	112.0	6.49	832	0.30	0.11	0.27	n.n.	87.5	296.8 ± 2.2	
302/16-2 **	2219.1	KRI	BAI	112.0	6.49	941	0.39	0.14	0.28	0.06	87.5	296.8 ± 2.2	
302/14-1 **	2267.0	KRI	BAI	112.0	7.38	7300	0.41	0.33	0.77	0.15	80.4	298.6 ± 2.8	
302/14-2 **	2267.0	KRI	BAI	112.0	7.38	1800	0.20	0.14	0.38	0.11	80.4	298.6 ± 2.8	
302/14-3 **	2267.0	KRI	BAI	112.0	7.38	941	0.39	0.14	0.28	0.06	80.4	298.6 ± 2.8	
RINIKEN		**(RIN)**											
303/1-1 **	515.8	km	AL	38.0	15.52	1180	1.09	2.07	3.66	0.33	77.4		
303/1-2 **	515.8	km	AL	38.0	15.52	12200	1.14	2.37	3.90	0.37	77.4		
303/1-3 **	515.8	km	AL	38.0	15.52	12700	1.24	2.42	3.96	0.37	77.4		
303/2-1	656.7	mo	AL	45.0	14.50	6790	6.55	5.95	13.04	1.74	58.7	303.0 ± 1.4	
303/2-2	656.7	mo	AL	45.0	14.50	3970	3.53	4.36	10.54	1.50	58.7	303.0 ± 1.4	
303/3-1	805.8	s-r	AL	50.0	10.50	110000	4.12	5.58	10.80	1.49	216.0	323.5 ± 2.0	
303/3-2	805.8	s-r	AL	50.0	10.50	111000	4.54	5.78	10.86	1.52	216.0	323.5 ± 2.0	
303/6-1 *	993.5	r	GTC	57.0	23.38	201000	2.40	4.00	7.50	1.07	212.9	356.8 ± 2.1	
303/6-2 **	993.5	r	GTC	57.0	23.38	43600	0.64	1.82	4.30	0.75	212.9	356.8 ± 2.1	
303/6-3 **	993.5	r	AL	57.0	23.38	41000	0.99	1.80	4.31	0.78	212.9	356.8 ± 2.1	
303/10c-1**	1709.3	r	BAI	84.0	9.20	5060	0.58	0.41	0.53	0.12	69.4	298.0 ± 2.7	
303/10c-2**	1709.3	r	BAI	84.0	9.20	2170	0.56	0.44	0.69	0.14	69.4	298.0 ± 2.7	
303/10c-3**	1709.3	r	BAI	84.0	9.20	2200	0.76	0.61	0.94	0.16	69.4	298.0 ± 2.7	
SCHAFISHEIM		**(SHA)**											
304/2-1 **	558.0	USM	GTC	30.0	8.71	466	0.14	0.30	0.92	0.22	79.4	357.3 ± 7.2	
304/2-2 **	558.0	USM	GTC	30.0	8.71	481	0.19	0.30	0.90	0.24	79.4	357.3 ± 7.2	
304/14-1 **	1251.2	mo	AL	57.5	18.24	661	0.39	0.73	2.22	0.61	43.9		
304/14-2 **	1251.2	mo	AL	57.5	18.24	1430	1.00	1.54	4.52	1.00	43.9		
304/3-1 **	1260.4	mo	AL	57.5	15.29	23700	4.18	6.46	12.26	1.72			
304/3-2 *	1260.4	mo	AL	57.5	15.29	10300	2.08	4.24	9.10	1.39			
304/3-3 **	1260.4	mo	AL	57.5	15.29	6920	2.30	3.59	8.07	1.33			
304/5-1	1488.2	s-KRI	GTC	68.0		145000	3.14	5.16	8.89	1.30	198.0	365.4 ± 3.1	
304/6-1 **	1488.2	s-KRI	BAI	68.0	16.05	5850	0.40	0.26	0.57	0.11	272.6		
304/6-2 **	1488.2	s-KRI	BAI	68.0	16.05	2030	0.57	0.31	0.50	0.08	272.6		
304/7-1	1488.2	s-KRI	GTC	68.0	14.85	148000	3.15	5.49	9.60	1.26			
304/8-1	1571.1	KRI	AL	72.0	8.33	202000	4.65	7.41	13.84	1.83	135.5	301.9 ± 1.3	
304/8-2	1571.1	KRI	AL	72.0	8.33	274000	6.64	10.65	19.07	2.41	135.5	301.9 ± 1.3	
304/9-1	1571.1	KRI	GTC	72.0	8.23	129000	3.42	5.77	11.22	1.64			
304/9-2	1571.1	KRI	GTC	72.0	8.23	149000	3.59	5.57	10.60	1.49			
304/10-1 **	1887.9	KRI	AL	85.0	8.58	28000	1.54	3.14	6.40	1.27	148.0	348.7 ± 2.3	
304/10-2 *	1887.9	KRI	AL	85.0	8.58	78100	2.54	4.74	8.37	1.45	148.0	348.7 ± 2.3	
304/10-3 *	1887.9	KRI	AL	85.0	8.58	91000	3.09	5.01	8.89	1.35	148.0	348.7 ± 2.3	
304/11-1 **	1887.9	KRI	GTC	85.0	8.33	175000	5.40	5.72	10.55	1.76	159.0		

Table 3.3.2: Noble gas concentrations of samples from the deep boreholes and the regional programme. (Page 3 of 4)

Sample/ NAGRA No.	Avg. Depth (m)	Geo- logy	Sam- pling Meth- od	GWT (°C)	Min. (g/kg)	He $\times 10^{-8}$	Ne $\times 10^{-7}$	Ar $\times 10^{-4}$	Kr $\times 10^{-8}$	Xe $\times 10^{-8}$	N2/Ar	Argon 40/36 Water Sample	Argon 40/36 Gas Sample
KAISTEN		(KAI)											
305/1-1	113.5	s	AL	16.0	6.85	29600	2.86	5.50	10.34	1.40	89.0	328.2 ± 2.7	
305/1-2	113.5	s	AL	16.0	6.85	30200	2.76	5.08	10.49	1.47	89.0	328.2 ± 2.7	
305/3-1 **	284.3	r	GTC	24.0	1.66	10100	17.44	5.30	10.88	1.69			
305/2-1	285.3	r	AL	24.0	1.64	9150	3.16	5.35	11.29	1.63	74.0*	308.6 ± 2.1	
305/2-2	285.3	r	AL	24.0	1.64	9290	3.37	5.29	11.04	1.60	74.0*	308.6 ± 2.1	
305/4-1	310.4	KRI	AL	25.0	1.58	11000	3.18	5.21	10.37	1.58	59.0	288.6 ± 1.6	
305/4-1	310.4	KRI	AL	25.0	1.58	11000	3.02	5.19	10.70	1.62	59.0	288.6 ± 1.6	
305/6-1	482.7	KRI	AL	30.0	1.57	10000	3.26	5.29	11.29	1.64	60.3	295.8 ± 1.3	
305/6-2	482.7	KRI	AL	30.0	1.57	10100	3.17	5.27	11.75	1.63	60.3	295.8 ± 1.3	
305/8-1	656.5	KRI	AL	36.0	0.00	9170	3.14	5.26	11.49	1.67	61.5	300.1 ± 0.7	
305/8-2	656.5	KRI	AL	36.0	0.00	9180	3.37	5.24	11.48	1.60	61.5	300.1 ± 0.7	
305/9-1	819.5	KRI	AL	47.5	1.45	10200	3.47	5.38	11.57	1.69	57.0	297.9 ± 1.8	294.1 ± 1.9
305/9-2	819.5	KRI	AL	47.5	1.45	9940	3.34	5.34	11.97	1.65	57.0	297.9 ± 1.8	294.1 ± 1.9
305/9-3	819.5	KRI	AL	47.5	1.45	9770	3.34	5.36	11.62	1.67	57.0	297.9 ± 1.8	294.1 ± 1.9
305/12-1 *	1031.0	KRI	AL	50.0	1.46	5140	3.27	5.52	9.06	1.74	80.2		
305/12-2 *	1031.0	KRI	AL	50.0	1.46	5220	3.37	5.44	10.20	1.68	80.2		
305/12-3 *	1031.0	KRI	AL	50.0	1.46	5290	3.28	5.37	10.95	1.65	80.2		
305/12-4 **	1031.0	KRI	AL	50.0	1.46	5080	3.24	5.43	11.34	1.71	80.2		
305/14-1 **	1153.3	KRI	BAI	54.0	1.55	157	1.55	0.98	2.27	0.36	74.5	293.8 ± 1.7	
305/15-1	1153.3	KRI	GTC	54.0	1.33	4580	3.25	5.37	10.30	1.86			
305/15-2	1153.3	KRI	GTC	54.0	1.33	1020	2.79	4.97	10.98	1.58			
305/16-1	1271.9	KRI	AL	58.0	1.46	5060	3.63	5.65	11.82	1.55	87.0	297.9 ± 0.7	
305/16-2	1271.9	KRI	AL	58.0	1.46	3530	2.75	4.84	9.98	1.95	87.0	297.9 ± 0.7	
305/16-3	1271.9	KRI	AL	58.0	1.46	2390	3.33	5.11	10.46	1.59	87.0	297.9 ± 0.7	
305/17-1	1271.9	KRI	GTC	58.0	1.48	955	3.47	5.41	11.72	1.48			
LEUGGERN		(LEU)											
306/1-1	75.0	mo	AL	13.0	1.16	465	2.54	4.12	9.41	1.32	43.4	296.6 ± 1.9	
306/1-2	75.0	mo	AL	13.0	1.16	453	2.54	4.06	9.03	1.42	43.4	296.6 ± 1.9	
306/2-1	217.9	s	AL	20.0	1.91	8820	3.05	5.07	10.65	1.66	54.1		
306/2-2	217.9	s	AL	20.0	1.91	8650	3.02	5.06	10.67	1.65	54.1		
306/2-3	217.9	s	AL	20.0	1.91	8290	3.01	4.97	9.59	1.56	54.1		
306/3-1 **	217.9	s	GTC	20.0		3740	4.33	5.35	11.33	1.49			303.4 ± 1.6
306/4-1 *	251.3	KRI	AL	20.0	1.25	2220	2.98	5.12	10.23	1.66	79.8		296.0 ± 1.6
306/4-2 **	251.3	KRI	AL	20.0	1.25	4820	33.30	20.91	29.95	2.90	79.8		296.0 ± 1.6
306/5-1 *	444.3	KRI	AL	25.0	1.12	4230	2.95	5.03	10.95	1.38	53.3	297.6 ± 1.2	
306/5-2 **	444.3	KRI	AL	25.0	1.12	4570	3.10	5.04	19.58	1.42	53.3	297.6 ± 1.2	
306/6-1	444.3	KRI	GTC	25.0	1.12	3590	4.20	5.25	11.17	1.54	57.0		
306/7-1	538.0	KRI	AL	28.0	1.07	4200	3.19	5.07	11.89	1.81	51.2	299.7 ± 0.6	
306/7-2	538.0	KRI	AL	28.0	1.07	6230	3.50	5.06	12.25	1.61	51.2	299.7 ± 0.6	
306/8-1 **	538.0	KRI	GTC	28.0	1.05	6940	33.96	4.24	12.44	1.64			
306/9-1	705.8	KRI	AL	34.0	1.16	8900	3.09	5.05	12.06	1.67	50.9	301.0 ± 1.3	
306/9-2	705.8	KRI	AL	34.0	1.16	7670	3.10	5.03	12.20	1.63	50.9	301.0 ± 1.3	
306/10-1 **	705.8	KRI	GTC	34.0	1.16	8310	8.95	4.24	12.31	1.55			
306/10-2 **	705.8	KRI	GTC	34.0	1.16	6470	15.13	5.03	12.06	1.61			
306/17-1 *	847.0	KRI	AL	40.0	1.10	5960	3.62	5.22	11.27	1.58	54.5		
306/17-2 **	847.0	KRI	GTC	40.0	1.10	5760	3.15	6.30	11.85	1.72	54.5		
306/11b-1**	923.0	KRI	BAI	42.0	4.33	552	6.46	3.92	5.67	0.59		292.3 ± 3.0	
306/11b-2**	923.0	KRI	BAI	42.0	4.33	572	1.12	0.77	1.21	0.19		292.3 ± 3.0	
306/12-1 **	923.0	KRI	BAI	42.0	4.59	1500	0.38	0.32	0.66	0.17			
306/12-2 **	923.0	KRI	BAI	42.0	4.59	1480	0.49	0.36	0.75	0.16			
306/13-1 **	923.0	KRI	BAI	42.0	4.61	5620	0.21	0.16	0.34	0.11	46.3		
306/20-1 **	1203.3	KRI	BAI	51.0	1.04	118	0.29	0.18	0.74	0.11	39.5		
306/20-2 **	1203.3	KRI	BAI	51.0	1.04	176	0.93	0.27	0.59	0.17	39.5		
306/21-1 **	1203.3	KRI	GTC	51.0	0.97	3760	3.76	6.12	8.35	1.22	69.5		
306/26-1 **	1433.4	KRI	BAI	59.0	1.45	288	0.38	0.31	0.68	0.17	53.7		
306/16-1	1643.4	KRI	AL	66.0	1.06	10900	3.18	5.26	11.47	1.79	68.7	298.4 ± 1.0	
306/16-2	1643.4	KRI	AL	66.0	1.06	11200	9.60	8.24	14.91	2.27	68.7	298.4 ± 1.0	
306/23-1	1665.6	KRI	GTC	66.0	1.04	5170	6.76	6.12	12.44	1.82			

Table 3.3.2: Noble gas concentrations of samples from the deep boreholes and the regional programme. (Page 4 of 4)

Site	(Nagra No.)	Geo-logy	GWT (°C)	Min. (g/kg)	He $\times 10^{-8}$	Ne $\times 10^{-7}$	Ar $\times 10^{-4}$	Kr $\times 10^{-8}$	Xe $\times 10^{-8}$	N2/Ar	Argon 40/36 Water Sample	Argon 40/36 Gas Sample
REGIONAL PROGRAMME												
Aqui	001/1-1*	tOMM	23.0	1.00	468	3.16	5.67	11.82	1.62			
Aqui	001/1-2**	tOMM	23.0	1.00	501	4.43	10.38	27.55	2.85			
Aqui	001/2-1*	tOMM			1350	3.12	5.32	11.73	1.57	58.6	296.3 ± 1.7	
Aqui	001/2-2*	tOMM			1360	3.05	5.25	11.37	1.52	58.6	296.3 ± 1.7	
Tiefenbrunnen	002/3-1	tOMM	25.4	3.64	5420	2.99	4.89	10.30	1.06	47.4		302.2 ± 1.5
Kaiseraugst	003/1-1**	s-r	18.4	12.84	16000	1.81	3.72	8.69	1.62			
Kaiseraugst	003/3-1*	s-r	17.0	12.96	27900	1.71	3.81	8.81	1.33	103.4		304.9 ± 1.2
Kaiseraugst	003 3-2*	s-r	17.0	12.96	27200	1.75	3.80	8.41	1.37	103.4		304.9 ± 1.2
Pratteln	003/2-1	mo	12.0	2.15	12	2.16	4.10	8.97	1.31	52.1	311.8 ± 3.0	295.0 ± 1.6
Frenkendorf	009/2-1	mo	18.4	0.73	17	2.73	4.63	10.14	1.36	60.6	313.2 ± 2.2	294.2 ± 0.8
Densbüren	014/3-1	mo	16.1	0.73	14	3.12	4.62	9.44	1.33	54.5		295.6 ± 1.0
Beznau	015/1-1	mo	20.5	6.48	4040	1.91	4.10	14.98	1.52			
Waldkirch	095/2-1	KRI	11.8	5.57	35100	2.65	5.04	10.96	1.54		308.8 ± 2.0	
Waldkirch	095/2-2	KRI	11.8	5.57	34700	2.70	5.03	11.15	1.57		308.8 ± 2.0	
Oberbergen	096/2-1**	tvu	19.2	0.41	27	2.11	3.36	7.54	1.04		295.2 ± 2.5	
Säckingen	125/1-1**	KRI	29.4	3.03	16700	3.31	5.16	13.49	2.02			
Rheinfeld.	159/4-1**	KRI	27.6	4.39	51700	3.25	4.43	7.79	1.43	102.4	317.1 ± 1.8	295.0 ± 4.1
Rheinfeld.	159/4-2**	KRI	27.6	4.39	53000	5.24	4.75	9.03	1.55	102.4	317.1 ± 1.8	295.0 ± 4.1
Rheinfeld.	159/4-3*	KRI	27.6	4.39	52900	3.48	5.10	11.05	1.61	102.4	317.1 ± 1.8	295.0 ± 4.1
Säckingen	161/1-1	KRI	28.0	7.33	27100	2.72	4.80	9.37	1.39	117.5	350.4 ± 3.4	318.5 ± 4.0
Säckingen	161/1-2	KRI	28.0	7.33	23900	2.35	4.60	9.27	1.38	117.5	350.4 ± 3.4	318.5 ± 4.0
Ennetbaden	201/1-1**	mo	47.2	4.53	1050	3.45	2.69	6.53	1.16			
Amsteg	231/1-1	KRI			56	2.35	3.95	9.02	1.26		295.5	
Amsteg	231/1-1**	KRI			93	8.61	6.31	11.99	1.42	60.4	300.0 ± 1.5	
Grimsel	233/1-2*	KRI			456	2.44	5.24	8.86	1.20	61.8	294.1 ± 0.9	
Lötschberg	234/1-1	KRI			4950	1.83	3.86	8.32	1.16	56.3	294.4 ± 1.1	
Magden Weiere	271/101-1	mo			56	2.99	4.38	10.08	1.55			
Magden Weiere	271/101-2	mo			79	3.51	4.42	10.41	1.56			
Magden Weiere	271/103-1	mo			26	2.65	4.41	9.86	1.66	43.6		
Magden Weiere	271/103-2	mo			25	2.61	4.38	9.87	1.59	43.6		
Magden Weiere	271/103-3	mo			31	2.75	4.41	9.86	1.96	43.6		
Magden Falke	272/102-1	mo			27	2.49	4.15	9.46	1.39		304.5 ± 1.6	
Magden Falke	272/102-2	mo			20	2.25	4.01	9.29	1.36		304.5 ± 1.6	
Reichenau	501/ 1)	Quaternary			131	2.06	5.40	12.70	1.75		302.0 2)	
Birnau	508/ 1)	Quaternary			26	2.94	4.96	11.50	1.68		302.0 2)	
Mainau	128/ 1)	OMM			949	3.47	5.28	12.50	1.73		302.0 2)	
Singen	097/ 1)	Malm			2490	4.00	5.88	12.70	1.66		302.0 2)	

1) Data from BERTLEFF (1986)
2) Value assumed for calculations

4. DATING BY RADIONUCLIDES

H. H. Loosli, B. E. Lehmann, and G. Däppen

4.1 Introduction

Groundwater dating based on radioisotope measurements makes use of the fact that the initial concentration c_0 in recharging groundwater decreases with time according to the decay-equation

$$c(t) = c_0 \cdot e^{-\ln 2 \, (t/T_{1/2})}$$

where $T_{1/2}$ is the half-life of the radioisotope.

In an ideal situation, the initial concentration c_0 is well known, no processes other than decay change the isotope concentration as the water mass moves through an aquifer, and the half-life of the selected isotope is of the same order of magnitude as the time that has passed since infiltration. Such a situation is hardly ever realized in nature.

Because each water molecule in a groundwater sample has its individual history, a single groundwater age does not strictly exist. That is, any given groundwater represents a mixture of waters with a distribution of ages, and it is generally not possible *a priori* to know that distribution. Therefore, only model ages based on hypothetical age distributions can be calculated.

Models for two extreme distributions are in common use. The piston flow model (PFM) assumes unmixed flow in a closed system, as in a pipe, so that all water at a given location has virtually the same age and the age of water in a given reservoir varies with space. The exponential model (EM) assumes that the age distribution is described by an exponential law so that young water is present more frequently than older water. A mean residence time can be calculated. It can be shown that such a exponential age distribution is also obtained if a perfectly well mixed reservoir is assumed. Other intermediate distribution models also exist, and usually it is difficult to decide which model is closest to reality (see, *e.g.*, ZUBER, 1986).

Each isotope method is most applicable within a certain time range and each has its limitations. The main reason for the limitation is that processes other than ageing may change the isotope concentration. These processes include:

- Underground production, which can add to the concentration originating from atmospheric sources. This process is particularly important for ^{36}Cl, ^{37}Ar and ^{39}Ar, as well as for such stable nuclides as ^4He and ^{40}Ar;

- Interactions of the water with the rock, such as dissolution, precipitation, sorption and exchange which can change isotopic concentrations. This process particularly affects ^{14}C and ^{36}Cl;

- Uncertainty about the input concentration c_0, which may be poorly known or vary with time as in the case of ^{3}H, ^{14}C, ^{36}Cl and ^{85}Kr.

Such limits of applicability for individual isotopes can be reduced by combining different dating methods. Furthermore, groundwater dating is not the only goal for using radioisotopes in hydrology. Results of isotope studies help to identify similarities or differences of water bodies. Valuable information about a variety of processes including infiltration conditions, mixing of different types of water, rock-water interactions, mechanisms of escape of nuclides from the solid phase into water and migration of radionuclides can be gained.

Furthermore, these different processes are important for evaluating possible transport mechanisms of isotopes originating from a nuclear waste disposal site into an aquatic system (analog studies). Again, a combination of different methods increases the reliability of conclusions.

Extensive information about the state of the art of the different groundwater dating methods is given in the literature (BALDERER, 1983; INTERNATIONAL ATOMIC ENERGY AGENCY, 1983a and b; FRITZ AND FONTES, 1980, 1986). In the following we summarize some of the main aspects of each isotope as a guideline to the more detailed discussions in the respective sections.

4.2   Overview of the Sources of the Individual Isotopes

There are three sources of radioisotopes and radiogenic stable isotopes to groundwaters. The two natural sources are production by cosmic rays in the atmosphere and by natural decay and activation processes in the subsurface. The third source is anthropogenic production, by reactor operation and weapons testing, for example. The relative importance of these three sources to concentrations measured in groundwater is indicated in Figure 4.2.1. The noble gas isotopes ^{3}He, ^{4}He, and ^{40}Ar, which are radiogenic but not radioactive, are also included. The qualitative classifications "main source", "possibly important source" or "negligible source" that are used may of course not hold for specific samples, but do give a general weight for an average situation.

Some general comments on the sources of individual isotopes are:

1)   Tritium ^{3}H (half-life 12.3 a) was produced in large amounts in the thermonuclear tests of the late 1950s and early 1960s; its concentration in the atmosphere has decreased since then. Local industrial sources may add some ^{3}H to the concentrations measured in precipitation and surface water. Subsurface production could take place by the reaction $^{6}Li(n,\alpha)^{3}H$, but the amount so produced is generally negligible (see Section 4.3);

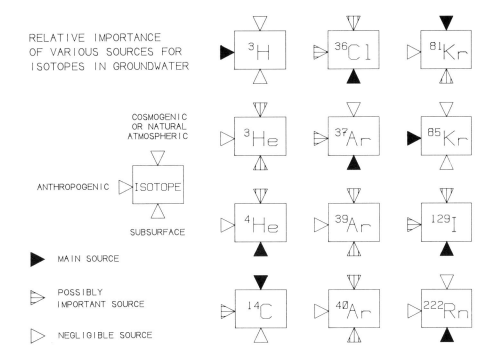

Figure 4.2.1: Relative importance of various sources of isotopes in groundwaters.

2) <u>Helium-3</u> (stable) is present in the atmosphere with a concentration of 1.38 ppm of the natural helium. It dissolves in groundwater in the recharge zone with the same ^{3}He to ^{4}He ratio. ^{3}He can be produced from the reaction ^{6}Li(n,α)^{3}H when ^{3}H decays by β-decay to ^{3}He. Another source of ^{3}He is gas coming from the earth's mantle. Groundwaters with elevated levels of ^{3}He may have accumulated helium over extended periods of time (see Section 6.4);

3) <u>Helium-4</u> (stable) dissolves in groundwater as a natural component of atmospheric air and is produced in the subsurface as neutralized α-particles from the natural decay series in the rock. Very often the measured ^{4}He concentrations in groundwater are orders of magnitude higher than the levels of saturation with modern air at recharge. Subsurface produced ^{4}He must then have been added locally or have migrated from other layers (see Section 6.4);

4) <u>Carbon-14</u> (half-life 5730 a) is produced by cosmic rays in the atmosphere by the reaction ^{14}N(n,p)^{14}C. Nuclear weapons tests increased the atmospheric level by a factor of about two, but it has decreased now to a level of approximately 120 per cent of that of pre-bomb times. Subsurface production of ^{14}C could occur mainly by the reaction ^{17}O(n,α)^{14}C in rocks, but is generally negligible. However, the carbonate system of groundwater may be changed considerably in the underground which makes necessary the application of correction models for dating purposes (see Chapter 5);

5) <u>Chlorine-36</u> (half-life $3 \cdot 10^5$ a) is produced in the atmosphere (by cosmic rays) in spallation reactions mainly from ^{40}Ar. Very large amounts were also produced by weapons testing. This transient bomb spike is highly specific for waters of this age, and is, therefore, used for their dating. In deep groundwaters, ^{36}Cl levels are very often dominated by ^{36}Cl originating from neutron activation of ^{35}Cl either in the rock or in dissolved chloride (see Section 6.2);

6) <u>Argon-37</u> (half-life 35 d) is produced by cosmic rays but generally only in negligible amounts. ^{37}Ar is produced in the subsurface by ^{40}Ca(n,α)^{37}Ar reactions and may escape to the atmosphere after underground nuclear weapons tests. Because of its short half-life, ^{37}Ar of atmospheric origin is usually not found in groundwaters. However, large amounts of ^{37}Ar have been measured in some groundwaters. Those from U-rich formations with high neutron fluxes, and, therefore, high production rates may have high ^{37}Ar contents. Ca-rich fracture fillings in close contact with water allow easy escape of ^{37}Ar from rock into water and also lead to high ^{37}Ar levels (see Section 6.3);

7) <u>Argon-39</u> (half-life 269 a) is produced by cosmic rays in spallation reactions in the atmosphere. In formations with low U-content (carbonate rocks, some sandstones) the decay of this atmospheric component may be used for dating. In groundwaters from such high-uranium rock as granite, elevated levels of ^{39}Ar have been found because it is produced underground by ^{39}K(n,p)^{39}Ar reactions. Anthropogenic sources are known to be below ten per cent of the cosmic ray produced level (see Sections 4.3 and 6.3.);

8) <u>Argon-40</u> (stable) is dissolved from the atmosphere during recharge. In subsurface waters, the measured ^{40}Ar/^{36}Ar ratio is sometimes higher than the atmospheric value of 295.5, indicating that ^{40}Ar from the decay of ^{40}K in rocks and water has accumulated in the water over possibly extended periods of time (see Section 6.5);

9) <u>Krypton-81</u> (half-life $2.1 \cdot 10^5$ a) is produced by cosmic rays in spallation reactions on stable Kr isotopes in the atmosphere and in n-capture by ^{80}Kr. These are considered to be the main sources of ^{81}Kr although subsurface produced ^{81}Kr could possibly contribute additional levels in certain U-rich formations. Although a number of samples have been taken, only a few results have been reported up to now (THONNARD AND OTHERS, 1987). Thus, a dating method for ^{81}Kr is not yet established;

10) <u>Krypton-85</u> (half-life 10.76 a) is released to the atmosphere mainly when reactor fuel rods are opened for reprocessing or plutonium separation. The atmospheric activity is increasing. Young waters that contain both ^{3}H and ^{85}Kr can be dated by taking advantage of the different input functions of the two radioisotopes. Cosmogenic production is negligible and subsurface production by spontaneous fission of ^{238}U has only been identified once in a U-rich granite (see Section 4.3);

11) <u>Iodine-129</u> (half-life $1.57 \cdot 10^7$ a) is produced by cosmic rays, and is also released in fuel-reprocessing plants. Levels in deep groundwaters seem to be dominated

by leaching ^{129}I from the rock, where it has been produced by spontaneous fission of ^{238}U (FABRYKA-MARTIN AND OTHERS, 1989). Because activity measurements are not easy and the interpretation of ^{129}I results is difficult, this method is only rarely used, and was not attempted in these Nagra studies;

12) Radon-222 (half-life 3.8 d) is a decay product of the natural ^{238}U-decay series. Levels in groundwaters are variable and depend on various factors such as the microgeometry, the porosity and the distribution of uranium and its daughter products in rock (see Section 7.3);

13) The ^{234}U/^{238}U method is discussed in Chapter 7.3.

4.3 Detection of Young Water Components With ^{3}H and ^{85}Kr

In this chapter, as elsewhere in this report, "young water" refers to water containing ^{3}H and ^{85}Kr as a signal of weapons testing and nuclear plant operation. "Very young water" is younger than two years.

The present ^{3}H concentration in precipitation is several orders of magnitude lower than its peak during the period of atmospheric weapons testing and it continues to decrease. On the other hand, the ^{85}Kr activity in the atmosphere is increasing (Figures 4.3.1 and 4.3.2). In the years when most samples from Nagra boreholes were collected, the ^{3}H activity in precipitation was about 20 to 50 TU (about 2 to 5 Bq/l) and the ^{85}Kr activity in the air was about 35 to 45 dpm/cc krypton (about 0.7 to 0.8 Bq/m^3 air). The average measured ^{3}H content of drilling fluid was about 90 TU (see Section 2.1.1). The application of ^{3}H becomes increasingly difficult because its activity showed a smaller decrease rate in the last years than previously, but a combination of both methods applying the ^{3}H/^{85}Kr ratio seems most promising. For example, this ratio increases with increasing mean residence time as shown in Figure 4.3.3 for the piston flow model and in Figure 4.3.4 for the exponential model. The ratio is about 0.5 to 1.5 TU/(dpm/cc Kr) for very young water. The gradient with time for the ratio is larger than the gradients of the activities of either isotope separately. This larger sensitivity is clearly visible for young waters with mean residence times of up to 30 years, based on the exponential model.

Measurement of ^{3}H and ^{85}Kr is not too difficult, although low ^{3}H activities (below about 10 Bq/l) make enrichment necessary (WOLF AND OTHERS, 1981). Samples for the measurement of ^{85}Kr activities are taken from the large gas samples collected for ^{39}Ar analyses. Therefore, there is enough krypton gas and ^{85}Kr activity available to measure levels as low as about 0.1 dpm/cc Kr with sufficient precision. Local spatial and temporal variations are larger than the analytical uncertainty, and increase the total uncertainty of recent input values to about 50 per cent for ^{3}H (KUER, 1989) and to about 20 per cent for ^{85}Kr (LUDIN, 1989).

Both isotopes can also be used for detecting possible contamination of a water sample by a very young component or, in combination, *e.g.*, with the O_2 content, to detect contamination of the extracted gases by ambient air (see Section 2.1). If the ^{3}H/^{85}Kr ratio is about one TU/(dpm/cc Kr) and (or) if the O_2 content in an extracted gas sample

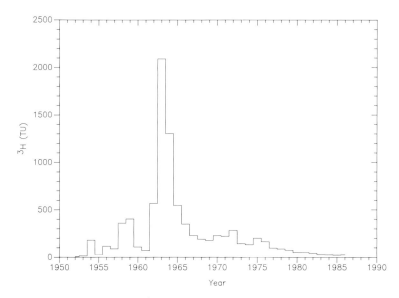

Figure 4.3.1: Measured ^{3}H activity in precipitation of Switzerland (SIEGENTHALER AND OTHERS, 1983; LUDIN, 1989). Practically all activity originates from nuclear weapons tests.

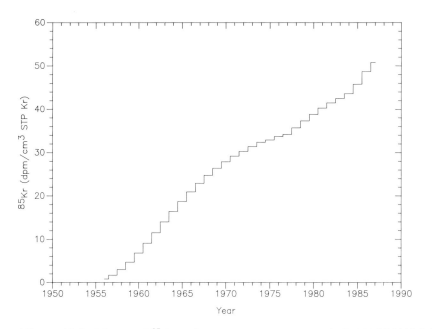

Figure 4.3.2: Measured ^{85}Kr activity in atmospheric air (ZIMMERMANN AND OTHERS, 1989; LUDIN, 1989). All activity is man-made, mainly originating from nuclear fuel reprocessing plants.

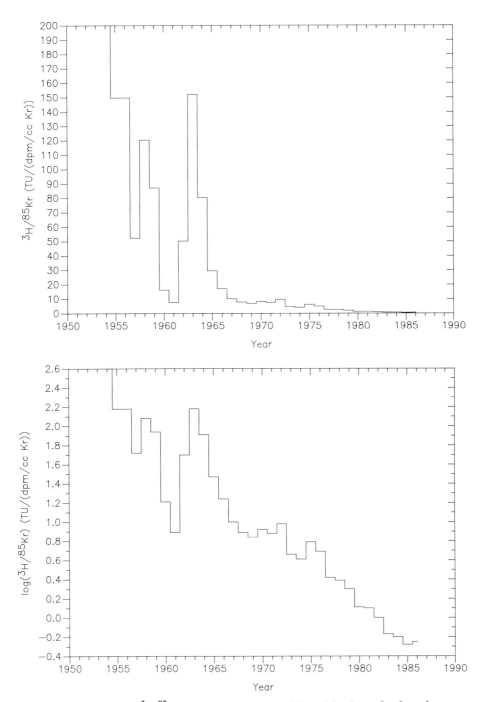

Figure 4.3.3: Calculated ^{3}H/^{85}Kr ratios on linear and logarithmic scales based on a piston flow model. Activities are corrected to 31-Dec-1984.

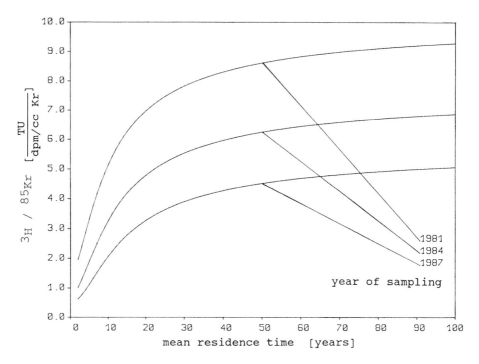

Figure 4.3.4: Calculated $^3H/^{85}Kr$ ratio as a function of the mean residence time of an exponential age distribution of water components. Calculation is done for three sampling dates.

is considerable, then contamination by or admixture with a very young water is probable. However, if the $^3H/^{85}Kr$ is larger than one TU/(dpm/cc Kr) and no O_2 is found, the assumption of a young but not very young water component may be appropriate (see Figure 4.3.4).

In addition, it is important to note that equally important conclusions about the <u>absence</u> of a very young or a young water component can be drawn from ^3H and ^{85}Kr measurements. Many samples showed ^3H activity values below one or two TU, and ^{85}Kr values below 0.3 dpm/cc Kr. Samples with admixtures of very young or young water below two to three per cent are given in Table 4.3.1 (see also Table 2.1.1). Most samples in Table 4.3.1 have ^3H values close to, or below the detection limit, although a few are included with low tritium values of around 2 to 3 TU. Most also have low ^{85}Kr activities ($\leq$ 0.2 dpm/cc Kr). Some samples included have considerable ^{85}Kr activities. This activity is due to contamination of the gas sample. The conclusion that they have a low young water component is based on ^3H only.

Table 4.3.1: Samples with ^{3}H and ^{85}Kr concentrations indicating very young or young water contents below two to three per cent.

Location Sample	Name	Depth (metres)	Aquifer	Date Collected
1/1	Aqui	300-500	tOMM	21. 7.81
2/3	Tiefenbrunnen	330-716	tOMM	30.10.82
9/2	Frenkendorf	230-250	mo	8. 2.84
15/1	Beznau	203-303	mo	22. 7.81
271/101	Magden Weiere	213-282	mo	13. 9.83
271/103	Magden Weiere		mo	13. 6.85
271/104	Magden Weiere		mo	28.10.86
272/102	Magden Falke	165.5-235	mo	22.11.85
272/103	Magden Falke		mo	24.10.86
273/101	Magden Eich	191-248	mo	10. 2.86
273/102	Magden Eich		mo	15. 5.86
273/103	Magden Eich		mo	26.10.86
301/2	Böttstein	123.2-202.5	mo	1.11.82
302/10a /10b	Weiach	822-896.1	mo mo	3. 4.83
303/2	Riniken	617.3-696	mo	17. 8.83
302/12	Weiach	981-989.6	s	19. 7.83
305/1	Kaisten	97-129.9	s	22. 2.84
306/2	Leuggern	208.2-227.5	s	8. 8.84
303/3	Riniken	793-820	s-r	16. 9.83
305/2	Kaisten	276-292.5	r-KRI	1. 3.84
301/5	Böttstein	305.6-327.6	s-KRI	16.11.82
302/19	Weiach	1109.2-1123.8	r	27. 6.84
303/6	Riniken	977-1010	r	1.11.83
301/18	Böttstein	782-802.8	KRI	16. 8.83
301/12b	Böttstein	618.5-624.1	KRI	22. 1.83
305/4	Kaisten	299.3-321.5	KRI	15. 3.84
306/16	Leuggern	1637.4-1649	KRI	15. 2.85
306/9	Leuggern	702-709.5	KRI	17.10.84

The following discussion addresses samples with very young or young water components present in amounts greater than a few per cent. These are listed in Table 4.3.2. The assumptions are:

1) The water consists of an old and a young component;

2) ^{3}H and ^{85}Kr activities result entirely from the young component and not from contamination by air or drilling fluid;

3) The age structure of the young component may be described by the exponential model (EM) or by piston flow (PFM).

In the following discussion there is no attempt to divide the very young component into drilling fluid and natural very young water. Waters showing sample contamination (of gas or water) are not treated.

To calculate the fraction of the young component one has to remember that the ^{3}H/^{85}Kr ratio is determined only by its mean residence time. From the input functions of both isotopes, the ^{3}H and ^{85}Kr activities of water can be calculated as a function of the mean residence time and the date of sampling (Figures 4.3.5 and 4.3.6 for the EM). These activities are then compared with the measured activities which allows determination of the fraction of the young component. It is important to recall that input values adopted include considerable uncertainties. These uncertainties are not included in the errors given below.

<u>306/1 Leuggern 53.5 to 96.4 m, Muschelkalk, 18-Jul-1984:</u>

The measured ^{3}H/^{85}Kr ratio of 3.5 (±6 per cent) TU/(dpm/cc Kr) indicates that some water is young with an EM mean residence time between 10 and 15 years (Figure 4.3.4). The amount of this young water is calculated as follows: If all the water were 10 to 15 years old, its ^{3}H activity would be about 75 TU and its ^{85}Kr activity about 20 dpm/cc Kr (Figures 4.3.5 and 4.3.6). The ^{3}H and ^{85}Kr concentrations actually measured are lower by a factor of seven. From this it can be estimated that the young component represents only about 15 per cent of the sample. The rest of the water must be older, with ^{3}H and ^{85}K contents below the detection limit.

The assumption that the young component can be described by a piston flow model yields an age of about 8 to 12 years; the amount would also be about 15 per cent (Figures 4.3.3, 4.3.1 and 4.3.2).

The amount of young waters may have been variable during sampling, because during this time an increase of the ^{3}H activity from 7 to 11 TU was observed.

<u>008/2 Pratteln 53 to 124 m, Muschelkalk, 7-Feb-1984:</u>

The ^{3}H/^{85}Kr ratio is 2.7 (±7 per cent) TU/(dpm/cc Kr). The EM leads to a young component with a mean residence time of about eight years, and a fraction

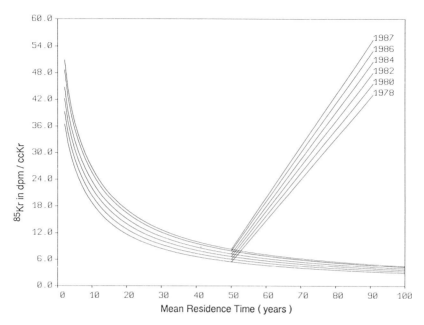

Figure 4.3.5: Calculated ^{85}Kr activity of water *versus* mean residence time based on the exponential model (LUDIN, 1989).

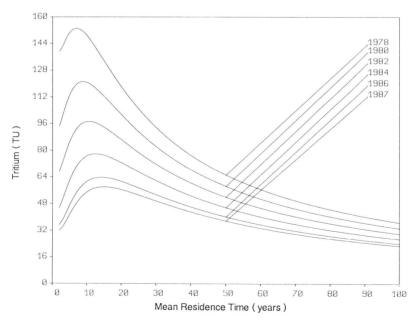

Figure 4.3.6: Calculated ^{3}H activity of water *versus* mean residence time based on an exponential model (LUDIN, 1989).

of about 12 per cent. If we assume that the measured O_2 content is due to a contamination by air, the $^3H/^{85}Kr$ ratio increases to a value of 3.2 TU/(dpm/cc Kr). However, this does not significantly change the amount and age distribution.

014/3 Densbüren 38 to 125 m, Muschelkalk, 10-Feb-1984:

The $^3H/^{85}Kr$ ratio of 9.9 (±7 per cent) TU/(dpm/cc Kr), is one of the largest ratios and is at the limit of applicability of the EM. Based on the lower limit of the ratio (value minus 2 σ), a mean residence time of more than 150 years is estimated; the amount of this water component would be 40 per cent or more. The ^{39}Ar result agrees with this conclusion and indicates that the measured ^3H, ^{85}Kr and ^{39}Ar activities may be explained by the same age distribution (see Section 4.4.2). However, a PFM description of the young component is probably more adequate, the age would be about 15 years and the fraction of young water about eight per cent. Waters with high $^3H/^{85}Kr$ values are discussed further at the end of this section.

123 Lostorf 3 530 to 584 m, Muschelkalk, 10-Oct-1979:

The ^{85}Kr result is very uncertain because in earlier days only activities in dpm and not in dpm/cc were measured. From estimates of the extraction and measuring yields, we can assume a probable range between 0.3 and 0.9 dpm/cc. The ^3H result also has a relatively high error so that no $^3H/^{85}Kr$ ratio can be given. If a young water component is assumed to add the ^3H and ^{85}Kr activity measured, its amount must be estimated to be smaller than about five per cent.

274/102 Magden Stockacher, Muschelkalk, 27-Oct-1986:

The $^3H/^{85}Kr$ ratio is 7.7 (±9 per cent) TU/(dpm/cc Kr) corresponding to an EM mean residence time larger than 50 years. Again the ^{39}Ar result indicates that the measured ^3H, ^{85}Kr and ^{39}Ar activities may be explained by assuming an age distribution with one mean residence time of about 500 years (see Section 4.4.2). If the PFM is adopted, a young component of about five per cent and an age of about 15 ± 5 years can be calculated.

301/23 Böttstein 649.0 m, Crystalline, 17-Oct-1983:

The $^3H/^{85}Kr$ ratio amounts to 11.4 (±13 per cent) TU/(dpm/cc Kr). This result is unexpected even if one takes into account that this sample is not perfectly reliable (Block 2, see Chapter 2) and that it contains an admixture of a young water used as drilling fluid. Only if one uses the calculated value minus 2-σ error is the applicability of the exponential model not exceeded. This limit allows the conclusion that the mean residence time of the younger EM water component is larger than 100 years and that its fraction is larger than ten per cent. Therefore, the PFM assumption may be more adequate. The PFM age reaches from 11 to 17 years and the corresponding amount of water is between 1.5 and 2.5 per cent.

Table 4.3.2: Isotope results for discussion of a possible young water component (more than a few per cent).

LOCATION SAMPLE	DATE	NAME	AQUI-FER	DEPTH (m)	39AR (% mod)	O2 IN EXTRACTED GAS SAMPLE (vol.-%)	85KR (dpm/cc Kr)	3H (TU)	AR IN EXTRACTED GAS SAMPLE (vol.-%)
306/1	18. 7.84	Leuggern	mo	53.5-96.4	53.0 ± 3.7	0.28 0.20	3.42 ± 0.13 3.27 ± 0.14	11.6 ± 1.1	1.84 1.94
008/2	7. 2.84	Pratteln	mo	53-124	255 ± 11 249 ± 11	0.80	3.42 ± 0.22	9.26 ± 0.8	1.82
014/3	10. 2.84	Densbüren	mo	38-125	63.6 ± 3.5	traces	0.79 ± 0.03	7.8 ± 0.9	1.71
123	10.10.79	Lostorf 3	mo	530-584	52.5 ± 4.2	traces	0.3 - 0.9	1.7 ± 2.0	---
274/102	27.10.86	Magden Stockacher	mo		41.9 ± 3.1	0.02	0.609 ± 0.037	4.7 ± 0.7	1.57
301/23	17.10.83	Böttstein	KRI	649.0	510 ± 12	0.10	0.28 ± 0.02	3.2 ± 0.7	1.65
125/1	22. 7.81	Säckingen Badquelle	KRI	81.9-201.3	224 ± 7	1.48	10.20 ± 0.30	46.5 ± 2.0	0.73

125/1 Säckingen Badquelle 81.9 to 201 m, Crystalline, 22-Jul-1981:

> The high ^{3}H and ^{85}Kr activities lead to the conclusion that the relatively high O$_2$ concentration in the extracted gas is due to young water and not caused by air contamination of the gas sample. This is in agreement with the observation that water from Säckingen contains a considerable amount of oxygen (see Table 2.1.1). Therefore, ^{85}Kr values need not be corrected for air contamination.
>
> The results of the EM concerning the young component are: $\tau \approx 6$ years, young water fraction ≈ 50 per cent.

It is important to consider the reliability of the high ^{3}H/^{85}Kr ratios measured in the Densbüren, Stockacher and Böttstein 301/23 samples. The errors given in the discussion of these samples include only the uncertainties in the measurements of the isotope activities and the krypton concentrations, but do not include such additional errors as those of the input function. For example, ^{3}H shows a considerable seasonal variation with values higher in the summer than averaged over the entire year. On the other hand, any contamination during collection or preservation of a sample would increase its ^{85}Kr activity, and could not be the reason for a high ^{3}H/^{85}Kr ratio.

The EM can be applied to samples with high ^{3}H/^{85}Kr ratios only if long mean residence times are assumed for the entire sample. The age distribution would then be very flat. It would include components as young as two or three years, but each would contribute only a small proportion of the total. For Densbüren and Stockacher, ^{39}Ar results permit the quantitative determination of the age distribution (Section 4.4).

On the other hand, applying the PFM to these three waters yields a mixture of two to ten per cent young water and an old component. The PFM model age of the old component can be determined with ^{39}Ar. From the results of these three isotopes, it is not possible to decide which model better describes reality even though the EM is at the limit of applicability. Many more models could be assumed between the idealized PFM and EM, which could also explain the measured isotope results (ZUBER, 1986). For instance, both models can be combined leading to a delayed exponential distribution. In such a model, the most probable age of the exponential distribution would not be very young waters of zero to two years but older waters of some assumed age. This model would increase the ^{3}H content and decrease the ^{85}Kr content, and could explain the high ^{3}H/^{85}Kr ratios. Also dispersion flow models could be used. However, this enlargement of possible model assumptions is not restricted to the above mentioned three water samples. No matter what modelling assumptions are made, it is certain that if some ^{3}H and (or) ^{85}Kr activities are measured, a component with relatively short residence time must be present and responsible for introducing these short-lived activities.

4.4 ^{39}Ar Results for Waters from Sedimentary Formations

Basic information about the application of ^{39}Ar for dating in hydrology is given by LOOSLI (1983), ANDREWS AND OTHERS (1983), and FORSTER AND OTHERS (1983). Application of this dating method is complicated by the fact that, in some rock formations, a considerable fraction of the measured ^{39}Ar activity is produced underground. This occurs mainly in crystalline rocks and is discussed in Section 6.3. In this chapter, only results obtained in samples from sedimentary formations down to the Muschelkalk are presented and discussed. The application of ^{39}Ar for studying mixing and ageing aspects seems possible in these zones. The ^{39}Ar results are combined with ^{3}H, ^{85}Kr, ^{14}C, δ^{13}C and other isotope results to reach a more consistent interpretation. Results are given in Table 4.4.1 at the end of this chapter.

4.4.1   General Conclusions from the ^{39}Ar Results

^{39}Ar in groundwater may originate partly from a young water component, which would be indicated by the presence of short-lived isotopes like ^{3}H and ^{85}Kr, as discussed above. Therefore, the measured ^{39}Ar activities need to be corrected for any contribution from young water before dating of an older PFM component can be attempted. This correction is based on a simple activity balance, assuming that the same amount of argon is dissolved per litre of water in the young and the old components (A = Activity):

$$A_{measured} = F_{young} \cdot A_{young} + (1 - F_{young}) \cdot A_{old}$$

where F_{young} is the fraction of young water and A_{young} the decay-corrected activity of this young fraction. A_{old} can be calculated from this equation.

Table 4.4.1 gives corrected ^{39}Ar values together with the ^{39}Ar activities originally measured and the available O_2, ^{3}H and ^{85}Kr results. Usually, application of these three tracers yields consistent estimates of possible contamination, or of a young water component. This is the case, for example, for the samples from the Muschelkalk at Weiach (302/10a, b), Leuggern (306/1), Pratteln (8/2) and Densbüren (14/3), where it is possible that the young water which introduces some ^{3}H and ^{85}Kr activities had already lost part or most of its O_2 content. For the other samples, these corrections are small and the ^{39}Ar values change little. A satisfactory, consistent correction is not possible for Frenkendorf (9/2). Apparently only the extracted gas is contaminated because the ^{3}H value in the water is very low. However, the correction using the O_2 result is larger than when applying the ^{85}Kr result. Therefore, for this sample, only an upper limit can be given for the corrected ^{39}Ar value (based on ^{85}Kr).

Underground production of ^{40}Ar by ^{40}K decay dilutes the specific ^{39}Ar activity in the water. The measured ^{39}Ar activity, which is given in per cent modern and which represents the ratio of ^{39}Ar to total argon in the atmosphere, is too low in waters with considerable excess ^{40}Ar. A correction is possible using the measured ^{40}Ar/^{36}Ar ratios, which reflect the dilution effect of ^{40}Ar produced underground. The ^{39}Ar/Ar ratio, A_0, of the unaffected dissolved argon can be calculated from the measured ^{39}Ar/Ar ratio A_m by the formula

$$A_0 = A_m \cdot \frac{R_m}{R_0} = A_m \cdot \frac{R_m}{295.5}$$

where R_m is the measured $^{40}Ar/^{36}Ar$ ratio.

The ^{39}Ar values given as "corrected" in Table 4.4.1 include the correction for ^{40}Ar produced underground, if an $^{40}Ar/^{36}Ar$ value is available.

For most samples, the differences between uncorrected and corrected ^{39}Ar results are small or negligible. Thus, most of the following conclusions depend very little on whether or not a correction is made. This shows that contamination of the gas samples by modern air is usually negligible for the interpretation of ^{39}Ar results, and that sampling from extremely large volumes of waters can be done carefully enough for this method to be useful.

An important observation is that of all the ^{39}Ar results given in Table 4.4.1, three of the Muschelkalk samples and both waters from the Upper Marine Molasse show activities below the present limit of measurement, i.e., below about seven per cent modern. This indicates that subsurface production in these geological formations can probably be neglected (compare Section 6.3 for underground production rates of ^{39}Ar and for waters from the crystalline, Permian and adjacent Buntsandstein). This conclusion is qualitatively supported by the Muschelkalk sample from Weiach (302/10) in which the ^{39}Ar activity is about 13 per cent modern and for which a rather short residence time or a considerable young water component must be deduced from the ^{14}C activity and the relatively high ^{13}C content. A similar conclusion may even be drawn from the ^{39}Ar, ^{14}C, (and $\delta^{13}C$) results of the Magden samples (271, 272, 273, 274). Therefore, dating by ^{39}Ar may be possible for waters from the Muschelkalk and Molasse. The number of samples, however, is still too small to absolutely exclude that ^{39}Ar activity produced underground can always be neglected, especially if the measured activity is below about 20 per cent modern (see Section 6.3).

An exceptional ^{39}Ar result is obtained for Pratteln (8/2). The activity of this sample clearly originates from subsurface production and would be an indication of an admixture of another water type to Muschelkalk water, e.g., from underlying Permian or crystalline. This assumption is supported for Pratteln by the exceptionally high Cl content of this water (compared with Frenkendorf (9), for example). In addition, its Mg content is too high to result from dolomite dissolution as in all other Muschelkalk samples. A recent duplicate sampling and analysis confirms the ^{39}Ar activity result and this conclusion.

To test the consistency of isotope results, the corrected ^{39}Ar results are compared with uncorrected ^{14}C values in Figure 4.4.1 and with ^4He concentrations measured in the water in Figure 4.4.2. From Figure 4.4.1 it can be seen that samples with very low ^{14}C values show also ^{39}Ar activities below ten per cent modern or close to the detection limit of five to seven per cent. Lostorf 3 (123), Densbüren (14) and Stockacher (274) contain considerable activities of both nuclides ($^{14}C \approx 38$ to 44 pmc, $^{39}Ar \approx 40$ to 60 per cent) and, therefore, are not very old waters. Samples between these extremes show similarly

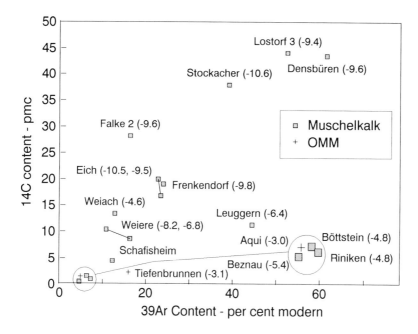

Figure 4.4.1: Corrected ^{39}Ar and uncorrected ^{14}C activities of samples from sediments. Measured δ^{13}C values are in parentheses.

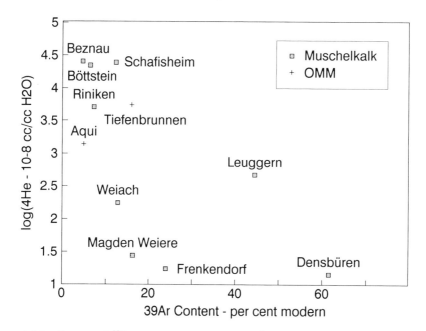

Figure 4.4.2: Corrected ^{39}Ar activity and measured ^{4}He concentrations of samples from sediments.

varying ^{14}C and ^{39}Ar activities. A more detailed discussion of the individual groundwaters follows in Section 4.4.2; the figure, however, presents a qualitatively consistent picture.

Figure 4.4.2 shows that, in general, samples with high ^{4}He contents have low ^{39}Ar activities. The situation of Leuggern (306/1) in this picture and to a smaller degree of Schafisheim (304/14) and Tiefenbrunnen (2) may indicate a mixing situation between an old and a younger component.

Some ^{40}Ar/^{36}Ar ratios were measured on the gas samples collected for ^{39}Ar measurements, and some on water samples collected in copper tubes. As explained in Section 6.5, the results obtained from water samples are more reliable and are to be used if both types of results are available. The ^{40}Ar/^{36}Ar results given in Table 4.4.1 do not show any systematic correlation with the corrected ^{39}Ar results and, therefore, cannot be used for a better age determination of the waters with ^{39}Ar.

4.4.2 Conclusions from Corrected ^{39}Ar Results about Residence Times of the "Old" Water Components

The mean residence times calculated from the corrected ^{39}Ar results are given and discussed in Table 4.4.2. These results refer only to an old component. Information about a young component has already been obtained from ^{3}H and ^{85}Kr results, and discussed in Section 4.3. It is assumed that all ^{3}H and ^{85}Kr in these samples are contributed by the young component. If a young component is present, its contribution to the ^{39}Ar activity, but not to the ^{14}C activity, has been subtracted. The influence of young water on the ^{14}C content of the sample is discussed in Chapter 5. The piston flow model is the only assumption appropriate for the remaining old component because an exponential age distribution would introduce an additional activity of short-lived isotopes. The residence times given in Table 4.4.2, therefore, are based on an assumed piston-flow model for the old component. If the ^{3}H and ^{85}Kr activity is explained by an age distribution with a high mean residence time, one must explore whether the same residence time could yield the total measured ^{39}Ar activity (see, *e.g.*, Densbüren and Stockacher). This, and a comparison with conclusions from other isotope methods (mainly ^{14}C) are discussed in a summary way in Table 4.4.2. For a detailed discussion of the ^{14}C data see Chapter 5. It is important to note that the residence times of the samples from Densbüren (14/3), Lostorf 3 (123) and Magden Stockacher (274) are in an age range on which information is available only using the ^{39}Ar method.

Table 4.4.1: Isotope results of waters from sedimentary rock; He analyses from Bath, except 1/1 from Weizmann.

Sample Location Number	Site	Formation	Depth (m)	39Ar (% mod.) (± 1 SD)	O2 (vol.-%)	85Kr (dpm/cc Kr) (± 1SD)	3H (TU ± 2SD)	39Ar corr. (% mod.) (±1 SD)
301/2	Böttstein	mo	123-202.5	<6.0	0.13±0.02	0.15±0.05	<1.0	<6.3
302/10a /10b	Weiach	mo	822-896 ---	16.5±3.6 15.5±1.6	tr tr	0.65±0.04 ---	3.2±0.7 ---	12.8±1.9 ---
303/2	Riniken	mo	617-696	<7.0	tr	---	<1.1	<7.2
304/14 /3	Schafisheim	mo	1241-1262	14.2±4.3	n.d.	---	1.4±0.7 <0.9	12.3±4.5
306/1	Leuggern	mo	535-96 ---	53.0±3.7 ---	0.28 0.20	3.42±0.13 3.27±0.14	11.6±1.1 ---	44.5±7 ---
1/1 /3	Aqui	tOMM	300-500	<4.7 <6.0	<0.03 n.d.	0.089 0.123±0.038	<0.6 ---	<4.9 ???
2/1 /3	Tiefenbrunnen	tOMM	330-716	--- 8±5	--- 0.05±0.01	--- <0.17	<0.7 <0.9	--- <16
8/2	Pratteln	mo	53-124	252±8*	0.80	3.42±0.22	9.2±0.8	293±10
9/2	Frenkendorf	mo	230-250	32.8±3.8	12.1	6.87±0.27	1.1	<24
14/3	Densbüren	mo	38-125	63.6±3.5	tr	0.79±0.03	7.8±0.9	61.5±5
15/1	Beznau	mo	203-303	<4.6	<0.03	0.09±0.01	<0.6	<4.6
123/110	Lostorf 3	mo	530-584	52.5±4.2	tr	0.3-0.9	<0.5	52.5±4.2
271/103 /104	Magden Weiere	mo mo	213-282	17.3±3.2 11.3±2.2	--- 0.01	0.27±0.11 0.15±0.01	0.8 <1.1	16.3±4 10.8±2.2
272/102 /103	Magden Falke 2	mo mo	165.5-235	17.3±4.5 ---	0.43 0.02	0.03±0.01 0.10±0.01	1.6 <0.8	16.3±4.5 ---
273/102 /103	Magden Eich	mo	192-248	23±2.4 23.7±2.4	0.007 0.01	0.072±0.006 0.101±0.008	1.9 <0.9	22.8±2.4 23.4±2.4
274/102	Magden Stockacher	mo	---	41.9±3.1	0.02	0.609±0.037	4.3±0.7	39.1±3

171

Table 4.4.1: Isotope results of waters from sedimentary rock; He analyses from Bath, except 1/1 from Weizmann.

Sample Location Number	Site	4He (10-8cc/ccH2O) 1)	40Ar/36Ar Gas	40Ar/36Ar Water	14C (pmc)	Delta 13C ‰	Delta 2H per mil ‰	Delta 18C ‰
301/2	Böttstein	21,700	309.9±1.5	---	1.49±0.12	- 4.8	-78.2	-11.16
302/10a /10b	Weiach	82 172	294.4±1.5 ---	--- ---	13.28±0.14	- 4.6	-82.2 -86.6	-11.87 -11.76
303/2	Riniken	6,800	303.0±1.4	---	0.9±0.2	- 4.8	-71.6	-10.14
304/14 /3	Schafisheim	1,400 23,700	--- ---	--- ---	--- 4.41±1.20	--- ---	-78.6 -80.0	-10.81 -11.02
306/1	Leuggern	465 ---	--- ---	296.6±1.9 ---	11.23±0.30	- 6.4 ---	-68.3 ---	- 9.96 ---
1/1 /3	Aqui	920 1,360	--- ---	291.6±3.5 296.3±1.7	<1.3 1.5	- 3.0 - 4.3	-88.0	-11.89
2/1 /3	Tiefenbrunnen	--- 5,420	--- ---	--- 302.2±1.5	<2.2 ---	- 3.1 ---	-73.6 -71.6	- 9.04 - 9.14
8/2	Pratteln	12	295.0±1.6	311.8±3.0	39.4±0.4	- 9.5	-66.7	- 9.29
9/2	Frenkendorf	17	294.2±0.8	313.2±2.2	18.97±0.30	- 9.8	-64.4	- 9.14
14/3	Densbüren	14	295.6±1.0	---	43.42±0.40	- 9.6	-73.8	-10.24
15/1	Beznau	25,000	---	---	<0.4	- 5.4	-81.2	-11.09
123/110	Lostorf 3	---	---	---	44.0±1.2	- 9.4	-70.9 *	-10.30 *
271/103 /104	Magden Weiere	27 ---	--- ---	--- ---	8.53±0.3 10.3±0.3	- 6.8 - 8.2	-74.8 -71.5	-10.49 - 9.84
272/102 /103	Magden Falke 2	27 20	--- ---	304.5±1 ---	28.1±0.4 5.5±0.2	- 9.6 - 9.3	-66.1 -72.3	- 9.22 -68.50
273/102 /103	Magden Eich	--- ---	--- ---	--- ---	19.9 16.7±0.6	-10.5 - 9.5	-69.8 -69.3	- 9.83 - 9.50
274/102	Magden Stockacher	---	---	---	37.8±0.7	-10.6	-68.5	- 9.30

Table 4.4.2: Summary of conclusions from ^{39}Ar results. (Page 1 of 2)

Sample/ Formation	No.	Mean Residence time of older component of sample from ^{39}Ar	Comments Especially comparisons with other isotope data (for ^{14}C interpretation see Chapter 5)
Böttstein	301/2	> 1000 a	Consistent with ^{14}C result. The very high δ^{13}C value requires a large correction to the measured ^{14}C. High ^{4}He and increased ^{40}Ar/^{36}Ar ratio in the gas sample may support a high residence time.
Weiach mo	302/10	800 ± 60 a	Consistent with the ^{14}C interpretation. The considerable ^{14}C activity points to a water with a residence time range as deduced from the ^{39}Ar activity. This value seems reliable, assuming negligible underground production.
Riniken mo	303/2	> 1000 a	Same comment as for 301/2.
Schafisheim mo	304/14	≥ 800 a	The ^{39}Ar result is preliminary and has a relatively large error. Since underground production cannot be completely excluded, the residence time is given as a lower limit. Since large quantities of H$_2$S are present in the water sample and since there is no measured δ^{13}C value, the ^{14}C value is very difficult to interpret.
Leuggern mo	306/1	≥ 300 a	The ^{3}H, ^{85}Kr, ^{39}Ar and ^{14}C activities cannot be explained by assuming an exponential model with one mean residence time only. Therefore, mixing has to be assumed. A very young component of about 15% introduces the ^{3}H and ^{85}Kr activities (Section 4.3). If only one older component leads to the measured ^{39}Ar activity, it would have a PFM model age of about 300 years. Evidence exists from hydrogeology that a young component is present since the Muschelkalk formation is in contact with superficial groundwaters through Quaternary gravels. This gives problems for the interpretation of the ^{14}C results. All results, however, can be explained by a mixing of three or more components. Proportion Age Activities introduced 15% 10-15 yrs All ^{3}H and ^{85}Kr All ^{39}Ar 30% 60 yrs and ^{14}C 60% very old No activities These numbers are rough estimates only; the uncertainties of the applied corrections would allow other combinations too. The mixing components may be variable with time, since even the very young water component was varying during sample collection (^{3}H increased from 7 TU to 11.6 TU).
Aqui tOMM	1/1	> 1100 a	Same comment as for 301/2.
Tiefenbrunnen tOMM	2/3	> 700 a	Same comment as for 301/2.

Table 4.4.2: Summary of conclusions from ^{39}Ar results. (Page 2 of 2)

Sample/ Formation	No.	Mean Residence time of older component of sample from ^{39}Ar	Comments Especially comparisons with other isotope data (for ^{14}C interpretation see Chapter 5)
Frenkendorf mo	9/2	≥ 550 a	Uncertain lower limit due to contamination of the gas sample.
Densbüren mo	14/3	190 ± 30 a EM ≈ 220 a	190 years are obtained from ^{39}Ar if PFM modelling is applied to the very young water component (see Chapter 4.3). An exponential distribution of components with a mean residence time of about 220 years explains the ^{39}Ar and short-lived isotope activities, although this model assumption is at the limits for application to ^{3}H and ^{85}Kr. Both time indications seem to be in agreement with the ^{14}C value, yielding a time of 400 ± 1100 years.
Beznau mo	15/1	> 1200 a	In agreement with ^{14}C (and ^{4}He) result.
Lostorf 3 mo	123	250 ± 30 a	For the uncorrected ^{39}Ar result. This result is in agreement with the ^{14}C result.
Magden Weiere and Magden Falke mo	271/103 and 104 272/102	≥ 700 a ≥ 700 a	Assuming no admixture of underground produced ^{39}Ar, the residence time is about 700 years, otherwise, it has to be considered as lower limit.
Magden Eich mo	273/102 and 103	≥ 570 a	Same as comment for Weiere and Falke.
Magden-Stockacher mo	274/102	about 350 a EM: 550 a	If a small component of young water of about 5% is assumed from the ^{3}H and ^{85}Kr results then the PFM age of the "old" component is about 350 years. If an EM age distribution is assumed, then the ^{39}Ar activity leads to a mean residence time of about 550 years. Each year in the last few decades would contribute about 2 per mil of the water; from this an ^{85}Kr activity of about 1 dpm/cc and a ^{3}H activity of about 4 to 8 TU can be estimated. These values agree well with the measured activities. Therefore, the assumption of an EM age distribution with <u>one</u> mean residence time of about 550 years explains the ^{3}H, ^{85}Kr, and ^{39}Ar activities. Local mixing has to be assumed for these waters from Magden since ^{39}Ar and ^{14}C values showed variations with time. In general, higher ^{14}C values are found for waters with higher ^{39}Ar concentrations. The measured activities can be explained if one admixed water component has low concentrations of both isotopes.

5. CARBONATE ISOTOPES

This chapter gives the results of isotope analyses on dissolved and mineral carbonates from the Nagra deep drilling and regional hydrochemical programmes. Selected results from work published by others on northern Switzerland and adjacent areas are also included to support the interpretation of the Nagra results. The presentation and discussion of these results is preceded by a summary of the principles which underlie the interpretation of carbonate isotope data.

Ratios of the stable carbon and oxygen isotopes reflect the origin of minerals and of dissolved carbonate species and may also provide insight into the sources of groundwater itself. The naturally-occurring radioisotope of carbon, ^{14}C, which is present in measurable amounts in some of the waters sampled, can provide information about flow rates and residence times of groundwater in systems with straightforward patterns of flow and geochemical evolution.

5.1 Overview of Groundwater Carbonate Evolution

F. J. Pearson, Jr.

The term "dissolved carbonate" will be used to refer to all forms in which carbon in the +4 oxidation state can be present in solution. These include the carbonate ion itself, CO_3^{-2}, as well as its hydrolysed forms bicarbonate, HCO_3^- and carbonic acid, H_2CO_3. Carbonic acid is equivalent to dissolved gaseous CO_2, $CO_{2(aq)}$, and the terms $CO_{2(aq)}$ and H_2CO_3 may be used interchangeably below. Dissolved carbonate may also be present as a constituent of ion pairs such as $NaCO_3^-$, $MgHCO_3^+$, and $CaCO_3^0$ (GARRELS AND CHRIST, 1965; DREVER, 1982, MATTHESS, 1982; STUMM AND MORGAN, 1981; NORDSTROM AND MUNOZ, 1985). The total dissolved carbonate concentration will be written $CO_{2(tot)}$. It is the sum of the concentrations of all dissolved species containing carbonate in any form.

Carbonate dissolved in groundwater may have several sources and sinks. Water falling as precipitation is exposed to the atmosphere which contains CO_2 at a concentration of about 350 ppm, equivalent to a CO_2 partial pressure P_{CO_2} of $10^{-3.5}$ bars. The atmosphere within the soil contains CO_2 at much higher concentrations than the normal atmosphere because of CO_2 produced by plant root respiration and the decay of plant debris. During recharge, groundwater is exposed to soil air and to CO_2 concentrations as high as 1 to 10 volume per cent or more, equivalent to P_{CO_2} values of 10^{-2} to 10^{-1} bars or greater. Because soil-air CO_2 originates from plant respiration and decay, it is isotopically similar to plant CO_2. In climates like those of northern Switzerland, it is depleted in ^{13}C by about -25 per mil relative to the marine carbonate PDB standard, and it contains modern ^{14}C.

Groundwaters are commonly also exposed to carbonate minerals, both in the soil and during flow within the saturated zone. The dominant rock forming minerals in dolomite and limestone formations are carbonates, but detrital carbonate or carbonate cement are also common in silicic clastic rocks as well. Marine carbonates have ^{13}C contents similar to those of the PDB standard and have $\delta^{13}C$ values in the range of 0 to +2 per mil. Mineral carbonates to which groundwaters are exposed generally contain no ^{14}C because of their age.

Under some conditions, groundwater can lose dissolved carbonate by precipitation of carbonate minerals within the aquifer. Such precipitation is generally brought about by reactions between the groundwater and other minerals. For example, a groundwater saturated with respect to calcite would begin to precipitate that mineral if its calcium content were increased by the dissolution of gypsum or anhydrite, or by exchange of calcium for some other cation. Minerals deposited by groundwater may be recognisable petrographically and are generally isotopically distinct from minerals of primary marine or early diagenetic origin.

Other sources and sinks for carbonate exist, but only rarely contribute significantly to the evolution of groundwaters. These include carbonate which can be produced by the oxidation of organic carbon within the aquifer, or the loss of gaseous CO_2 within the aquifer.

An understanding of the evolution of the carbonate chemistry of a groundwater can provide information on the conditions of its recharge, the types of formation through which it has passed to reach the sampling point, and the geochemical reactions to which it has been subjected. Such an understanding is prerequisite to an interpretation of the ^{14}C contents of groundwaters.

Two approaches are possible to develop such an understanding. The first is employed when studying aquifer systems in which the position of each sample in the groundwater flow field is known. The evolution of such samples along flow paths can be described with relative ease. This approach is exemplified by studies such as those of THORSTENSON AND OTHERS (1979) and PLUMMER AND OTHERS (1983).

An alternate approach must be used in studies such as this one, where the details of flow among the many sampling points may not be known. This technique makes use of water chemical and isotopic data, knowledge of the types of geochemical reactions which occur generally in groundwater systems, and certain assumptions about initial conditions to develop an understanding of how particular water samples evolved. From this understanding, it is often possible to place them in correct flow relationships with other samples.

The reactions considered in interpreting the carbonate evolution of the Nagra samples will be discussed in Section 5.1.2 following a discussion of carbonate isotope equilibrium.

5.1.1 Isotope Equilibria

Both carbon and oxygen isotope exchange reactions are important when interpreting the evolution of the carbonate chemistry of groundwaters. Oxygen in carbonate ions exchanges rapidly with the oxygen in the water molecule. Thus, the oxygen isotopic composition of a carbonate mineral reflects the oxygen isotopic composition of the water from which the mineral precipitated. Carbon isotope exchange between the various dissolved forms of carbonate, $CO_{2(aq)}$, HCO_3^-, and CO_3^{-2}, as well as between gaseous CO_2 ($CO_{2(g)}$), dissolved carbonate, and solid carbonate minerals must also be considered.

The conventions and nomenclature used to describe stable isotopic compositions and reactions are given in Section 1.3.

5.1.1.1 Oxygen Isotopes

As pointed out in Section 1.3, oxygen isotopic compositions are reported relative to one of two standards. The compositions of solid carbonates are usually referred to the carbonate PDB standard as $\delta^{18}O_{PDB}$ values. Water oxygen isotopic compositions are usually reported relative to the water SMOW standard as $\delta^{18}O_{SMOW}$ values. $\delta^{18}O$ values on the two scales are related by the equation (FRIEDMAN AND O'NEIL, 1977, FRITZ AND FONTES, 1980):

$$\delta^{18}O_{SMOW} = 1.03086 \; \delta^{18}O_{PDB} + 30.86 \qquad (5.1.1)$$

Isotope fractionation factors (α and ϵ) vary with temperature. The equation used here to describe the temperature variation of oxygen isotope fractionation between calcite and water is that published in 1969 by O'Neil, Clayton and Mayeda, as given by FRIEDMAN AND O'NEIL (1977, Figure 13):

$$1000 \ln \alpha^{18}_{(CaCO_3 - H_2O)} = \frac{2.78 \cdot 10^6}{T^2} - 2.89 \qquad (5.1.2)$$

where T is in Kelvin.

Another widely used equation was given by Epstein and others in 1953 for fossil paleo-temperature work. Equation 5.1.2 agrees with the version of the Epstein equation given by FRIEDMAN AND O'NEIL (1977, Figure 2) within 0.2 per mil between 10° and 90°C.

In Section 5.3, the oxygen isotopic composition of calcite in equilibrium with water is calculated using combined equations 5.1.2 and 5.1.1 and the definition of $\delta^{18}O$ from Section 1.3.

5.1.1.2 Carbon Isotopes

The dominant dissolved carbonate species are carbonate, CO_3^{-2}, bicarbonate, HCO_3^-, and $CO_{2(aq)}$. Carbon isotope fractionation takes place among these species and with $CO_{2(g)}$

or carbonate minerals. This fractionation must be taken into account in calculating the carbon isotopic composition of calcite or $CO_{2(g)}$ in equilibrium with a water.

The isotopic composition of carbonate dissolved in water is generally reported as $\delta^{13}C_{CO2(tot)}$, the $\delta^{13}C$ value of the total dissolved carbonate in the water, relative to the PDB carbonate standard. To calculate carbon isotope fractionation between dissolved carbonate and calcite or $CO_{2(g)}$, it is necessary to know the $\delta^{13}C$ value of a single dissolved carbonate species. Bicarbonate is a convenient choice because it is the most prevalent dissolved carbonate species in most groundwaters.

The $\delta^{13}C$ value of dissolved bicarbonate is:

$$\delta^{13}C_{HCO_3^-} = \left[\frac{(\delta^{13}C_{CO_{2(tot)}} + 1000)}{\alpha_{(CO_{2(tot)} - HCO_3^-)}} \right] - 1000 \qquad (5.1.3)$$

In this equation,

$$\alpha_{(CO_{2(tot)} - HCO_3^-)} = \qquad (5.1.4)$$

$$\Sigma Q_{HCO_3^-} + \alpha_{(CO_{2(aq)} - HCO_3^-)} \cdot Q_{CO_{2(aq)}} + \alpha_{(CO_3^{-2} - HCO_3^-)} \cdot \Sigma Q_{CO_3^{-2}}$$

where $Q_{C_i} = m_{C_i}/m_{CO_{2(tot)}}$, so that $\Sigma Q = 1$, and m_{C_i} and $m_{CO_{2(tot)}}$ are the molal concentrations of carbonate species i and of the total dissolved carbonate, respectively.

Factors for carbon isotope fractionation among the major dissolved carbonate species $CO_{2(aq)}$, HCO_3^-, and CO_3^{-2}, and $CO_{2(g)}$ and carbonate minerals are known. However, no data are available on fractionation factors for ion pairs such as $CaHCO_3^+$ or $NaCO_3^-$. To evaluate equation 5.1.4, it is generally assumed that the fractionation factors for HCO_3^- and CO_3^{-2} complexes are the same as for free HCO_3^- and CO_3^{-2} ions. Thus, the ΣQ terms in the equation represent the sums of the proportions of both free and complexed HCO_3^- and CO_3^{-2} species, respectively.

The carbon isotopic composition of calcite in equilibrium with the solution can be calculated using the fractionation factor $\alpha_{(CaCO_3 - HCO_3^-)}$:

$$\delta^{13}C_{CaCO_3} = \left(\frac{\alpha_{(CaCO_3 - HCO_3^-)}}{\alpha_{(CO_{2(tot)} - HCO_3^-)}} \right) \cdot \left(1000 + \delta^{13}C_{CO_{2(tot)}} \right) - 1000 \quad (5.1.5)$$

Likewise, the isotopic composition of $CO_{2(g)}$ in equilibrium with the solution equals:

$$\delta^{13}CO_{2(g)} = \left(\frac{\alpha_{(CO_{2(g)} - HCO_3^-)}}{\alpha_{(CO_{2(tot)} - HCO_3^-)}} \right) \cdot \left(1000 + \delta^{13}C_{CO_{2(tot)}} \right) - 1000 \quad (5.1.6)$$

Three sets of equations for carbon isotope fractionation factors as functions of temperature are commonly considered. DEINES AND OTHERS (1974) fitted equations to a compilation of experimental results from the literature. Another set was developed on theoretical grounds by Bottinga for fractionation between $CO_{2(g)}$ and calcite (FRIEDMAN AND O'NEIL, 1977, Figure 31). The set selected for this work was developed by MOOK (1983, 1986), and adopted by FONTES (1983). The fractionations in per mil are:

$$\epsilon(CO_{2(g)} - HCO_3^-) = \frac{-9483}{T} + 23.89$$

$$\epsilon(CO_{2(aq)} - HCO_3^-) = \frac{-9866}{T} + 24.12$$

$$\epsilon(CO_3^{-2} - HCO_3^-) = \frac{-867}{T} + 2.52$$

$$\epsilon(\text{Calcite} - HCO_3^-) = \frac{-4232}{T} + 15.10$$

Values of α required for equations 5.1.4, 5.1.5, and 5.1.6 are found using these factors and the definition of the additive fractionation factor:

$$\epsilon \text{ (A-B) } (^o/oo) = (\alpha_{(A-B)} - 1) \cdot 10^3 \quad (1.3.5)$$

5.1.2 Evolution of the Isotopic Composition of Dissolved Carbonate

The isotopic composition of carbonate dissolved in groundwater is determined by the sources of carbonate and by water-rock reactions. If the isotopic composition of the carbonate sources and the type and extent of water-rock reactions are known, the isotopic composition of dissolved carbonate can be calculated. Agreement between calculated and measured ^{13}C contents of the total carbonate dissolved in a groundwater validates the sources chosen and the reactions used to model the groundwater. In addition, any difference between calculated and measured ^{14}C values can be attributed to ^{14}C decay and interpreted in terms of groundwater age or residence time.

In practice, it is rarely possible to describe the evolution of a groundwater and its carbonate chemistry so completely that an unambiguous calculation of its carbon isotopic composition can be made. Instead, the chemistry of a groundwater and the general character of its water-bearing unit are considered to indicate the overall reactions by which the water evolved. These reactions are then described quantitatively using chemical mass balance calculations similar to those of PLUMMER AND BACK, (1980) and PLUMMER AND OTHERS (1983). The mass balance is refined until calculated and measured ^{13}C values agree. The refined mass balance provides a basis for adjusting the initial ^{14}C content of the water for the effects of water-rock reactions. If such complicating hydraulic phenomena as mixing of waters or matrix diffusion can be ruled out, a difference between the adjusted initial ^{14}C content and the measured ^{14}C content of a water can be interpreted as a model age of the water.

In the five sections which follow, the procedures used to develop an understanding of the carbonate evolution of Nagra groundwaters are described in some detail. The reactions which determine the carbonate chemistry of groundwaters are discussed in Section 5.1.2.1, along with the mass balance equations which describe them. Section 5.1.2.2 discusses the equations used to describe the carbon isotopic changes accompanying chemical evolution. Section 5.1.2.3 treats the practical application of the mass balance and isotope evolution equations. Section 5.1.2.4 presents an analysis of the uncertainty in the results of the calculations due to uncertainties in various parameters of the mass balance and isotope evolution equations. Finally, in Section 5.1.2.5, the approach used here is compared with other models used to interpret the ^{14}C contents of groundwater carbonate.

5.1.2.1 Chemical Mass Balances

In studies of regional aquifer systems in which flow patterns are well known, it is often possible to deduce the reactions by which the groundwater has evolved from observed chemical changes along flow paths. A number of such studies have been reported including those of PEARSON AND WHITE (1967), PLUMMER (1977), and BACK AND OTHERS (1983).

Unfortunately, the flow systems from which the Nagra samples were taken are not well enough defined to permit the determination of chemical change along specific flow paths. Therefore, it has been necessary to assume that these waters evolved as a result of certain general reactions typical of many groundwater systems. The reactions by which waters in the sediments evolved are summarized as follows:

- In the recharge zone, ^{14}C-bearing CO_2 from the soil air dissolves together with some mineral carbonate from calcite and (or) dolomite;

- In the zone of saturated flow, carbonate mineral solution continues and a calcium sulphate mineral, gypsum or anhydrite, may dissolve;

- Calcite will precipitate if the carbonate and calcium added from dolomite and calcium sulphate solution lead to calcite oversaturation;

- Additional dissolved anions are present which can be assumed to come from solution of fictive "NaCl". The actual source and identity of the anions and their balancing cations are unimportant to the carbonate balance. Only the sum of their ionic charges matters;

- The concentration of dissolved calcium and magnesium are generally lower than the amount of carbonate and sulphate present from carbonate and sulphate mineral dissolution. This is assumed to result from cation exchange of dissolved Ca^{+2} for Na^+ and other cations. The calcium deficiency is made up by an equivalent amount of the general "Na^+" cation.

These reactions lead to the following mass balance equations for the amount of ^{14}C-bearing CO_2 dissolved from the soil air, and the amounts of minerals which have dissolved and precipitated to produce the chemistry of a given water sample:

Calcium sulphate dissolved (mol) = SO_4^{-2} concentration (mol)

Dolomite dissolved (mol) = Mg^{+2} concentration (mol)

"NaCl" dissolved (mol) = Mean ion concentration (eq)
 - 2 · SO_4^{-2} concentration (mol) - carbonate alkalinity (eq)

Ca^{+2} exchanged for 2 "Na^+" = 0.5 · [Mean ion concentration (eq)
 - 2 · (Mg^{+2} + Ca^{+2} (eq) - "NaCl" dissolved (mol)]

Calcite dissolved or precipitated (mol) = Ca^{+2} concentration (mol)
 - $CaSO_4$ dissolved (mol) - dolomite dissolved (mol) - Ca^{+2} exchanged (mol)

^{14}C-bearing CO_2 dissolved (mol) = Total dissolved carbonate (mol)
 - 2 · dolomite dissolved (mol) ± calcite precipitated or dissolved (mol)

The mean ion concentration appears in these equations. It is the mean of the sums of the charges of the analysed cations and anions, and is included so that the mass balance will be made on a charge-balanced solution.

The actual evolution of dissolved carbonate is more complicated than this simple mass balance suggests. A particularly troublesome oversimplification is neglect of the fact that in many aquifers, calcite both dissolves and precipitates. Because of its ubiquity and rapid rate of dissolution, calcite solution occurs early in the evolution of most waters. However, during later solution of dolomite, and (or) calcium sulphate minerals, calcite may precipitate. Solution and reprecipitation will affect the isotopic composition of dissolved carbonate in the same way as would isotope exchange between dissolved and mineral carbonate. The amount of solution and reprecipitation or of isotope exchange, if any, cannot be estimated from the mass balances alone. However, adjusting the mass balances until there is agreement between the calculated and measured ^{13}C content of the dissolved carbonate provides an estimate of the amount of mineral solution and reprecipitation or isotope exchange.

A second oversimplification is the assumption that all gaseous CO_2 dissolved is ^{14}C-bearing and from the soil zone, and that all ^{14}C-free carbon is dissolved from mineral carbonate. Groundwater may also dissolve CO_2 within the saturated zone. If this CO_2 is produced by the oxidation of organic matter or dissolved methane in the formation, it would be ^{14}C-free. If it results from the oxidation of dissolved organic carbon, it could contain ^{14}C.

Subsurface production of CO_2 from methane or other reduced organic carbon is an oxidation process. This oxidation requires a coupled reduction reaction which, in many groundwaters, has been found to be the reduction of sulphate to sulphide (PEARSON AND HANSHAW, 1970). To produce 1 mmol of CO_2 (44 mg/l) from CH_4 requires the reduction of SO_4^{-2} to produce 1 mmol of H_2S (36 mg/l). Of the samples on which ^{14}C measurements were made, only three contained more than 4 mg H_2S/l. This corresponds to about 0.1 mmol dissolved carbonate. As Tables 2.1.2 and 2.1.3 show, the total dissolved carbonate of most samples is greater than about 4 mmol, so the amount of carbonate which could be associated with H_2S production is less than 2.5 per cent. This is well within the uncertainties included in the modelling calculations described below.

The three samples with more than 4 mg H_2S/l are from Schinznach Bad S2 (120, c. 60 to 70 mg), the Permian at Riniken (303/6, 22 mg/l), and the Muschelkalk at Schafisheim (304/3, c. 750 mg/l). The Schinznach samples are mixed waters with ^{14}C contents of about 25 pmc. Virtually all the ^{14}C was probably supplied by the young component of the mixture, which is a modern ^3H-bearing water (Tables 2.1.3 and 5.2.2). The ^{14}C content of the Riniken sample (303/6) is below detection, and so any CO_2 introduced as a by-product of H_2S production was not ^{14}C-bearing. The H_2S content of the Schafisheim sample (304/3) corresponds to c. 21 mmol CO_2, nearly twice the total dissolved carbonate in this sample (Table 2.1.1). Thus, all the H_2S cannot be associated with CO_2 production. On the other hand, this sample has a ^{14}C content of 4.4 pmc which could well have been introduced as a by-product of H_2S production. The ^{14}C content of this sample is treated as an upper limiting value in the discussion in Section 5.2.

5.1.2.2 Isotope Evolution Model

Many quantitative models have been proposed to describe the evolution of the carbon isotopic composition of groundwater and to assist in the interpretation of groundwater ^{14}C measurements. While most of these models consider carbonate sources and sinks of different isotopic compositions, they generally do not consider mixing of waters of different origins.

The model proposed by WIGLEY AND OTHERS (1978) has been adopted here. It is a general description of carbon isotope evolution in groundwaters in piston flow or in an elemental volume of flowing groundwater in which mixing can be neglected.

The "exponential" model suggested by SIEGENTHALER (1972), and applied frequently to ^{14}C measurements is of a different type. This model is mathematically equivalent to a description of a completely mixed flow system. Thus, it is not comparable to the model of WIGLEY AND OTHERS (1978), or to most other models in the literature for groundwater ^{14}C, which assume piston flow of groundwater.

The model of WIGLEY AND OTHERS (1978) describes the change in the carbon isotopic composition of a water resulting from carbonate dissolving in or otherwise entering the solution, or precipitating from or otherwise leaving the solution. It is based on an isotope balance equation stating that the change in the ^{13}C (or ^{14}C) content of a mass of groundwater equals the sum of the amounts of the heavy isotope entering the water from all carbon sources minus the sum of the amounts leaving to all sinks. The equation is written:

$$d(R_S m_{CO_{2(tot)}}) = \sum_{i=1}^{N} d(R_{E_i} m_{CO_{2(E_i)}}) - \sum_{j=1}^{M} d(R_{L_j} m_{CO_{2(L_j)}}) \qquad (5.1.7)$$

In this equation,

$m_{CO_{2(tot)}}$ = total dissolved carbonate in mols/kg H_2O,

$m_{CO_{2(E_i)}}$ = mols of carbonate entering the water from source i,

$m_{CO_{2(L_j)}}$ = mols of carbonate leaving the groundwater to sink j.

R_S, R_E, and R_L are the ratios of ^{13}C or ^{14}C to ^{12}C (or total CO_2) in solution, in substances entering, and in substances leaving the solution, respectively.

It is assumed that carbonates leaving groundwater do so slowly enough to maintain isotopic equilibrium with the solution. Thus,

$$R_{L_j} = R_S \cdot \alpha_{(L_j - S)}$$

where $\alpha_{(L_j - S)}$ is the fractionation factor between substance j leaving the solution, and the total dissolved carbonate.

WIGLEY AND OTHERS (1978, equation 17) provide a closed-form solution to equation 5.1.7 for the general N-input, M-output case. Restated in the notation of this report, that equation is:

$$\beta R_S - R_E = (\beta R_S^0 - R_E) \cdot \left[\frac{m_{CO_{2(tot)}}}{m^0_{CO_{2(tot)}}} \right]^\Phi \qquad (5.1.8)$$

where $m^0_{CO_{2(tot)}}$ and $m_{CO_{2(tot)}}$ are the total dissolved carbonate concentrations of the starting water and of the sample, respectively. The parameters β and Φ include the amounts of carbonate entering and leaving the solution and the fractionation factors between the dissolved carbonate and the substances leaving the solution.

It is sufficient for the samples to be discussed here to consider systems with no more than two sources or sinks for carbonate, that is, with $N \leq 2$ and $M \leq 2$. For such systems, the total carbonate entering and leaving the solution, respectively, will be:

$$\Sigma_E = dm_{CO_{2(E_1)}} + dm_{CO_{2(E_2)}} \text{ and,}$$

$$\Sigma_L = dm_{CO_{2(L_1)}} + dm_{CO_{2(L_2)}}$$

In 5.1.8, β, the composite fractionation factor, equals:

$$\beta = 1 + \frac{dm_{CO_{2(L_1)}}(\alpha_{(L_1 - S)} - 1) + dm_{CO_{2(L_2)}}(\alpha_{(L_2 - S)} - 1)}{\Sigma_E} \qquad (5.1.9)$$

R_E, the weighted-average isotopic composition of the carbonate entering the solution is:

$$R_E = \frac{R_{E_1} dm_{CO_{2(E_1)}} + R_{E_2} dm_{CO_{2(E_2)}}}{\Sigma_L - \Sigma_E} \quad (5.1.10)$$

The exponent Φ equals

$$\Phi = \frac{\Sigma_E + dm_{CO_{2(L_1)}} (\alpha_{(L_1 - S)} - 1) + dm_{CO_{2(L_2)}} (\alpha_{(L_2 - S)} - 1)}{\Sigma_L - \Sigma_E} \quad (5.1.11)$$

If the amount of carbonate dissolving during a step of water evolution equals the amount leaving the solution, no change takes place in the dissolved carbonate content of the groundwater, and $\Sigma_E = \Sigma_L$. Under these conditions, equation 5.1.8 is not a solution for equation 5.1.7 because the exponent Φ is undefined. The closed-form solution under these conditions is:

$$\beta R_S - R_E = (\beta R_S^0 - R_E) \exp\left[-\frac{\beta \Sigma_E}{m_{CO_{2(tot)}}^0}\right] \quad (5.1.12)$$

Carbonate can both enter and leave a solution if a dissolving mineral such as dolomite is simultaneously replaced by precipitation of a similar mineral such as calcite. This process is formally called incongruent dissolution. Only the minimum extent to which it may have occurred can be estimated from a chemical mass balance, but with $\delta^{13}C$ information, its full extent can be determined.

The masses of mineral carbonate dissolving and reprecipitating are expressed by the Σ_E and Σ_L terms in the equations above. The change in the total dissolved carbonate content of groundwater, however, reflects only the difference $\Sigma_E - \Sigma_L$, and thus may be much smaller than the amounts of carbonate which actually dissolved in and precipitated from a groundwater during its evolution. When the values of Σ_E and Σ_L approach each other, no change will be evident in the dissolved carbonate content of the water, and it may appear that the water is simply undergoing isotopic exchange with the carbonate minerals of the aquifer. This condition is described by equation 5.1.12, with the amount of "exchange" described by the quotient in the exponent:

$$\frac{\beta \Sigma_E}{m^0_{CO_{2(tot)}}}$$

Waters in very fine grained formations such as chalks may show isotopic evidence for reactions between dissolved and solid carbonate, but give no indication by changes in water chemistry what these reactions might be. Such waters appear to be undergoing isotopic exchange with formation carbonate. However, their carbonate isotope evolution can still be modelled with the techniques used for modelling incongruent dissolution.

The model equations above are written in terms of isotope ratios, R, and are applicable to either the stable or radioactive carbon isotopes. These equations are more useful to describe stable isotope evolution if they are restated in terms of $\delta^{13}C$ values. From the definition of δ values, equations 5.1.8 and 5.1.12 can be transformed to:

$$\beta (\delta^{13}C_S + 1000) - (\delta^{13}C_E + 1000) =$$
$$\{\beta (\delta^{13}C_S^0 + 1000) - (\delta^{13}C_E + 1000)\} \Omega \tag{5.1.13}$$

When $\Sigma_E \neq \Sigma_L$,

$$\Omega = \left[\frac{m_{CO_{2(tot)}}}{m^0_{CO_{2(tot)}}}\right]^\Phi \tag{5.1.14}$$

and when $\Sigma_E = \Sigma_L$,

$$\Omega = \exp - \left[\frac{\beta \Sigma_E}{m^0_{CO_{2(tot)}}}\right] \tag{5.1.15}$$

The terms β, and Φ, are defined by equations 5.1.9, and 5.1.11, respectively, and Σ_E and Σ_L are the total amounts of carbonate dissolving and precipitating. $\delta^{13}C_E$ is the weighted average $\delta^{13}C$ value of the carbonate entering the solution. It is analogous to R_E, defined by equation 5.1.10, and equals:

$$\delta^{13}C_E = \frac{\delta^{13}C_{E_1} dm_{CO_{2(E_1)}} + \delta^{13}C_{E_2} dm_{CO_{2(E_2)}}}{\Sigma_E} \quad (5.1.16)$$

The isotope model equations above are also applicable to the ^{14}C content of carbonate dissolved in groundwater. Carbonate dissolving from an aquifer is likely to be ^{14}C free because of the age of the dissolving minerals. Except in unusual circumstances, even secondary minerals will probably also have residence times before redissolution which render them virtually ^{14}C free as well. Thus, from equation (5.1.10),

$$^{14}R_E = 0,$$

and equations 5.1.8 and 5.1.12 become:

$$^{14}R_S \; (pmc) = \;^{14}R_S^0 \; (pmc) \cdot \Omega \quad (5.1.17)$$

with Ω defined by equation 5.1.14 or 5.1.15. In this equation, $^{14}R°_S$ is the ^{14}C content of the dissolved carbonate of the initial solution and $^{14}R_S$ is that of the dissolved carbonate after the chemical evolution of the water but without considering radioactive decay of ^{14}C.

Employing the same values of Ω in both ^{13}C and ^{14}C equations follows the usage of WIGLEY AND OTHERS (1978, p. 1131; X in their paper is equivalent to Ω in this report). Strictly speaking, the values of Ω appropriate for the $\delta^{13}C$ equation, 5.1.13, are not identical to those of the ^{14}R equation, 5.1.17, because the fractionation factors, α, which appear in Ω, are not the same for ^{14}C as for ^{13}C.

It is generally assumed that $\alpha^{14} = (\alpha^{13})^2$. Values of α^{13} for calcite precipitating from waters of interest in this report will generally not differ from 1.00 by more than 0.01. Thus, differences between α^{13} and α^{14} should be less than one per cent, and no significant errors should result from using the same Ω values in the $\delta^{13}C$ and ^{14}R equations.

Equations 5.1.13 through 5.1.16 will be used in this report to calculate the ^{13}C effects accompanying carbonate evolution. The value of Ω from equation 5.1.14 or 5.1.15 will then be used in equation 5.1.17 to adjust the ^{14}C content of groundwater. The practical use of these equations is discussed in Section 5.1.2.3, and the uncertainties in the numerical results obtained are discussed in Section 5.1.2.4.

The model of WIGLEY AND OTHERS (1978), to which these equations correspond, is less widely used in isotope hydrology than are certain other approaches. Section 5.1.2.5 discusses several other models commonly used to interpret groundwater ^{14}C data, and shows that they are subsets of this model.

5.1.2.3 Application of Model Equations

If the chemical evolution of a groundwater is known, from regional knowledge of its aquifer system, for example, the $\delta^{13}C$ value calculated using the chemical description and equations 5.1.13 through 5.1.16 should be the same as the measured $\delta^{13}C$ value of the water. To evaluate equations 5.1.13 through 5.1.16 and solve for $\delta^{13}C_S$ requires knowledge of the isotopic composition, $\delta^{13}C_{(E)}$, and amounts, $dm_{CO_2(E)}$, of each source of carbonate to the solution, and the amounts, $dm_{CO_2(L)}$, of carbonate leaving solution. The amounts of carbonate entering and leaving the solution could be taken from chemical mass balances discussed in Section 5.1.2.1.

The total dissolved carbonate contents of the initial and final solutions, $m_{CO_2^o(tot)}$ and $m_{CO_2(tot)}$, respectively, and $\delta^{13}C^o{}_S$, the isotopic composition of the initial solution, must also be known. The chemistry, temperature, and isotopic composition of the final solution are the values measured for the groundwater being interpreted. The initial solution for most of the samples discussed below can be thought of as recharge water in the soil before any reaction with mineral carbonate has occurred.

The composition and temperature of the solution from which carbonate loss occurs must be known to calculate the fractionation factors, α, required in equation 5.1.9 to evaluate β and in equation 5.1.11 to evaluate Φ. If carbonate loss occurs during the entire evolutionary step being modelled, fractionation factors averaged from those of the starting and ending solutions are used. If losses occur only as the solution composition approaches that of the final solution, it is reasonable to choose the fractionation factor of the final solution.

The general procedure when examining a groundwater is to use the chemical mass balances of Section 5.1.2.1 to estimate the quantities of minerals dissolving and precipitating during its evolution, and the initial dissolved carbonate content of the water. This chemical information is used in equation 5.1.13 to find $\delta^{13}C_S$, the ^{13}C content of the final solution. Lack of agreement between the calculated value and the measured ^{13}C content of the groundwater may result from errors in the mass balance describing the chemical evolution of the water. Mass balances may underestimate the amount of dissolution balanced by precipitation, or isotope exchange, which has occurred. This situation is evident when the values of $dm_{CO_2(E)}$ and $dm_{CO_2(L)}$ required by the measured $\delta^{13}C$ values are larger than those derived from the mass balance alone.

As will be shown in Section 5.1.2.4, the calculated $\delta^{13}C$ value is a sensitive indicator of the amount of dissolution and precipitation. However, that sensitivity decreases as the isotopic composition of the dissolved carbonate approaches equilibrium with that of the mineral carbonate. The $\delta^{13}C$ values of dissolved carbonate in equilibrium with solid formation carbonate depend on the $\delta^{13}C$ of the formation and the temperature and chemistry of the groundwater. Generally, concern about possible exchange equilibria begins with waters $\delta^{13}C$ values more positive than about -4 per mil.

The evolution of the chemistry of the Nagra groundwaters has been modelled using a starting water containing no carbonate except $CO_{2(g)}$ dissolved from the soil air. Two types of behaviour may control the relative isotopic compositions of CO_2 in the soil air and in solution. First, the water may dissolve all of the $CO_{2(g)}$ with which it comes in contact. In this case, there will be no isotope fractionation, and

$$\delta^{13}C_E = \delta^{13}C_{CO_{2(g)}}$$

Second, the water may be in contact with and reach isotopic exchange equilibrium with an excess of soil-air $CO_{2(g)}$. In this case,

$$\delta^{13}C_S = \left[\frac{(\delta^{13}C_{CO_{2(g)}} + 1000)}{\alpha_{(CO_{2(g)} - S)}} \right] - 1000 \qquad (5.1.18)$$

The first type is known as **Closed System** behaviour in the ^{14}C literature, while the second is **Open System** behaviour (*e.g.*, DEINES AND OTHERS, 1974; REARDON AND FRITZ, 1978).

The range of isotopic compositions likely for soil-air $CO_{2(g)}$ is limited, as are the conditions of temperature and water chemistry, which determine the amount of fractionation between $CO_{2(g)}$ and dissolved carbonate. Thus, it is possible to define ranges of $\delta^{13}C$ values representative of starting waters.

As pointed out in Section 5.1, soil-air $CO_{2(g)}$ is derived primarily from plant root respiration and decay (PEARSON AND HANSHAW, 1970). Plants of the type dominant in regions like northern Switzerland, and therefore the soil-air CO_2 in that region as well, have $\delta^{13}C$ values of about -25 per mil (DEINES, 1980). At a temperature of 10°C, ^{13}C fractionation between gaseous and dissolved CO_2 is about +1 per mil, and that between gaseous CO_2 and bicarbonate about -10 per mil. The $\delta^{13}C$ value of carbonate dissolved in water from soil CO_2 should therefore be about -25 per mil for closed system conditions, or between about -26 and -15 per mil for open system conditions.

The dominant mineral carbonate in most aquifers is of marine sedimentary origin. Therefore, the $\delta^{13}C$ value of carbonate dissolving during the evolution of the Nagra groundwaters has been taken as that of marine carbonate, from about 0 per mil to +2 per mil. The isotopic composition of a number of minerals from the Nagra deep boreholes are discussed in Section 5.3. As pointed out there, the oxygen isotopic contents of many of the minerals suggest that they are in equilibrium with, and probably precipitated from, water like that presently found in the formations. Carbonate minerals with ^{18}O contents clearly indicating a marine origin were sampled from the Muschelkalk of Böttstein, Schafisheim, and Leuggern. These have $\delta^{13}C$ values from 0.2 to 1.3 per mil.

Section 5.1.2.4 discusses the effects of variations of initial ^{13}C values and of mineral isotopic compositions on calculated $\delta^{13}C$ values of waters. Many of the Nagra groundwaters are at a relatively advanced stage of chemical evolution. For these waters, uncertainties in the isotopic composition of dissolving minerals are more important than uncertainties in the isotopic composition of the initial dissolved CO_2.

The source of ^{14}C to groundwater is soil CO_2, which is a product of the decay of plant debris and root respiration. Therefore, the ^{14}C content of plants and of soil CO_2 will be from 100 to perhaps 140 pmc. The higher values reflect the inclusion of significant amounts of ^{14}C generated during the nuclear era. At present, the ^{14}C values of plants are about 120 pmc.

The point of zero age for a groundwater can be taken as the time at which it entered the zone of saturated flow and was no longer exposed to soil-air CO_2 as a source of ^{14}C. If water moved from the recharge area to the sampling point by piston flow (that is, with no admixture of waters from other sources), and if the ^{14}C content of the water changed only by radioactive decay, the model age of a sample could be calculated using the conventional ^{14}C age equation:

$$t(years) = - \frac{t_{1/2}(years)}{\ln 2} \ln\left(\frac{^{14}R_M}{^{14}R^0}\right) \qquad (5.1.19)$$

where $t_{1/2}$ is the half-life of ^{14}C, and $^{14}R^0$ and $^{14}R_M$ are the ^{14}C contents at the time of recharge and measured at time t, respectively.

Because the isotopic composition of dissolved carbonate commonly is effected by water-aquifer reactions of the types described above, equation 5.1.19 cannot be applied directly to groundwater ^{14}C measurements. This equation is valid, however, if the ^{14}C content has been adjusted for chemical effects using equation 5.1.17. Under these conditions, equation 5.1.19 becomes:

$$t(years) = - \frac{t_{1/2}(years)}{\ln 2} \ln\left(\frac{^{14}R_M}{^{14}R_S}\right) \qquad (5.1.20)$$

where $^{14}R_S$ is from equation 5.1.17 and is the ^{14}C content of the water adjusted for chemical dilution, but before radioactive decay.

5.1.2.4 Uncertainties in Results of Modelling

This section examines the uncertainties which are to be expected in the calculated ^{13}C and ^{14}C values and the model ages derived from them. These uncertainties result from uncertainties in values of the parameters of the chemical and isotope mass balance equations and in the properties assumed to describe the initial solution. Additional uncertainties brought about by possible shortcomings of the conceptual model used to develop the equations are also discussed.

The section begins with a description of the evolution of an hypothetical, but realistic, groundwater. Reasonable values of the parameters required to model its evolution are selected and the equations are solved to illustrate the behaviour of the system as it evolves.

The complexity and the number of uncertain parameters in the model equations preclude an analysis of error propagation through them by direct techniques. The procedure adopted has been to model the evolving hypothetical groundwater as if all parameters were not uncertain to illustrate the changes taking place in the ^{13}C and ^{14}C contents of the dissolved carbonate with water evolution. Next, several points on the path of evolution were selected, and a risk analysis procedure applied to evaluate the uncertainty in the values calculated at those points resulting from uncertainties in the parameters of the model equations.

The risk analysis technique involves repetitive solution of the model equations for the $\delta^{13}C$ and ^{14}C values of the final solution. The actual value of each parameter used during a calculation is chosen randomly from the distribution of values assigned to it. The statistical distributions of the calculated values are then examined as an indicator of the overall uncertainty in the calculations. The distributions of parameter values are chosen to be consistent with the accuracy to be expected in the chemical and isotopic data available on most samples, and with the uncertainties in the values of such essentially unmeasurable parameters as the initial isotopic composition of carbonate.

The evolutionary path modelled in the uncertainty analysis leads to a water chemically similar to those found in the Muschelkalk of the Böttstein and Beznau boreholes (SCHMASSMANN AND OTHERS, 1984, Section 6.4.5) The water is saturated with respect to calcite, dolomite and gypsum at a P_{CO_2} of about $10^{-1.2}$ bars. In addition to calcium, magnesium, sulphate and dissolved carbonate, it contains 37 millimoles (mmol) of chloride (*ca.* 1450 mg/l) and 50 mmol of sodium (*ca.* 1150 mg/l).

Table 5.1.1 shows the chemistry of two of the waters along the path modelled. An initial solution containing small concentrations of magnesium, calcium, sulphate, and chloride, and with dissolved carbonate due only to solution of soil CO_2 comes to equilibrium with calcite and dolomite. During this process 2.89 mmol of calcite and 0.14 mmol of dolomite are dissolved. The solution then dissolves sodium chloride, raising its chloride content to 37 mmol, while also being brought to gypsum saturation and maintaining saturation with respect to calcite and dolomite. To reach the final state, 26.15 mmol of gypsum and 6.45 mmol of dolomite dissolve, 13.26 mmol of calcite precipitate, and 6.4 mmol of Ca^{+2} are exchanged for 12.8 mmol of Na^+. All modelling assumed a temperature of 25°C.

Table 5.1.1: Chemistry of modelled solutions used to illustrate the carbon isotope evolution of groundwaters.

		Initial Solution	Calcite and Dolomite Saturated	Calcite, Dolomite, and Gypsum Saturated
Sodium	(mmol)	0.4	0.400	50.000
Magnesium	(mmol)	1.8	1.940	8.393
Calcium	(mmol)	0.5	3.528	16.470
Alkalinity	(meq/kg)	0.0	6.335	5.624
Sulphate	(mmol)	2.4	2.400	28.551
Chloride	(mmol)	0.2	0.200	37.000
Total Ions	(meq/kg)	5.0	11.336	99.726
$CO_{2(aq)}$	(mmol)	4.857	1.699	2.055
HCO_3^-	(mmol)		6.316	5.604
CO_3^{2-}	(mmol)		0.010	0.010
Total CO_2	(mmol)	4.857	8.025	7.669

Reaction: (mmol dissolved [+] or precipitated [-])

Calcite			2.888	-13.262
Dolomite			0.140	6.453
Gypsum				26.151

Saturation Indices

Calcite		-5.40	0.00	0.00
Dolomite		-10.18	-0.20	-0.20
Gypsum		-1.81	-1.12	0.00

Isotopic Composition of Dissolved Carbonate

^{14}C	(pmc)	100.0	60.5	11.6

With ^{13}C Dolomite and Calcite = 0.0 per mil:

$\delta^{13}C$	(per mil)	-25.00	-15.13	-5.64
$\delta^{13}C$	(per mil)	-15.00	-9.08	-4.48

With ^{13}C Dolomite and Calcite = 2.0 per mil:

$\delta^{13}C$	(per mil)	-25.00	-13.34	-3.88
$\delta^{13}C$	(per mil)	-15.00	-8.29	-2.72

The isotopic composition of the solution at points along the evolutionary path was calculated using the equations of Section 5.1.2.2. The composition of the CO_2 dissolved in the initial solution was taken as -25 or -15 per mil for $\delta^{13}C$, and as 100 pmc for ^{14}C. A value of -25 is equivalent to the composition of soil-air CO_2 and implies formation of the initial solution by closed system dissolution of soil-air CO_2. A value of -15 would result from open system dissolution as discussed below.

Calculations were made with calcite and dolomite $\delta^{13}C$ values of 0.0 and of +2.0 per mil. The isotopic composition of the total dissolved carbonates in the modelled solutions are given in Table 5.1.1 and their evolution is illustrated in Figures 5.1.1 through 5.1.5.

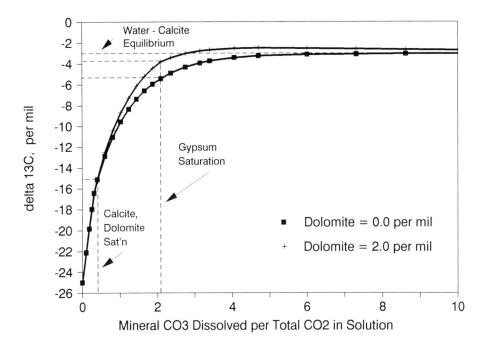

Figure 5.1.1: Change of ^{13}C of total dissolved carbonate with dissolution of mineral carbonate.

Figures 5.1.1, 5.1.2, and 5.1.3 show the carbon isotopic composition of a solution as a function of its carbonate chemical evolution. The ratio of the mols of mineral carbonate dissolved to the mols of total dissolved carbonate is used as the measurement of the stage of evolution. The figures include indications of the points along the path at which calcite and dolomite saturation and gypsum saturation occur. In the reaction scheme described in Section 5.1.2.1, water chemical evolution would end when gypsum saturation is reached. However, the carbon isotopic composition of a water could continue to change even after gypsum saturation if carbonate mineral solution and reprecipitation continued, or if isotope exchange between dissolved and mineral carbonate took place. Thus, in Figures 5.1.1 through 5.1.3 the carbon isotope evolution is shown well beyond gypsum saturation. The solution has an initial $\delta^{13}C$ value of -25 per mil and the $\delta^{13}C$ of the dissolving calcite is 0 per mil. The effects of dissolving dolomite with $\delta^{13}C$ values of both +2 and 0 are shown.

These figures and Table 5.1.1 show that during the evolution of the initial water to calcite and dolomite saturation, the ^{14}C content of the total dissolved carbonate decreases from 100 to 60.5 pmc. The apparent age of a conventional ^{14}C sample containing 60.5 pmc is 4.2 ka. The $\delta^{13}C$ value of the calcite and dolomite saturated water will be from -15.13 to -8.29 per mil depending on the $\delta^{13}C$ values of the initial carbonate and the dissolving minerals.

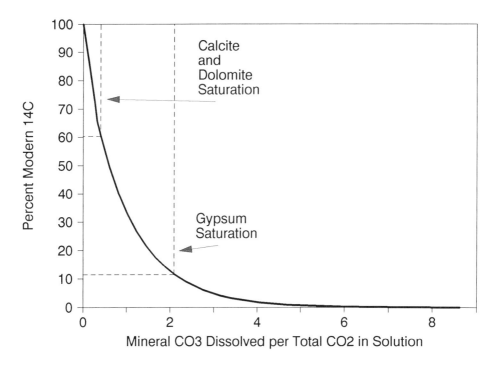

Figure 5.1.2: Change of ^{14}C content of total dissolved carbonate with dissolution of mineral carbonate.

During the evolution of the calcite- and dolomite-saturated water to gypsum saturation, the ^{14}C content of the water decreases to 11.6 pmc, an apparent age of 17.8 ka. The ^{13}C content of the water at gypsum saturation will be from -5.64 to -2.72 per mil, depending again on the $\delta^{13}C$ values of the initial and mineral carbonate.

As these figures also show, the effects of continuing dissolution and reprecipitation, or isotope exchange, occurring beyond gypsum saturation are to continue to lower the ^{14}C content of the water as ^{14}C-free mineral carbonate dissolves. The ^{13}C content of the water also continues to approach that of the rock, but when the $\delta^{13}C$ values of both calcite and dolomite equal 0.0 it levels off at about -3 per mil. This represents isotopic equilibrium between the total dissolved carbonate of water of this composition and calcite with $\delta^{13}C$ of 0 per mil.

The $\delta^{13}C$ of water in which dolomite of +2 per mil dissolves reaches values more positive than -3 per mil, but then becomes more negative and also approaches -3. This is because the calculations assume that calcite is the mineral which is dissolving and reprecipitating (or with which isotope exchange occurs) after gypsum saturation is reached. If this solution were to equilibrate isotopically with dolomite, the $\delta^{13}C$ value of its total dissolved carbonate would be about five per mil more negative than that of the dolomite itself because at the 25° temperature of this solution the fractionation between dolomite and calcite is about two per mil (FRIEDMAN AND O'NEIL, 1977, Figure 33).

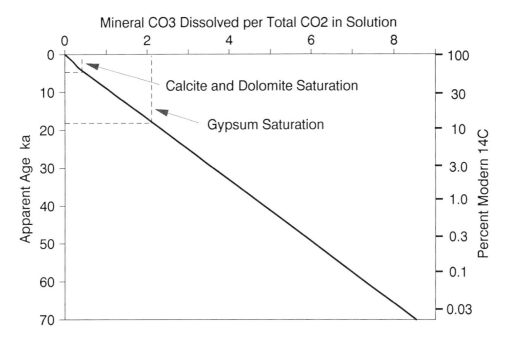

Figure 5.1.3: Change of apparent age calculated from ^{14}C content of total dissolved carbonate with dissolution of mineral carbonate.

Figures 5.1.4 and 5.1.5 display the relationship between the ^{13}C and ^{14}C contents of the total dissolved carbonate during the evolution of the water. As these figures show, during the evolution of the initial water to calcite and dolomite saturation, there is a linear relationship between $\delta^{13}C$ and ^{14}C. During this stage of evolution, only mineral dissolution is occurring. As the water evolves toward gypsum saturation and mineral precipitation occurs, the variation of ^{14}C with $\delta^{13}C$ begins to curve slightly and has a different slope than during the dissolution-only stage.

A particularly useful implication is that if a group of groundwaters from the same aquifer or flow system describe a straight line like that drawn in Figure 5.1.4, all are of the same age. In addition, the ^{14}C value where the line crosses the $\delta^{13}C$ value of the soil CO_2 can be used in the conventional ^{14}C-age equation to find the age of the group of waters. Such lines are defined most commonly by groups of recent waters. The ^{14}C-intercept of such groups at the soil $\delta^{13}C$ values is the ^{14}C content of the soil CO_2.

When the dissolving dolomite has a $\delta^{13}C$ value of 0 per mil like that of the dissolving calcite, the rate of change of ^{14}C with $\delta^{13}C$ is the same during solution/reprecipitation (or isotope exchange) occurring after gypsum saturation as it is during the evolution to gypsum saturation. When the $\delta^{13}C$ value of the dissolving dolomite is +2 per mil, the $^{14}C/\delta^{13}C$ ratio changes as calcite solution and reprecipitation occur beyond gypsum

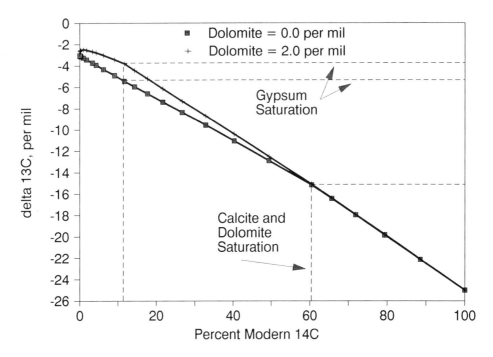

Figure 5.1.4: $\delta^{13}C$ values plotted against ^{14}C contents of total dissolved carbonate in waters dissolving and reprecipitating mineral carbonate.

saturation. Figure 5.1.5 also shows that as the $\delta^{13}C$ value of the water approaches equilibrium with that of the calcite, the ^{14}C content becomes infinitesimally small.

The fact that the relationship between the ^{13}C and ^{14}C contents of groundwaters is linear and has a similar slope over a wide range of $\delta^{13}C$ values means that the errors in ^{14}C contents introduced by assuming a simple linear model for the variation of ^{14}C with ^{13}C will be relatively small, even in waters as evolved as those saturated with gypsum. In Figure 5.1.4, for example, a water with a measured $\delta^{13}C$ of -5 per mil, which had evolved chemically like the hypothetical water, could have a ^{14}C content ranging from about 8 pmc if the dissolving dolomite had a $\delta^{13}C$ value of 0 per mil, to about 20 pmc if a simple linear $^{14}C/^{13}C$ ratio were assumed. This corresponds to a total uncertainty of about 7.6 ka which is relatively small, particularly in waters with measured ^{14}C contents of only a few pmc. For waters with more negative $\delta^{13}C$ values, the uncertainty will be less.

To this point, analysis of the evolution of a hypothetical water neglects additional sources of uncertainty which are important in interpreting ground-water samples. These include uncertainties in the:

- Isotopic composition of the carbonate dissolved in the initial water;

- Isotopic composition of mineral or other carbonate dissolved in the subsurface;

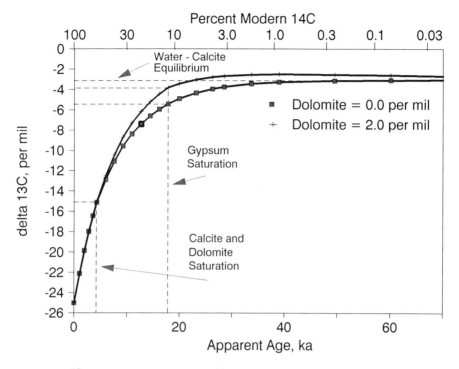

Figure 5.1.5: $\delta^{13}C$ values plotted against ^{14}C contents as pmc and as apparent ages in waters dissolving and reprecipitating mineral carbonate.

- Precision of the chemical and isotopic analytical data; and

- Path of chemical evolution of the water sample.

In the remainder of this section, the range of variation to be expected in each of these variables is discussed and they are combined using a risk analysis technique to estimate the cumulative uncertainty in quantitative interpretation of groundwater carbon isotope data (PEARSON, 1989).

The isotopic composition of the initial dissolved carbonate depends on the isotopic composition of the ^{14}C-bearing soil-air gas to which the water was last exposed and on whether that gas dissolved under open or closed system condition. The fact that soil-air CO_2 is the source of initial carbonate is usually not in dispute. The carbonate chemistry of most groundwaters cannot be explained except by early exposure to a source of CO_2 at high partial pressures found only in the soil. Soil-air CO_2 is isotopically similar to plant carbon and in northern Switzerland will have $\delta^{13}C$ values of within a few per mil of -25 per mil and a ^{14}C content of about 100 pmc. Waters recharged since the beginning of the nuclear era may have been exposed to soil CO_2 with ^{14}C contents of up to 140 pmc. However, such waters should be easily recognizable by their tritium contents.

The initial dissolved CO_2 in water recharged under closed system conditions will be the same as that of the soil-air, while in water recharged under open system conditions, isotope fractionation between dissolved and gaseous CO_2 will be important. As pointed out in Section 5.1.2.3, this fractionation would lead to $\delta^{13}C$ values between -26 and -15 per mil and ^{14}C values of 100 to 102 pmc.

For the isotopic composition of the dissolving minerals, a range of 0 to +2 per mil typical of marine carbonates is reasonable. For the uncertainty analysis, the $\delta^{13}C$ values of both calcite and dolomite have been given a uniform distribution ranging from 0 to +2 per mil. This analysis does not consider the situation in which the dissolving carbonate had been previously deposited by groundwater or had some other non-marine origin.

Uncertainties in the analysis both of dissolved constituents and of carbon isotopes must also be considered. Errors in the concentrations of dissolved constituents will cause errors in the mass balance results, and errors in the analysis of individual carbonate species will cause errors in the calculated isotope fractionation factors. For the uncertainty analysis, the chemical results are assumed to be normally distributed, with standard deviations of ten per cent for the carbonate species, and of five per cent for all others.

Errors in ^{14}C determinations are usually given explicitly by the analytical laboratories. For this uncertainty analysis, the standard deviation of the ^{14}C value is assumed to be 0.3 pmc.

$\delta^{13}C$ values may also be in error. Although analytical errors for ^{13}C measurements should be only on the order of a few tenths per mil, the sample collection procedure may also introduce error. Thus, for the uncertainty analysis, ^{13}C values are assumed to be normally distributed, with a standard deviation of 0.5 per mil.

The amount of uncertainty in the path of chemical evolution modelled is difficult to quantify. The principles of groundwater chemistry are well enough known to choose the reactions included in the chemical mass balance. The effects of uncertainties in the amounts of reacting minerals have been analysed here by varying the chemical parameters in the mass balance equations from which the paths are calculated. The mass balance equations are those described in Section 5.1.2.1.

When interpreting a real groundwater, the data available include the concentrations of its dissolved constituents and the isotopic composition of its dissolved carbonate. The quantities of minerals dissolving and reprecipitating can be found from the chemistry of the water and the mass balance equations, provided the composition of the initial water is known. To make the calculations on the evolution of the hypothetical water discussed in the first part of this section, an initial water of the composition shown in Table 5.1.1 was assumed.

The importance of the mass balance equations in this context is to interpret the isotopic chemistry of the dissolved carbonate, rather than to accurately describe the actual chemical evolution of the water. Thus, some simplifications are possible. In particular, the calculated quantities of carbonate dissolving in, precipitating from, and initially present in the water do not depend on whether the sodium, sulphate and chloride in the final solution were present in the initial solution or were added during its underground evolution.

The carbonate balance calculated is strongly influenced by the sources of the calcium and magnesium in the sample. For example, if the magnesium in the sample is the product of dolomite dissolution, a corresponding amount of carbonate from dolomite must also have entered the water. On the other hand, if magnesium was present in the initial solution, or is a product of reactions with non-dolomitic minerals, no corresponding carbonate need be considered.

When interpreting the isotopic composition of waters from a known flow system, it is generally possible to estimate the chemistry of the initial water and the effects of all evolution reactions, and to know directly how much of the magnesium in a given sample represents dolomite solution. When examining samples from unknown positions in their flow systems, the amount of magnesium from non-dolomite sources cannot be found without also considering the isotope chemistry of the system.

If the isotopic composition of the dissolving dolomite and of the initial dissolved carbonate are specified, a corresponding amount of dolomite dissolved can be calculated. In general, only the range, not precise values of the two isotopic compositions, will be known. Any pair of isotopic compositions corresponds to a given amount of dolomite dissolution, so the range of possible isotope values gives rise to a range of possible amounts of dolomite dissolution. In this uncertainty analysis, and in that made on each of the Nagra samples described in the following section, the $\delta^{13}C$ values of the initial carbonate and of the dissolving minerals were assigned ranges. During the risk analyses, values are chosen randomly from the given distributions of initial $\delta^{13}C$ and mineral $\delta^{13}C$ values, each pair of which corresponds to an amount of dolomite dissolution.

By this procedure, uncertainty in the path of water evolution is also included in the overall uncertainty analyses.

To estimate the uncertainty in the $\delta^{13}C$ and ^{14}C values calculated using the chemical and isotope mass balances, 1000 trial calculations were made. The values of the uncertain input parameters for each trial were selected at random from the distributions described. The results are shown in Figures 5.1.6 and 5.1.7.

Figure 5.1.6 is a cumulative probability distribution of the calculated $\delta^{13}C$ values of the solution. The distribution has a mean of -4.80 per mil with a standard deviation of 0.57. This is virtually identical with the $\delta^{13}C$ of the test solution, -4.76 per mil. The uncertainty of the measured $\delta^{13}C$ of dissolved carbonate may be about 0.5 per mil or more, if artifacts of sampling and precipitation are considered. A curve illustrating a 1-σ error of 0.5 per mil is also shown in Figure 5.1.6. There is good agreement between the distribution from the uncertainty analysis and the probable experimental uncertainty. This agreement suggests that the uncertainty in the ^{14}C results would not be lowered by rewriting the equations of Section 5.1.2.3 so that ^{14}C could be found directly from ^{13}C values.

Figure 5.1.7 is a histogram of the calculated ^{14}C contents of the water. The distribution has a mean and standard deviation of 13.9 ± 3.6 pmc. The ^{14}C content of the water calculated from its actual evolutionary path is 11.6 pmc, 2.3 pmc lower than the mean of the distribution. This difference is equivalent to a ^{14}C age error of 1.2 ± 2.3 ka.

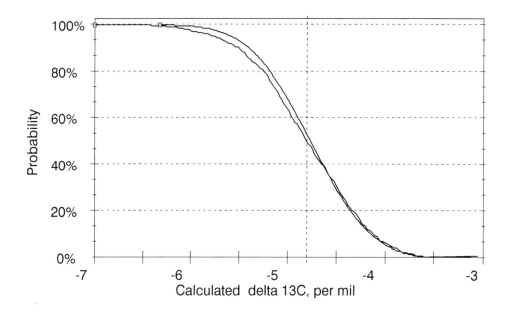

Figure 5.1.6: Cumulative probability of occurrence of $\delta^{13}C$ values.

The uncertainty analysis was made assuming chemical evolution in a single step from the initial water to the measured sample. The hypothetical water was developed by a two-step evolution, which more realistically represents real groundwater. The difference between the ^{14}C content of the hypothetical solution and the mean value from the uncertainty analysis results from the simplified evolutionary model used in the uncertainty analysis.

Interpretation of the Nagra results given in Section 5.2 includes uncertainty analyses similar to the one described here. Thus, the errors associated with the adjusted ^{14}C contents and model ages of the samples include uncertainties in the chemical evolution model, and in the parameters required for the isotope evolution model.

5.1.2.5 Comparison with Other Models for ^{14}C Correction

^{14}C measurements are expressed as activities relative to the total carbonate dissolved in a groundwater. Dissolution of ^{14}C-free carbon increases the total carbonate of a water, but not its ^{14}C content. This decreases the measured ^{14}C concentration and makes the water appear falsely old. Two approaches are possible for correcting or adjusting for this effect.

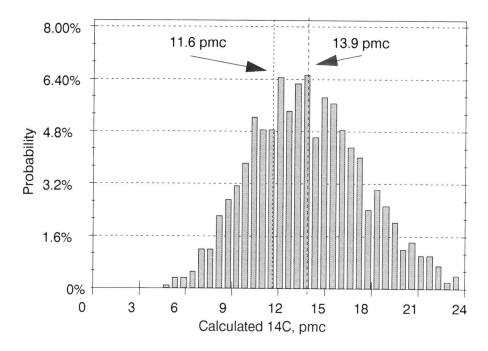

Figure 5.1.7: Histogram showing the distribution of modelled ^{14}C values.

The method generally used attempts to correct the initial ^{14}C content of dissolved carbonate for dilution with ^{14}C-free carbon. The adjusted ^{14}C content is then used as ^{14}R° in the age equation 5.1.19. Correction techniques used with this method range from the simple use of 85 pmc instead of 100 pmc for ^{14}R° by VOGEL (1967), to the use of detailed chemical mass balances and stable carbon isotope ratios as done in this report.

A second approach, which was suggested by OESCHGER (1974), considers the ^{14}C activity of a groundwater per unit volume or mass of water rather than per unit mass of dissolved carbonate. The addition of ^{14}C-free carbonate does not change the activity per unit of water as it does the activity per mass of carbon. This method is particularly useful if the distribution of ^{14}C in a groundwater system is to be simulated using conventional radionuclide transport models because isotope concentrations in such models are expressed per unit of water. To use this approach for absolute ages requires knowledge of the initial ^{14}C content of the water, a value which may be difficult to determine (PEARSON AND OTHERS, 1983).

FONTES (1983) and BALDERER (1985b, p. 63-84) have reviewed and give references to a number of models for the interpretation of groundwater ^{14}C data. Only the model of WIGLEY AND OTHERS (1978) includes carbonate both dissolving in and precipitating from a solution. The others describe dissolution processes alone.

To describe dissolution, the $dm_{CO_{2(L)}}$ values in the equations of Section 5.1.2.2 are set to zero. Thus, in equation 5.1.9, $\beta = 1$, and equation 5.1.13 reduces to:

$$\frac{\delta^{13}C_S - \delta^{13}C_E}{\delta^{13}C_S^0 - \delta^{13}C_E} = \Omega \tag{5.1.21}$$

Equation 5.1.21 is equivalent to equation 10 of INGERSON AND PEARSON (1964) and equation 4 of PEARSON AND WHITE (1967), with Ω equal to P of the earlier papers.

When dissolution only is occurring, $\Phi = -1$ in equation 5.1.11 so that equation 5.1.14 reduces to:

$$\frac{m^0_{CO_{2(tot)}}}{m_{CO_{2(tot)}}} = \Omega \tag{5.1.22}$$

Equation 5.1.22 is equivalent to equation 6 of INGERSON AND PEARSON (1964) and equation 8 of PEARSON AND WHITE (1967). Combining equations 5.1.21 and 5.1.22 to eliminate Ω leads to the "two inputs, zero output" equation of WIGLEY AND OTHERS (1978, Errata).

Several approaches explicitly consider how isotope exchange between ^{14}C-bearing soil-air CO_2 and dissolved CO_2 influences the initial ^{14}C content of a groundwater. Mook first addressed this problem because of his concern with the carbon isotopic chemistry of surface waters (see references in MOOK, 1980). DEINES AND OTHERS (1974) and WIGLEY (1975) analysed the effects on groundwaters of closed and open system dissolution of soil CO_2. These effects are included in the model used here in the values chosen for the $\delta^{13}C$ and ^{14}C of the dissolved carbonate of the initial solution. In the uncertainty analysis of Section 5.1.2.4, the full range of conditions from complete exchange between dissolved and gaseous CO_2 (open system) to dissolution only (closed system) are allowed by including initial $\delta^{13}C$ values between -15 and -26 per mil.

Several models which include water chemistry as well as carbon isotope data have also been used. INGERSON AND PEARSON (1964) proposed the use of the proportion of $CO_{2(aq)}$ to total dissolved carbonate as a measure of the proportion of soil-air CO_2 to total dissolved CO_2. This model was adopted by Tamers (see references in FONTES, 1983). MOOK (1980) and FONTES (1983) explicitly include isotope fractionation effects in their version of this model.

The concentrations of other dissolved constituents have been used to indicate the extent and identity of dissolving minerals in several studies. FONTES AND GARNIER (1979) include chemical balance considerations like those outlined in Section 5.1.2.1 in developing their global exchange model. More sophisticated models including aqueous speciation calculations were described by PLUMMER (1977) and by REARDON AND FRITZ (1978).

In summary, there are no processes included in other models for ^{14}C in groundwater which are not also included in the model used here. Some concerns of other models, such as the distinction between open and closed system control on the isotopic composition of the initial dissolved carbonate, are not explicitly present in this model. However their effects are accounted for in the ranges of input parameter values which have been chosen for adjusting the measured ^{14}C values.

5.2 Age Interpretations of Dissolved Carbonate Isotopes

F. J. Pearson, Jr.

Carbon isotope data on samples from the Nagra borehole and regional programmes and on other samples from northern Switzerland and adjacent areas were discussed in Section 2.1.2 of this report. That section addressed the quality of the samples and the analytical results and gave carbonate chemical data as well as the isotope results, in Tables 2.1.2 (borehole samples) and 2.1.3 (regional programme and other samples). This section presents and discusses the model ages which can be developed from these carbon isotope data.

5.2.1 Calculation of Adjusted ^{14}C Ages

The principles of the calculations used to develop model ages from dissolved carbon isotope data were given in Section 5.1. The chemistry of the water samples is required for the calculations. The chemical compositions of all samples from the borehole programme were taken from PEARSON AND OTHERS (1989); the compositions of the other samples were taken from NAGRA (1989).

As pointed out by PEARSON (1985) and by PEARSON AND OTHERS (1989), water equilibrium calculations using the analytical data suggest that many samples are oversaturated with respect to calcite. Because of the relative rapidity with which calcite can precipitate, it is unlikely that the groundwaters represented by these samples actually are oversaturated. The probable explanation for the apparent oversaturation is the loss of $CO_{2(g)}$ during the sampling or analytical processes, leading to falsely high pH values. To correct for this, the pH values and carbonate contents of all samples were adjusted for calcite saturation.

The average correction required was less than five per cent of the total dissolved carbonate concentration, and for only 12 samples was the correction greater than 15 per cent. These samples were of relatively high temperature waters of high dissolved carbonate content from the Muschelkalk at Weiach (302/10a, b) and the springs of Baden and Ennetbaden (201, 203, 206 and 219), from the Buntsandstein at Weiach (302/12), the Buntsandstein and Permian of Riniken (303/3 and 303/5b, c), and the Buntsandstein and crystalline of Schafisheim (304/5, 6 and 304/10). The corrections to these samples are described by PEARSON AND OTHERS (1989) and amounted to as much as 35 per cent of the total dissolved carbonate content. Two samples from the Muschelkalk at Beznau (174 and 175) required 17 per cent adjustments. The uncertainty in the ^{14}C model ages introduced by these corrections are included in the calculations.

There appears to be no correlation between the size of the carbonate correction required and the extent of sample outgassing deduced from the noble gas measurements discussed in Section 3.3. The noble gas analyses are made on small samples collected under pressure, while the measurements from which total dissolved carbonate concentrations are calculated are made on samples after exposure to the atmosphere. There is little reason to expect correlation between noble gas loss and erroneously low carbonate contents.

The procedure for calculating model ^{14}C ages described in Section 5.1 can be summarized as follows:

- reconciliation of the chemical analyses to best represent groundwater conditions;

- use of chemical mass balances to describe reaction paths by which the groundwater chemistry may have developed;

- calculation of the ^{14}C composition of the dissolved carbonate consistent with possible developmental reaction paths and the measured ^{13}C content of the water; and

- comparison of the calculated ^{14}C content with the measured value to arrive at the model age of the water.

The uncertainties given with the calculated ages are based on estimated or measured uncertainties in the chemical and isotopic composition of the samples. These uncertainties were propagated through the calculations using risk analysis procedures. Several hundred repetitive calculations were made for each sample. In each repetition, different parameter values were chosen using the Latin hypercube (stratified Monte Carlo) sampling technique. The uncertainties given in the table correspond to one standard deviation (σ) of the resulting distribution of calculated ages. When only limiting measured ^{14}C values were available, or when the age is that of one component of a mixed sample, limiting ages are given based on 2-σ values of the calculated distribution.

The uncertainty values used for the parameters required for the adjustment calculations were chosen as discussed in Section 5.1 and were as follows:

- chemical composition of groundwater: concentration values are normally distributed with standard deviation, σ, of ten per cent for dissolved carbonate species and five per cent for all other dissolved constituents. Ten per cent is about twice the correction to the total carbonate content required to bring the average sample to calculated calcite saturation;

- isotopic composition of dissolved carbonate of groundwater: both δ^{13}C and ^{14}C values are normally distributed with σ_{14C} as reported by the laboratory and $\sigma_{\delta^{13}C} = 0.5$ per mil;

- measured isotopic composition of initial dissolved carbonate: δ^{13}C and ^{14}C were assigned triangular distributions with minimum, most likely, and maximum values of 100, 100, and 102 pmc, and -26, -25, and -15 per mil, respectively. These distributions result from the assumption that, at times greater than a few hundred years ago when samples sensitive to the ^{14}C method would have been recharged, normal soil-air CO_2 in Switzerland had a δ^{13}C value of -25 ± 1 per mil and a ^{14}C content of 100 pmc. Carbonate dissolved from gaseous CO_2 under closed system conditions will have the same isotopic composition as the CO_2 itself. Carbonate dissolving under open system conditions, however, could be enriched by as much as 10 per mil in ^{13}C and 2 pmc in ^{14}C over that of the gas. Closed system dissolution is considered far more likely than open system dissolution, so the open system values (-25 per mil and 100 pmc) were taken as the most probable values, and the closed system limits (-15 per mil and 102 pmc) as the least probable values;

- isotopic composition of dissolving mineral carbonate: ^{14}C = 0, δ^{13}C uniformly distributed with minimum and maximum values of 0 and +2 per mil, respectively. This range assumes that the mineral carbonate is of marine origin. For waters from the non-marine USM, a triangular distribution between -7 and +2 per mil, with a most probable value of 0 per mil, was used.

For most samples, several chemical reaction paths could be chosen which would lead to the measured ^{13}C contents of the groundwater. Four interdependent parameters determine the calculated ^{13}C content. They are the δ^{13}C value of the initial carbonate, the δ^{13}C value of the dissolving carbonate minerals, the proportion of the total dissolved magnesium which is assumed to come from dolomite solution, and the amount of calcite which has been dissolved and reprecipitated. To determine how these parameters were related for each sample modelled, a preliminary calculation was made using fixed values for the uncertain parameters as follows:

- the δ^{13}C values of the initial and mineral carbonate were set to the heaviest values in their chosen range (-15 per mil and +2 per mil, respectively);

- the amount of magnesium derived from dolomite solution was set to the value which would yield the measured groundwater δ^{13}C value with the minimum of calcite dissolution. Often it was not possible to match the measured δ^{13}C value unless the δ^{13}C value of the heaviest initial carbonate was taken to be more negative than -15 per mil. For many samples, this restricted the initial δ^{13}C values which were possible to ranges as narrow as -20 or even -23 to -25 per mil. DEINES AND OTHERS (1974) and REARDON AND FRITZ (1978) have previously used the ^{13}C content of dissolved carbonate and chemical evolution modelling to calculate the range of the initial δ^{13}C values possible for given groundwaters; and

- with the amount of non-dolomitic magnesium fixed at this value, several values for the δ^{13}C of the initial and mineral carbonate were chosen, and the amount of dissolving calcite required to match the measured δ^{13}C value of the water was determined. From these sets of values, linear regression was used to calculate the coefficients of the equation

$$\text{dissolving calcite} = X_1 + X_2 \cdot \delta^{13}C_{mineral} + X_3 \cdot \delta^{13}C_{initial\ carbonate}$$

The coefficients X_i in this equation are specific for a given sample and express how the amount of calcite dissolved is related to the δ^{13}C values of the initial carbonate and dissolving mineral carbonate. In the uncertainty modelling, these δ^{13}C values are chosen from their defined distributions as discussed above, and the amount of dissolving carbonate is calculated using this equation.

The model ages in thousands of years (ka) for samples for the deep boreholes are given in Table 5.2.1 and for those regional programme and other studies in Table 5.2.2.

5.2.2 Discussion of Results

As pointed out in Section 2.1.2 and illustrated in Figure 2.1.1, there is a strong relationship between the ^{3}H and the ^{14}C contents of these waters. All lie along, or on the high-^{14}C side of the dashed line on Figure 2.1.1, which passes through the origin. Therefore, the age results are discussed as if the samples were mixtures between a young ^{3}H-bearing water containing modern ^{14}C and an older ^{3}H-free component. The proportions of young water in many samples can be determined from their ^{3}H and (or) ^{85}Kr contents or, in the case of the Nagra borehole samples, from their drill fluid tracer contents, as described in Sections 2.1 and 4.2. Estimates of the ^{14}C contributions of young water to all samples, and the data on which these estimates are based are given in Tables 2.1.2, 2.1.3, and 4.3.2.

In principle, if the amount of ^{14}C contributed by the young component is known, it could be subtracted from the measured ^{14}C content and the difference used to estimate the age of the old component. This procedure is rarely possible in practice for several reasons:

- The ^{3}H content of young water--water recharged since the start of the nuclear era--has varied from less than 10 to over 1000 TU in northern Switzerland (Figure 4.3.1). The ^{14}C content of plants, which is equivalent to the ^{14}C content of recharging groundwaters, has varied only between a nominal 100 pmc and a maximum of about 140 pmc. Thus, the ^{14}C to ^{3}H ratio of young groundwaters has varied so greatly that the precision with which the ^{14}C content of an old component can be estimated is too low to be useful for purposes of estimating the age of that old component;

- Second, mixing between old and young waters affects not only their ^{14}C contents, but their ^{13}C values and chemistry as well. To correctly adjust the ^{14}C content of the old component of a mixture would require the δ^{13}C value and chemistry of that component as well as its ^{14}C content;

- Third, many samples from the Nagra boreholes contain dissolved organic carbon (DOC) which may well be derived from material introduced into the borehole during drilling. As discussed in Section 2.1.2, this carbon is probably ^{14}C-bearing, and its oxidation could add ^{14}C-bearing dissolved carbonate to the samples. Although the DOC contents of many samples were measured and are given in Tables 2.1.2 and 5.2.1, the ^{14}C content of the DOC is not known, nor is the amount which might have oxidized to carbonate in a given sample. Thus, the presence of DOC is only a qualitative indication that some ^{14}C contamination may be present.

These considerations give rise to the conventions followed in reporting the model ^{14}C ages of the Nagra samples in Tables 5.2.1 and 5.2.2, and to the discussion of these ages. In particular, the samples are divided on the basis of their ^3H content into those which are likely to contain significant ^{14}C contributed by a young water component, and those with little or no young ^{14}C.

5.2.2.1   Samples Containing Modern ^{14}C

Samples containing more than a few per cent of young water, or ^{14}C of modern origin, are discussed in this section. Two groups are distinguished: one of samples containing virtually 100 per cent young ^{14}C; the other of samples containing smaller, but still significant amounts of young water or ^{14}C of modern origin.

As shown in Figure 2.1.1, the ^3H and ^{14}C contents of samples from Kaisten Felsbohrung (12), Windisch BT 2 (16), Meltingen 2 (19), Beuren (100), and Görwihl Bohrloch 30 (167) suggest they are virtually all young groundwater. The ^{14}C contents of their initial dissolved carbonate were calculated assuming model ages of zero. These concentrations, which are given in the comments to Table 5.2.2, range from 104 ± 8 to 132 ± 8 pmc. These are in the range of atmospheric ^{14}C contents during the latter part of the nuclear era.

The ^{14}C content of the Eptingen sample (23) is the highest measured, and indicates a source of nuclear-era ^{14}C virtually undiluted by geochemical effects. The ^3H content of this sample is also high, and supports a young water age. The $\delta^{13}C$ values of samples with dissolved ^{14}C contents approaching those of plants or the atmosphere are generally in the range of -20 to -25 per mil, as discussed in Section 5.1.2. The sample from Görwihl Bohrloch 30 (167) with the next highest measured ^{14}C content, 108 ± 1 pmc and $\delta^{13}C = -21$, is an example of such a sample. The Eptingen sample, however, has an enriched $\delta^{13}C$ value of -3.6 per mil, which is typical of a highly evolved water, and which would normally be associated with a relatively low ^{14}C content, even in a sample with a low water age. The Weiach Muschelkalk sample (302/10a, b), which contains water of low age and has $\delta^{13}C = -4.6$ per mil and $^{14}C = 13.28$ pmc, exemplifies this type of sample. CO_2 gas evolution from water tends to enrich the remaining dissolved carbonate in the heavier isotope. Such CO_2 loss could be driven by gypsum dissolution in a calcite-saturated water, for example, or simply by the presence of the water in an environment with a P_{CO_2} lower than the one in which the water evolved. Processes such as these may account for the ^{13}C and ^{14}C contents of the Eptingen sample. Whatever may have brought about the ^{13}C enrichment in this sample, however, its ^{14}C content alone indicates that it is modern.

Two samples have low or undetectable ^3H contents, indicating that they include little or no young water, yet have ^{14}C contents which adjust to initial values ≥ 100 pmc. These are the samples from Schönenbuch (5), with an adjusted initial ^{14}C content of 119 ± 13 pmc and from Schafisheim (304/1, 2) with 125 ± 23 pmc. Both these samples are from the non-marine USM so there is considerable uncertainty in the range of $\delta^{13}C$ values used for dissolving mineral carbonate in the calculation of their model ages. The initial ^{14}C contents calculated for both samples exceed 100 pmc by less than 2 σ, so these samples are considered to have modern ^{14}C model ages, but to have been recharged before the start of the nuclear era.

The ^{14}C content of the Schafisheim sample is inconsistent with its high ^{40}Ar/^{36}Ar ratio and saline Na-Cl character, both of which suggest a water of considerable age. BALDERER (1990) proposes that this sample represents a mixture between a deep, old Na-Cl water of the sort represented in the Na-Cl waters from the Tertiary and Malm, and a modern, but ^{3}H-free, water.

Several samples are probably mixtures of young and prenuclear waters. All contain ^{3}H, indicating the presence of some young water, but have ^{14}C contents corresponding to recharge before the nuclear era. However, the uncertainties in their ^{14}C model ages are such that they could contain all nuclear era ^{14}C. These samples are from Pratteln (8), Densbüren Felsbohrung (14), Oberbergen (96), Schinznach Bad Alt (119), Lostorf 3 (123), Magden Stockacher (274), and Sauldorf (502). These samples are given definite ages (±1σ) in Table 5.2.2, even though their 2-σ range includes ages of zero. ^{39}Ar data on the Densbüren Felsbohrung (14), Lostorf 3 (123), and Magden Stockacher (274) samples suggest ages of a few hundred years for the waters, consistent with the ^{14}C results (Table 4.4.2). The sample from Sulzburg Waldhotel (166) has the ^{3}H content of a young water, but a model age older than zero by more than 2 σ.

The samples from the Kimmeridgian (302/6) and Muschelkalk (302/10a, b) at Weiach are probably also best considered members of this group. They have ^{3}H contents of < 5.7 and 3.2 ± 0.7 TU, respectively, and are highly evolved chemically, as shown by their δ^{13}C values of -4.2 and -4.6 per mil. Most of the Nagra samples with δ^{13}C values more positive than -5 per mil also have ^{14}C contents which are below detection, or can be attributed entirely to mixing with young water or to contamination. Only these samples from Weiach and the one from Eptingen (23) have both δ^{13}C values more positive than -5 per mil and significant ^{14}C contents. The sample from the Muschelkalk (302/10a, b) has a ^{14}C content which, after adjustment, is equivalent to an initial ^{14}C concentration of 150 ± 70 pmc. The conclusion from the ^{14}C, that the older component of this sample is modern or at least relatively young, is consistent with its ^{39}Ar age of 800 ± 60 years (Table 4.4.2).

The model ^{14}C age content of the Kimmeridgian sample (302/6) is given as > 1 ka rather than as modern. This is done because the upper limit of the ^{3}H content of this sample, 5.7 TU, neither includes the presence of young water, as does the measurable ^{3}H content of the Muschelkalk sample, nor does it exclude it. Thus, in the absence of a definitive ^{3}H value, the ^{14}C value can only be interpreted as a limiting age. If the ^{3}H information is neglected, the model ^{14}C age for this sample would be 2.8 ± 1.8 ka. BALDERER (1990) interprets this sample as he does that from the USM of Schafisheim (304/1, 2) to be a mixture of a deep, ^{14}C-free Na-CL water and a modern, ^{3}H-free water.

The remaining samples in this group also contain low but measurable ^{3}H contents, but have modelled initial ^{14}C contents which are more than 2 σ below 100 pmc. These waters are mixtures of pre-nuclear and young waters. As mentioned above, the ^{14}C age of a mixture is not a linear function of the age and proportion of its two components. To estimate the residence times of the old components of mixtures would require knowledge of the isotopic composition of both the young and the old components. Without such information, the only safe interpretation one can make of data on mixed samples is that the age of their old component is greater than the model age calculated from the adjusted

^{14}C content of the mixture. Thus, only limiting ages are given for these samples in Tables 5.2.1 and 5.2.2. These should be attributed to the old components.

Several locations were sampled more than once and provide insight into the nature of mixed samples. Those with measurable ^{3}H, and which, therefore, appear in Figure 2.1.1, include two samples each from Densbüren Felsbohrung (14), Beuren (100), Schinznach Bad alt (119), Schinznach Bad S2 (120), Lostorf 3 (123), and three samples from Säckingen Badquelle (125). The ^{3}H and ^{14}C contents of samples from three of these locations vary sympathetically, qualitatively at least, suggesting mixing between a low ^{14}C component and varying amounts of a younger water. For these locations, it can be stated that the model age of the old component is at least as old as the greatest limiting age found. Thus, the old component at Schinznach Bad S2 (120) has a model age > 2.3 ka, while that from Säckingen Badquelle (125) is > 4.3 ka. The uncertainties in the adjusted ages of the samples from Densbüren (14) are such that even the old component could well be modern, so it is included with a group of samples discussed previously.

The variation of ^{14}C and ^{3}H in the samples from Beuren (100), Schinznach Bad alt (119), and Lostorf 3 (123) differs from that of the samples discussed in the previous paragraph. The model initial ^{14}C contents of the Beuren samples are > 100 pmc, and they are shown as modern in Table 5.2.2. The ^{14}C contents of the Schinznach Bad alt samples lead to model ages of greater than zero but with errors which overlap zero. These samples are, therefore, included in groups which were discussed above.

The sample from the Muschelkalk at Leuggern (306/1) is unique in that the combination of its ^{3}H, ^{85}Kr, and ^{39}Ar contents can be accounted for only by a three-component mixture, as shown in Table 4.4.2. Because of lack of knowledge of the ^{14}C content of young water and of the chemical character of the mixing components, and because of the uncertainty in the mixing proportions themselves, the measured ^{14}C of this sample cannot be distributed among the sample components. Therefore, as for the other mixed samples, the only conclusion possible is that the age of the old component is greater than the model age of the mixture itself. This age is shown in Table 5.2.1 as > 5.5 ka.

Some samples with young water have an old component of significant age. The four samples from the thermal springs of Baden and Ennetbaden (201, 203, 206 and 219) are mixtures of waters of model ages of at least 4 ka with from five to ten per cent young water. The sample from Beznau 104.5 to 109.5 m (171), although its ^{14}C yields only an upper limiting age, also contains some young water. The 649.0 m sample from the crystalline at Böttstein (301/23) is in this category, as is the sample from Reichenau (501).

Several additional samples from the Nagra boreholes have ^{14}C concentrations greater than a few pmc, some or all of which may have been contributed by a young fluid component or by the oxidation of dissolved organic carbon (DOC) containing ^{14}C. The sample from the lower crystalline at Böttstein (301/21) has a ^{14}C content of 154 pmc. Drilling fluid makes up nearly 50 per cent of this sample. Exchange between this drilling fluid and the atmosphere would account for all the ^{14}C in the sample. The 1116.5 m sample from the Permian at Weiach (302/19) contains 13 to 14 pmc, but also has a relatively high DOC. The samples from the deeper Permian and the crystalline at Weiach (302/18 and 302/16) may have as much as 12 pmc from young water, which is higher than the ^{14}C content measured for either of them. The $\delta^{13}C$ values for these three deeper Weiach samples are so unlike values to be expected from groundwaters of this type, or have such large associated errors, that not even limiting ages could be calculated for these samples.

The samples from the Permian at Riniken (303/5b, c and 303/6) have significantly more ^{14}C than could have been contributed from contamination with young fluid. The high DOC, particularly of the deeper sample, suggests that at least some of the ^{14}C present could be from this source of contamination, however, so only limiting adjusted ages are given. The Muschelkalk sample from the Schafisheim borehole (304/3) has a ^{14}C concentration of about 4 pmc, larger than the amount likely to have been introduced by fluid contamination. The DOC concentration of this sample is low, but still could have been a source of ^{14}C. No interpretation of the ^{14}C content, even as a limiting age, is possible because of the lack of a $\delta^{13}C$ value. The samples from the 1203.2 m and 1433.4 m intervals of the crystalline at Leuggern (306/20 and 306/26) may have ^{14}C contents larger than can be attributed to fluid contamination (although the DOC of the deeper sample is high for this borehole). Limiting ages of the older component of these samples are given as well. These were small samples and the ^{14}C and ^{13}C measurements were made using the AMS technique. The ^{14}C model ages were calculated using $\delta^{13}C$ values of -9.5 per mil, similar to the $\delta^{13}C$ of the intervals immediately overlying and underlying these samples.

5.2.2.2 Samples with Virtually No Modern ^{14}C

The ^3H and (or) ^{85}Kr contents of the remaining samples from the regional programme and other sources, and the drilling fluid content of the remaining borehole samples, are so low that they are unlikely to contain more than 2 or 3 pmc from mixing with young water. These remaining samples can be considered in two groups: samples with ^{14}C contents less than about 3 pmc; and samples containing significantly higher ^{14}C contents. The ^{14}C in the samples of the first group could have been a result of mixing with small amounts of young water, so the model ages calculated for them are given as younger limiting values. The higher ^{14}C contents of the other group of samples may well represent their residence times.

Samples with such low ^{14}C contents that only limiting model ages can be calculated include: Zürich Aqui (1), Zürich Tiefenbrunnen (2), Kaiseraugst (3), Beznau (15 and 172 through 175), Neuwiller (77), Waldkirch (95), Singen (97), Konstanz (98), Lottstetten (127), Mainau (128), Rheinfelden (159), Säckingen Stammelhof (161), Birnau (508), and Ravensburg (509). Nagra borehole samples of this type include those from the Keuper at Riniken (303/1), the Muschelkalk at Riniken and Böttstein (303/2 and 301/2), and the Buntsandstein and Permian from Weiach (302/12), Riniken (303/3), Kaisten (305/1 and 305/2), and Leuggern (306/2). The ^{14}C measured in the samples from the Buntsandstein and crystalline at Schafisheim (304/5, 6 and 304/10), the sample from the Böttstein crystalline at 621.3 m (301/12b), and all samples from the crystalline of Kaisten and Leuggern except those previously mentioned, can also be attributed to the presence of contaminating borehole fluid.

Samples with relatively high ^{14}C contents yielding model ages of up to 10 ka include Frenkendorf (9, 7.2 ± 1.0 ka), Lostorf 4 (124, three samples of which have a mean model age of 5.7 ± 0.4 ka), and Magden Eich (273/102 and 273/103, which have overlapping model ages of 6.8 ± 1.0 and 7.6 ± 1.2 ka). The only Nagra borehole samples of this type are from the Buntsandstein-weathered crystalline zone at Böttstein (301/25b and 301/5) which yield overlapping model ages of 9.0 ± 0.9 and 7.6 ± 1.4 ka.

Several samples with no ^3H and measurable ^{14}C contents represent mixtures and so their adjusted ^{14}C contents are expressed only as limiting ages. The samples from Eglisau (141 to 144) represent mixtures of a water of relatively low mineralization and ^2H and ^{18}O contents similar to those of young waters, with a more highly mineralized water enriched in ^2H and ^{18}O (Sections 1.5.3 and 3.2.1). The sample from Eglisau 1 (141/1) is one of the least mineralized of the Eglisau group. Its model ^{14}C age is 9.0 ± 3.0 ka. In Table 5.2.2, the younger limiting age of its old component and the older limiting age of its young component are given.

Samples from Magden Weiere (271) and Magden Falke (272) were taken at several times and had different measured ^{14}C contents. These are interpreted as the product of mixing between two waters. Thus, the model ^{14}C ages of the samples from these locations are given as a younger limiting age for the old component and an older limiting age for the young component.

Samples from Zurzach 1 and 2 (131 and 132) and three samples from the crystalline of Böttstein (301/8c, 301/16 and 301/18) have ^{14}C contents ranging from about 4 to 8 pmc, and yield model ages of 13 to 21 ka. These ^{14}C contents are too high to have been introduced entirely by mixing with young, nuclear-era water because these samples all have undetectable ^3H contents. These and other samples yielding definite ^{14}C ages > c. 12 ka are discussed in the following section.

5.2.2.3 Samples Yielding Definite ^{14}C Ages Greater than c. 12 ka.

The hydrologic implications of the model ^{14}C ages are discussed in the synthesis chapter, Chapter 8, of this report. However, certain of the ages given in Tables 5.2.1 and 5.2.2 may indicate groundwater mixing or reflect on the processes by which the raw ^{14}C data were interpreted as ages. This section addresses samples of this type.

Definite, or real, model ages (as opposed to limiting ages) > c. 12 ka would represent waters recharged during glacial times. Such waters should also be characterized by, for example, more negative ^2H and ^{18}O values and lower noble gas temperatures than young waters. Model ^{14}C ages > 12 ka could also result from mixing between ^{14}C-free, pre-glacial water and water younger than 10 to 12 ka. The samples from Magden illustrate the possibilities of mixing. One sample from Weiere (271/101) and one from Falke (272/103) have measured ^{14}C concentrations which correspond to model ages of 18 ± 1.5 and 16.5 ± 1.2 ka, respectively. As discussed in the previous section, however, these samples are not all of waters of one age, but are mixtures of waters older than 18 ka and younger than c. 5 ka. The fact that the waters were mixtures could be detected because several samples were available. It is possible that undetected mixing may have affected the apparent ages of other samples as well.

Water from the crystalline at Kaisten (305) and Leuggern (306) boreholes, from the upper part of the crystalline at Böttstein (301), and from the thermal boreholes at Zurzach (131 and 132) are chemically similar. They also have similar ^{18}O and ^{2}H contents, and have ^{14}C contents which range from below detection in most samples from the Nagra boreholes, through 3 to 7 pmc in Zurzach samples, to 14 pmc in the Buntsandstein and one deeper crystalline zone at Böttstein. The ^{14}C in the Böttstein crystalline sample may have been due to mixing downward of Buntsandstein water during drilling and sampling (PEARSON, 1985), so it can be discounted. The Zurzach samples, however, lead to real model ages between 15 ± 1 and 21 ± 1 ka.

The ^{2}H and ^{18}O contents of these crystalline waters are identical with those of high-^{3}H waters in the same region, suggesting that the climate of recharge of these groundwaters is similar to that of the present (BALDERER AND OTHERS, 1987a). If so, it would mean that the waters were not recharged during glacial periods, but are mixtures of much older waters with more recent recharge. On the other hand, the noble gas contents of waters from this group would result from recharge temperatures lower than the present mean annual temperature in the region, and lower than those calculated from two samples with high-^{14}C contents from Germany (see Figure 3.3.8a and RUDOLPH AND OTHERS, 1983). A lower recharge temperature would be consistent with recharge during glacial times, so that it would not be necessary to consider mixing in the interpretation of the ^{14}C content of these samples.

It is also possible to assert that these crystalline waters were entirely recharged during present climatic cycle, and that their low ^{14}C contents results from diffusive loss of ^{14}C from flowing groundwater into essentially stagnant matrix water. Such diffusion has been proposed to account for disparate ages of groundwaters from the Lincolnshire Limestone (DOWNING AND OTHERS, 1977), as a matter for general concern in groundwater dating by FRITZ (1982), and as a particular concern for ^{14}C studies by NERETNIEKS (1981). Matrix diffusion has also been considered as a mechanism capable of retarding the movement of radionuclides in groundwater in granites (NERETNIEKS, 1981 and HADERMANN AND JACOB, 1987).

In the interpretation of these samples in the remainder of this report, the noble gas results are accepted as indicating recharge during a cooler period than the present. The model ^{14}C ages given in Tables 5.2.1 and 5.2.2 are also accepted at their face values.

^{2}H and ^{18}O isotope concentrations and noble gas contents are useful to evaluate the model ^{14}C ages of several other samples. The ^{2}H and ^{18}O contents of the sample from the Keuper at Riniken (303/1) are much higher than those of other Keuper samples from northern Switzerland (Table 3.1.3 and Figure 3.1.5). All but the Riniken sample are modern, so the fact that the latter sample has a high model age is not unreasonable. Likewise, the ^{2}H and ^{18}O contents of the Buntsandstein samples from Weiach (302/12) and Riniken (303/3) are different from those of modern waters (Figure 3.1.9) and are consistent with their great ^{14}C model ages.

214

Table 5.2.1: Adjusted 14C ages of samples from Nagra deep boreholes. (Page 1 of 5)

Sample Location Number	Source Formation	Mean Depth (m)	Quality Block (WITTWER, 1986)	3H (TU) (GSF)	Dissolved Organic Carbon (mg/l)	14C Contamination from Fluid (pmc) (Table 2.1.2)	CARBON ISOTOPES Delta 13C (per mil)	14C (pmc)	Adjusted 14C Age ± One Sigma (ka)	– – – – – C o m m e n t s – – – – –
	BÖTTSTEIN									
301/2	mo	162.9	1	1.0	3.5	< 1.4	-4.8	1.49 ± 0.12	> 17	All measured 14C could be from fluid contamination; limiting age given. Sample 15/1 from nearby Muschelkalk with same chemistry has < 0.4 pmc; 39Ar undetectable this sample and in 015/1.
301/25b	s-KRI	312.5	2	0.2 ± 0.2	2.90	< 0.2	-9.3	14.03 ± 0.13	9.0 0.9	Negligible 14C from fluid contamination in either sample; Ages of both samples overlap; No effect evident from higher DOC in 301/5.
301/5	s-KRI	316.6	2	0.7	11.10	< 0.9	-6.3	12.13 ± 0.25	7.6 1.4	
301/8c	KRI	399.5	1	< 1.0	0.90	< 1.1	-9.3	8.25 ± 0.13	13.2 1.0	samples 8c, 16 and 18 have overlapping ages and negligible 14C from fluid contamination; Ages are improbable; samples may be mixtures of 12b and 25b-5-23 types.
301/16	KRI	618.4	1	1.0 ± 0.7	0.74	< 1.1	-9.0	7.89 ± 0.09	13.4 1.0	
301/12b	KRI	621.3	1	1.0 ± 0.7	0.82	< 1.1	-8.8	1.30	> 27	14C below detection; Limiting age given; Age is distinct from other Böttstein crystalline samples but similar to those from Kaisten and Leuggern crystalline.
301/23	KRI	649.0	2	3.2 ± 0.7	– –	< 2.6	-9.1	13.80 ± 0.13	> 7	Calculated with chemistry of 12b; 14C resembles samples 25b and 5 from s/KRI rather than other crystalline samples; some 14C from fluid contamination, limiting age given.
301/18	KRI	792.4	1	0.9 ± 0.9	0.53	< 1.0	-9.1	7.64 ± 0.07	13.8 1.0	
301/21	KRI	1326.2	2	57 ± 5	0.20		-6.3 ± 2.9	154.08 ± 2.16		All 14C from young fluid contamination.

Table 5.2.1: Adjusted 14C ages of samples from Nagra deep boreholes. (Page 2 of 5)

Sample Location Number	Source Forma- tion	Mean Depth (m)	Qual- ity Block (WITTWER, 1986)	3H (TU) (GSF)	Dis- solved Organ- ic Car- bon (mg/l)	14C Con- tami- na- tion from Fluid (pmc) (Table 2.1.2)	CARBON ISOTOPES Delta 13C (per mil)	14C (pmc)	Ad- justed 14C Age ± One sigma (ka)	Comments
WEIACH										
302/6	joki	255.0	2	< 5.7	18.00	< 4.4	-4.2	12.47 ± 0.25	> 1	Some 14C probable from fluid contamination and (or) DOC; Limiting age given.
302/10a,b	mo	859.1	1	3.2 ± 0.7	34.00	< 4.3	-4.6 ± 0.5	13.28 ± 0.14	Mod- ern	14C negligible from fluid contamination but possible from DOC; Adjusted initial 14C = 150 ± 70 pmc; Consistent with 39Ar result of 0.8 ka.
302/12	s	985.3	1	0.8	3.90	< 0.9	-8.3	3.63 ± 0.15	> 17	14C negligible from fluid contamination but possible from DOC; Limiting age given.
302/19	r	1116.5	2	2.2 ± 0.6	74.00	< 1.4	-25.4 2.6	12.60 ± 0.30 13.96 ± 0.50		14C negligible from fluid contamination but possible from DOC; 13C values suspect; No 14C adjustment attempted.
302/18	r	1408.3	2	6.4 0.9	21.50	11.3	-4.4 +4.9	0.40		13C values suspect; No 14C adjustment attempted.
302/16	KRI	2218.1	2	2.7 ± 0.7	7.80	11.5	-26.7 ± 2.0	10.10 ± 0.40		All 14C from young fluid contamination.
RINIKEN										
303/1	km	515.7	1	1.2	1.4	< 2.1	-8.2	3.35 ± 0.30	> 17	Much of measured 14C could be from fluid contamination; Limiting age given; 2H and 18O contents suggest recharge during much warmer period than present, consistent with a great age: Sect. 3.2.3.
303/2	mo	656.7	1	1.1	1.0	< 1.8	-4.8	0.90 ± 0.20	> 19	All measured 14C could be from fluid con- tamination; Limiting age given; 39Ar undetectable; 2H, 18O and noble gas tem- perature suggest recharge during period like the present: Sect. 3.3.5.
303/3	s-r	806.6	1	0.7 ± 0.7	0.68	< 0.5	-9.2	0.55 ± 0.20	> 30	All measured 14C could be from fluid con- tamination; Limiting age given.

Table 5.2.1: Adjusted 14C ages of samples from Nagra deep boreholes. (Page 3 of 5)

Sample Location Number	Source Forma-tion	Mean Depth (m)	Qual-ity Block (WITTWER, 1986)	3H (TU)	Dis-solved Organ-ic Car-bon (mg/l)	14C Con-tami-na-tion from Fluid (pmc) (Table 2.1.2)	CARBON ISOTOPES Delta 13C (per mil)	CARBON ISOTOPES 14C (pmc)	Ad-justed 14C Age ± One Sigma (ka)	Comments
	RINIKEN (continued)									
303/5b,c	r	965.5	2	1.1	9.2	< 0.7	-7.6	10.85 ± 0.20	> 6.8	14C negligible from fluid contamination but possible from DOC; Limiting age given.
303/6	r	993.5	2	2.6 ± 1.0	144	< 2.7	-14.6	40.81 ± 0.60	> 3	14C negligible from fluid contamination but negative 13C and 14C possible from DOC; Limiting age given.
	SCHAFISHEIM									
304/1,2	USM	558.0	1	1.2	2.0	< 3.5	-8.5	40.46 ± 0.70	Modern	Some 14C probable from fluid contamina-tion; Adjusted 14C = 125 ± 23 pmc; Prob-ably Modern water.
304/3	mo	1260.4	1	0.9	4.0	< 0.5		4.41 ± 1.20		14C negligible from fluid contamination possible from DOC; No adjustment possible without 13C value.
304/5,6	s-KRI	1488.2	2	1.8 ± 0.7	95.0	< 0.6	-21.2 ± 2.0	1.40 ± 0.40	> 21	All measured 14C could be from fluid con-tamination and (or) high DOC; Limiting age given; Adjustment made using 13C of 304/10.
304/10	KRI	1887.9	2	1.5 ± 0.9	0.3	< 0.4	-9.2	0.70	> 31	All measured 14C could be from fluid con-tamination; Limiting age given.
	KAISTEN									
305/1	s	113.5	1	0.4 ± 0.2	0.34	< 1.6	-17.2 -9.8	0.50 1.64	> 25	All measured 14C could be from fluid con-tamination; Limiting age given; Calcu-lations made with more positive 13C value.
305/2	r	284.3	1	0.7 ± 0.3	0.69	< 0.7	-17.8 -9.7	0.30 0.80 ± 0.20 2.52 ± 0.30	> 22	All measured 14C could be from fluid con-tamination; Limiting age given; Calcula-made with more positive 13C value.
305/4	KRI	310.4	1	1.1 ± 0.7	0.33	< 1.1	-9.8	0.98 ± 0.20	> 29	All measured 14C could be from fluid con-tamination; Limiting age given.

Table 5.2.1: Adjusted 14C ages of samples from Nagra deep boreholes. (Page 4 of 5)

Sample Location Number	Source Formation	Mean Depth (m)	Quality Block (WITTWER, 1986)	3H (TU)	Dissolved Organic Carbon (mg/l)	14C Contamination from Fluid (pmc) (Table 2.1.2)	CARBON ISOTOPES Delta 13C (per mil)	14C (pmc)	Adjusted 14C Age ± One Sigma (ka)	Comments
	KAISTEN (continued)									
305/6	KRI	482.6	1	1.3 ± 0.7	0.55	< 0.9	-10.4	0.86 ± 0.20	> 31	All measured 14C could be from fluid contamination; Limiting age given.
305/9	KRI	819.4	1	0.9 ± 0.7	0.58	< 0.6	-9.9	0.63 ± 0.20	> 32	All measured 14C could be from fluid contamination; Limiting age given.
305/12	KRI	1031.0	1	0.8	0.10	< 0.5	-9.2	0.84 ± 0.20	> 30	All measured 14C could be from fluid contamination; Limiting age given.
305/16	KRI	1271.9	1	1.2 ± 0.7	0.95	< 0.8	-10.1	1.56 ± 0.30	> 26	All measured 14C could be from fluid contamination; Limiting age given.
	LEUGGERN									
306/1	mo	74.9	1	11.6 ± 1.1	0.30	16.0	-15.8 / -6.4	12.00 ± 0.30 / 11.23 ± 0.30	> 5.5	Adjustment made with 13C = -6.4. 3H, 85Kr, 39Ar, and 14C contents of sample consistent only with 3-component mixture; Limiting age is that of oldest component; See Table 4.4.2 and text.
306/2	s	217.9	2	0.8	0.91	< 2.4	-9.3	1.56 ± 0.20	> 26	All measured 14C could be from fluid contamination; Limiting age given.
306/4	KRI	251.2	1	0.8 ± 0.7	0.22	< 0.8	-10.1	1.36 ± 0.30	> 27	All measured 14C could be from fluid contamination; Limiting age given.
306/5	KRI	444.2	1	1.2	0.70	< 1.4	-10.1	3.19 ± 0.20	> 20	All measured 14C could be from fluid contamination; Limiting age given.
306/7	KRI	538.0	1	1.0	0.28	< 1.4	-10.2	2.41 ± 0.20	> 23	All measured 14C could be from fluid contamination; Limiting age given.
306/9	KRI	705.7	1	1.3	0.30	< 2.3	-10.2	1.40 ± 0.30	> 27	All measured 14C could be from fluid contamination; Limiting age given.
306/17	KRI	847.0	2	3.5 ± 0.7	2.20	< 4.1	-9.8	2.18 ± 0.50	> 23	All measured 14C could be from fluid contamination; Limiting age given.
306/20	KRI	1203.2	2	1.3 ± 0.7	0.93	< 1.5	-26 ± 5	4.50 ± 0.30	> 16	Possibly some 14C above fluid contamination; Limiting age given; Adjustment made with 13C = -9.5.

Table 5.2.1: Adjusted 14C ages of samples from Nagra deep boreholes. (Page 5 of 5)

Sample Location Number	Source Formation	Mean Depth (m)	Quality Block (WITTWER, 1986)	3H (TU)	Dissolved Organic Carbon (mg/l)	14C Contamination from Fluid (pmc) (Table 2.1.2)	CARBON ISOTOPES Delta 13C (per mil)	14C (pmc)	Adjusted 14C Age ± One Sigma (ka)	Comments
	LEUGGERN (continued)									
306/26	KRI	1433.4	2	1.4 ± 0.9	4.00	< 1.9	-23 ± 5	11.70 ± 0.30	> 8	14C could be from high DOC for borehole; Limiting age given; Adjustment made with 13C = -9.5.
306/16	KRI	1643.4	1	0.7 ± 0.7	0.34	< 1.8	-9.3	1.19 ± 0.20	> 27	All measured 14C could be from fluid contamination; Limiting age given.

Table 5.2.2: Adjusted 14C ages of samples from the regional programme and other sources. (Page 1 of 5)

Sample Location Number	Source Formation	Tritium (TU)	CARBON ISOTOPES		Adjusted Ages ± one Sigma (ka)	Comments
			Delta 13C (per mil)	14C (pmc)		
1/1	tOMM	< 0.7	-3.0	< 1.3	> 19	< 2% young water: Tab. 2.1.3. Chemistry of 1/1 used to adjust both samples; 39Ar below detection.
1/3			-4.28	< 1.3	> 21	
2/1	tOMM	< 0.7	-3.1	< 2.2	> 13	< 2% young water: Tab. 2.1.3; 39Ar below detection.
3/1	s-r	< 0.7	-8.3	2.85 ± 0.28	> 18	< 2% young water: Tab. 2.1.3; 2H, 18O and noble gas temperature and total mineralization suggest recharge under conditions different from present: Sect. 3.2.5, 3.3.5.
3/3	s-r	< 0.7	-9.2	2.55 ± 0.30	> 20	
5/1	tUSM	0.9 ± 0.7	-14.2	67.7 ± 0.9	Modern	Adjusted 14C = 119 ± 13 pmc.
				68.5 ± 1.1		
8/2	mo	9.2 ± 0.8	-9.5	39.4 ± 0.4	1.4 ± 1.6	ca. 15% young water: Sect. 4.3.
9/2	mo	1.1	-9.8	18.97 ± 0.3	7.2 ± 1.0	< 2% young water: Tab. 2.1.3; 39Ar gives > 0.55 ka; Tab. 4.4.2.
12/1	mo	69.8 ± 4.8	-14.0	75.6 ± 0.6	Modern	Adjusted 14C = 125 ± 7 pmc.
14/1	mo	5.2 ± 0.6	-8.9	33.3 ± 0.4	2.0 ± 1.2	Consistent with 39Ar age of 0.2 ka: Tab. 4.4.2.
14/3	mo	7.8 ± 0.9	-9.6	43.4 ± 0.4	0.4 ± 1.1	
15/1	mo	1.0	-5.4	< 0.4	> 29	< 2% young water: Tab 2.1.3; 39Ar below detection; Tab. 4.4.1; Noble gas recharge temperature and water isotopes consistent with recharge under cooler conditions and in agreement with nearby similar Böttstein mo (301/2).
16/1	mo	83.7 ± 5.6	-13.4	77.3 ± 0.6	Modern	Adjusted 14C = 132 ± 8 pmc; NTB 84-21 suggests young, mixed water.
19/1	km(+?m)	103.5 ± 6.8	-15.4	83.2 ± 0.8	Modern	Adjusted 14C = 127 ± 8 pmc; NTB 84-21 suggests 10 a residence time.

Table 5.2.2: Adjusted 14C ages of samples from the regional programme and other sources. (Page 2 of 5)

Sample Location Number	Source Formation	Tritium (TU)	Delta 13C (per mil)	CARBON ISOTOPES 14C (pmc)	Adjusted Ages ± One sigma (ka)	Comments
23/1	km	129.8 ± 8.5	-3.6	122.0 ± 0.9 121.9 ± 1.0	Modern	3H suggests modern water: Fig. 2.1.1; Enriched 13C inconsistent with young 3H and 14C; No 14C adjustment attempted; see text.
77/1	jmHR	< 0.7	-4.4	1.3 ± 0.2 1.2 ± 0.3	> 26	All 14C could be from young water: Tab. 2.1.3; Chemistry and slight 18O enrichment consistent with long residence time: Sect. 3.2.2.
95/2	KRI	< 0.9	-8.0	0.69 ± 0.2	> 31	All 14C could be from young water; 2H, 18O contents and noble gas temperature slightly above Böttstein (301), Leuggern (306), and Kaisten (305) upper crystalline and within range of present conditions: Sect. 3.3.5.
96/2	tvu	4.9 ± 0.7	-10.5	50.67 ± 0.5	0.2 ± 1.0	3H indicates ca. 5% young water: Fig. 2.1.1; Remainder also virtually modern.
97/1	jo	< 0.7	-5.2	2.2 2.2	> 18	2H, 18O and noble gas temperatures consistent with cooler conditions: Sect. 3.3.5. 10.7 pmc value grossly inconsistent with other results.
97/103 97/105	jo jo	0.6	-6.1	10.7 ± 0.4 0.5	> 31	
98/1	tOMM	< 0.7	-4.3	1.3 1.3	> 13	
98/102	tOMM	0.5	-4.0	0.6	> 26	
100/1	jo(-q)	17.0 ± 1.3	-11.2	52.4 ± 0.5 51.2 ± 0.5 54.3 ± 1.0	Modern	Adjusted 14C = 104 ± 8 pmc; 3H and chemistry of sample 149/1 used for 100/1; No 13C or chemistry available for 100/101.
100/101	jo(-q)	5.1 ± 1.1				
119/118 119/119	mo mo	46.9 ± 5.1 39.8 ± 4.5	-12.0 -10.4	41.4 ± 0.5 42.0	1.6 ± 0.9 0.4 ± 1.2	3H indicates ca. 60% young water: Fig. 2.1.1; NTB 84-21 suggests more young water than sample from location 120.
120/1 120/118	mo mo	18.0 ± 2.1	-9.3 -10.4	27.1 ± 0.4 24.5 ± 0.3	> 0.6 > 2.3	Samples are varying mixtures of 3H-bearing and 3H-free waters; The model age of the 3H-free component is > 2.3 ka; see text.
120/119	mo	24.9 ± 4.2	-9.4	28.0	Modern	

Table 5.2.2: Adjusted 14C ages of samples from the regional programme and other sources. (Page 3 of 5)

Sample Location Number	Source Formation	Tritium (TU)	Delta 13C (per mil)	CARBON ISOTOPES 14C (pmc)	Adjusted Ages ± One Sigma (ka)	Comments
123/110	mo		-9.4	44.0 ± 1.2	Modern	Adjusted 14C = 101 ± 13 pmc using 123/111 chemistry.
123/111	mo	5.8 ± 0.7	-9.6	44.1 ± 0.5	Modern	Adjusted 14C = 106 ± 14 pmc.
123/113	mo	7.1 ± 3.3	-9.3	41.0	0.6 ± 1.3	14C at age = 0.0 is 93 ± 15 pmc, overlapping other results; Samples contain < 5% young water: Sect. 4.2; Fig. 2.1.1; 39Ar gives 0.25 ± 0.03 ka, Tab. 4.4.2, consistent with 14C indication that all water is virtually modern.
124/109	mo	< 1.0	-6.8	14.8 ± 0.8	5.7 ± 2.1	
124/110	mo	< 0.8	-6.8	15.5 ± 0.2	5.3 ± 1.7	
124/111	mo		-9.3	15.0	6.1 ± 1.8	
125/1	KRI	46.5 ± 2.0	-13.0	36.9 ± 0.4	Modern	Samples are varying mixtures of 3H-bearing and 3H-free waters; The 14C age of the 3H-free component is > 4.3 ka; See text.
125/108	KRI	39.4 ± 4.2	-14.2	34.0 ± 0.4	> 3.9	
125/109	KRI	35.3 ± 4.3	-12.4	28.0	> 4.3	
127/1	jo	< 0.7	-3.8	2.1 ± 0.4	> 13	2H and 18O contents similar to young waters, but chemistry consistent with considerable age: Sect. 3.1.2.
				2.0 ± 0.3		
127/104	jo	0.8 ± 0.7	-3.9	3.5	> 11	
128/1	tOMM	< 0.9	-5.5	< 2.2	> 19	2H, 18O and noble gas temperature consistent with recharge under cooler conditions.
			-5.7	< 2.2		
128/101	tOMM	0.3		0.5	31	
131/105	KRI		-9.6	4.62 ± 0.16	19 ± 1	Waters from locations 131 and 132 closely resemble Böttstein upper crystalline, with model ages from 7 to > 27 ka.
131/115	KRI	< 2.0	-9.5	6.78 ± 0.64	15 ± 1	
132/103	KRI		-9.4	4.67 ± 0.16	18 ± 1	
132/131	KRI	< 1.0	-9.4	3.50	21 ± 1	

Table 5.2.2: Adjusted 14C ages of samples from the regional programme and other sources. (Page 4 of 5)

Sample Location Number	Source Formation	Tritium (TU)	Delta 13C (per mil)	CARBON ISOTOPES 14C (pmc)	Adjusted Ages ± One Sigma (ka)	Comments
141/1	tUSM	< 0.7	-8.7	11.1 ± 2.1	> 3 < 15	Sample is mixture of low-mineralized water with 2H and 18O values like those of young waters with highly mineralized, 2H- and 18O-enriched water; Ages are younger limiting age of old component and older limiting age of young component; See text.
159/106	KRI	< 0.9	-7.1	< 2.2	> 16	2H and noble gas recharge temperature within range of present conditions; Enriched 18O consistent with considerable age: Sect. 3.2.5, 3.3.5.
161/1	KRI	1.5 ± 0.7	-10.9	2.9 ± 0.2	> 22	3H indicates 14C could be from young water; Tab. 2.1.1; Limiting age of old component given.
166/1	KRI	23.8 ± 1.8	-13.2	45.0 ± 0.5	1.5 ± 0.5	3H indicates significant young water: Fig. 2.1.1; Remainder also virtually modern.
167/1	KRI	57.3 ± 4.0	-21.2	107.8 ± 1.1	Modern	Adjusted 14C = 124 ± 6 pmc.
171/101	km3b	5.6 ± 3.9	-6.4	< 4.7	> 12	3H indicates 14C could be from young water; Tab. 2.1.1; Limiting age of old component given.
172/101	mo3		-6.4	0.7 ± 0.3 0.7 ± 0.3	> 19	The low 14C contents of samples 172 to 174 could well be from young water contamination, so only limiting ages are given; These are consistent with sample 15/1 from same unit in this borehole.
173/101	mo3		-6.9	1.1 ± 0.2 1.1 ± 0.2	> 21	
174/101	mo1		-6.4	1.7 ± 0.3 1.0 ± 0.3	> 16	
175/101	mm4		-6.7	1.0 ± 0.2 0.5 ± 0.3	> 23	

Table 5.2.2: Adjusted 14C ages of samples from the regional programme and other sources. (Page 5 of 5)

Sample Location Number	Source Formation	Tritium (TU)	Delta 13C (per mil)	14C (pmc)	Adjusted Ages ± One Sigma (ka)	Comments
201/1	mo	3.3 ± 0.8	-7.0	6.3 ± 0.5	> 4	Locations 201 to 219 are the thermal springs of Baden and Ennetbaden; NTB 84-21 and their 3H contents indicate 5 to 10% young water, Fig 2.1.1. This could supply all the 14C measured; Limiting ages of old component are given.
201/3	mo	3.4 ± 0.6				
203/112	mo	6.2 ± 0.6	-8.5	7.2 ± 0.3	> 8	
206/102	mo	3.6 ± 0.5	-9.9	6.0	> 12	
219/112	mo	5.8 ± 0.7	-8.0	7.0	> 7	
271/101	mo	0.8	-8.1	4.0 ± 0.3	> 15	Locations 271 and 272 contain < 3% young water, Sect. 4.2; Changing 14C indicates varying mixtures of waters of > 15 and < 5 ka; Noble gas recharge temperatures and water isotopes consistent with post-glacial waters; 39Ar gives ≥ 0.7 ka, Tab. 4.4.2.
271/103	mo <	1.1	-6.8	8.5 ± 0.3		
271/104	mo <	1.1	-8.2	10.3 ± 0.3	< 7	
272/102	mo <	1.6	-9.6	28.1 ± 0.4	< 5	
272/103	mo <	1	-9.3	5.5 ± 0.2	> 14	
273/102	mo <	1.4	-10.5	19.9	6.8 ± 1.0	Sample contains < 3% young water: Sect. 4.2; 39 Ar age of ≥ 0.57 ka only marginally consistent with 14C.
273/103	mo	0.9	-9.5	16.7 ± 0.6	7.6 ± 1.2	
274/102	mo	4.7 ± 0.7	-10.6	37.8 ± 0.7	1.5 ± 0.9	Sample contains ca. 5% young water: Fig. 2.1.1; 14C is consistent with 39Ar ages of 0.4 to 0.6 ka; Tab. 4.4.2.
501/102	q	24.0	-15.2	24.0	> 5.3	3H indicates ca. 30% young water: Fig. 2.1.1; Limiting age of 3H-free component given; 2H, 18O and noble gas temperatures suggest cooler conditions and are similar to samples from locations 97, 128 and 508 (Sect. 3.2.1, 3.3.5); This is consistent with great age for old component.
502/101	tOMM	4.0 ± 0.6	-10.8	39.6 ± 2.4	1.5 ± 0.8	3H indicates < 10% young water; 3H component has virtually modern 14C.
508/101	tOMM	0.5	-6.4	1.9	> 21	Noble gases and water isotopes in samples 508 and 509 indicate recharge temperatures colder than at present, consistent with great 14C model ages.
509/101	tOMM	0.4 ± 0.6	-4.7	0.8 ± 0.8	> 16	

5.3 Isotopic Composition of Carbonate Minerals and Water

F. J. Pearson, Jr. and W. Balderer

This section examines possible isotopic equilibria between carbon and oxygen of carbonate minerals selected from core and groundwaters sampled from the boreholes. This is done by comparing the measured calcite isotopic compositions with calculated compositions at equilibrium with each water sampled. The ^{18}O composition of calcite depends on the ^{18}O content of the water from which it precipitates and its temperature of precipitation. The ^{13}C composition of the calcite depends on the ^{13}C content of the total dissolved carbonate, the distribution of dissolved carbonate species, and the temperature. TULLBORG (1989) reports a similar comparison between dissolved and mineral carbonate in crystalline rock fractures in Sweden.

In the following sections, a table of the results of analyses and calculations is given for each Nagra borehole. The first three columns of each table show the depth and formation from which each water or calcite samples was taken, and the temperature at that depth. The depths given for water samples are the mean depths of the intervals sampled. The temperatures were taken from equations fit by a least-squares procedure to data for each borehole given by WITTWER (1986). These equations are given in the text of the following sections.

The fourth and fifth columns give $\delta^{18}O$ values for calcite. These are given relative to the PDB-standard, as discussed in Sections 1.3 and 5.1.1.1. The fourth column includes values measured on calcite samples. The fifth gives values calculated for calcite at equilibrium with the waters analysed at the temperature given in the third column. The procedure for this calculation and the value of the fractionation factor used are discussed in Section 5.1.1.1.

The sixth and seventh columns give the $\delta^{18}O$ and $\delta^{2}H$ values of the water samples. These are the same as are given in the tables in Section 3.2 and are expressed relative to the SMOW-standard.

The last four columns are $\delta^{13}C$ values. All are expressed relative to the PDB-standard. The eighth column gives the values measured on calcite samples, while the ninth gives values calculated for calcite in equilibrium with the dissolved carbonate in the water samples at the temperature given in the third column. The tenth column gives the measured $\delta^{13}C$ values of the total dissolved carbonate of the water samples, and the last column is the calculated $\delta^{13}C$ value of the bicarbonate in the water samples. The procedure for calculating the isotopic composition of bicarbonate and calcite in equilibrium with the water and the fractionation factors used are described in Section 5.1.1.2. The distributions of dissolved carbonate species required to make these calculations were taken from modelled values for each water sample given by PEARSON AND OTHERS (1989).

5.3.1 Böttstein Samples

Results for the Böttstein borehole are given in Table 5.3.1. Two dolomite samples were also measured from this borehole. These are denoted by a "D" following the depth in the tables.

Table 5.3.1: Isotopic composition of water and carbonate minerals from the Böttstein borehole.

SAMPLE			CALCITE $\delta^{18}O_{PDB}$		WATER $\delta^{18}O_{SMOW}$	WATER δ^2H_{SMOW}	CALCITE $\delta^{13}C_{PDB}$		WATER $\delta^{13}C_{PDB}$	
Depth (metres)	Unit	Temp (C)	Measured	Calculated			Measured	Calculated	Total CO_2	Bicarbonate
122.65	mo	20.7	-12.71				-1.24			
131.35	mo	21.0	-13.07				-1.20			
137.20 D	mo	21.2	1.72				1.33			
162.90	mo	21.0		-12.30	-11.16	-78.2		-1.5	-4.8	-2.2
192.90	mo	23.0	-9.93				0.51			
192.90 D	mo	23.0	-8.58				1.05			
312.50	KRI	27.0		-12.48	-10.08	-73.2		-8.0	-9.3	-9.0
316.60	KRI	27.0		-12.55	-10.15	-73.7		-4.8	-6.3	-5.8
399.50	KRI	29.5		-12.95	-10.05	-72.7		-8.0	-9.3	-9.1
544.33	KRI	34.7	-16.48				-8.80			
618.40	KRI	38.5		-14.61	-10.01	-74.1		-7.3	-9.0	-8.8
618.72	KRI	37.2	-8.87				-8.26			
619.90	KRI	37.2	-8.82				-7.90			
621.30	KRI	37.3		-14.42	-10.04	-72.7		-7.1	-8.8	-8.6
638.00	KRI	37.8	-14.38				-7.97			
649.00	KRI	39.0		-14.66	-9.97	-74.6		-7.4	-9.1	-8.9
649.10	KRI	39.0		-14.68	-9.99	-74.3			(- -)	
649.21	KRI	39.0		-14.64	-9.95	-73.7			(- -)	
722.32	KRI	40.7	-15.09				-9.45			
789.90	KRI	42.9	-11.16				-9.47			
792.40	KRI	43.0		-15.36	-9.96	-73.7		-7.3	-9.1	-9.0
792.50	KRI	43.0	-18.52				-10.73			
816.70	KRI	43.8	-17.03				-9.04			
865.25	KRI	45.4	-20.15				-6.52			
1000.35	KRI	49.9	-16.71				-8.58			
1048.86	KRI	51.5	-18.11				-9.40			
1117.73	KRI	53.8	-19.81				-8.91			
1149.95	KRI	54.9	-17.12				-10.11			
1324.27	KRI	60.7	-16.93				-10.91			
1326.30	KRI	60.8		-16.15	-7.90	-49.0				
1487.98	KRI	66.1	-19.30				-9.16			
Upper Crystalline: Mean					-10.02	-73.6			-9.1	
Sigma					0.06	0.6			0.2	

The temperature equation used was:

$$\text{Temp (°C)} = 16.6 + 0.0333 \cdot \text{Depth (metres)}$$

The isotopic composition of all nine water samples from depths between 300 and 800 metres (Buntsandstein and upper crystalline waters) are virtually identical ($\delta^{18}O$ = -10.02 ± 0.04 per mil; δ^2H = -73.6 ± 0.6 per mil). The sample from the Muschelkalk (162.9 m) is relatively depleted in the heavy isotopes ($\delta^{18}O$ = -11.16 per mil; δ^2H = -78.2 per mil), while the sample from the lower crystalline (1326.3 m) is enriched ($\delta^{18}O$ = -7.9 per mil; δ^2H = -49.0 per mil).

Six of the seven water samples from the Buntsandstein and upper crystalline also have similar carbon isotopic compositions ($\delta^{13}C$ = -9.1 ± 0.2 per mil). The sample from the Muschelkalk (162.9 m) is enriched in ^{13}C ($\delta^{13}C$ = -4.8 per mil), while the remaining Buntsandstein sample, from 316.6 m, has a $\delta^{13}C$ value of -6.3 per mil. There is no obvious explanation for the enrichment of this sample relative to the other sample from the Buntsandstein. Both Buntsandstein samples lost CO_2 gas during collection (PEARSON AND OTHERS, 1989), and under some conditions, loss of gaseous CO_2 can increase the ^{13}C content of the remaining dissolved carbonate (PEARSON AND OTHERS, 1978). Possibly, the CO_2 loss accompanying the collection of the 316.6 m sample caused fractionation, while that of the 312.5 m sample did not.

Figures 5.3.1 and 5.3.2 compare the measured $\delta^{18}O$ and $\delta^{13}C$ values of carbonate minerals with values calculated for carbonate at equilibrium with the waters. The lines on Figure 5.3.1 illustrate the influence of temperature on the calculated isotopic composition of calcite. The solid line shows $\delta^{18}O$ values in equilibrium with Buntsandstein/upper crystalline waters. This line becomes more negative with depth as a result of the increase with temperature of the fractionation factor for ^{18}O between water and calcite. The dashed lines show the $\delta^{18}O$ values for calcite at equilibrium with the groundwaters at temperatures five degrees above and five degrees below formation temperature.

No trend line is drawn in the $\delta^{13}C$ plot, Figure 5.3.2. However, this figure shows that the calculated $\delta^{13}C$ values of calcite in equilibrium with water from the upper crystalline become slightly more positive with depth. This results from the decrease with temperature of the fractionation factor for ^{13}C between total dissolved carbonate and calcite.

Based on both their ^{18}O and ^{13}C contents, it appears that the carbonate minerals from depths of 122.65 and 131.35 m could have formed from present-day Muschelkalk water. Within the crystalline section, carbonate from 638.00 m also could have formed at equilibrium with present groundwater, as could minerals from 722.32, 1000.35, 1149.95 and 1324.27 m. The ^{13}C content of the 722.32 m sample suggests equilibrium with water enriched in ^{13}C by about two per mil relative to present water. No ^{13}C data are available on waters from the lower crystalline, so no conclusions can be drawn from the ^{13}C contents of minerals from below about 1000 m.

The positive $\delta^{13}C$ values of the dolomite from 137.2 m ($\delta^{13}C$ = 1.33 per mil) and of the dolomite and calcite samples from 192.9 m ($\delta^{13}C$ = 1.05 and 0.51 per mil, respectively) suggest that all are primary depositional minerals. The relatively positive $\delta^{18}O$ value for the 137.2 m sample confirms this origin. The more negative $\delta^{18}O$ values of the samples from 192.9 m could be a result of a higher depositional temperature or from deposition from (or exchange with) water depleted in ^{18}O relative to normal sea water.

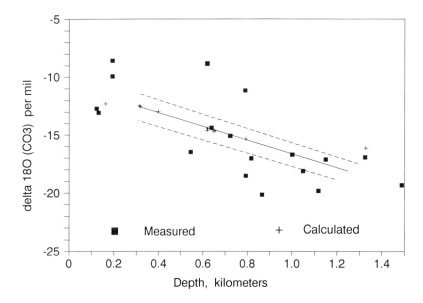

Figure 5.3.1: Oxygen isotopic composition of calcite at various depths in the Böttstein borehole.

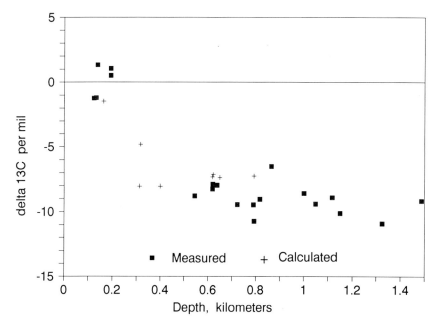

Figure 5.3.2: Carbon isotopic composition of calcite at various depths in the Böttstein borehole.

The remaining minerals, all from the crystalline, were apparently formed under different temperature conditions and (or) from waters of different isotopic compositions than exist at present.

5.3.2 Weiach Samples

Results from the Weiach borehole are given in Table 5.3.2. The temperature equation used was:

$$\text{Temp (°C)} = 7.4 + 0.0466 \cdot \text{Depth (metres)}$$

Table 5.3.2: Isotopic composition of water and carbonate minerals from the Weiach borehole.

SAMPLE			CALCITE $\delta^{18}O_{PDB}$		WATER $\delta^{18}O_{SMOW}$	$\delta^{2}H_{SMOW}$	CALCITE $\delta^{13}C_{PDB}$		WATER $\delta^{13}C_{PDB}$	
Depth (metres)	Unit	Temp (C)	Measured	Calculated			Measured	Calculated	Total CO_2	Bicarbonate
255.00	joki	19.0		-6.47	-5.76	-61.8		-2.3	-4.2	-2.9
859.10	mo	50.0		-18.32	-11.76	-86.6		-0.1	-4.6	-2.1
985.30	s	53.5		-15.26	-8.12	-60.5		-4.4	-8.3	-6.6
1116.50	r	59.0		-13.33	-5.32	-37.8			(-25.4,	2.6)
1408.40	r	70.5		-14.26	-4.61	-31.5		0.3	-4.4	-2.5
2055.58	KRI	103.2	-20.66				-10.20			
2211.60	KRI	110.4	-23.28				-14.87			
2218.10	KRI	112.0		-22.49	-8.17	-62.0			(26.7)	
2224.95	KRI	111.1	-20.37				-11.67			
2265.84	KRI	113.0	-23.33				-13.39			
2267.00	KRI	112.5		-22.43	-8.06	-62.7			(- -)	
2272.07	KRI	113.3	-16.10				-13.01			

Only minerals from the crystalline in this borehole were analysed. The oxygen isotopic composition of the minerals from 2211.60 and 2265.84 m are within one per mil (10°C) of those calculated for calcite at equilibrium with the waters from 2218.1 and 2267.0 m. The samples from 2055.58 and 2224.95 m are two per mil more positive than those calculated, and could have precipitated either:

- From present water at a temperature about 20° cooler than at present; or

- At the present temperature from waters about two per mil enriched in ^{18}O relative to present waters.

The only carbon isotope measurement for waters from the crystalline is an AMS determination on the sample from 2218.1 m. As discussed in Chapter 2, AMS measurements tend to give too negative $\delta^{13}C$ values and are unreliable, so the value is given in parentheses in Table 5.3.2 and no calcite value was calculated from it. A $\delta^{13}C$ value calculated for total dissolved carbonate in water from this depth assuming equilibrium with the calcite would be from -16 to -19 per mil.

5.3.3 Riniken Samples

Results from the Riniken borehole are given in Table 5.3.3. The temperature equation used was:

$$\text{Temp (°C)} = 19.6 + 0.0373 \cdot \text{Depth (metres)}$$

Table 5.3.3: Isotopic composition of water and carbonate minerals from the Riniken borehole.

SAMPLE			CALCITE $\delta^{18}O_{PDB}$		WATER $\delta^{18}O_{SMOW}$	$\delta^{2}H_{SMOW}$	CALCITE $\delta^{13}C_{PDB}$		WATER $\delta^{13}C_{PDB}$	
Depth (metres)	Unit	Temp (C)	Measured	Calculated			Measured	Calculated	Total CO$_2$	Bicarbonate
515.70	km	38.0		-10.96	-6.43	-41.2		-5.7	-8.2	-7.2
656.70	mo	45.0		-15.88	-10.14	-71.6		-0.8	-4.8	-2.6
805.00	s	49.6	-16.90				-6.68			
806.60	s	50.0		-13.51	-6.92	-65.4		-4.4	-9.2	-6.4
965.50	r	56.0		-14.27	-6.73	-57.5		-3.4	-7.6	-5.7
993.50	r	57.0		-14.35	-6.65	-53.4		-3.1	-7.4	-5.3
									-10.7	-14.6 -10.3
1024.74	r	57.8	-9.81				-8.90			
1045.45	r	58.6	-13.76				-7.30			

Two values are given for the measured $\delta^{13}C$ content of water from the 993.5 m sample. The value of $\delta^{13}C$ = -14.6 per mil was measured. This sample also had a ^{14}C content of 41 pmc. A major loss of fluid in this zone took place during drilling, and to minimize further loss, a high viscosity mud was introduced. From the drilling record it appears that cellulose-based high viscosity "pills" were added to the mud as well as mica and additional bentonite (WITTWER, 1986, Section 5.3.5). Oxidation of that cellulose in the formation could have produced the high ^{14}C content of the sample and would also have added ^{13}C-depleted carbonate to the water.

Modern cellulose can be assumed to have a ^{14}C content of about 100 pmc and a $\delta^{13}C$ value of about -25 per mil. The measured ^{14}C content of this sample, 41 pmc, indicates that 41 per cent of the dissolved carbonate results from cellulose oxidation. From mass balance considerations, the $\delta^{13}C$ value of the uncontaminated water would then be -7.4 per mil. This is the other value given in Table 5.3.3 for the 993.5 m sample.

Only three samples of carbonate were analysed from the Riniken borehole. As Table 5.3.3 shows, none of these appears to be in either carbon or oxygen isotope equilibrium with present groundwaters.

5.3.4 Schafisheim Samples

Results from the Schafisheim borehole are given in Table 5.3.4. The temperature equation used was:

$$\text{Temp (°C)} = 6.8 + 0.0411 \cdot \text{Depth (metres)}$$

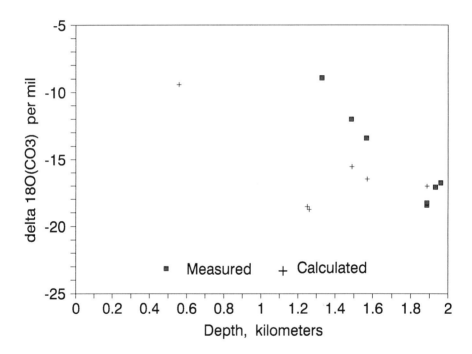

Figure 5.3.3: Oxygen isotopic composition of calcite at various depths in the Schafisheim borehole.

Figure 5.3.3 compares the measured $\delta^{18}O$ values of mineral carbonates with values calculated for calcite at equilibrium with groundwaters. As the figure shows, minerals from 1932.10 and 1960.77 m appear to be in oxygen isotope equilibrium with water of the type sampled at 1887.9 m. The two mineral samples from 1887.19 m would be at equilibrium with the 1887.9 m water at a temperature of about 95°C, 10° warmer than the present formation temperature. The oxygen isotopic composition of the remaining carbonate minerals would require their precipitation from present day waters at much lower temperatures, or their precipitation at present temperatures from waters with higher ^{18}O contents than the present ones.

Only two reliable ^{13}C measurements are available for waters from Schafisheim. The third $\delta^{13}C$ value reported is shown in parenthesis in Table 5.3.4, because it is an AMS measurement, and, therefore, not likely to be reliable. Calcite in equilibrium with carbonate in the water sample from 1887.9 m would have δ^{13} = -4.5 per mil. The mean $\delta^{13}C$ of the minerals sampled below 1887 m is -7.9 per mil, suggesting that they are not at carbon isotope equilibrium with the present water.

Table 5.3.4: Isotopic composition of water and carbonate minerals from the Schafisheim borehole.

SAMPLE			CALCITE $\delta^{18}O_{PDB}$		WATER $\delta^{18}O_{SMOW}$	$\delta^{2}H_{SMOW}$	CALCITE $\delta^{13}C_{PDB}$		WATER $\delta^{13}C_{PDB}$	
Depth (metres)	Unit	Temp (C)	Measured	Calculated			Measured	Calculated	Total CO_2	Bicarbonate
558.00	USM	30.0		-9.42	-6.41	-61.7		-6.9	-8.5	-8.1
1251.10	mo	57.5		-18.55	-10.81	-78.6			(- -)	
1260.40	mo	57.5		-18.76	-11.02	-80.0			(- -)	
1329.20	mo	61.3	-8.92				0.15			
1485.75	s	67.8	-12.01				-7.45			
1488.20	S-KRI	68.0		-15.53	-6.24	-61.6			(-21.2)	
1566.24	KRI	71.1	-13.4				-6.48			
1571.10	KRI	72.0		-16.49	-6.66	-64.8			(- -)	
1887.19	KRI	84.4	-18.44				-7.99			
1887.19	KRI	84.4	-18.28				-7.96			
1887.90	KRI	85.0		-17.03	-5.55	-60.8		-4.5	-9.2	-7.7
1932.10	KRI	86.2	-17.10				-7.96			
1960.77	KRI	87.4	-16.78				-7.80			

5.3.5 Kaisten Samples

Results from the Kaisten borehole are given in Table 5.3.5. The temperature equation used was:

$$\text{Temp (°C)} = 13.1 + 0.0355 \cdot \text{Depth (metres)}$$

The seven water samples from the crystalline (depths below 280 metres) have virtually identical ^{18}O and ^{2}H isotopic compositions ($\delta^{18}O$ = -10.40 ± 0.10 per mil; $\delta^{2}H$ = -73.2 ± 0.7 per mil). This mean $\delta^{2}H$ value is within one standard deviation of the means of the measured $\delta^{2}H$ values of waters from the upper crystalline at Böttstein (-73.6 ± 0.6 per mil; Table 5.3.1), and at Leuggern (-72.6 ± 0.9 per mil; Table 5.3.6). The mean $\delta^{18}O$

value of waters from Kaisten (given above) and Leuggern (-10.36 ± 0.14 per mil) also agree within one standard deviation, but is slightly depleted in ^{18}O relative to the upper crystalline water from Böttstein (-10.02 ± 0.06 per mil).

The mean $\delta^{13}C$ value of this same group of samples is -9.9 ± 0.4 per mil, the same as the mean of the Crystalline samples from Leuggern. Samples from the upper crystalline at Böttstein have slightly more ^{13}C ($\delta^{13}C$ = -9.1 ± 0.2 per mil).

Table 5.3.5: Isotopic composition of water and carbonate minerals from the Kaisten borehole.

SAMPLE			CALCITE $\delta^{18}O_{PDB}$		WATER $\delta^{18}O_{SMOW}$	δ^2H_{SMOW}	CALCITE $\delta^{13}C_{PDB}$		WATER $\delta^{13}C_{PDB}$	
Depth (metres)	Unit	Temp (C)	Meas- ured	Calcu- lated			Meas- ured	Calcu- lated	Total CO_2	Bicar- bonate
113.50	s	16.0		-9.06	-9.03	-68.3		-7.4	-9.8	-7.8
144.48	r	18.2	-10.28				-5.09			
284.30	r	24.0		-12.14	-10.36	-73.6		-8.4	-9.7	-9.3
310.40	KRI	25.0		-12.28	-10.29	-73.5		-8.5	-9.8	-9.4
482.60	KRI	30.0		-13.34	-10.34	-72.7		-8.9	-10.4	-10.0
656.40	KRI	36.0								
664.27	KRI	36.7	-16.24				-7.64			
672.13	KRI	36.9	-14.73				-7.20			
799.88	KRI	41.5	-16.20				-7.89			
819.40	KRI	42.0		-15.54	-10.32	-73.8		-7.6	-9.9	-9.2
819.67	KRI	42.2	-15.98				-7.18			
834.28	KRI	42.7	-16.69				-7.37			
846.88	KRI	43.1	-15.38				-7.30			
918.72	KRI	45.7	-17.37				-7.69			
929.49	KRI	46.1	-10.45				-7.51			
1031.00	KRI	50.0		-17.18	-10.61	-74.1		-6.5	-9.2	-8.5
1048.80	KRI	50.3	-15.79				-6.64			
1153.30	KRI	54.0		-17.61	-10.40	-71.7				(- -)
1205.11	KRI	55.8	-18.56				-7.50			
1271.90	KRI	58.0		-18.28	-10.46	-73.2		-7.2	-10.1	-9.5
1273.94	KRI	58.3	-19.46				-7.61			
1273.94	KRI	58.3	-15.88				-5.00			
		Crystalline: Mean			-10.40	-73.2			-9.9	
		Sigma			0.10	0.7			0.4	

The long-term analytical uncertainty of $\delta^{18}O$ measurements of a large number of samples is about 0.1 per mil, and that associated with $\delta^{13}C$ measurements on dissolved carbonate are about 0.5 per mil. The standard deviation of the means of the Kaisten $\delta^{13}C$ and $\delta^{18}O$ values are within these ranges, so that the differences among the individual values may well represent analytical uncertainties rather than real variations.

Figures 5.3.4 and 5.3.5 compare the $\delta^{18}O$ and $\delta^{13}C$ values measured on the carbonate minerals with values calculated for calcite in equilibrium with formation waters. The solid line in Figure 5.3.4 shows the $\delta^{18}O$ values of calcites in equilibrium with waters from the

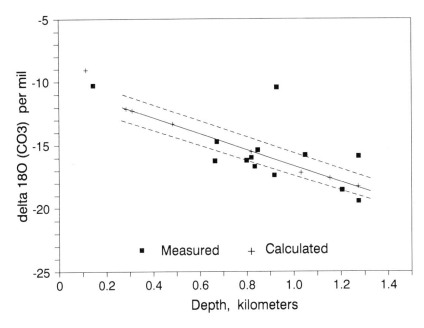

Figure 5.3.4: Oxygen isotopic composition of calcite at various depths in the Kaisten borehole.

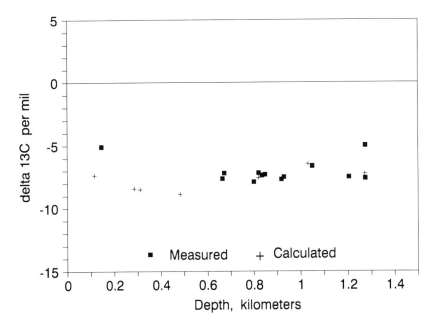

Figure 5.3.5: Carbon isotopic composition of calcite at various depths in the Kaisten borehole.

crystalline section. The dashed lines show the $\delta^{18}O$ values of calcite in equilibrium with water at temperatures 5° above and 5° below the present formation temperatures.

The water and mineral samples from the Buntsandstein and Permian (113.5 and 144.48 m, respectively) appear to be neither in oxygen nor in carbon isotopic equilibrium. Oxygen isotopic equilibrium with present water would require a formation temperature about 10° above the present temperature, or would require water depleted in ^{18}O by 1.2 per mil relative to present water.

Most of the samples from the crystalline appear to be at equilibrium with respect to both ^{18}O and ^{13}C with present formation water. Those which are in clear ^{18}O disequilibrium include the 929.49 and the second of the 1273.94 m samples, and perhaps the 644.27, 834.27, 918.72, 1048.80 and second 273.94 m samples as well. The remaining samples are within about one per mil (±5°C) of the calculated values. Clear ^{13}C disequilibrium is evident only for the second 1273.94 m sample.

5.3.6 Leuggern Samples

Results from the Leuggern borehole are given in Table 5.3.6. The temperature equation used was:

$$\text{Temp (°C)} = 10.6 + 0.0336 \cdot \text{Depth (metres)}$$

The ten water samples from depths between 200 and 1650 m (Buntsandstein and crystalline waters) have virtually identical isotopic compositions ($\delta^{18}O$ = -10.36 ± 0.14 per mil and δ^2H = -72.6 ± 0.9 per mil). This composition also encompasses the composition of waters from the crystalline in the Kaisten borehole (Table 5.3.5), and has the same ^2H content and only 0.4 per mil less ^{18}O than water from the upper crystalline at Böttstein (Table 5.3.1). The sample from the Upper Muschelkalk at 74.9 m is slightly enriched in ^{18}O and ^2H relative to the crystalline samples. The standard deviation of the mean of the Leuggern $\delta^{18}O$ value is slightly above, and the $\delta^{13}C$ values are within the probable experimental uncertainty.

$\delta^{18}O$ and $\delta^{13}C$ values measured on carbonate minerals and calculated from the water sample compositions are compared in Figures 5.3.6 and 5.3.7. The solid line in Figure 5.3.6 shows the $\delta^{18}O$ values in calcite in equilibrium with waters like those from the crystalline section. The dashed lines show $\delta^{18}O$ values of calcite in equilibrium with these waters at temperatures 5° above and below the present formation temperatures. Values for the carbonate sample from 234.34 m are not included in the following discussion because they are so different from any other carbonate sample measured for this study as to suggest some difficulty in the analyses.

As Figure 5.3.6. shows, most of the samples do not appear to be at equilibrium with the present waters at the present temperatures. Three samples, from 704.44, 846.29 and 1437.44 m, would have precipitated either from water more enriched in ^{18}O than the

Table 5.3.6: Isotopic composition of water and carbonate minerals from the Leuggern borehole.

SAMPLE			CALCITE $\delta^{18}O_{PDB}$		WATER		CALCITE $\delta^{13}C_{PDB}$		WATER $\delta^{13}C_{PDB}$	
Depth (metres)	Unit	Temp (C)	Measured	Calculated	$\delta^{18}O_{SMOW}$	δ^2H_{SMOW}	Measured	Calculated	Total CO_2	Bicarbonate
54.08	mo	12.8	-12.77				0.16			
74.90	mo	13.0		-9.30	-9.96	-68.3		-4.6	-6.4	-4.9
217.90	S-KRI	18.0		-10.87	-10.39	-72.9		-8.6	-9.3	-9.1
218.61	s	18.3	-14.65				-7.68			
234.34	KRI	18.8	-9.17				-13.70			
251.20	KRI	19.5		-11.30	-10.49	-73.3		-9.4	-10.1	-10.0
444.20	KRI	25.0		-12.34	-10.35	-73.0		-9.1	-10.1	-10.0
458.34	KRI	26.3	-14.88				-8.04			
517.64	KRI	28.2	-13.67				-7.15			
538.00	KRI	28.0		-12.97	-10.37	-71.9		-9.1	-10.2	-10.1
704.44	KRI	34.5	-10.07				-8.07			
705.70	KRI	34.0		-14.29	-10.52	-74.6		-8.8	-10.2	-10.1
846.29	KRI	39.2	-13.10				-10.39			
847.00	KRI	40.0		-15.23	-10.36	-72.6		-8.1	-9.8	-9.6
863.35	KRI	39.7	-17.95				-8.82			
923.00	KRI	42.0		-15.67	-10.45	-71.3			(- -)	
925.15	KRI	41.8	-16.43				-9.83			
949.88	KRI	42.6	-16.25				-11.02			
1046.05	KRI	45.8	-19.02				-10.55			
1188.00	KRI	50.6	-17.97				-8.31			
1197.62	KRI	50.9	-18.41				-8.12			
1203.20	KRI	51.0		-17.15	-10.42	-72.3			(-26.0)	
1212.70	KRI	51.4	-19.07				-8.37			
1298.18	KRI	54.2	-19.51				-9.01			
1433.40	KRI	59.0		-17.99	-10.02	-72.1			(-23.0)	
1437.44	KRI	58.9	-16.92				-9.60			
1458.92	KRI	59.6	-17.80				-9.08			
1643.40	KRI	66.0		-19.23	-10.25	-71.9		-6.5	-9.3	-9.1
Crystalline:		Mean			-10.36	-72.6			-9.9	
		Sigma			0.14	0.9			0.4	

present formation water or at temperatures more than 5° below present temperatures. Four samples, 517.64, 922.15, 1188.00 and 1485.92 m, appear to be at equilibrium with present formation water, at temperatures within 5° of those measured. The remaining samples, including that from the Muschelkalk at 54.08 m, show precipitation either from water with a lower ^{18}O content or at higher temperatures than the present waters.

The carbon isotopic composition of carbonate mineral samples and values calculated from the carbon isotopic composition of the groundwaters are compared in Figure 5.3.7. The carbonates from the Buntsandstein and crystalline have measured $\delta^{13}C$ values between -7.7 and -10.6 per mil, while the range of values calculated for calcite in equilibrium with present groundwater is from -6.5 to -9.4 per mil. The overlap in these two ranges suggests that the minerals could well have formed from fresh water similar to the formation waters now present. There is sufficient disagreement in detail between carbonate-water pairs from nearly the same depth, however, to make it possible that either the temperature or the chemistry of the waters from which some of the minerals precipitated were not identical with those of today.

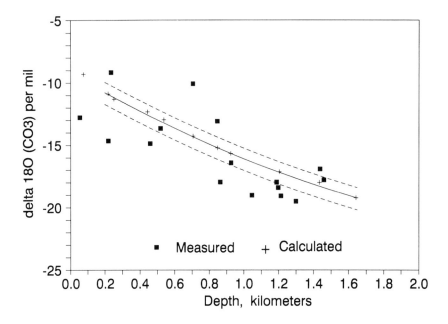

Figure 5.3.6: Oxygen isotopic composition of calcite at various depths in the Leuggern borehole.

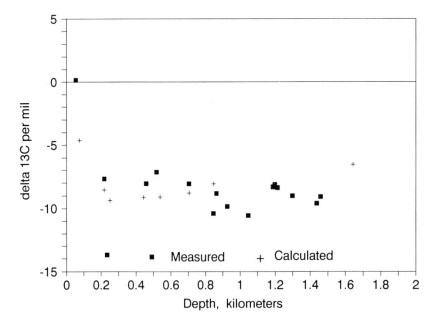

Figure 5.3.7: Carbon isotopic composition of calcite at various depths in the Leuggern borehole.

5.3.7 Summary

Most of the carbonate minerals from the Nagra boreholes analysed for $\delta^{13}C$ and $\delta^{18}O$, are from the Buntsandstein, Permian and crystalline. The few samples from the Muschelkalk have $\delta^{13}C$ values between 1.3 and -1.2 per mil, consistent with marine origin or with precipitation from a water containing relatively enriched dissolved carbonate, such as was sampled from the Muschelkalk at Böttstein. The $\delta^{18}O$ values of the carbonates are from +1.7 per mil for a dolomite from Böttstein, to between -9 and -13 per mil in calcites from Böttstein and Schafisheim. The dolomite has a typical marine value, but the calcites have ^{18}O contents indicating exchange with, or precipitation from, water more depleted in ^{18}O than sea water. Water now present in the Muschelkalk at Böttstein and Schafisheim would precipitate calcite with $\delta^{18}O$ values between -12 and -19 per mil. Some of the calcites measured are within this range, and could have formed from the water now present. The higher ^{18}O contents of the other calcites indicate precipitation either at lower temperatures than prevail at the present, or from more ^{18}O-enriched waters.

The $\delta^{13}C$ values of calcites from the crystalline of the Böttstein, Kaisten, and Leuggern boreholes range from -5 to -11 per mil, and the range of $\delta^{13}C$ values calculated for calcites at equilibrium with the waters now present in the boreholes is from -5 to -9 per mil. The range of $\delta^{18}O$ values measured on the calcites is also generally the same as that calculated for calcite at equilibrium with the waters. Therefore, many of the calcites in the crystalline were formed by waters, and under conditions like those prevailing at the present time. There is little evidence of calcites formed under different conditions, such as might have existed at earlier times. The petrographic observation that many of the calcites are fresh, and not eroded, suggests that the calcites are forming from today's waters rather than that the character of today's waters is the result of the dissolution of previously formed calcites.

The $\delta^{18}O$ and $\delta^{13}C$ values of calcites from the deep crystalline, Permian, and Buntsandstein of the Weiach, Riniken, and Schafisheim boreholes are also strongly negative and entirely inconsistent with a marine origin. There is also general agreement between the mineral values and those calculated for calcite in equilibrium with the waters from these formations. The agreement is not as striking in detail as is that between minerals and waters from the crystalline of the Böttstein, Kaisten, and Leuggern boreholes, but it is evident that the minerals in the deeper crystalline, Permian, and Buntsandstein also formed from waters and under conditions not very different from those of the present.

6. ISOTOPES FORMED BY UNDERGROUND PRODUCTION

6.1. Introduction

B. E. Lehmann and H. H. Loosli

Various processes in the rock matrix can produce isotopes, which eventually escape into groundwater. As outlined in Section 4.1, these subsurface sources may be <u>negligible</u> compared to atmospheric sources, but they may also be <u>dominant</u>. The possible pathways for subsurface production are summarized in Figure 6.1.1.

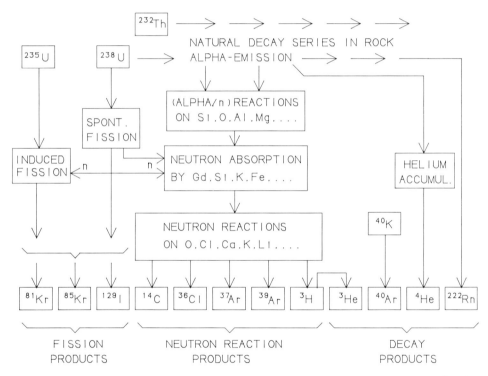

Figure 6.1.1: Schematic presentation of the various channels for subsurface production of isotopes in rock.

Three types of products can be distinguished:

- <u>Fission decay products</u> (^{81}Kr, ^{85}Kr, ^{129}I). These are produced by spontaneous or neutron induced fission of ^{235}U or ^{238}U atoms in the rock;

- <u>Neutron reaction products</u> (^{3}H, ^{14}C, ^{36}Cl, ^{37}Ar, ^{39}Ar). These are produced when neutrons are absorbed by the target isotopes ^{6}Li, ^{17}O, ^{35}Cl, ^{40}Ca and ^{39}K in the rock.

The neutrons originate either from (α,n) reactions between the major elements in the rock (Si, O, Al, Mg, ...) and α-particles emitted in the natural decay series of ^{238}U and ^{232}Th, or they are produced in fission processes. The neutron flux in rock depends on the U and Th concentration and is controlled by the concentration of neutron absorbers. Elements present only in very low concentrations, such as Gd, may have a major influence because of their large absorption cross section;

- Decay products (^{3}He, ^{4}He, ^{40}Ar, ^{222}Rn). ^{3}He is produced by β-decay of ^{3}H, ^{40}Ar by β-decay of ^{40}K, ^{4}He accumulates when α-particles neutralize and ^{222}Rn is a daughter product of the ^{238}U decay series.

Although all of the isotopes measured in this study can be produced in the subsurface, the amounts so produced are quite variable (ANDREWS AND OTHERS, 1989). As qualitatively indicated in Figure 4.2.1, subsurface production is generally:

- negligible for ^{3}H, ^{14}C, ^{85}Kr and probably ^{81}Kr;

- dominant for ^{37}Ar and ^{222}Rn and often also for ^{4}He, ^{36}Cl and ^{129}I.

The other isotopes in Figure 6.1.1, ^{3}He, ^{39}Ar and ^{40}Ar, must be discussed for each individual sample.

In the following sections, it will be important to compare the measured concentrations of isotopes in groundwater with calculated production rates in rocks. Table 6.1.2 includes theoretical calculations of concentrations in closed volumes of rock typical of the various formations from which water samples were taken. All eight isotopes for which subsurface production is or can be important are included, although currently measurements in groundwater are not available for all of them (*e.g.*, there are no measurements of ^{129}I).

For the four isotopes that are produced by neutron reactions (^{3}He, ^{36}Cl, ^{37}Ar, ^{39}Ar), the local neutron flux must be known. This *in situ* flux in the different rock formations was not measured, but it can be calculated if the concentrations of all important elements in a rock are known. These include U and Th, which produce neutrons, and elements that strongly absorb neutrons either because they are very abundant (Si, K, Fe...) or because they have very high absorption cross sections (Gd, Sm, Li, B,...). Except as noted in the Addendum to Section 6.3, this study considers subsurface production by thermal neutrons only.

In order to calculate production rates, one also needs the cross sections for the individual reactions and the concentrations of the respective target nuclides (^{6}Li, ^{35}Cl, ^{39}K, ^{40}Ca). If fast neutrons also contribute to the production of nuclides, as is the case for ^{37}Ar and ^{39}Ar, the product of cross-section and neutron flux has to be integrated over the energy spectrum of the neutrons. The status of such calculations is summarized in the Addendum to Section 6.3.

Since the elemental compositions of rocks are by no means homogeneous on a microscopic scale, one has to work with average "typical" values in each rock formation.

Table 6.1.1: Elemental composition of rock used for the calculations of subsurface production rates and calculated *in situ* neutron flux.

```
Depth     : Mean depth of water sampling interval
U, Th, Li : Uranium, Thorium and Lithium concentration in ppm
K, Ca     : Potassium and Calcium concentrations in per cent
RHO       : Rock density in g/cm³
PHI       : Calculated in situ neutron flux in neutrons/(cm² · s)
            (see text) (· 10⁻⁴)
```

Site	Sample	Mean Depth (m)	Formation	RHO	U	Th	K	Ca	Li	PHI
SHA	304/1	558	USM	2.3	2.5	8.0	1.8	8.80	37	0.268
BOE	301/2	163	mo	2.2	1.8	0.6	0.5	26.00	11	0.209
WEI	302/10b	859	mo	2.2	2.0	1.3	0.5	23.00	11	0.250
RIN	303/2	656	mo	2.2	2.0	1.5	0.5	27.00	11	0.256
SHA	304/3	1233	mo	2.2	1.5	1.0	0.3	24.00	11	0.189
LEU	306/1	75	mo	2.2	2.0	1.0	0.5	21.00	11	0.242
BOE	301/6	317	s	2.3	2.0	5.0	1.9	0.19	37	0.192
WEI	302/12	985	s	2.3	2.0	5.0	1.2	5.50	37	0.192
RIN	303/3	807	s	2.3	3.0	6.0	2.5	0.72	37	0.264
SHA	304/7	1488	s	2.3	3.0	12.0	2.5	2.00	37	0.360
KAI	305/1	113	s	2.3	3.0	6.0	2.0	1.80	37	0.264
LEU	306/2	218	s	2.3	2.0	6.0	1.5	2.00	37	0.208
WEI	302/19	1116	r	2.3	2.0	14.0	3.2	0.19	37	0.336
WEI	302/18	1408	r	2.3	6.0	16.0	4.0	0.34	37	0.592
RIN	303/5	965	r	2.3	5.0	27.0	5.9	2.90	37	0.712
RIN	303/6	993	r	2.3	2.0	20.0	5.2	2.50	37	0.432
KAI	305/2	284	r	2.3	3.0	9.0	4.0	1.90	37	0.312
BOE	301/8	399	KRI	2.6	2.5	13.0	4.4	1.00	113	0.605
BOE	301/12	621	KRI	2.6	2.5	18.0	4.5	1.00	113	0.755
BOE	301/23	649	KRI	2.6	7.0	17.5	5.1	1.00	113	1.127
BOE	301/18	792	KRI	2.6	5.0	18.0	4.5	1.00	113	0.970
BOE	301/21	1326	KRI	2.6	6.0	28.0	5.0	1.00	113	1.356
WEI	302/16	2218	KRI	2.6	5.0	19.0	4.5	0.67	110	1.000
WEI	302/14	2267	KRI	2.6	4.5	20.0	3.8	1.10	110	0.987
SHA	304/9	1571	KRI	2.6	2.0	56.0	4.5	1.20	60	1.852
SHA	304/10	1888	KRI	2.6	6.0	38.0	5.5	3.20	60	1.656
KAI	305/4	310	KRI	2.6	2.0	17.0	4.0	0.52	109	0.682
KAI	305/6	482	KRI	2.6	3.0	17.0	4.0	0.76	109	0.768
KAI	305/9	819	KRI	2.6	3.0	16.0	4.0	0.39	109	0.738
KAI	305/12	1031	KRI	2.6	3.0	18.0	4.0	0.42	109	0.798
KAI	305/16	1272	KRI	2.6	5.0	14.0	3.5	0.50	109	0.850
LEU	306/4	251	KRI	2.6	3.0	20.0	4.0	0.78	114	0.858
LEU	306/7	538	KRI	2.6	3.5	16.0	5.0	1.00	114	0.781
LEU	306/9	706	KRI	2.6	3.5	6.0	3.0	2.70	114	0.481
LEU	306/13	923	KRI	2.6	3.5	14.0	4.3	0.18	114	0.721
LEU	306/20	1203	KRI	2.6	3.5	16.0	3.5	0.12	114	0.781
LEU	306/16	1643	KRI	2.6	10.0	17.0	5.8	0.12	114	1.370

The mean free path for thermal neutrons is approximately 0.5 m. Therefore, inhomogeneities and anomalies on a smaller scale are generally not important. Fluctuations in the elemental compositions over a larger area must be averaged in order to get representative production rates for a particular stratigraphic section.

In Table 6.1.1, the average elemental concentrations which are used in our calculations are presented.

The samples are identified by site, sample number, average depth of sampling interval and formation and the following abbreviations are used:

Sites: BOE = Böttstein, WEI = Weiach, RIN = Riniken, SHA = Schafisheim, KAI = Kaisten, LEU = Leuggern

Formations: tUSM = Untere Süsswassermolasse (Tertiary): sandstones, siltstones, shales, marls
mo = Oberer Muschelkalk (M Triassic): calcareous and dolomitic limestones, dolomites
s = Buntsandstein (L Triassic): sandstones
r = Rotliegend (Permian): sandstones, siltstones, shales
KRI = Kristallin (Crystalline rocks of the Variscan Basement): mainly granites and gneisses with aplitic and lamprophyric dikes

Detailed information on uranium, thorium and potassium is available from the continuous γ logs of the individual boreholes (WEBER AND OTHERS, 1986). Concentrations over the length of each sampling interval have been averaged. Fluctuations of 20 per cent to 50 per cent around the average value occur within these intervals.

The values for calcium in Table 6.1.1 are based on individual rock samples from within the respective water sampling intervals as published in the various tables "Haupt- und Spurenelementanalyse" for the deep boreholes (MATTER AND OTHERS, 1987, 1988a, 1988b; PETERS AND OTHERS, 1986, 1988a, 1988b). If such samples are not available, an average of two or three nearby samples was used.

Only a fairly limited set of rock data is available for lithium. The values are extremely variable, sometimes even over short distances. For example:

             Kaisten Crystalline   1156.17 m: 11 ppm
             Kaisten Crystalline   1158.66 m: 185 ppm

The following average lithium concentrations are used for the calculations:

Muschelkalk:	11 ppm (±60%);	3 values between 6-18 ppm
Sandstone/Perm:	37 ppm (±70%);	4 values between 14-73 ppm
Crystalline/BOE:	113 ppm (±60%);	11 values between 24-281 ppm
WEI:	110 ppm (assumption, no rock data available)	
SHA:	60 ppm (±66%);	8 values between 17-145 ppm
KAI:	109 ppm (±83%);	9 values between 11-309 ppm
LEU:	114 ppm (±75%);	9 values between 49-333 ppm

Table 6.1.1 also includes the *in situ* neutron flux (PHI) calculated as proposed by FEIGE AND OTHERS (1968) for average compositions of standard rock types. Furthermore, we use the following integrated neutron absorption cross sections σ_m for various rock types of typical elemental composition, as compiled by BENTLEY (1978):

	P (n/(g · a))	σ_m (mol · barn/g)
Granite	2.04[U] + 0.70[Th]	0.0125
Sandstone	1.31[U] + 0.38[Th]	0.0123
Limestone	1.11[U] + 0.29[Th]	0.0055

P = neutron production (neutrons per gram of rock per year), where [U] and [Th] are uranium and thorium concentrations in ppm (FEIGE AND OTHERS, 1968)

σ_m = weighted total neutron absorption cross section (mol · barn per gram of rock) (BENTLEY, 1978)

The neutron flux in a rock matrix can be calculated using (ANDREWS AND OTHERS, 1986):

$$\Phi = 5.3 \cdot 10^{-8} \, (P/\sigma_m)$$

Combining the values above, one gets:

$$\Phi = 10^{-5} \, (\alpha[U] + \beta[Th]) \, (n/(cm^2 \cdot s))$$

with the values of α and β

	α	β
Granite	0.86	0.30
Sandstone	0.56	0.16
Limestone	1.07	0.28

The coefficients of "granite" are used for all crystalline rocks, those of "sandstone" for all Buntsandstein samples, all Permian samples (Rotliegend) and for the sample from the Molasse, and those of "limestone" for the Muschelkalk.

One must keep in mind that such an approach works with "typical" elemental compositions of the rock under consideration. The hope is that all local variations average out over a larger volume and that the rock compositions used by FEIGE AND OTHERS (1968) and BENTLEY (1978) are indeed representative for the geological formations of this study.

Based on the available elemental analyses, we believe that this is correct within errors of approximately 30 to 50 per cent, unless critical elements such as gadolinium in crystalline rock or boron in sandstone are much different than assumed by the above authors.

The neutron flux (PHI) in Table 6.1.1 is calculated with the average values for P and σ_m as outlined. The actual concentrations of K, Ca and Li have not been used for a more detailed calculation because this refinement would only change the neutron flux by a few per cent.

The actual values for K, Ca and Li are, of course, used for the target element concentrations in the calculation of underground production of ^{39}Ar, ^{37}Ar and ^{3}He.

The results of our calculations are presented in Table 6.1.2.

For stable isotopes which would accumulate in a closed rock volume with time (^{3}He, ^{4}He, ^{40}Ar), the production rates in cm^3 STP per cm^3 rock per year are listed. For radioactive isotopes, the equilibrium values that would be reached in a closed rock volume after approximately five half-lives are listed. This would be within approximately three weeks for ^{222}Rn, six months for ^{37}Ar, 1350 years for ^{39}Ar, 1.5 million years for ^{36}Cl and 80 million years for ^{129}I. It is assumed that no losses occur during these times. The realism of this assumption will be discussed when comparing measured values in water with calculated maximum levels in rock in the following sections.

The calculations are based on the following numerical values:

Half-lives/Decay Constants

Chlorine-36	$T_{1/2}$	$= 3.01 \cdot 10^5$ a	τ_{36}	$= 7.30 \cdot 10^{-14}$ s^{-1}
Argon-37	$T_{1/2}$	$= 34.8$ d	τ_{37}	$= 2.31 \cdot 10^{-7}$ s^{-1}
Argon-39	$T_{1/2}$	$= 269$ a	τ_{39}	$= 8.17 \cdot 10^{-11}$ s^{-1}
Potassium-40	$T_{1/2}$	$= 1.28 \cdot 10^8$ a	τ_{40}	$= 1.72 \cdot 10^{-17}$ s^{-1}
Iodine-129	$T_{1/2}$	$= 1.59 \cdot 10^7$ a	τ_{129}	$= 1.38 \cdot 10^{-15}$ s^{-1}
Radon-222	$T_{1/2}$	$= 3.82$ d	τ_{222}	$= 2.10 \cdot 10^{-6}$ s^{-1}
Thorium-232	$T_{1/2}$	$= 1.4 \cdot 10^{10}$ a	τ_{232}	$= 1.57 \cdot 10^{-18}$ s^{-1}
Uranium-235	$T_{1/2}$	$= 7.04 \cdot 10^8$ a	τ_{235}	$= 3.12 \cdot 10^{-17}$ s^{-1}
Uranium-238	$T_{1/2,\alpha}$	$= 4.47 \cdot 10^9$ a	$\tau_{238,\alpha}$	$= 4.92 \cdot 10^{-18}$ s^{-1}
	$T_{1/2 sf}$	$= 8 \cdot 10^{15}$ a	$\tau_{238,sf}$	$= 2.75 \cdot 10^{-24}$ s^{-1}

Cross Sections/Fission Yields

^{6}Li(n,α)^{3}H	σ_3	= 940 barns = $9.4 \cdot 10^{-22}$ cm^2
^{35}Cl(n,γ)^{36}Cl	σ_{36}	= 44 barns = $4.4 \cdot 10^{-23}$ cm^2
^{39}K(n,p)^{39}Ar	σ_{39}	= 16 millibarns = $1.6 \cdot 10^{26}$ cm^2
^{40}Ca(n,α)^{37}Ar	σ_{37}	= 2.4 millibarns = $2.4 \cdot 10^{-27}$ cm^2

^{40}K $\xrightarrow{\beta^+}$ ^{40}Ar	y_{40}	= 0.11
^{238}U $\xrightarrow{sf}$ ^{129}I	y_{129}	= $3 \cdot 10^{-4}$

Natural Abundance of Isotopes (0 < x < 1)

[^{6}Li]	= 0.075		[^{40}K]	= $1.17 \cdot 10^{-4}$
[^{35}Cl]	= 0.758		[^{235}U]	= $7.2 \cdot 10^{-3}$
[^{39}K]	= 0.933		[^{238}U]	= 0.933
[^{40}Ca]	= 0.969		[^{232}Th]	= 1.0

Molecular Weights m_x

Li:	6.94 g/mol	Ca:	40.1 g/mol
Cl:	35.4 g/mol	Th:	232 g/mol
K:	39.1 g/mol	U:	238 g/mol

Numerical Constants/Notation

N_L = Avogadro's Number = $6.02 \cdot 10^{23}$ atoms/mol
ρ = rock density in g/cm^3
t = accumulation time in seconds
Φ = calculated neutron flux in neutron/(cm$^2 \cdot$ s) (from Table 6.1.1)
[Li], [K], [Ca], [U], [Th] = concentrations of the respective elements in rock (Table 6.1.1) used as fractions x, with 0 < x < 1.

Table 6.1.2: Calculated production rates or equilibrium concentrations in rock.

^{36}Cl : $^{36}Cl/Cl$ equilibrium ratio in rock ($\cdot$ 10^{-15})
^{3}He : production rate in cm^3 STP/cm^3 rock $\cdot$ year ($\cdot$ 10^{-21})
^{4}He : production rate in cm^3 STP/cm^3 rock $\cdot$ year ($\cdot$ 10^{-13})
^{40}Ar : production rate in cm^3 STP/cm^3 rock $\cdot$ year ($\cdot$ 10^{-13})
^{37}Ar : equilibrium concentration (atoms/cm^3 rock), production by thermal neutrons only
^{39}Ar : equilibrium concentration (atoms/cm^3 rock), production by thermal neutrons only
^{129}I : equilibrium concentration (atoms/cm^3 rock)
^{222}Rn : equilibrium concentration (atoms/cm^3 rock)

Site	Sample	Mean Depth (m)	Form.	^{36}Cl calc	^{3}He calc	^{4}He calc	^{40}Ar calc	^{37}Ar calc	^{39}Ar calc	^{129}I calc	^{222}Rn calc
SHA	304/1	558	tUSM	12.2	16.4	12.3	1.7	0.00082	3.11	8625	33925
BOE	301/2	163	mo	9.5	3.6	5.2	0.4	0.00181	0.64	5940	23364
WEI	302/10b	859	mo	11.4	4.3	6.2	0.4	0.00191	0.77	6600	25960
RIN	303/2	656	mo	11.6	4.4	6.3	0.4	0.00230	0.79	6600	25960
SHA	304/3	1233	mo	8.6	3.3	4.6	0.3	0.00150	0.35	4950	19470
LEU	306/1	75	mo	11.0	4.2	6.0	0.4	0.00169	0.75	6600	25960
BOE	301/6	317	s	8.7	11.7	8.9	1.7	0.00001	2.29	6900	27140
WEI	302/12	985	s	8.7	11.7	8.9	1.1	0.00037	1.48	6900	27140
RIN	303/3	807	s	12.0	16.1	12.4	2.3	0.00007	4.25	10350	40710
SHA	304/7	1488	s	16.4	22.0	16.4	2.3	0.00025	5.80	10350	40710
KAI	305/1	113	s	12.0	16.1	12.4	1.8	0.00017	3.40	10350	40710
LEU	306/2	218	s	9.5	12.7	9.6	1.4	0.00014	2.01	6900	27140
WEI	302/19	1116	r	15.3	20.5	14.9	2.9	0.00002	6.92	6900	27140
WEI	302/18	1408	r	26.9	36.1	27.4	3.7	0.00007	15.25	20700	81420
RIN	303/5	965	r	32.4	43.4	31.9	5.4	0.00072	27.05	17250	67850
RIN	303/6	993	r	19.7	26.4	18.9	4.8	0.00038	14.47	6900	27140
KAI	305/2	284	r	14.2	19.0	14.4	3.7	0.00021	8.04	10350	40710
BOE	301/8	399	KRI	27.5	127.4	17.7	4.6	0.00024	19.38	9750	38350
BOE	301/12	621	KRI	34.4	159.0	21.5	4.7	0.00030	24.73	9750	38350
BOE	301/23	649	KRI	51.3	237.4	35.2	5.3	0.00044	41.84	27300	107380
BOE	301/18	792	KRI	44.1	204.3	29.3	4.7	0.00038	31.78	19500	76700
BOE	301/21	1326	KRI	61.7	285.6	40.0	5.2	0.00053	49.36	23400	92040
WEI	302/16	2218	KRI	45.5	205.1	30.1	4.7	0.00026	32.76	19500	76700
WEI	302/14	2267	KRI	44.9	202.4	29.3	4.0	0.00043	27.30	17550	69030
SHA	304/9	1571	KRI	84.3	207.1	48.6	4.7	0.00087	60.67	7800	30680
SHA	304/10	1888	KRI	75.3	185.2	47.6	5.7	0.00208	66.31	23400	92040
KAI	305/4	310	KRI	31.0	138.6	19.1	4.2	0.00014	19.86	7800	30680
KAI	305/6	482	KRI	34.9	156.1	22.3	4.2	0.00023	22.36	11700	46020
KAI	305/9	819	KRI	33.6	150.0	21.5	4.2	0.00011	21.49	11700	46020
KAI	305/12	1031	KRI	36.3	162.2	23.0	4.2	0.00013	23.24	11700	46020
KAI	305/16	1272	KRI	38.7	172.7	26.3	3.6	0.00017	21.66	19500	76700
LEU	306/4	251	KRI	39.0	182.3	24.5	4.2	0.00026	24.98	11700	46020
LEU	306/7	538	KRI	35.5	166.0	23.1	5.2	0.00031	28.43	13650	53690
LEU	306/9	706	KRI	21.9	102.2	15.5	3.1	0.00051	10.51	13650	53690
LEU	306/13	923	KRI	32.8	153.2	21.6	4.5	0.00005	22.57	13650	53690
LEU	306/20	1203	KRI	35.5	166.0	23.1	3.6	0.00004	19.90	13650	53690
LEU	306/16	1643	KRI	62.3	291.2	44.3	6.0	0.00006	57.85	39000	153400

The following expressions are used to calculate the production rates of the various isotopes. To convert the calculated rates from atoms/cm^3 rock into cm^3 STP gas/(cm^3·a), multiply $1.17 \cdot 10^{-12}$.

Helium-3

This isotope is produced by the reactions:

$$^6Li(n, \alpha)\ ^3H \xrightarrow{\beta^-}\ ^3He$$

N_3, the number of ^{3}He atoms produced per cm^3 rock, is calculated from:

$$N_3 = \rho \cdot [Li] \cdot [\,^6Li\,] \cdot (N_L/m_{Li}) \cdot \sigma_3 \cdot \Phi \cdot t$$

$$N_3 = \rho \cdot [Li] \cdot \Phi \cdot t \cdot 6.12$$

Helium-4

^{4}He is produced in the natural decay series of ^{238}U, ^{235}U and ^{232}Th, when α-particles neutralize. A total of 8, 7 and 6 α-particles, respectively, are emitted in each of the three series until a stable Pb isotope is created. At equilibrium, the production rate is controlled by the half-lives of the three mother isotopes.

N_4, the number of ^{4}He atoms produced per cm^3 of rock, is:

$$N_4 = \rho \cdot t \cdot N_L\,[\,8 \cdot [U] \cdot [\,^{238}U\,]\,(\tau_{238}/m_U)\ +$$

$$7 \cdot [U] \cdot [\,^{235}U\,]\,(\tau_{235}/m_U) + 6 \cdot [Th] \cdot [\,^{232}Th\,]\,(\tau_{232}/m_{Th})\,]$$

$$N_4 = \rho \cdot t \cdot (1.03 \cdot 10^5\,[U] + 2.45 \cdot 10^4\,[Th])$$

Chlorine-36

This isotope is produced by the reaction:

$$^{35}Cl(n,\gamma)^{36}Cl$$

N_{36}, the number of ^{36}Cl atoms per cm^3 of rock at equilibrium, is calculated from:

$$N_{36} = \rho \cdot \Phi \cdot [Cl] \cdot [^{35}Cl] \cdot (N_L/m_{Cl}) \cdot \sigma_{36}/\tau_{36}$$

Because AMS measurements give the ratio $^{36}Cl/Cl$, the calculated values for Table 6.1.2 are also given as such a ratio. They are independent of the chlorine content of the rock and depend only on the neutron flux Φ.

$$^{36}Cl/Cl = \Phi \cdot 4.55 \cdot 10^{-10}$$

Argon-37

This isotope is produced in the reaction:

$$^{40}Ca(n,\alpha)^{37}Ar$$

N_{37}, the number of ^{37}Ar atoms per cm^3 of rock at equilibrium, is calculated from:

$$N_{37} = \rho \cdot [Ca] \cdot [^{40}Ca] \cdot [N_L/m_{Ca}] \cdot (\sigma_{37}/\tau_{37}) \cdot \Phi$$

$$N_{37} = \rho \cdot [Ca] \cdot \Phi \cdot 151$$

Except as mentioned in the Addendum to Section 6.3, this study considers subsurface production by thermal neutrons only.

Argon-39

This isotope is produced by neutron capture of ^{39}K:

$$^{39}K(n,p)^{39}Ar$$

N_{39}, the number of ^{39}Ar atoms per cm^3 of rock at equilibrium, is calculated from:

$$N_{39} = \rho \cdot [K] \cdot [^{39}K] \cdot [N_L/m_K] \cdot (\sigma_{39}/\tau_{39}) \cdot \Phi$$

$$N_{39} = \rho \cdot [K] \cdot \Phi \cdot 2.8 \cdot 10^6$$

Except as mentioned in the Addendum to Section 6.3, this study considers subsurface production by thermal neutrons only.

Argon-40

^{40}Ar accumulates as a β-decay product of ^{40}K. N_{40}, the number of ^{40}Ar atoms produced per cm^3 of rock, is calculated from:

$$N_{40} = \rho \cdot [K] \cdot [^{40}K] \cdot [N_L/m_K) \cdot \tau_{40}/y_{40} \cdot t$$

$$N_{40} = \rho \cdot [K] \cdot 3.4 \cdot t$$

Iodine-129

In the spontaneous fission decay of ^{238}U a fraction y_{129} is converted to ^{129}I. N_{129}, the number of ^{129}I atoms per cm^3 of rock at equilibrium, is calculated from:

$$N_{129} = \rho \cdot [U] \cdot [^{238}U] \cdot [N_L/m_U) \cdot (\tau_{238,sf}/\tau_{129}) \cdot y_{129}$$

$$N_{129} = \rho \cdot [U] \cdot 1.5 \cdot 10^9$$

Radon-222

^{222}Rn is a daughter product in the ^{238}U decay series. N_{222}, the number of ^{222}Rn per cm^3 of rock in equilibrium with ^{238}U, is calculated from:

$$N_{222} = \rho \cdot [U] \cdot [^{238}U] \cdot [N_L/m_U) \cdot (\tau_{238,\alpha}/\tau_{222})$$

$$N_{222} = \rho \cdot [U] \cdot 5.9 \cdot 10^9$$

(To convert from atoms ^{222}Rn/cm^3 rock into Bq/cm^3, multiply by $2.1 \cdot 10^{-6}$.)

6.2 Chlorine-36

B. E. Lehmann, H. H. Loosli, W. Balderer, J. Ch. Fontes, J. L. Michelot, and S. Soreau

^{36}Cl has a half-life of 301,000 years and should be a very useful radionuclide for estimating groundwater residence times in a range of approximately 50,000 to 1,500,000 years. Measurements in natural water samples have become possible only in recent years due to the progress in accelerator mass spectrometry (AMS). ^{36}Cl/Cl ratios are very small, typically in the range of 1 to 100 · 10^{-15}. From the measured ratio R_{36} (= ^{36}Cl/Cl) and the chloride content C (in mg/l) the number N_{36} of ^{36}Cl atoms per litre of water can be calculated using:

$$N_{36} = 1.7 \cdot 10^{19} \cdot R_{36} \cdot C \qquad (6.2.1)$$

6.2.1 Results

The results of ^{36}Cl analyses on Nagra groundwater samples are presented in Table 6.2.1. All samples were processed at the Universite de Paris-Sud (Prof. J. Ch. Fontes) and measured at the Nuclear Structure Research Laboratory at the University of Rochester (Dr. D. Elmore). The table includes the measured values for R_{36}, the measured chloride contents, C, taken from WITTWER (1986), and the calculated number of ^{36}Cl atoms per litre of water, N_{36}. The same abbreviations are used for sample identification as introduced in Section 6.1. The calculated values resulting from subsurface production are also listed in the last column. These calculations were discussed in Section 6.1; the results are taken from Table 6.1.2 and will be discussed later in this chapter.

6.2.2 Origin of ^{36}Cl in Groundwater Samples

Several articles in the last few years review the sources of ^{36}Cl in the terrestrial environment and draw conclusions about the use of this isotope in groundwater studies (BENTLEY AND OTHERS, 1986; FONTES AND OTHERS, 1984; ANDREWS AND OTHERS, 1986). Four different sources for ^{36}Cl in groundwater samples are generally considered:

Atmospheric input:

a) ^{36}Cl produced by <u>cosmic rays</u> in the spallation of ^{40}Ar and the neutron activation of ^{36}Ar in the atmosphere.

b) ^{36}Cl produced in <u>nuclear weapons testing</u> in the atmosphere mainly in the years 1952 to 1959.

Table 6.2.1: Results of ^{36}Cl analyses of groundwater samples.

R_{36} (meas) = measured ratio of ^{36}Cl/Cl in groundwater ($\cdot\ 10^{-15}$)
C = measured chloride concentration of groundwater (mg/litre)
N_{36} = calculated number of ^{36}Cl atoms per litre of water ($\cdot\ 10^8$)
R_{36} (calc) = calculated ^{36}Cl/Cl ratio in rock (from Table 6.1.2)($\cdot\ 10^{-15}$)

Site	Sample	Mean Depth (m)	Formation	R_{36} meas	± 1σ	C (mg/l)	N_{36}	R_{36} calc
SHA	304/1	558	tUSM	18	5	5200	16	12.2
BOE	301/2	163	mo	8	4	1300	1.8	9.5
WEI	302/10b	859	mo	17	4	53	0.15	11.4
SHA	304/14	1251	mo	23	11	6400	25	8.6
SHA	304/3	1260	mo	5	2	8100	6.9	8.6
BOE	301/6	317	s	14	7	640	1.5	8.7
WEI	302/12	985	s	19	3	2900	9.4	8.7
RIN	303/3	807	s	34	7	4100	24	12
SHA	304/7	1488	s	19	4	6500	21	16.4
LEU	306/2	218	s	69	10	490	5.7	9.5
WEI	302/19	1116	r	20	8	18000	61	15.3
WEI	302/18	1408	r	29	4	60000	300	26.9
RIN	303/5	965	r	34	6	7900	46	32.4
KAI	305/2	284	r	32	6	100	0.54	14.2
BOE	301/8	399	KRI	13	10	120	0.27	27.5
BOE	301/12	621	KRI	43	9	130	0.95	34.4
BOE	301/23	649	KRI	48	19	120	0.98	51.3
BOE	301/18	792	KRI	29	10	140	0.69	44.1
BOE	301/21	1326	KRI	53	10	6600	59	61.7
WEI	302/16	2218	KRI	40	5	3400	23	45.5
SHA	304/9	1571	KRI	27	5	2400	11	84.3
SHA	304/10	1888	KRI	38	5	3600	23	75.3
KAI	305/6	482	KRI	27	6	60	0.28	34.9
KAI	305/12	1031	KRI	53	8	63	0.57	36.3
KAI	305/16	1272	KRI	54	8	73	0.67	38.7
LEU	306/4	251	KRI	73	14	180	2.2	39
LEU	306/9	706	KRI	170	10	110	3.2	21.9
LEU	306/13	923	KRI	40	6	190	1.3	32.8
LEU	306/20	1203	KRI	35	5	130	0.77	35.5
LEU	306/16	1643	KRI	38	6	125	0.81	62.3

Subsurface input:

c) ^{36}Cl produced <u>near the surface</u> through secondary particles of cosmic radiation, especially in spallation reactions, neutron activation of ^{35}Cl, μ capture by ^{40}Ca and (n,α) reactions on ^{39}K.

d) ^{36}Cl produced <u>deep underground</u> by neutron irradiation of ^{35}Cl either in the rock matrix or in the chloride dissolved in the groundwater.

If the groundwater forms a closed system with respect to chloride so that chloride is neither lost nor gained from the rock matrix, then ^{36}Cl can be used for the determination of groundwater residence times according to

$$N_{36}(t) = N_0 \cdot e^{-\tau t} + N_{eq}(1-e^{-\tau t}) \qquad (6.2.2)$$

N_0 = initial concentration at recharge
τ = decay constant = $7.3 \cdot 10^{-14}$ s^{-1}
N_{eq} = equilibrium concentration

The first term describes the decay of the ^{36}Cl atoms that were introduced into the aquifer in the recharge area. N_0, the initial concentration of ^{36}Cl atoms, may be measured in samples taken near the surface or estimated from a detailed discussion of processes (a), (b)--including the fallout pattern--and, possibly, (c).

The second term describes production by *in situ* irradiation of ^{35}Cl in solution, process (d). The equilibrium concentration, N_{eq}, which is reached after approximately 1.5 million years (about five half-lives), depends on the local neutron flux.

If the salinity changes during groundwater evolution so that the closed system criterion is not satisfied, dating of groundwater is not possible without a detailed knowledge of the chloride evolution.

In the next section we will estimate the ^{36}Cl content of groundwater at recharge and argue that none of the processes (a), (b) or (c) significantly contributes to the levels of ^{36}Cl measured in the Nagra samples and that, therefore, the large majority of ^{36}Cl atoms originate from deep subsurface sources.

6.2.3 ^{36}Cl in Recharge

a)   Cosmic ray produced ^{36}Cl

The ^{36}Cl concentration in near-surface groundwater can be estimated from the fallout rate, F (atoms m^{-2} s^{-1}), the annual precipitation, P (mm a^{-1}) and the evapotranspiration, E (%), according to

$$N_{36} = 3.2 \cdot 10^7 \cdot \frac{F \cdot 100}{P \cdot (100-E)} \qquad (6.2.3)$$

For an estimated value of F ≈ 26 atoms m^{-2} s^{-1} (BENTLEY AND OTHERS, 1986) at the latitude of Switzerland, and choosing P = 2000 mm a^{-1} and E = 30 per cent, one calculates $N_{36} = 6 \cdot 10^5$ atoms/l H$_2$O.

This number is one to three orders of magnitude lower than the values found in the Nagra samples (see Table 6.2.1) and, therefore, cosmic ray produced ^{36}Cl in these samples is negligible.

Even though the value of F could well vary within a factor of two to four (SUTER AND OTHERS, 1987), even if a maximum value were chosen, the concentration in groundwater would still be far below the measured values. Thus, a more detailed discussion is not necessary.

b) Thermonuclear ^{36}Cl

Several nuclear weapons tests in the South Pacific in the years 1952 to 1959 produced large amounts of ^{36}Cl by neutron activation of ^{35}Cl in sea water. The radioactive clouds penetrated into the stratosphere and the global fallout rate of ^{36}Cl increased by about three orders of magnitude to approximately $3 \cdot 10^4$ atoms/$m^{-2} s^{-1}$ (± 50 per cent) in mid-latitudes (BENTLEY AND OTHERS, 1986). Using equation 6.2.3, one calculates a maximum of $7 \cdot 10^8$ atoms of ^{36}Cl per litre of water during the test period. For an assumed chloride content of a typical surface water of 10 mg/l, the $^{36}Cl/Cl$ ratio would then be $4000 \cdot 10^{-15}$. A fraction of such a relatively young groundwater component could be present in a sample either as a natural admixture or by contamination of the groundwater with drilling fluid during sampling.

Table 6.2.2 summarizes the extent of a possible admixture of drilling fluid to the samples for which ^{36}Cl results are available. The values given (K_{max}, in per cent) are based either on drilling-fluid tracer analysis as, discussed in WITTWER (1986), or are estimated from the ^3H content.

A value of 100 TU (± 50 per cent) is taken as representative of the water supply used in the drilling operations (see WITTWER, 1986, and Chapter 2 of this report). Therefore, the measured ^3H contents (in TU) are simply converted (1 TU = one per cent contamination) to get an estimate of the extent of drilling fluid contamination whenever the tracer analysis does not give a good value.

Several types of drilling fluids were used including deionized water (D), fresh water (F), and fresh water to which clay (C) or clay and salt (S) had been added.

To estimate the possible ^{36}Cl contribution from drilling fluid, the following Cl$^-$ concentrations and $^{36}Cl/Cl$ ratios (maximum values) were used for the various types of drilling fluids:

Contamination with deionized water (D) or fresh water (F) could add ^{36}Cl of thermonuclear origin to the groundwater; the clay (C) and salt (S) additives to the drilling fluid could add ^{36}Cl of subsurface origin. The assumptions for fresh water lead to an upper limit for possible contamination. A maximum value for N_{36} of $6.8 \cdot 10^8$ atoms $^{36}Cl/l$ H_2O was reached in precipitation only in the years of nuclear testing and most likely has been diluted since then.

Type		Cl⁻ (mg/l)	^{36}Cl/Cl ($\cdot 10^{-15}$)	N_{36} ($\cdot 10^8$)
Deionized water	D	1	4000	0.68
Fresh water	F	10	4000	6.8
With clay	C	2000	20	6.8
With clay and salt	S	200 000	0.1	3.4

For an admixture of clay it is assumed that 2 g/l are dissolved from the clay and that ^{36}Cl is in equilibrium with production by neutron activation at a level of $R_{36} = 20 \cdot 10^{-15}$ (see next section for details). The salt added, from large, deep salt formations, should contain essentially no ^{36}Cl; we take $R_{36} = 10^{-16}$ as an upper limit.

Note that no ^{36}Cl analyses are available for any drilling fluids and that it is, therefore, necessary to keep all estimates rather conservative. As can be seen in the last column of Table 6.2.2, even with these conservative estimates, none of the samples--with the exception of the Weiach 859 m--contain a significant fraction of thermonuclear ^{36}Cl or of ^{36}Cl from drilling fluid. Any corrections would well be within the errors of the ^{36}Cl/Cl ratio measurements by AMS.

For the Weiach 859 m sample, however, one cannot rule out the possibility that all, or most of the ^{36}Cl found originates from contamination with drilling fluid or natural young water.

c) ^{36}Cl produced near the surface

As summarized by KUBIK AND OTHERS (1984), several processes induced by secondary particles from cosmic radiation can produce ^{36}Cl in the top layers of the earth's surface. Spallation reactions or neutron activation of ^{35}Cl may be important to depths of about five metres. In limestones, however, μ capture of ^{40}Ca dominates production to depths of up to 100 metres. Measured ^{36}Cl/Cl ratios are in the range of 1 to $6 \cdot 10^{-13}$. Although the Cl⁻ content of limestone is fairly small (30 to 60 ppm by weight; KUBIK AND OTHERS, 1984), one can, in principle, produce any of the measured ^{36}Cl concentrations in Table 6.2.1 by mixing:

- a small fraction of chloride released by weathering from a near surface limestone, with;

- chloride from a salt formation that does not contain any ^{36}Cl.

Table 6.2.2: Maximum possible ^{36}Cl contamination from nuclear weapons test or drilling fluid.

^3H = Tritium concentration (TU)
K_{max} = maximum contamination estimated from
  (a) drilling fluid tracer analysis (WITTWER, 1986), or
  (b) tritium content
DF = Drilling Fluid Type (see text)
A_{36} = maximum number of ^{36}Cl atoms per litre of water ($\cdot 10^8$)

Site	Sample	Mean Depth (m)	Formation	^3H (TU)	K_{max} (%)	Ref.	DF	A_{36}	A_{36}/N_{36} (%)
SHA	304/1	558	tUSM	<1.2	2.4	a	FC	33	2.1
BOE	301/2	163	mo	1	1	b	FC	14	7.8
WEI	302/10b	859	mo	3.2 ± 0.7	3.9	b	FC	53	350
SHA	304/14	1251	mo	1.4 ± 0.7	1	a	FCS	17	2.5
SHA	304/3	1260	mo	<0.9	0.8	a	F	5.4	0.22
BOE	301/6	317	s	0.2 ± 0.2	0.4	b	FCS	6.8	4.5
WEI	302/12	985	s	<0.8	0.8	b	FCS	14	1.5
RIN	303/3	807	s	0.7 ± 0.7	0.05	a	D	0.03	0.0012
SHA	304/7	1488	s	1.8 ± 0.7	2	a	FCS	34	1.6
LEU	306/2	218	s	<0.8	0.2	a	FCS	3.4	0.6
WEI	302/19	1116	r	2.2 ± 0.6	2.8	b	FC	38	0.62
WEI	302/18	1408	r	6.4 ± 0.9	7.3	b	FC	99	0.33
RIN	303/5	965	r	<1.1	0.8	a	FCS	14	0.3
KAI	305/2	284	r	0.7 ± 0.3	1.1	a	F	7.5	13.9
BOE	301/8	399	KRI	<1	0.8	a	D	0.54	2
BOE	301/12	621	KRI	1 ± 0.7	3.5	a	D	2.4	2.5
BOE	301/23	649	KRI	3.2 ± 0.7	8	a	D	5.4	5.5
BOE	301/18	792	KRI	0.9 ± 0.9	0.1	a	D	0.07	9.9
BOE	301/21	1326	KRI	69 ± 5	55	a	D	37	0.63
WEI	302/16	2218	KRI	2.7 ± 0.7	2	a	D	1.4	0.06
SHA	304/9	1571	KRI	2.5 ± 0.7	2.9	a	D	2	0.18
SHA	304/10	1888	KRI	1.5 ± 0.9	1.4	a	D	0.95	0.04
KAI	305/6	482	KRI	1.3 ± 0.7	0.9	a	D	0.61	2.2
KAI	305/12	1031	KRI	<0.8	0.3	a	D	0.2	0.35
KAI	305/16	1272	KRI	1.2 ± 0.7	1.3	a	D	0.88	1.3
LEU	306/4	251	KRI	0.8 ± 0.7	1.1	a	D	0.75	0.34
LEU	306/9	706	KRI	<1.3	0.3	a	D	0.2	16
LEU	306/13	923	KRI	2.8 ± 0.7	4.6	a	D	3.13	2.4
LEU	306/20	1203	KRI	1.3 ± 0.7	1.5	a	D	1	0.77
LEU	306/16	1643	KRI	1.2 ± 0.7	1.3	a	D	0.88	1.1

Although such an infiltration of a small fraction of chloride with a high $^{36}Cl/Cl$ ratio from a shallow depth of five to ten metres cannot be ruled out, it is considered an unlikely scenario for the interpretation of the ^{36}Cl concentrations found in the deep subsurface samples, all of which originate from depths below 150 m.

6.2.4 Deep Subsurface Production of ^{36}Cl

As was shown in the previous section, none of the sources in the atmosphere or near the surface can contribute enough ^{36}Cl to explain the high levels of ^{36}Cl found in the groundwater samples of the deep drilling programme. Most ^{36}Cl must have been produced deep underground. All measured values listed in Table 6.2.1 are plotted in Figure 6.2.1. As can be seen in this figure, there is a strong correlation between the concentrations of ^{36}Cl and stable Cl atoms.

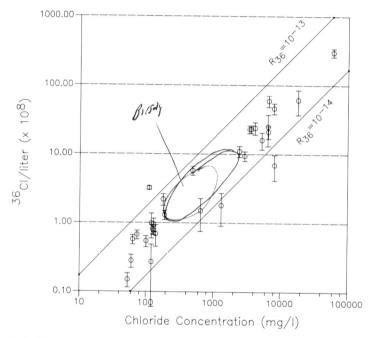

Figure 6.2.1: ^{36}Cl concentration *versus* chloride content of all samples from Table 6.2.1.

The main path for producing ^{36}Cl deep underground is neutron capture by ^{35}Cl (FONTES AND OTHERS, 1984, BENTLEY AND OTHERS, 1986, ANDREWS AND OTHERS, 1986). The reaction ^{35}Cl(n,γ)^{36}Cl has a cross section of 44 barns for thermal neutrons. It occurs on chlorine atoms incorporated in the rock matrix, in minerals and fluid inclusions, and on chloride dissolved in the groundwater. Aqueous ^{36}Cl of deep origin was first noticed at Stripa (MICHELOT AND OTHERS, 1984).

The ratio of ^{36}Cl atoms to total stable chlorine atoms, R_{36}, will increase with time according to

$$R_{36}(t) = 4.55 \cdot 10^{-10} \cdot \Phi \cdot (1-e^{-\tau t}) \tag{6.2.4}$$

Φ is the neutron flux (neutrons/(cm$^2 \cdot$ s))
τ is the decay constant of ^{36}Cl ($7.3 \cdot 10^{-14}$s^{-1})
(see *e.g.,* ANDREWS AND OTHERS, 1986)

Equilibrium between ^{36}Cl formation and decay is established after about 1.5 My (about five half-lives of ^{36}Cl).

As was discussed in Section 6.1, the calculation of subsurface production rates is based on the estimated *in situ* neutron flux derived from the local uranium and thorium content. The results of these calculations are included in Table 6.1.2. Figure 6.2.2 shows all measured ratios R_{36} in water (from Table 6.2.1) *versus* the calculated equilibrium ratio in the corresponding rock formation (from Table 6.1.2).

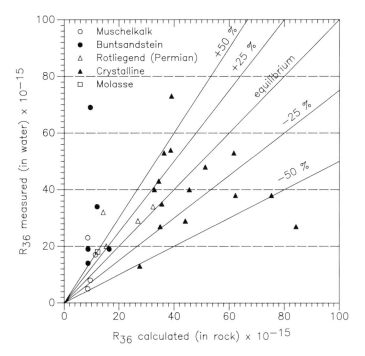

Figure 6.2.2: Measured ^{36}Cl/Cl ratios in groundwater *versus* calculated equilibrium ^{36}Cl/Cl ratios in rock.

To get a first overview no error bars are plotted. The 45° line represents a situation where the measured ^{36}Cl/Cl ratio in water agrees with the calculated equilibrium ratio in the formation from which the water was sampled.

Samples above the 45° line contain more ^{36}Cl than can be produced in the surrounding rock; samples below the 45° line contain less ^{36}Cl than the equilibrium value and could, therefore, be dated if the closed system criterion is satisfied (see Section 6.2.2). The lines of -25 per cent and -50 per cent deviation from the equilibrium line would represent model ages of 600,000 a and 300,000 a according to equation 6.2.2.

Figure 6.2.2 illustrates that more than half of the samples are above the equilibrium line and that all others would be at least several hundred thousand years old. As was pointed out earlier, such ^{36}Cl model ages are based on the assumption that only ^{36}Cl-free chloride (*e.g.,* from large salt formations) has been added to the groundwater, and that the ^{36}Cl build-up occurs in water according to the second term in equation 6.2.2 while the chloride content remains constant.

Alternatively, the ^{36}Cl concentrations found in groundwater can be explained by a dissolution of chloride with a $^{36}Cl/Cl$ ratio in equilibrium with the local neutron flux. No dating is then possible, but information about the origin of the chloride may possibly be gained.

6.2.5   Origin of Aqueous Chloride

Indications of the origin of aqueous chloride can be obtained by studying the behaviour of several natural ionic tracers (especially Br^- and Cl^-, which are usually assumed to be conservative), and the isotopic composition of certain dissolved species (especially sulphates), provided they have the same origin as chloride. ^{36}Cl contents themselves can sometimes be used to indicate the origin of chloride in groundwater samples.

For the water samples taken from sedimentary formations, most tracers indicate that salinity is derived from interactions with the sedimentary rocks themselves. Therefore, one would expect that the $^{36}Cl/Cl$ ratio in water would correspond to the equilibrium ratio in the surrounding rock formation. This is generally confirmed by the available data (see next section).

In the crystalline two types of water can be distinguished:

- waters of low mineralization, of the $Na-(Ca)-SO_4-HCO_3-(Cl)$ type, with Cl^- concentrations of less than 200 mg/l. These include 12 of the 16 samples for which $^{36}Cl/Cl$ data are available (see Table 6.2.1). These samples could have derived their Cl^- content from the crystalline rock, although no definitive conclusions about the origin can be drawn from chemical or isotope data;

- water of high mineralization, of the $Na-(Ca)-Cl-(SO_4)$ type, with Cl^- concentrations of several grams per litre. These are the four samples:

Böttstein	1304 m	with	6600 mg Cl⁻/litre
Schafisheim	1571 m	with	2400 mg Cl⁻/litre
Schafisheim	1888 m	with	3600 mg Cl⁻/litre
Weiach	2218 m	with	3400 mg Cl⁻/litre

Chemical and isotopic data strongly suggest an external (sedimentary) source for the salinity in the waters from the Böttstein and Schafisheim boreholes even though they were sampled in the crystalline. For example, all ion ratios (Na^+/Cl^-, Ca^{2+}/Cl^-, Mg^{2+}/Cl^-, SO_4^{2-}/Cl^-, Br^-/Cl^-, Li^+/Cl^-, F^-/Cl^-) fall into the ranges measured for waters of sedimentary formations. Furthermore, these three samples contain sulphates which are isotopically similar to sedimentary sulphates. The 2218 m sample from Weiach differs in its lower sulphate content. Finally, the ^{18}O and ^2H contents of these waters are higher than in the other crystalline waters (SCHMASSMANN AND OTHERS, 1984), and similar to those of the sedimentary waters.

In the first three cases, ^{36}Cl contents can possibly be used to estimate migration times from sedimentary, *e.g.*, Permian formations, into the crystalline (see Section 6.2.7).

The individual geological formations are discussed in the following sections. The data points of Figure 6.2.2 are shown separately in Figures 6.2.3 to 6.2.6 according to the individual formations, and error bars are included. For the measured values, these error bars represent the analytical uncertainty given by the AMS measurement; for the calculated values in rock an uncertainty of ±30 per cent is used.

6.2.6 ^{36}Cl in Water from Sedimentary Formations

Upper Muschelkalk (Carbonate Rock) (Figure 6.2.3): This unit has low uranium and thorium concentrations leading to a low neutron flux and correspondingly small $^{36}Cl/Cl$ ratios of about $10 \cdot 10^{-15}$. As Figure 6.2.3 shows, all measured values in water are in agreement with the calculated local equilibrium values, if the error bars are taken into account. The sample with the highest $^{36}Cl/Cl$ ratio (Schafisheim 1251 m) has a fairly large analytical error in the AMS measurement, and is not considered to be in disagreement with the other three Muschelkalk values.

Buntsandstein (Figure 6.2.4): All Buntsandstein waters appear to be influenced by the underlying formations. These are the crystalline in the Böttstein, Leuggern and Schafisheim boreholes and the Permian sediments in Weiach, Riniken and Kaisten. Such an influence from a region where the equilibrium $^{36}Cl/Cl$ ratio is higher would move horizontally to the right.

The following conclusions can be drawn: Two of the three highly mineralized waters (WEI 985 m and RIN 807 m) are in equilibrium with the neutron flux in the underlying Permian formation. The third (SHA 1488 m) is far from equilibrium with the underlying crystalline, where due to the high thorium content, a very high neutron flux and a correspondingly high $^{36}Cl/Cl$ ratio are to be expected. However, as was outlined in Section 6.2.5, the highly mineralized waters in the crystalline of Schafisheim have derived their chloride from a sedimentary formation, most likely from Permian sediments. The $^{36}Cl/Cl$

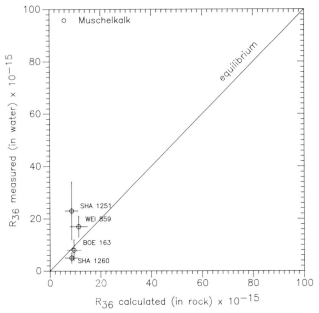

Figure 6.2.3: Measured ^{36}Cl/Cl ratios in water from the Muschelkalk *versus* calculated ^{36}Cl/Cl ratios in rock.

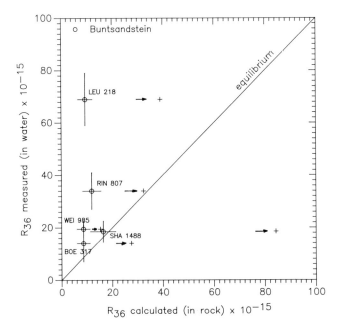

Figure 6.2.4: Measured ^{36}Cl/Cl ratios in water from the Buntsandstein *versus* calculated ^{36}Cl/Cl ratios in rock.

ratio in the crystalline waters has not yet reached the local equilibrium value indicating a short residence time in the crystalline since transfer from the Permian (see next section for numerical values). The fact that the ^{36}Cl/Cl ratio in the Buntsandstein (SHA 1488 m) is also not in equilibrium with the underlying crystalline indicates that the transfer time from the original Permian sediment through the Crystalline into the Buntsandstein is fast compared to the half-life of ^{36}Cl. Alternatively, a direct transfer from Permian sediments into the Buntsandstein is also in agreement with the ^{36}Cl results.

The two other Buntsandstein samples ("Böttstein 317 m" and "Leuggern 218 m") are waters of low mineralization. An influence of the crystalline would shift the points in Figure 6.2.4 towards the equilibrium line. This shift appears to be somewhat too large for Böttstein and somewhat too small for Leuggern (see Figure 6.2.4).

Permian (Rotliegend) (Figure 6.2.5): The three highly mineralized samples from the Weiach and Riniken boreholes have ^{36}Cl/Cl ratios that agree within the error bars with the calculated local equilibrium ratio. The fourth, Kaisten 284 m, with a low mineralization of 100 mg Cl/litre is known to be a crystalline water. The equilibrium ^{36}Cl/Cl ratio in the underlying crystalline is about $31 \cdot 10^{-15}$ (e.g., sample "Kaisten 310 m" in Table 6.1.2) very close to the measured value in Kaisten 284 m. The residence time in the Permian formation was too short for the ^{36}Cl/Cl ratio to have shifted the local equilibrium value.

6.2.7 ^{36}Cl in Water from the Crystalline

All data from crystalline rocks are plotted in Figure 6.2.6. Most of the crystalline values are higher than those of the other geologic formations due to the higher U and Th concentrations of the Crystalline. The measured values in only three water samples (SHA 1571 m, SHA 1888 and LEU 251) disagree within the error bars with the calculated equilibrium values in rock.

Sample LEU 251 is a water of low mineralization. Its high measured ^{36}Cl/Cl ratio of $(73 \pm 14) \cdot 10^{-15}$ confirms the high value found in the Buntsandstein water of Leuggern. It appears that the chloride originates from an area with a higher neutron flux than used for calculating the local subsurface production rate. A different elemental concentration (of, e.g., uranium, thorium or gadolinium) would explain such a difference.

As discussed in Section 6.2.5, three of the samples must have derived their chloride content from a sedimentary source (BOE 1304, SHA 1571 and SHA 1888). If this source was Permian with an estimated average equilibrium ^{36}Cl/Cl ratio of $25 \cdot 10^{-15}$, the following numbers can be calculated for the time that passed since the chloride was transferred from the (Permian) sediment into the crystalline:

Böttstein	1304 m:	t ≥ 290 000 years
Schafisheim	1571 m:	t ≤ 60 000 years
Schafisheim	1888 m:	t = 135 000 ± 60 000 years

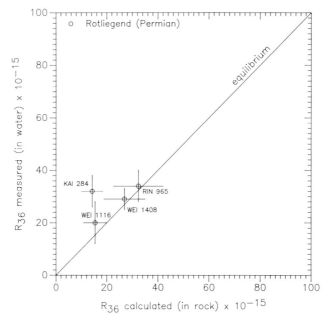

Figure 6.2.5: Measured ^{36}Cl/Cl ratios in water from the Permian *versus* calculated ^{36}Cl/Cl ratios in rock.

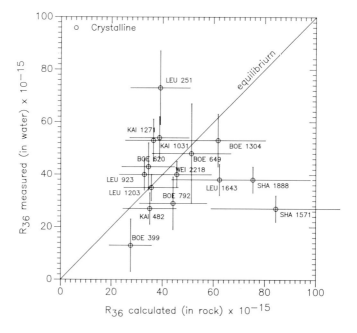

Figure 6.2.6: Measured ^{36}Cl/Cl ratios in water from the Crystalline *versus* calculated ^{36}Cl/Cl ratios in rock.

The measured value of $R_{36} = 170 \cdot 10^{-15}$ in the sample LEU 705 (see Table 6.2.1) has not been included in Figure 6.2.6. This ratio is extremely high, and we cannot offer any explanation. However, we would recommend a remeasurement.

6.2.8 ^{36}Cl Measurements on Rock Samples

To check the validity of the neutron flux calculations one can directly analyse the ^{36}Cl/Cl ratios in rock. A first set of samples has been processed. The rocks were crushed to a fine powder from which AgCl targets were produced for AMS analysis. The measurements were performed at the Institut für Mittelenergiephysik (ETH, Zürich: Prof. Wölfli). U and Th concentrations on a fraction of each sample were measured at PSI. The results are given in Table 6.2.3 together with the values for R_{36} calculated using the relations of Section 6.1.

There is general agreement between the measured values in rock and the calculated values based on measured U and Th concentrations using the approach discussed in Section 6.1. Muschelkalk values (mo) are low (around $10 \cdot 10^{-15}$), Crystalline values (KRI) are higher (between 30 and $93 \cdot 10^{-15}$), and Buntsandstein(s) and Permian values (r) are intermediate (10 to $20 \cdot 10^{-15}$). However, the correspondence for individual samples varies, probably because small samples of approximately 100 g are not always representative of the surrounding layers.

Table 6.2.3: ^{36}Cl measurements on rock samples.

R_{36} (meas) = measured ^{36}Cl/Cl ratio ($\cdot 10^{-15}$) by AMS

R_{36} (calc) = calculated equilibrium ratio in rock ($\cdot 10^{-15}$) (according to Section 6.1)

Site	Depth (m)	Formation	R_{36} meas	Uran. (ppm)	Thor. (ppm)	R_{36} calc
BOE	142	mo	<10	1.28	1.2	7.8
WEI	832	mo	<10	1.87	0.45	9.7
SHA	1253	mo	14 ± 6	1.1	0.6	6.1
BOE	311	s	<10	1.72	4.35	7.5
WEI	983	s	17 ± 7	0.48	1.78	2.5
LEU	218	s	<10	3.61	2.28	10.9
WEI	1369	r	20 ± 6	15.1	23.7	55.7
BOE	699	KRI	57 ± 10	4.75	11.7	34.6
BOE	1059	KRI	93 ± 11	20.7	15	102
BOE	1489	KRI	41 ± 7	7.75	18.3	55.3
LEU	1202	KRI	46 ± 7	2.14	9.35	21.1
LEU	1642	KRI	30 ± 7	3.81	12.5	32

6.2.9 Summary

- Thirty water samples from the deep drilling programme have been analysed for ^{36}Cl by accelerator mass spectrometry (Table 6.2.1). Most $^{36}Cl/Cl$ ratios are within the fairly narrow range from about $10 \cdot 10^{-15}$ to $60 \cdot 10^{-15}$. (One sample has a value of $170 \cdot 10^{-15}$, which requires further confirmation).

- The Cl content of the same water samples, on the other hand, varies by a factor of more than 1000. Therefore, the concentrations of ^{36}Cl and stable Cl atoms in all samples are strongly correlated (Figure 6.2.1).

- We conclude that the ^{36}Cl atoms in all water samples (with the possible exception of Weiach 859 m) originate mainly from deep subsurface sources. Atmospheric input, thermonuclear ^{36}Cl or ^{36}Cl produced near the surface can be neglected (Section 6.2.3).

- The dominant path for ^{36}Cl production is neutron capture by ^{35}Cl. This process occurs on chloride dissolved in groundwater and on chlorine atoms within the rock matrix. After approximately 1.5 million years, the $^{36}Cl/Cl$ ratio reaches equilibrium. Its magnitude is proportional to the local neutron flux.

- Models for calculating the neutron flux in a given rock formation require a detailed knowledge of the elemental composition. If such data are not available, one must work with "typical" concentrations of the elements that dominate neutron production and neutron absorption (Section 6.1). Therefore, all such models have a basic uncertainty and do not take into account any anomalies or local inhomogeneities. On the other hand, one can reasonably assume that over a larger area these variations average out to a certain degree (The mean free path for thermal neutrons is approximately 0.5 m).

- The general result of the neutron flux calculations is that one would expect low $^{36}Cl/Cl$ ratios of about $10 \cdot 10^{-15}$ in carbonate rock (Muschelkalk), medium values in the range of 20 to $30 \cdot 10^{-15}$ in Sandstones (Buntsandstein and Permian) and higher values of approximately 40 to $50 \cdot 10^{-15}$ in crystalline rock (see Table 6.1).

- As can be seen in Figures 6.2.3 to 6.2.6, there is generally good correspondence between the calculated equilibrium $^{36}Cl/Cl$ ratios in the rock and the measured $^{36}Cl/Cl$ ratio in the water samples taken from the respective formations. This shows that:

  ○ The production of ^{36}Cl did actually occur deep underground as was discussed in Sections 6.2.3 and 6.2.4. (If any other sources were important one would not observe such a correspondence).

  ○ The neutron flux calculations seem to be fairly accurate in spite of the basic uncertainties related to variable elemental compositions. Error bars of about ±30 per cent are therefore considered to be realistic.

- Such a correspondence would result if:

  o ^{36}Cl production occurred in water with chloride initially containing no ^{36}Cl atoms and continuing for several hundred thousand years (Figure 6.2.2); or

  o ^{36}Cl production occurred in rock and the fact that the ratios in water are the same as in the surrounding rock simply reflects that the chloride has been dissolved from the rock formation in which the water flows. Dating of groundwater is then not possible.

- The first hypothesis cannot be ruled out as a possible explanation for the ^{36}Cl data. However, for all sedimentary samples and for the samples of low mineralization from the crystalline (see Section 6.2.5), an origin of the chloride from the local rock matrix is quite possible, and a correspondence between measured values in water and calculated ratios in rock would result. One can estimate residence times within the crystalline for the three highly mineralized samples discussed in Section 6.2.7.

- Finally, the ^{36}Cl/Cl results confirm the influence of the underlying Permian or crystalline formation on the waters sampled in Buntsandstein (Section 6.2.6).

6.3 Argon-39 and Argon-37

H. H. Loosli and B. E. Lehmann

6.3.1 Introduction

As outlined in Section 6.1, the two noble gas radioisotopes ^{39}Ar (half-life 269 a) and ^{37}Ar (half-life 35 d) can be produced underground in neutron reactions:

$$^{39}K(n,p)^{39}Ar$$
$$^{40}Ca(n,\alpha)^{37}Ar$$

It has been demonstrated in various studies (LOOSLI, 1983, LOOSLI AND OTHERS, 1989) that in granitic groundwaters especially, the amount of ^{39}Ar produced underground can exceed that of atmospheric origin. Concentrations of up to 1600 per cent have been measured in extreme cases (100 per cent = specific activity of atmospheric ^{39}Ar = 1.78 · 10^{-3} Bq/litre of Ar).

In this section, the production rates in the granite, sandstone and limestone rock types (from Section 6.1) are compared with concentrations measured in groundwater. The possibilities and limits of ^{39}Ar for dating groundwater, as well as for understanding rock-water interactions and transport of radionuclides are discussed.

The first measurements of ^{37}Ar in groundwater have been made only recently (LOOSLI AND OTHERS, 1989), so only a limited number of results are available from the Nagra programme. Using ^{39}Ar/^{37}Ar ratios, one can discuss underground production and escape from rock into groundwater even when the *in situ* neutron flux is not well known. As mentioned in Section 6.1, fast neutrons may contribute to the subsurface production of ^{37}Ar and ^{39}Ar; recent estimates show that the contribution is probably dominant (ANDREWS AND OTHERS, 1990). Since the investigation about the energy spectrum of the neutrons and the energy dependence of the cross section is not yet finished, the discussion in this section is based on production by thermal neutrons only.

Therefore, the fractional loss coefficients calculated in this section are preliminary; they will have to be reduced for ^{37}Ar and for ^{39}Ar by the same factor as their production rates will turn out to be higher. Recent data on production by fast neutrons is given in an Addendum to this section.

6.3.2 ^{39}Ar Results

All ^{39}Ar results from the Nagra deep drilling programme are presented in Table 6.3.1. Results on additional samples from the regional programme were given in Table 4.4.1. Table 6.3.1 also gives measured argon gas concentrations of groundwater and the calculated number of ^{39}Ar atoms per cm^3 of water. The last column gives the number of

Table 6.3.1: Results of ^{39}Ar analyses of groundwater samples from Nagra boreholes. Additional results from the regional programme are given in Table 4.4.1.

(1) ^{39}Ar activity in per cent corrected for subsurface ^{40}Ar (100 per cent = activity of modern air = $1.78 \cdot 10^{-3}$ Bq/litre of Ar)
(2) Measured argon gas concentration in cm^3 STP/cm^3 H$_2$O ($\cdot 10^{-4}$)
(3) Concentration in atoms/cm^3 of water (N_W). Calculated from (1) and (2)
(4) Calculated equilibrium concentration in rock taken from Table 6.1.2 in atoms/cm^3 of rock (N_R). (*For Siblingen the average of all Muschelkalk values is used)
Depth Mean depth of water sampling interval

Site	Sample	Mean Depth (m)	Formation	(1)	± 1σ	(2)	(3)	(4)
BOE	301/2	163	mo	<6.3		4.7	<0.65	0.64
WEI	302/10b	859	mo	12.8	1.9	3.8	1.06	0.77
RIN	303/2	656	mo	<7.2		5.2	<0.82	0.79
SHA	304/3	1233	mo	12.3	4.5	4.2	1.13	0.35
LEU	306/1	75	mo	44.5	7	4.1	3.98	0.75
SIB	307/1	164	mo	5.7	2.5	4.4	0.55	0.66 *
BOE	301/6	317	s	149	4	5.4	17.54	2.29
WEI	302/12	985	s	44.8	3.8	4.8	4.69	1.48
RIN	303/3	807	s	106.5	6.6	5.7	13.23	4.25
KAI	305/1	113	s	57.4	3.5	5.3	6.63	3.4
LEU	306/2	218	s	173	7	5.1	19.23	2.01
WEI	302/19	1116	r	114	6	4	9.94	6.92
RIN	303/5	965	r	270	15	4.5	26.49	27.05
KAI	305/2	284	r	192	6	5.3	22.18	8.04
BOE	301/12	621	KRI	450	10	5.1	50.03	24.73
BOE	301/23	649	KRI	510	12	5	55.59	41.84
BOE	301/18	792	KRI	523	11	5.1	58.15	31.78
KAI	305/4	310	KRI	183	8	5.2	20.74	19.86
LEU	306/9	706	KRI	168	5	5	18.31	10.51
LEU	306/16	1643	KRI	586	10	5.3	67.71	57.85

atoms of ^{39}Ar per cm^3 of rock produced underground according to the calculations presented in Table 6.1.2.

For a first overview, all measured ^{39}Ar concentrations are plotted in Figure 6.3.1 *versus* the equilibrium concentrations in the formation from which the groundwater sample was taken.

6.3.3 Escape of Argon Atoms from Rock into Water

The equilibrium concentrations in rock were calculated in Section 6.1 from the known local U, Th and K concentrations under the assumption that no ^{39}Ar atoms escape from the rock matrix into groundwater. Equilibrium in rock will be reached in approximately five half-lives of ^{39}Ar or 1350 years. As noted in Section 6.1, only production by thermal neutrons is considered except in the Addendum to this section.

In order to estimate maximum possible concentrations in a volume of groundwater, N_M (atoms/cm^3 of water), the equilibrium concentration in rock, N_R (atoms/cm^3 of rock) is divided by the local open porosity p (cm^3 of water/cm^3 of rock). This estimate implies that all atoms produced in rock are transferred to the water in a time that is short compared to the half-life of ^{39}Ar. Furthermore, it permits no water movement which would remove ^{39}Ar from the volume under discussion.

It is to be expected that only a fraction of the available ^{39}Ar will escape from the rock to the groundwater before decaying. Therefore, a fractional loss coefficient, e, is defined as the ratio of the measured concentration in groundwater, N_W, to the maximum possible concentration, N_M. Thus:

$$N_W = e \cdot N_M = e \cdot (N_R/p)$$

These fractional loss coefficients can be estimated from the results in Table 6.3.1 by introducing appropriate values for the porosity of the individual formations.

6.3.4 Fractional Loss Coefficients of ^{39}Ar for the Individual Formations

a) Crystalline Rock

For the six measurements on groundwaters from the crystalline, an average value of e/p = 1.5 ± 0.4 is calculated. This is represented by the straight line in Figure 6.3.2. For values of open porosity between 0.5 per cent and two per cent, fractional loss coefficients in a range of 0.6 per cent to 3.8 per cent result. These numbers are in accordance with values of 0.2 per cent to 0.9 per cent found at the Stripa site (LOOSLI AND OTHERS, 1989), where the local neutron flux is higher by about a factor of three. No groundwater dating is possible for these samples.

b) Muschelkalk

From the six measurements on water from the Muschelkalk, two (BOE 163 and RIN 656) have ^{39}Ar concentrations below and one (SIB 164) has a value close to the detection limit. Two other samples (LEU 75 and WEI 859) contain a young water component as discussed in Section 4.4, which can explain all of the ^{39}Ar activity in these samples. These two samples most likely contain no detectable ^{39}Ar produced underground. The result for the remaining sample (SHA 1233) is not very reliable due to the small amount of

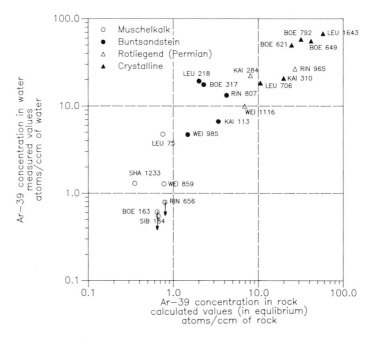

Figure 6.3.1: Measured ^{39}Ar concentrations in groundwater *versus* calculated equilibrium concentrations in rock.

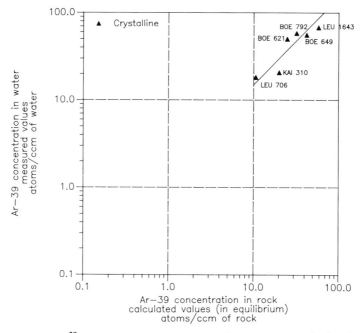

Figure 6.3.2: Measured ^{39}Ar concentrations in groundwater *versus* calculated equilibrium concentrations in rock for the Crystalline.

argon obtained after a difficult sample preparation caused by the high H_2S content of this water.

An upper limiting ratio of e/p = 1 is represented by the straight line in Figure 6.3.3. We conclude that fractional loss coefficients in the Muschelkalk, with a typical open porosity of ten per cent, are smaller than ten per cent. Subsurface production in the Muschelkalk is probably not important. Such a statement is further confirmed by the ^{39}Ar value of < 4.6 per cent modern measured in the Beznau borehole (15/1) (see Table 4.4.1). Therefore, if measurable ^{39}Ar activities occur in this formation, they either represent atmospheric ^{39}Ar and can be used for groundwater dating, or they indicate an admixture of water from, *e.g.*, the crystalline. The second situation would explain the value of 293 per cent modern measured in Pratteln (8). Other ^{39}Ar measurements in Table 4.4.1, from Densbüren (14), Magden (271 to 274), and Lostorf 3 (123), can most likely be used for groundwater dating.

c)   <u>Buntsandstein and Permian</u>

The interpretation of the ^{39}Ar results in groundwaters from the Buntsandstein and from the Permian formations is more difficult for three reasons:

- The ^{39}Ar activities are between the low values in Muschelkalk, where subsurface production is not important, and the high values of the crystalline, where subsurface production is dominant. They cover a range from 45 per cent modern (Weiach 985 m) to 270 per cent modern (Riniken 965 m). For at least some of the samples, a large fraction of the ^{39}Ar activity must, therefore, originate from subsurface sources.

- The calculated equilibrium concentrations in Buntsandstein samples are based on an average elemental composition of "sandstone" and are probably fairly reliable (see Section 6.1). However, it is not known if the same average elemental composition is adequate for Permian rocks.

- According to chemical evidence, most waters from this group are influenced by other formations.

All Buntsandstein waters have an admixture of either crystalline water or water from the underlying Permian aquifers. Higher U, Th and (or) K concentrations than in the Buntsandstein would lead to higher production rates of ^{39}Ar. This effect is illustrated in Figure 6.3.4. A mixing of two water types (*e.g.*, Buntsandstein and underlying crystalline) would shift points along the horizontal line between the two end-members calculated for the respective rock formations. A section extending 50 m below the water sampling interval was selected to be representative for the underlying formation.

One should keep in mind that the calculated values always have error bars of approximately ±30 per cent because of the basic uncertainties outlined in Section 6.1. Furthermore, it should not be expected that all points in such a figure would lie on a straight line representing a constant e/p. Either of the two parameters, e and p, can vary locally. In view of these reservations the correlation between calculated concentrations in rock and measured concentrations in water is surprisingly good.

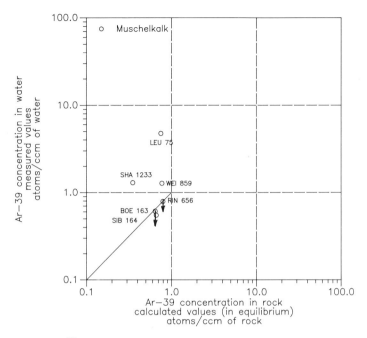

Figure 6.3.3: Measured ^{39}Ar concentrations in groundwater *versus* calculated equilibrium concentrations in rock for the Muschelkalk.

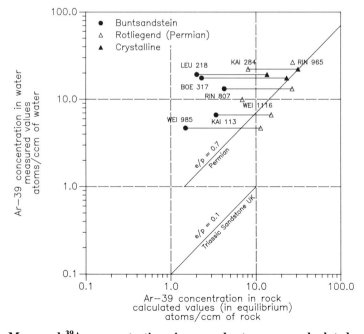

Figure 6.3.4: Measured ^{39}Ar concentrations in groundwater *versus* calculated equilibrium concentrations in rock for Buntsandstein and Permian samples.

The assumption that the production of ^{39}Ar in Buntsandstein/Permian waters is determined by production in Permian sediments leads to five data points in Figure 6.3.4. These are the open triangles shown for Riniken 965 and 807 m, Weiach 1116 and 985 m, and Kaisten 113 m. A best fit to these data yields e/p = 0.7 ($\pm$0.4). For possible porosities in a range of 8 to 12 per cent, one calculates fractional loss coefficients for ^{39}Ar between approximately five per cent and nine per cent.

These values are higher than those measured in an earlier study of a UK Triassic Sandstone formation (ANDREWS AND OTHERS, 1983). This formation, which is better defined than the Permian aquifers of this study, has a porosity of 30 per cent and fractional loss coefficients of e $\leq$ 3 per cent observed in a number of samples. The straight line of e/p = 0.1 from this study is included in Figure 6.3.4 for comparison. With e $\leq$ 10 per cent and p = 10 per cent, the average *in situ* production of 3.2 ($\pm$1.6) atoms/cm^3 rock (from Table 6.1.2) in the Buntsandstein yields a maximum ^{39}Ar activity in water of 30 ($\pm$15) per cent for this formation.

It is important to keep in mind that the calculated ^{39}Ar equilibrium concentrations in rock (Table 6.1.2) are based on production by thermal neutrons only. If the contribution by fast neutrons is large (as preliminary calculations discussed in the Addendum seem to indicate) the values of e/p given in this section will have to be adjusted. The same reservation holds for the discussion of ^{37}Ar results.

6.3.5 ^{37}Ar Measurements

Only a limited number of ^{37}Ar results are available because the short 35 day half-life of this nuclide requires that measurements be made within a few months of sample collection. This has been possible only since 1985. A first set of results from the Stripa programme illustrated the use of this isotope in studying rock-water interactions (LOOSLI AND OTHERS, 1989). The ^{37}Ar results from the Nagra programme are presented in Table 6.3.2, together with the number of ^{37}Ar atoms produced in the subsurface as calculated in Section 6.1.

All measured ^{37}Ar activities in groundwater are produced in the subsurface (see Sections 4.1 and 6.1). The atoms originate from the rock in contact with the water in the last few months before sampling. Factors that influence the ^{37}Ar concentration in water include: the local neutron flux, the concentration and spatial distribution of Ca atoms, and the fractional loss coefficient, e, that quantifies the probability of ^{37}Ar escape from the rock matrix to the liquid phase. Fractional loss coefficients may be estimated as described for ^{39}Ar in Section 6.3.3. This is done by comparing the measured concentrations in water (N_W) with calculated equilibrium concentrations in rock (N_R) for a range of representative (open) porosity values (p) according to

$$e = p \cdot (N_W/N_R)$$

Table 6.3.2: Results of ^{37}Ar analyses of groundwater samples.

(1) Measured activity in dpm/l of argon.

(2) Calculated concentration of ^{37}Ar in groundwater (atoms/cm^3) based on (1) and assuming an argon concentration of $5 \cdot 10^{-4}$ cm^3 STP/cm^3 of water

(3) Calculated equilibrium concentration in rock taken from Table 6.1.2 or from an assumed average elemental composition of the respective formation (atoms/cm^3 rock)

Site	Sample	Formation	Mean Depth (m)	Date	(1)	± 1σ	(2)	(3)
Aqui	1/3	tOMM	200	6/05/87	0.146	0.01	0.0053	0.00080
Stockacher	274/102	mo		27/10/86	0.02	0.004	0.0007	0.00180
Eich	273/103	mo		26/10/86	0.013	0.0039	0.0005	0.00180
Falke	272/102	mo	200	22/11/85	0.031	0.011	0.0011	0.00180
Weiere	271/103	mo	248	13/06/85	0.037	0.006	0.0013	0.00180
Weiere	271/104	mo	248	28/10/86	0.035	0.006	0.0013	0.00180
Siblingen	307/1	mo	184	26/09/88	0.056	0.007	0.0020	0.00180
Leuggern	306/9	KRI	705	17/10/86	<0.081		0.0029	0.00051
Leuggern	306/16	KRI	1643	15/02/85	0.029	0.0035	0.0010	0.00006

For waters from the Muschelkalk with an assumed typical open porosity of the rock of p = 10 per cent, the data in Table 6.3.2 leads to fractional loss coefficients, e_{37}, from 3 to 11 per cent. For waters from the crystalline at Leuggern 705 m, an assumed open porosity of one per cent yields $e_{37} \leq 6$ per cent and for Leuggern 1643 m, 20 per cent.

The range of these estimated loss coefficients is similar to that of the ^{39}Ar coefficients. However, the spatial distribution of Ca is likely to be much less uniform than that of K because of fracture fillings, so there will generally be larger uncertainties associated with the values for e_{37} than with those for e_{39}.

At Stripa (LOOSLI AND OTHERS, 1989), for instance, calcite fracture fillings clearly increased the probability for ^{37}Ar to escape to the water phase as compared to ^{39}Ar which was more uniformly distributed throughout the bulk rock. The approach used for calculating fractional loss coefficients from bulk production rates and bulk porosities had to be modified in such a situation.

6.3.6   Comparison of ^{39}Ar and ^{37}Ar

It is interesting to compare the activity ratios of the pair of radioactive argon isotopes. This ratio is independent of the local neutron flux and of the local porosity, neither of which is usually very well known. The remaining factors that influence the ^{37}Ar/^{39}Ar ratio are the concentrations and local spatial distributions of the target elements Ca and K, and possible differences in the fractional loss coefficients, e. Unfortunately, a very small number of samples from the Nagra programme are available for such a comparison, since the ^{39}Ar activities for several samples originate at least partially from remaining atmospheric ^{39}Ar as discussed in Section 4.4. Some measured and calculated activity ratios are given in Table 6.3.3.

Table 6.3.3: ^{39}Ar/^{37}Ar **activity ratios calculated in rock and measured in water samples.**

		Calculated in rock (1)	Measured in water (2)
Stripa	bulk	2.20E-05	7.00E-03
Stripa	fracture fillings	2.50E-03	
Leuggern	705 m	1.50E-04	<4.70E-04
Leuggern	1643 m	3.40E-06	5.00E-05
Muschelkalk	average	8.00E-03	>3.00E-03
Aqui			2.50E-02

There is general agreement between the calculated ratios in rock and the measured ratios in groundwater, particularly when the following factors are considered:

1) In the Stripa samples the measured activity ratio in water is closer to the activity ratio calculated for fracture filling material (15 per cent Ca and 1 per cent K) than to the one calculated for bulk rock (0.5 per cent Ca and 3.8 per cent K). This supports the conclusion that at Stripa the fracture fillings are very important for the chemistry of the groundwater.

2) The sample from Leuggern 1643 m originates from a zone with hydrothermally strongly altered granite with exceptionally low Ca content (0.1 per cent used in the calculation, PETERS AND OTHERS, 1988b). It is unknown if this is indeed representative for the calculation of *in situ* ^{37}Ar production.

3) For the samples from Leuggern 705 m and for the Muschelkalk samples only lower or upper limits for the activity ratios could be determined because either the ^{37}Ar or the ^{39}Ar activity was below the detection limit.

At present, there are not enough data to determine if the fractional loss coefficients for ^{37}Ar and ^{39}Ar are of the same magnitude. There may be a tendency for ^{37}Ar to have a higher fractional loss coefficient. Another basic uncertainty relates to the values of the reaction cross sections used for calculating the subsurface production rates. The values for both isotopes are not very well known. The value for ^{37}Ar has been measured once (MÜNNICH, 1958); the one for ^{39}Ar is still a best theoretical estimate (ALDER, personal communication, 1983). Preliminary results from irradiation experiments suggest that a larger value for the ^{37}Ar product may be correct (FORSTER, personal communication,

1988). Finally, the extent to which fast neutrons promote the production of these two isotopes is presently unknown.

ADDENDUM:

During the final preparation of this report, improvements were made in calculating ^{37}Ar and ^{39}Ar production by fast neutrons. As was first pointed out by FLORKOWSKI (1989), these nuclides are produced in amounts that cannot be neglected when the primary fast neutrons from (α,n)-reactions in the rock are slowed down to thermal energies. Calculations of the energy distribution of neutrons in various rock formations, however, are quite complicated. We have used a PC-version of a neutron slowing-down programme developed by CZUBEK (1988) together with cross-section data from the ENDF/B-neutron data base. Preliminary results indicate that the production rates of Section 6.1, where only thermal neutrons were considered, have to be raised by factors from 5 to 25.

The following table summarises the current status:

	Production by thermal neutrons only (Table 6.2.1)		Production by fast neutrons included	
	^{37}Ar	^{39}Ar	^{37}Ar	^{39}Ar
Muschelkalk	0.0018	0.66	0.016	3.2
Buntsandstein	0.00017	3.2	0.0033	24
Crystalline Rock	0.00035	28	0.0086	290

(all numbers are average values of (atoms/cm^3 rock) at equilibrium)

All estimates given in Section 6.3 on escape factors must therefore be lowered according to the values given in the above table. As already pointed out by ANDREWS AND OTHERS (1990), the ^{37}Ar and ^{39}Ar release efficiencies from rock to water now appear to be very similar for both isotopes, indicating that the process is surface controlled. Matrix diffusion cannot be involved because of the much shorter half-life of ^{37}Ar. Furthermore, there seems to be a connection to ^{222}Rn release rates.

6.4 ^{3}He and ^{4}He

W. Balderer and B. E. Lehmann

6.4.1 Introduction

Subsurface production of ^{3}He and ^{4}He in rocks and subsequent accumulation in groundwater leads to concentrations of both isotopes that very often are several orders of magnitude above the values in samples that have reached equilibrium with the atmosphere during infiltration.

Atmospheric helium has a ^{3}He/^{4}He ratio of $1.38 \cdot 10^{-6}$. At 10°C only $4.75 \cdot 10^{-8}$ cm^3 STP helium are dissolved per cm^3 water (in equilibrium). The concentration of ^{3}He is $6.56 \cdot 10^{-14}$ cm^3 STP/cm^3 water under these conditions.

In contrast, groundwaters from the Nagra deep drilling programme have helium concentrations between 10^{-6} and 10^{-3} cm^3 STP/cm^3 water. ^{3}He/^{4}He ratios are in the fairly narrow range of about $5 \cdot 10^{-8}$ to $2 \cdot 10^{-7}$. Most ^{4}He and ^{3}He concentrations are two or more orders of magnitude above the equilibrium value with air, clearly indicating their subsurface origin.

By comparing production rates and isotopic ratios calculated for the various rock formations with values measured in water samples one can, in principle, study groundwater movements and mixing processes. A review of such a use of radiogenic helium in hydrology was published by ANDREWS (1985). Helium accumulation in water of the Great Artesian Basin in Australia is discussed in detail by TORGERSON AND CLARKE (1985).

6.4.2 Results and First Interpretation

Because of the extreme oversaturation of most groundwaters with helium, relative to equilibrium with the atmosphere, one must be very careful to avoid gas losses during sampling. As discussed in more detail in Section 2.3, not all sampling techniques are equally adequate. Best results were achieved with copper tubes with a volume of about 5 cm^3 sealed with pinch clamps as described in BALDERER (1985b, p. 29).

Noble gas concentrations were measured by Prof. J. Andrews, University of Bath, and ^{4}He concentrations and ^{3}He/^{4}He ratios on selected samples by Prof. K. O'Nions, University of Cambridge. The results are presented in Table 6.4.1. Twenty six of the 31 values listed are for samples which suffered no gas losses during collection based on the criteria used in Section 3.2 in calculating recharge temperatures.

Five values are marked with an asterisk (*) in Table 6.4.1. These were measured at PSI and should be considered as upper limits because one cannot rule out the possibility that gas bubbles were trapped in the cylinders used for the PSI samples, leading to an overestimate of the noble gas content of the water.

Table 6.4.1: ^{3}He/^{4}He ratios and ^{4}He concentrations in waters from the Nagra boreholes.

^{4}He = measured ^{4}He concentrations (cm^3 STP/cm^3 water) measured at Bath and (or) Cambridge

* = PSI-measurement

Site	Sample	Mean Depth (m)	Formation	^{3}He/^{4}He meas	^{4}He meas
SHA	304/1	558	tUSM	5.46E-08	5.60E-05 *
BOE	301/2	163	mo	1.03E-07	2.11E-04
WEI	302/10b	859	mo	1.67E-07	1.72E-06
RIN	303/2	656	mo		5.00E-05
SHA	304/3	1233	mo		2.40E-04
LEU	306/1	75	mo		4.70E-06
BOE	301/6	317	s	6.51E-08	1.33E-04
WEI	302/12	985	s	1.12E-07	1.81E-03
RIN	303/3	807	s	9.24E-08	1.10E-03
SHA	304/7	1488	s		1.45E-03
KAI	305/1	113	s	1.01E-07	3.02E-04
WEI	302/19	1116	r	1.16E-07	1.59E-03
WEI	302/18	1408	r	9.38E-08	4.48E-03 *
RIN	303/6	993	r		2.01E-03
KAI	305/2	284	r		9.30E-05
BOE	301/8	399	KRI		2.24E-04
BOE	301/12	621	KRI	6.06E-08	2.11E-04
BOE	301/23	649	KRI	7.14E-08	1.88E-04
BOE	301/18	792	KRI	6.72E-08	1.47E-04
WEI	302/16	2218	KRI	7.14E-08	2.58E-03 *
WEI	302/14	2267	KRI		2.46E-03 *
SHA	304/9	1571	KRI	9.52E-08	2.74E-03
SHA	304/10	1888	KRI	8.40E-08	9.10E-04
KAI	305/4	310	KRI	1.47E-07	1.10E-04
KAI	305/6	482	KRI	2.03E-07	1.01E-04
KAI	305/9	819	KRI	1.95E-07	9.90E-05
KAI	305/16	1272	KRI		5.06E-05
LEU	306/7	538	KRI	5.46E-08	6.20E-05
LEU	306/9	706	KRI		8.90E-05
LEU	306/13	923	KRI		2.24E-04 *
LEU	306/16	1643	KRI		1.12E-04

All results of Table 6.4.1 are illustrated in Figure 6.4.1. As the figure shows, most ^{4}He concentrations are between $5 \cdot 10^{-5}$ and $3 \cdot 10^{-3}$ cm^3 STP/cm^3 of water, and there is no clear grouping of values by source formations.

The lowest values for the ^{4}He concentrations were measured in the samples from the Muschelkalk waters of Weiach and Leuggern. Waters with the highest ^{4}He concentrations from $9 \cdot 10^{-4}$ to $4.5 \cdot 10^{-3}$ cm^3 STP/cm^3 H$_2$O are found in the Buntsandstein, Permian and crystalline of the Weiach, Riniken and Schafisheim boreholes.

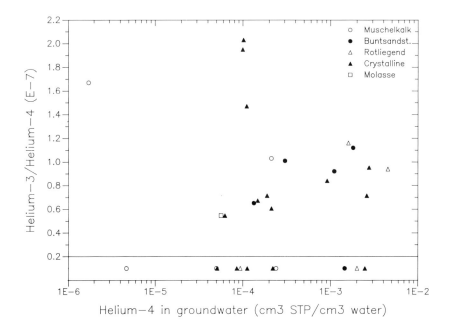

Figure 6.4.1: ^{3}He/^{4}He ratios and ^{4}He concentrations in Nagra samples. Samples plotted below the line lack ^{3}He data.

6.4.3 Subsurface Production

As outlined in Section 6.1, ^{4}He is formed in rock formations by neutralisation of α-particles produced in the U and Th decay series nuclides. ^{3}He is produced by:

$$^6Li(n, \alpha) \; ^3H \xrightarrow{\beta^-} \; ^3He$$

Table 6.1.2 gives calculated production rates for both isotopes from which ^{3}He/^{4}He ratios can be calculated.

As also mentioned in Section 6.1, the key unknown parameter for ^{3}He production is the lithium concentration of the various rock types. Although measurements on a few selected samples show large variations even over short distances, useful ^{3}He/^{4}He ratios can still be calculated. In the Muschelkalk, an average lithium concentration of 11 ppm yields a ^{3}He/^{4}He ratio of $7 \cdot 10^{-9}$. In the Buntsandstein and Permian rocks, an assumed average Li concentration of 37 ppm yields a ^{3}He/^{4}He ratio of $1.4 \cdot 10^{-8}$.

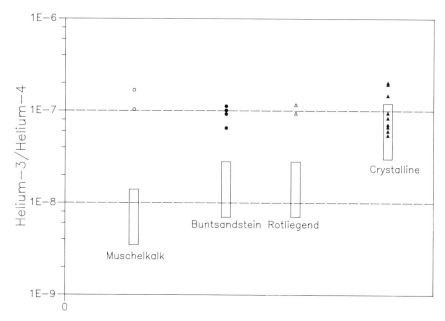

Figure 6.4.2: Measured ^{3}He/^{4}He ratios in water (symbols) compared with the possible range of ^{3}He/^{4}He ratios (boxes) produced in the various formations.

Both these theoretical ratios are clearly lower than the lowest measured values, about $6 \cdot 10^{-8}$, and would remain so even if an uncertainty as large as a factor of four is admitted for the Li concentration.

Theoretical ratios of rates of production in crystalline rocks range from about $3 \cdot 10^{-8}$ to $1.2 \cdot 10^{-7}$, if a factor of four uncertainty in Li concentration is used again. This clearly illustrates that all measured helium samples show a crystalline signature. In particular, they are not produced *in situ* in Muschelkalk, Buntsandstein or Permian rocks. This situation is illustrated in Figure 6.4.2. The rectangles indicating the ranges of calculated ratios represent an uncertainty of a factor of two around the average Li concentrations.

6.4.4   Helium Accumulation Times and Water Residence Times

^{3}He and ^{4}He isotopes are stable and, therefore, should accumulate with time in a closed volume of rock. Some or all of the gas will dissolve in the volume of water corresponding to the local open porosity (cm^3 of water per cm^3 of rock) and, if the water does not move, high concentrations of dissolved helium may eventually result. Generally, high He concentrations should indicate long residence times of water in a particular rock formation. However, it is not easy to quantify such a dating technique.

As a starting point, one can calculate helium accumulation times in a closed volume of rock based on the production rates given in Table 6.1.2, and use various values of porosity to calculate the resulting helium concentrations in water. Because such a model does not allow for any losses of helium, by diffusion for example, it is likely that calculated accumulation times based on *in situ* production alone will not be particularly realistic.

Figure 6.4.3 shows the ages one would get from such a model. The average production rates were taken from Table 6.1.2 and are:

$5.7 \cdot 10^{-13}$ cm^3 STP He/cm^3 rock and year in the Muschelkalk
$11.5 \cdot 10^{-13}$ cm^3 STP He/cm^3 rock and year in the Buntsandstein
$20 \cdot 10^{-13}$ cm^3 STP He/cm^3 rock and year in the Rotliegend
$28 \cdot 10^{-13}$ cm^3 STP He/cm^3 rock and year in the Crystalline

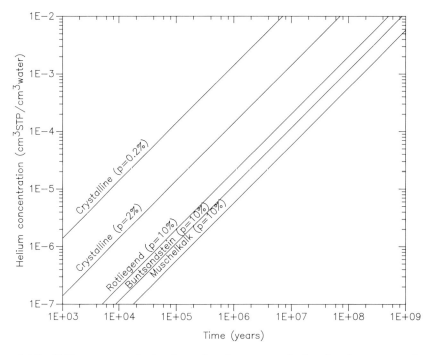

Figure 6.4.3: Helium accumulation in a closed system calculated using average *in situ* production rates (Table 6.1.2) and the porosities indicated.

The porosities, 10 per cent for the Muschelkalk, Buntsandstein and Rotliegend and either 2 per cent or 0.2 per cent for the crystalline, are representative values for the interconnected or open porosity used to quantify the pore volume which can be reached by He atoms through convective water transport or by diffusion along water-filled channels.

To reach a ^{4}He concentration of 10^{-4} cm^3 STP/cm^3 water, a value typical for many of the samples (see Figure 6.4.1), *in situ* accumulation times of 5-20 million years would be necessary for samples from sedimentary formations (Muschelkalk, Buntsandstein, Rotliegend). For the group of samples containing around 10^{-3} cm^3 STP/cm^3 water, these times would have to be ten times longer. However, as discussed in the previous section, the ^{3}He/^{4}He ratio clearly indicates a crystalline origin of the helium in all sedimentary formations. Therefore, model ages calculated from *in situ* production are meaningless for these formations.

For crystalline waters it would take 80,000 to 800,000 years to reach a He concentration of 10^{-4} cm^3 STP/cm^3 water, depending on whether one assumes a porosity of 0.2 per cent or 2 per cent. As was pointed out by several authors (*e.g.*, ANDREWS, 1985, TORGERSON AND CLARKE, 1985) ages based on models of *in situ* production alone tend to overestimate water residence times by large factors. This is demonstrated when cross calibration either with ^{14}C ages or with reliable hydrodynamic model ages (as for example in the Great Artesian Basin in Australia) is possible. The main reason is that long range transport of helium is not taken into account in this simple first-order approach.

6.4.5   Helium Transport Models

It is difficult to create a complete description of the pathways for helium atoms from their sites of production inside a source rock deep below the surface, through a complex geologic environment with many different layers of quite different materials up to the surface. Not only must one understand release mechanisms from the solid phase into the pore fluids, but one also needs to know how all the water-filled channels are interconnected or if there are, for instance, pathways for helium gas that do not lead through the liquid phase at all. Furthermore, it is not even clear whether a helium source deep below the surface would continuously release helium, or whether the gas accumulates over longer periods and is released in relatively short periods of, for instance, rock alteration or tectonic stress.

The most simple model to relate the residence time, t_R, of water in an aquifer to the flux, F, of ^{4}He into the aquifer at the lower boundary and to the measured concentration, [^{4}He], of helium can be written:

$$[^4\text{He}] = F \cdot t_R \cdot d^{-1} \cdot p^{-1} \qquad (6.4.1)$$

where d is the thickness of the aquifer and p its porosity.

This model treats the aquifer as a trap for helium gas. No diffusive loss at the upper boundary is considered and *in situ* production inside the aquifer is neglected.

As demonstrated by TORGERSON AND CLARKE (1985) for the J-aquifer in Australia, for instance, the flux of ^{4}He into a confined sandstone aquifer can be estimated from the *in situ* production rate in the underlying crystalline basement assuming a steady-state situation where, for every atom produced in the upper crust of approximately 10 km, one atom is removed at the upper boundary.

With an average ^{4}He production rate for crystalline rocks of $2.8 \cdot 10^{-12}$ cm^3 STP per cm^3 rock per year (from Table 6.1.2) and a thickness of 10 km for the upper crust in northern Switzerland, a flux of $F = 2.8 \cdot 10^{-6}$ cm^3 STP He/(cm$^2 \cdot$ a) is calculated for such a steady-state situation. This value is similar to the values of TORGERSON AND CLARKE (1985) of $F = 3.6 \cdot 10^{-6}$ cm^3 STP/(cm$^2 \cdot$ a), deduced from measurements in Australia and to the average continental steady-state flux of $F = 3.3 \cdot 10^{-6}$ cm^3 STP/(cm$^2 \cdot$ a) given by O'NIONS AND OXBURGH (1983) to explain the global atmospheric helium budget.

ANDREWS (1985) used a different approach, a diffusion model, to calculate a flux of ^{4}He towards the surface. As a solution of the diffusion equation for a thick crust with uniform radioelement content ANDREWS (1985) gets

$$c(z,t) = Gt(1 - \exp(-2\pi^{-1/2}z(Dt)^{-1/2})) \qquad (6.4.2)$$

where $c(z,t)$ is the helium concentration in rock at time t and at a distance z below the surface, G is the rate of generation of He atoms per unit volume of rock and D is a diffusion constant. The flux F at the surface ($z = 0$) is obtained by differentiation with respect to z:

$$F = 2 \cdot G \cdot (Dt/\pi)^{1/2} \qquad (6.4.3)$$

While numerical values for the helium generation rate, G, can be taken from Table 6.1.2 with reasonable accuracy, it is more difficult to choose values for the diffusion coefficient, D, and the total time, t, for the build-up of a concentration profile.

As a maximum value for the latter, one may use the rock age of 300 million years. If diffusion occurs in the liquid phase only, the diffusion coefficient D is $= 0.18$ m$^2 \cdot$ a^{-1}.

The production or generation rate, G, calculated above for crystalline rock is $2.8 \cdot 10^{-12}$ cm^3 STP/(cm$^3 \cdot$ a). Using this rate, the flux of He at $z = 0$ (top of the crystalline) is

$$F = 2 \cdot G \cdot (Dt/\pi)^{1/2} = 2.3 \cdot 10^{-6} \text{ cm}^3 \text{ STP/(cm}^2 \cdot \text{a)}$$

This value is very close to the one calculated above based on an average production in the crust and a steady-state release.

Therefore, a fast local transfer of ^{4}He from rock into pore fluids and undisturbed diffusion along water-filled channels to the top of the crystalline would supply enough ^{4}He to match the values required by the global atmospheric helium budget.

However, diffusion in porous media most likely occurs more slowly. The porosity, p, for instance, can be used to calculate an effective diffusion coefficient D_e in a porous medium according to

$$D_e = D_o \cdot (p/(1+0.5(1-p)))$$

For a crystalline rock with a porosity of p = 0.01, for example, the diffusion coefficient D_o in water of 0.18 m$^2 \cdot$ a^{-1} would be lowered to

$$D_e = 1.2 \cdot 10^{-3} \text{ m}^2 \cdot \text{a}^{-1} = 3.8 \cdot 10^{-11} \text{ m}^2 \cdot \text{s}^{-1}$$

The flux F in the diffusion model would accordingly be lowered to $1.9 \cdot 10^{-7}$ cm^3 STP/(cm$^2 \cdot$ a), eight per cent of the previous number. This seems to indicate that diffusion is not effective enough to transport helium through crystalline rocks.

This conclusion is supported by a comparison of the concentrations c(z,t) predicted by equation 6.4.2 with values measured in groundwater.

For a porosity, p, of one per cent one would expect the following ^{4}He concentrations at various depths, z, in a diffusion profile ($D_e = 1.2 \cdot 10^{-3}$ m$^2 \cdot$ a^{-1}, t = 3.10^8 a, G = 2.8 $\cdot$ 10^{-12} cm^3 STP/(cm^3 rock $\cdot$ a):

10^{-4} cm^3 STP He/cm^3 water at z = 0.64 m
10^{-3} cm^3 STP He/cm^3 water at z = 6.4 m
10^{-2} cm^3 STP He/cm^3 water at z = 67 m
$8.4 \cdot 10^{-2}$ cm^3 STP He/cm^3 water at z = ∞

This shows that subsurface steady-state concentrations of nearly 10^{-1} cm^3 He/cm^3 water would be required to drive a diffusive flux sufficient to remove at the surface all the helium produced at depth. Measured values are at most from 10^{-3} to 10^{-4} cm^3 STP/cm^3 water, which means that some mechanism in addition to diffusion must be responsible for the helium flux to the surface.

In accord with TURCOTTE AND SCHUBERT (1987), we conclude that diffusion processes are unlikely to be important for the transport of helium on a planetary scale. More likely, helium is transported by water circulation in the near surface part of the continental crust.

6.4.6 Correlation Between Helium and Chloride Concentrations

Figure 6.4.4 shows that there is a certain correlation between the concentrations of chloride and helium in groundwater from the crystalline, Buntsandstein and Permian. The group of crystalline waters centred around values of 100 mg Cl/litre and 10^{-4} cm^3 STP He/cm^3 water are the waters of low mineralisation from the Böttstein, Kaisten and Leuggern boreholes. The highly mineralised waters from the crystalline of Weiach and Schafisheim have concentrations of both Cl and He that are larger by factors of about 30. The He/Cl ratio, R, is, therefore, about constant and has a value of approximately R = 1/630 = 0.0016. This is represented by the straight line "Cryst.-NAGRA" in Figure 6.4.4.

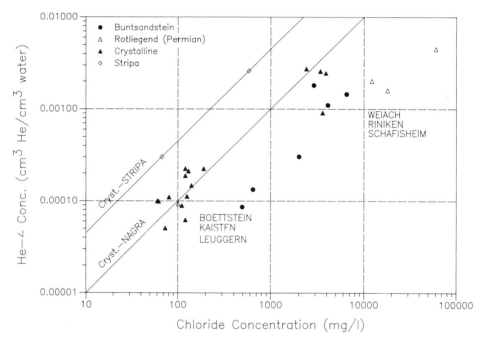

Figure 6.4.4: Helium concentrations *versus* chloride concentrations for the Crystalline, Buntsandstein and Permian aquifers.

It is interesting to note that samples from the Stripa granite in Sweden show a similar behaviour: values of 580 mg Cl/litre of water with $2.6 \cdot 10^{-3}$ cm^3 STP of He/cm^3 of water were measured in the deep borehole V1; values of 67 mg Cl/litre of water with $3 \cdot 10^{-4}$ cm^3 STP of He/cm^3 of water in borehole N1 (NORDSTROM AND OTHERS, 1985). These two points are also included in Figure 6.4.4 for comparison. They define a second straight line, "Cryst.-STRIPA," which has the same slope as the Nagra line. The He/Cl ratio equals R = 1/140 = 0.0071, higher than the Nagra ratio by a factor of 4.5.

The Stripa granite has higher uranium (44 ppm) and thorium (53 ppm) concentrations than an average granite in the crystalline of northern Switzerland. Application of the ^{4}He production rate expression derived in Section 6.1 leads to a ^{4}He production rate in Stripa which is higher by a factor of 6.3, close to the differences between the He/Cl ratios observed in the two granites.

These facts seem to indicate that Cl and He originate from a common source. One possible source of chloride is the fluid inclusions and other water-soluble chloride present in crystalline rock. Whereas the chloride concentration of the crystalline is expected to stay constant, ^{4}He concentrations would increase with time if no losses occur. Based on the ^{4}He production rate derived in Section 6.1, one can calculate the time required for the He/Cl ratio to reach the groundwater value of $R = 0.0016$. The average content of water-extractable chloride in the Böttstein granite measured by PETERS (1986), for example, is 60 ppm. This corresponds to $2.8 \cdot 10^{18}$ atoms of Cl per cm^3 of rock. From the known production rate in average granite of $2.8 \cdot 10^{-12}$ cm^3 STP He/(cm^3 rock · a) (see Section 6.1), a time of 60 million years results.

The crystalline in northern Switzerland was formed about 300 million years ago. At least two hydrothermal events are known to have occurred since then. The first some 250 million years ago; the second, between 50 million years and 100 million years ago (MULLIS, 1987). At that time, fluids of up to 120°C flushed the rock. If a large fraction of the ^{4}He that had accumulated prior to this event was removed at that time, then the ^{4}He clock would have been set to zero. As outlined above, ^{4}He could have accumulated since then to just about the present He/Cl ratio in groundwater, assuming that no losses occurred during this time. In this model, He and Cl would simultaneously be released from rock into groundwater wherever fluid inclusions are opened, or fresh grain boundaries exposed by rock alteration or tectonic stress. High ^{4}He concentrations would not be a direct measure of long water residence times except for the fact that the leaching process itself probably requires a certain time: higher mineralisation requires a longer time for rock/water interactions to become effective. Speculations along the model outlined above are supported by a correlation between ^{4}He, radiogenically produced ^{40}Ar and potassium concentration in groundwater (see next section).

Figure 6.4.5 is a graph of helium concentration *versus* chloride concentration for samples from all rock types. As the figure shows, samples from the Muschelkalk and Molasse have He/Cl ratios similar to or lower than those of samples from the Buntsandstein and Rotliegend, and significantly lower than those from the crystalline. In principle, the lower He/Cl ratios in the samples from sedimentary formations could be due to either lower helium or higher chloride sources to these waters. As the ^{3}He/^{4}He ratios demonstrate, the source of helium dissolved in the waters from the sediments is not the sediments themselves, but crystalline rock. Therefore, the lower He/Cl ratio is better attributed to the higher chloride content of the sediments. This is consistent with the observation that chloride is more readily leached into groundwater from sedimentary rock than from crystalline rock (PEARSON AND FISHER, 1971, Figure 7).

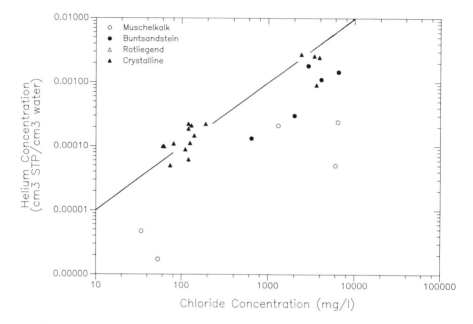

Figure 6.4.5: Helium concentrations *versus* chloride for samples from all units.

6.4.7 Summary

- The ^{3}He and ^{4}He in Nagra groundwaters are of subsurface origin. An atmospheric contribution is negligible.

- The ^{3}He/^{4}He ratio indicates a crystalline origin for the helium in all samples, including those from sedimentary aquifers.

- Because *in situ* production is not important in sediments, the measured ^{4}He concentrations must be explained by long range transport from deeper parts of the continental crust.

- The flux estimated for a steady-state situation, $F = 3 \cdot 10^{-6}$ cm^3 STP/cm$^2 \cdot$ a ($= 2.6 \cdot 10^6$ atoms/cm$^2 \cdot$ s), is in fairly good agreement with measurements in the Great Artesian Basin in Australia (TORGERSON AND CLARKE, 1985) and with the global atmospheric helium budget (O'NIONS AND OXBURGH, 1983).

- Diffusion appears to be too slow to maintain such a helium flux. Therefore, He transport by convective water movement in the upper crystalline is likely.

- A correlation between He and Cl concentrations exists. For the crystalline, the atomic ratio of the two substances in groundwater appears to be approximately 1:630. A similar value of 1:140 was found in Stripa granite. The relative enrichment in helium in Stripa results from the higher uranium and thorium concentrations in the granite there.

- Such a correlation can be explained if the two substances have a common origin. Calculations indicate that accumulation in a closed volume of granite followed by transfer of ^{4}He into the fluid inclusions would yield the measured He/Cl ratio after an accumulation time of about 60 million years. This might coincide with the last observed hydrothermal event. One could speculate that most of the ^{4}He accumulated prior to this event was removed from the rock during that event.

- If such a model is correct, ^{4}He cannot be used as a direct measurement of water residence times. Instead, ^{4}He and Cl have a common source, possibly fluid inclusions. Since it can be assumed that the responsible water/rock interaction process requires some time to become effective, the general statement that waters with high ^{4}He concentrations are comparatively old or contain an old component (and are more highly mineralised) still holds. However, ^{4}He cannot be used to quantify the release rate from rock. The ^{4}He concentrations in groundwater are controlled by the release rate from rock, on the one hand, and the removal rate by water movement, on the other hand.

6.5 $^{40}Ar/^{36}Ar$ Ratios

D. Rauber, H. H. Loosli and B. E. Lehmann

6.5.1 Introduction

The $^{40}Ar/^{36}Ar$ ratio of atmospheric argon, R_o, is 295.5. However, in groundwater this ratio may be different because of radiogenic argon production in the subsurface (see Section 6.1). Production of ^{40}Ar from long-lived ^{40}K can be significant, but production of ^{36}Ar by decay of ^{36}Cl is so low as to be negligible. Therefore, the $^{40}Ar/^{36}Ar$ ratio of argon dissolved in groundwater is 295.5 or higher.

The precision of a $^{40}Ar/^{36}Ar$ measurement is generally around one per cent, but, in some cases, may be as high as five per cent. Groundwaters usually contain $(4 \text{ to } 5) \cdot 10^{-4}$ cm^3 STP Ar per cm^3 of water from the solution of atmospheric argon. Subsurface-produced ^{40}Ar can be detected unambiguously only if at least an additional 10^{-5} cm^3 STP Ar has been accumulated per cm^3 of water. Thus, the minimum detectable $^{40}Ar/^{36}Ar$ ratio is higher than about 303.

6.5.2 Results

Measured $^{40}Ar/^{36}Ar$ results are given in Table 3.3.2. Most analyses were made on water samples taken for the measurement of noble gas concentrations. However, some water samples had no $^{40}Ar/^{36}Ar$ measurements. For these, $^{40}Ar/^{36}Ar$ analyses were made on the large volume gas samples extracted in the field for ^{39}Ar and ^{85}Kr measurements. Measurements were made on both water and gas from a few samples. Systematically lower values were obtained for the gas samples, as shown in Figure 6.5.1. In addition, all gas samples have relatively low ratios, between 291 and 319. A qualitative explanation may be that fractionation accompanies the degassing during sampling. The diffusion coefficients of ^{40}Ar and ^{36}Ar differ by about five per cent. It can be assumed that fractionation does not effect the $^{40}Ar/^{36}Ar$ ratios of the water samples, provided no gas losses occur during sampling. Thus, the analyses of the water samples are probably more reliable, and they are used in the discussion which follows.

Figure 6.5.2 gives an overview of the results from the deep boreholes. Most of the ratios are between 290 and 305, although several are between 310 and 395. The highest values are found in the samples from the Permian of the Weiach and Riniken boreholes, and in samples from the Molasse and deep crystalline of the Schafisheim borehole.

6.5.3 Subsurface Production of ^{40}Ar

As discussed in Section 6.1, the production rate of ^{40}Ar in rock, P_{40}, is proportional to the potassium content in rock:

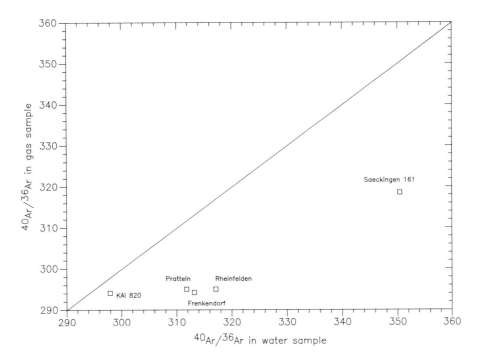

Figure 6.5.1: $^{40}Ar/^{36}Ar$ ratios measured in large-volume gas samples *versus* those in water samples.

$$P_{40} = p \cdot [K] \cdot 3.4 \text{ (atoms/(cm}^3 \text{ rock} \cdot \text{s))}$$

or

$$P = p \cdot [K] \cdot 4 \cdot 10^{-12} \text{ (cm}^3 \text{ STP } ^{40}Ar/(cm^3 \text{ rock} \cdot \text{year)})$$

In these equations,

p = rock density in g/cm^3
[K] = potassium concentration (0 < [K] < 1)

Since ^{40}Ar is a stable nuclide, its concentration would increase with time in a closed volume of rock. Under the assumption that all ^{40}Ar produced *in situ* is transferred to the volume of water contained in the local open porosity p, the resulting $^{40}Ar/^{36}Ar$ ratio in water after an accumulation time t can be calculated:

$$R = R_o + (P \cdot t/(p \cdot V_{36}))$$

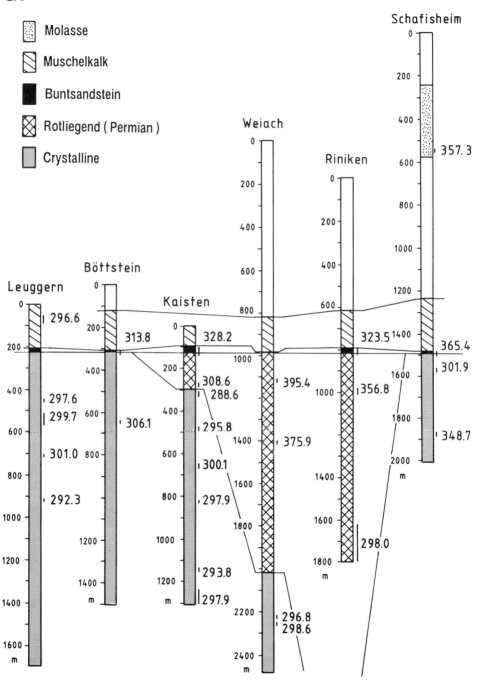

Figure 6.5.2: Columnar sections of deep boreholes with $^{40}Ar/^{36}Ar$ ratios of dissolved argon.

P_{40} = production rate (cm³ STP/cm³ · a))
t = accumulation time (a)
p = open porosity (cm³ of water/cm³ of rock)
V_{36} = volume of dissolved ³⁶Ar in water (cm³ STP, including ³⁶Ar from excess air)
R_o = 295.5 (atmospheric $^{40}Ar/^{36}Ar$ ratio)

The influence of the various parameters is illustrated by the following examples, which were calculated with a constant $V_{36} = 1.7 \cdot 10^{-6}$ cm³ STP/cm³ H₂O:

Formation	p (g/cm³)	[K] rock	T (a)	p	R calc.
Molasse	2.3	0.018	10^6	0.1	296.7
Muschelkalk	2.2	0.005	10^6	0.1	295.7
Rotliegend	2.3	0.04	10^6	0.1	297.8
Crystalline	2.6	0.04	10^6	0.01	323.8

As can be seen in the last column, the $^{40}Ar/^{36}Ar$ ratio should show only small changes even after the very long accumulation time of one million years, and with an unrealistic escape rate of 100 per cent from rock into water. Such a total transfer of underground produced ^{40}Ar from the rock matrix into water is unrealistic because the well-established K-Ar rock dating method works with the assumption that most radiogenic ^{40}Ar is trapped in the rock matrix.

The $^{40}Ar/^{36}Ar$ ratios of a few samples are so high that they cannot be a product of *in situ* production followed by transfer to the local pore water. The water from the tUSM at Schafisheim (558 m, 304/1), for instance, has a measured value of R = 357. To reach this ratio would require an accumulation time of approximately 40 million years, twice the age of the formation itself. Calculated *in situ* accumulation times turn out to be very high for most samples with elevated ^{40}Ar contents, even if the (unrealistic) hypothesis of a complete release of all ^{40}Ar from rock into water is maintained:

Magden Falke 272/101	mo,	$^{40}Ar/^{36}Ar$ = 304.5	t ≈ 3	$\cdot 10^7$y
Böttstein 301/2, 164 m,	mo,	$^{40}Ar/^{36}Ar$ ≥ 310	t ≥ 5	$\cdot 10^7$y
Weiach 302/19, 1116 m,	r,	$^{40}Ar/^{36}Ar$ = 395.4	t ≈ 4.5	$\cdot 10^7$y
Weiach 302/18, 1408 m,	r,	$^{40}Ar/^{36}Ar$ = 375.9	t ≈ 8	$\cdot 10^6$y
Riniken 303/6, 993 m,	r,	$^{40}Ar/^{36}Ar$ = 356.8	t ≈ 1.7	$\cdot 10^6$y

Therefore, it seems that at least some radiogenic ^{40}Ar is transported from neighbouring rock strata outside the aquifer.

6.5.4 Correlations Between $^{40}Ar/^{36}Ar$ Ratios and the Potassium Contents of Rock and Water

Figure 6.5.3 shows the $^{40}Ar/^{36}Ar$ ratios in water samples plotted against the potassium concentrations in the rock from which the water was taken. Figure 6.5.4 shows these ratios plotted against the dissolved potassium contents of the waters. As shown in Figure 6.5.3, there is no correlation between the measured $^{40}Ar/^{36}Ar$ ratio in water and the potassium content of the rocks from which the water sample was taken. However, Figure 6.5.4 shows a certain correlation between potassium dissolved in the water and the $^{40}Ar/^{36}Ar$ ratio. The most obvious explanation, that ^{40}Ar is produced in water by decay of the ^{40}K in water, is completely inadequate.

If a rather high potassium content of 100 mg per litre is assumed, more than 10^9 years would be required to change the $^{40}Ar/^{36}Ar$ ratio from 295.5 to 296 by decay of dissolved ^{40}K. The same relatively small change would also occur if dissolving potassium brought with it into solution all the ^{40}Ar which had accumulated in the rock from its decay for 10^9 years.

The general correspondence between high dissolved potassium concentrations and large amounts of accumulated ^{40}Ar can be interpreted as a result of intensive rock-water interactions. Such interactions qualitatively indicate long water residence times. This interpretation has also been given to the correlation between ^4He and chloride concentrations of groundwater presented for correlations between ^4He and ^{40}Ar concentrations.

6.5.5 Correlations Between He Concentrations and $^{40}Ar/^{36}Ar$ Ratios

Figure 6.5.5 shows that a certain correlation exists between the accumulated helium and ^{40}Ar. Most waters with helium concentrations above 10^{-4} cm^3 STP/cm^3 of water have significantly increased $^{40}Ar/^{36}Ar$ ratios. This follows from the consideration mentioned above that subsurface-produced ^{40}Ar cannot be detected below $\approx 10^{-5}$ cm^3 STP per cm^3 of water. In other words, it is most likely that all samples contain a small fraction of radiogenic ^{40}Ar, as well as subsurface-produced ^4He, but while ^4He can easily be detected, radiogenic ^{40}Ar is masked by dissolved atmospheric argon in most of the samples.

Except for the 558 m sample from the Molasse of Schafisheim (304/1), all samples with increased ^{40}Ar are from Permian or crystalline rock. As mentioned in several previous chapters, waters from the Buntsandstein show clear signatures of the underlying Permian or crystalline. The relatively low helium concentration of Schafisheim 558 m (304/1) and Weiach 1116 m (302/19) are explained by significant gas losses during sampling. This is evident from the other noble gas measurements (see Table 3.3.2). These two measured points should be shifted by at least a factor of ten toward higher He concentrations in Figure 6.5.5. They are then in the vicinity of measured points of other samples with increased noble gas contents and ratios. The Schafisheim 1571 m sample (304/8) has a strikingly high helium content, but a rather low $^{40}Ar/^{36}Ar$ ratio. Part of the relatively high helium concentration can be explained by the high thorium concentrations measured in

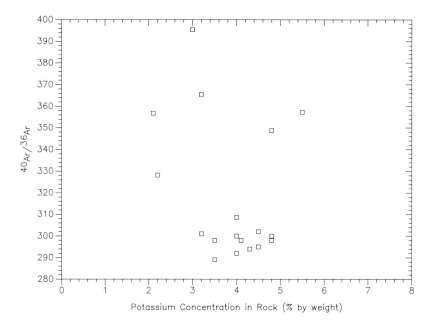

Figure 6.5.3: $^{40}Ar/^{36}Ar$ ratios in water *versus* potassium concentrations in surrounding rock.

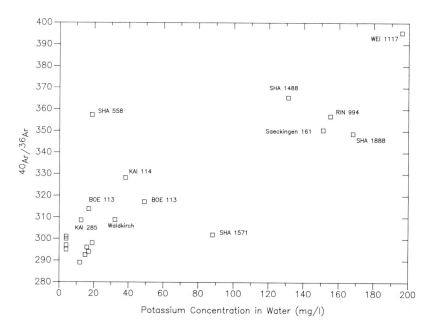

Figure 6.5.4: $^{40}Ar/^{36}Ar$ ratios in water *versus* dissolved potassium concentrations.

the sampling depth interval (see Table 6.1.1). However, this measured point still remains outside the range of all others.

The correlation between ^{4}He and ^{40}Ar can be quantitatively linked to the production rates of the two isotopes that were summarised in Table 6.1.2. For all crystalline and Permian formations in Table 6.1.2, the theoretical production rate ratio ^{4}He/^{40}Ar equals 6 ± 1.5. Figure 6.5.6 shows the calculated ratios on a graph of the measured values. Curve A in Figure 6.5.6 represents a ratio of ^{4}He/^{40}Ar = 6 and Curve B, a ratio of 19, which would account for a preferential escape of ^{4}He over ^{40}Ar by a factor of 3.2. This factor is roughly the difference in the diffusion constants of ^{40}Ar and ^{4}He, and equals $(40/4)^{1/2}$. Most data points are within the area defined by the two curves. Apparently, the ^{4}He/^{40}Ar ratios measured in groundwater closely reflect the production of the two isotopes in rock over concentration ranges of at least one to two orders of magnitude.

6.5.6 Summary

- Elevated ^{40}Ar/^{36}Ar ratios have been measured in a number of groundwater samples. They reflect addition of subsurface produced ^{40}Ar to argon of atmospheric origin.

- Accumulation times calculated from *in situ* production in a closed volume of rock, followed by transfer of ^{40}Ar to the corresponding volume of water, yield "ages" that generally appear to be too large even when assuming a complete release of ^{40}Ar. This contradicts the K-Ar rock dating method. This situation is very similar to the one discussed in Section 6.4 for ^{4}He.

- Long-range transport processes most likely are responsible for the observed ^{40}Ar levels. The measured ^{4}He/^{40}Ar ratios in groundwater closely reflect the production-rate ratios in rock which are summarised in Table 6.1.2. It is believed that this is a general effect even though it could only be demonstrated because radiogenic ^{40}Ar is masked by atmospheric ^{40}Ar in most samples.

- A certain correlation between the ^{4}He, ^{40}Ar, chloride and potassium concentrations (Figures 6.4.4, 6.5.4 and 6.5.5) has been observed. The samples with the highest salinity are the ones which have accumulated the largest amount of subsurface-produced noble gases. This seems to reflect the degree of rock-water interaction which would transfer the above constituents from rock into groundwater.

- Since such processes are expected to take some time, it appears that the water samples with the highest salinity, ^{4}He and ^{40}Ar concentrations are also the oldest samples. Therefore, the following waters have a relatively high residence time or contain an old component (Figure 6.5.2):

    Weiach 1116 m (302/19) and 1408 m (302/18)
    Schafisheim 558 m (304/1), 1488 m (304/5) and 1888 m (304/10)
    Riniken 993 m (303/6) and 806 m (303/3)
    Kaisten 114 m (305/1)

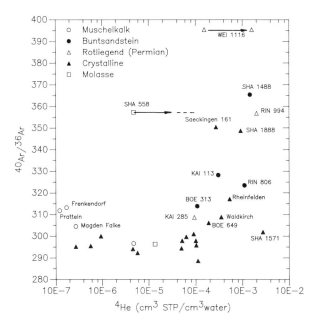

Figure 6.5.5: $^{40}Ar/^{36}Ar$ ratios in water *versus* dissolved helium concentrations.

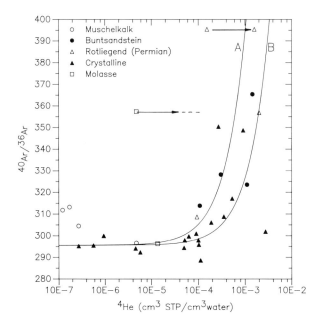

Figure 6.5.6: Ratios of theoretical production rates of ^{40}Ar and ^{4}He with measured $^{40}Ar/^{36}Ar$ and helium concentrations.

Less markedly, but nevertheless significant, increased $^{40}Ar/^{36}Ar$ values appear in the following samples:

Böttstein 313 m (301/25) and 649 m (301/23)
Rheinfelden (159)
Waldkirch (95)
Frenkendorf (9)
Pratteln (8)
Magden Weiere (271)
Magden Falke (272)

Therefore, these cannot be uniformly young waters.

- A more quantitative evaluation of $^{40}Ar/^{36}Ar$ ratios (and of elevated ^4He levels) in terms of "water residence time" will only become possible when the global transport processes from the upper continental crust to the atmosphere are better understood. Such research is in progress (*e.g.*, TORGERSON, 1989; TORGERSON AND OTHERS, 1989). These studies in particular need to synthesise noble gas data such as the ones presented in this chapter with the geochemical evolution of rock salts (chloride, potassium, bromide and their ratios). A better understanding of rock weathering processes and the role of fluid inclusion is required.

7. FORMATION-SPECIFIC CHARACTERISTICS OF GROUNDWATERS

Certain isotopic characteristics of groundwater and its solutes reflect properties of rock with which the water is, or has been, in contact. Examination of these characteristics can yield information on the flow paths followed by water to reach the sampling point, and on the nature of water-rock geochemical reactions.

This chapter presents data on the isotopic composition of dissolved and mineral sulphate and sulphide, dissolved and mineral strontium, and dissolved and mineral uranium, thorium and their daughter products. From these data, it is possible to draw conclusions about the flow history of water sampled from the test boreholes and from regional aquifer systems.

7.1 Sulphur and Oxygen Isotopes in Sulphate and Sulphide

W. Balderer, F. J. Pearson, Jr., and S. Soreau

Sulphur has four naturally-occurring stable isotopes, ^{32}S, ^{33}S, ^{34}S and ^{36}S, with abundances of 95.0, 0.77, 4.2 and 0.017 per cent, respectively. Ratios of the most common sulphur isotopes, $^{34}S/^{32}S$, are measured and used to describe the stable isotopic composition of sulphur-bearing substances. Little or no additional information of use in hydrochemical studies would be gained by measuring the ratios of other sulphur isotope pairs (NIELSEN, 1979).

$^{34}S/^{32}S$ ratios are expressed using the δ-notation, as are other stable isotopes ratios. This notation is described in Section 1.3. The standard to which sulphur isotope measurements are referred is sulphur from the mineral troilite, FeS, found in the Canyon Diablo meteorite. Thus, sulphur isotopic compositions are often written $\delta^{34}S_{CD}$. Here, the shorter form, $\delta^{34}S$, will be used.

The isotopic composition of oxygen in dissolved and mineral sulphate is also of interest. Oxygen isotopic compositions of sulphates are referred to the SMOW standard and should formally be written $\delta^{18}O_{SMOW}$. In this chapter, the shorter form, $\delta^{18}O$, will be used, with the understanding that it refers to SMOW both for sulphate and water.

The principles underlying the use of sulphate and sulphide isotopes in hydrology and geochemistry are reviewed in several publications (KROUSE, 1980; PEARSON AND RIGHTMIRE, 1980; INTERNATIONAL ATOMIC ENERGY AGENCY, 1983a, b, 1987; SCHMASSMANN AND OTHERS, 1984, Section 5.6; FONTES AND MICHELOT, 1985; and BALDERER, 1985b, Section 6.12), and are not repeated here.

The $\delta^{34}S$ and $\delta^{18}O$ values of marine evaporite sulphate minerals vary with their geologic age. This is generally interpreted to result from changes in the isotopic composition of oceanic sulphate with time. The observed ranges of $\delta^{34}S$ and $\delta^{18}O$ values of sulphates from rocks of the same ages as prominent water-bearing units in northern Switzerland are given in Table 7.1.1. These ranges are taken from the literature (PILOT AND OTHERS, 1972, NIELSEN, 1979, CLAYPOOL AND OTHERS, 1980). This table also indicates the ranges of isotopic compositions of sulphate minerals analysed as part of the Nagra deep drilling programme. The ranges given in Table 7.1.1 are compared with the results of Nagra analyses of dissolved sulphate later in this section.

As Table 7.1.1 shows, there is overlap but not absolute agreement between the $\delta^{34}S$ ranges given by CLAYPOOL AND OTHERS (1980), and by NIELSEN (1979) and PILOT AND OTHERS (1972). The ranges of CLAYPOOL AND OTHERS (1980) are for marine evaporite minerals from formations worldwide, while the NIELSEN (1979) and PILOT AND OTHERS (1972) data are from formations in Germany which include both marine and non-marine sulphate minerals. Therefore, the agreement is generally good between the $\delta^{34}S$ ranges of PILOT AND OTHERS (1972) and NIELSEN (1979), while the data of CLAYPOOL AND OTHERS (1980) differ from those of the other authors. NIELSEN (1979) gives only $\delta^{34}S$ values. The $\delta^{34}S$ of modern oceanic sulphate is +20.0 per mil, and modern evaporites (gypsum) are enriched by one to two per mil over dissolved sulphate (CLAYPOOL AND OTHERS, 1980). The $\delta^{18}O$ of modern oceanic sulphate is +9.6 per mil. The $^{18}O/^{16}O$ ratio of oceanic sulphate is not in equilibrium with that of ocean water, however. At equilibrium, oceanic sulphate would be considerably enriched in ^{18}O (PEARSON AND RIGHTMIRE, 1980).

Table 7.1.2, at the end of Section 7.1, gives the concentrations and $\delta^{18}O$ and $\delta^{34}S$ values of dissolved sulphate and sulphide, and the $\delta^{18}O$ values of groundwaters from northern Switzerland and adjacent areas. The $\delta^{18}O$ values of these waters are as given in Tables 3.2.1 through 3.2.5. As evident from Table 7.1.2, many more samples were analysed for ^{34}S than for ^{18}O.

A number of samples of sulphate and sulphide minerals taken from core from the Nagra deep boreholes were also analysed for their sulphur and oxygen isotopic compositions. The results of the analyses of sulphate minerals are given in Tables 7.1.3 and 7.1.4, and of sulphide minerals in Table 7.1.5. These tables also appear at the end of Section 7.1.

A description and interpretation of the systematics of the sulphate and sulphide isotopes is given in the following sections after a brief discussion of the uncertainties in the $\delta^{34}S$ and $\delta^{18}O$ values.

7.1.1 Comparison of Sulphur and Oxygen Isotope Analytical Results

Several laboratories participated in the preparation and analysis of sulphate samples.

All analyses of the ^{18}O content of sulphates were made at the laboratory of Professor J. Ch. Fontes, Université Paris-Sud, Orsay. The sulphate isotope analytical procedure begins with a solid sulphate, so sulphate in solution must be converted to a solid ($BaSO_4$) before analysis. Preparation of solids from water samples was done for some samples by Institut Fresenius and for others at Orsay. Both laboratories prepared solids from a few water samples.

Table 7.1.1: Range of delta 34S and delta 18O values of evaporite sulphate minerals from the literature and Nagra borehole.

		Delta 34S in Middle Europe (1)	Delta 18O in Middle Europe (1)	Delta 34S in German Evaporites (2)	Delta 34S in Marine Evaporites Worldwide (3)	Delta 18O in Marine Evaporites Worldwide (3)	Delta 34S in Mineral Sulphates (4)	Delta 18O in Mineral Sulphates (4)	Delta 34S resulting range used in Figures in this report	Delta 18O resulting range used in Figures in this report
Tertiary	Miocene				20.5-23.5	11-13			20.5-23.5	11-13
Jurassic	Malm Dogger Lias	15-18	14-17	15-20					15-20	14-17
	Keuper	10-17	14-18	13-16	14-17	10-14	14-16	12-14	10-17	10-18
Triassic	Muschelkalk	17-20	15-18	17-21	16-18 *	10-14 *	18-20	13-18	17-21	15-18
	Röt	25-29	14-16	18-29	25-29				18-29	14-16
	Lower Buntsandstein	9-15	11-16	7-10	9-15				9-15	11-16
Permian	Zechstein	7-13	10-13	8-13	8-14	8-12			7-14	8-13
	Rotliegend	5-9	13-17	7-10			11, 19	14	5-10	13-17

(1) Mineral samples from formations in Middle Europe (GDR). Data from PILOT AND OTHERS (1972), Figure 4.
(2) Water and mineral samples from formations in Germany. Data from NIELSEN (1979), Figure 8.
(3) Worldwide samples of marine evaporite minerals. Data from CLAYPOOL AND OTHERS (1980), Figures 3, 4, 5 and text.
* Muschelkalk evaporites from Israel (CLAYPOOL AND OTHERS (1980)).
(4) Minerals from cores from Nagra deep boreholes. See Table 7.1.3, this report.

The results of the analyses of the duplicate preparations are shown in Figure 7.1.1. Agreement between the several pairs is good but not perfect. The regression equation for the data is:

$$\delta^{18}O_{\text{Orsay Prep.}} = -(0.25 \pm 0.90) + (0.990 \pm 0.043) \cdot \delta^{18}O_{\text{Fres. Prep.}}$$

The slope of the regression is essentially unity and the intercept is the origin, suggesting that there are no systematic differences between the data sets. However, there appear to be random differences with a standard deviation of 0.9 per mil between the two data sets. Therefore, duplicate $\delta^{18}O$ results were averaged.

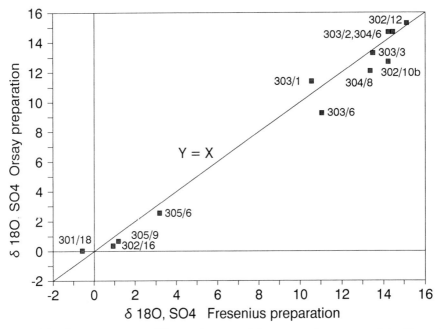

Figure 7.1.1: **Comparison of $\delta^{18}O$ results on sulphate samples prepared by Fresenius with those prepared at Orsay.**

^{34}S analyses were done in different laboratories. ^{34}S measurements on samples taken as part of the regional programme were made at the laboratory of Dr. H. Nielsen, Geochemisches Institut der Universität Göttingen on solid sulphates prepared by Institut Fresenius.

Samples taken from the Nagra deep boreholes were analysed at Orsay. Most of these samples were prepared at Orsay. Sulphate samples from the first five samples from Böttstein were prepared by Institut Fresenius, and preparations by both Institut Fresenius and Orsay were made on a further seven samples. Consistency between the analyses on the seven duplicate preparations is good. The average difference between them is 0.2 ± 0.5 per mil, and so the results were averaged.

Only three samples were analysed for ^{34}S by both the Göttingen and Orsay laboratories and there is poor agreement between them. The average difference between the three pairs is 1.4 ± 4.6 per mil. It is not known whether this discrepancy between the two laboratories is a result of calibration differences or of inhomogeneity in the samples analysed.

7.1.2 Samples from the Tertiary and Jurassic

Figure 7.1.2 shows δ^{34}S values of dissolved sulphate in waters from Tertiary and Jurassic aquifers plotted against their sulphate δ^{18}O values. Figure 7.1.3 shows the δ^{34}S values of samples from the same zones plotted against their dissolved sulphate concentrations. The ranges of isotopic compositions from the literature (see Table 7.1.1) for Miocene and Malm sulphate minerals are also included in these figures.

The samples from the Upper Freshwater Molasse at Schwarzenberg (42), the Malm at Durlinsdorf (80), and the Hauptrogenstein at Hauensteinbasis 6.785 (76) and 3.135 (68) are superficial Ca-HCO$_3$ waters (SCHMASSMANN AND OTHERS, 1984, Sections 6.1 and 6.2). The samples from Wildegg (30) and Weissenstein 1.480 (56) are considered special cases from the Malm and Lias, respectively.

As Figures 7.1.2 and 7.1.3 show, the Hauensteinbasis 6.785 (76), Schwarzenberg (42), and Durlinsdorf (80) samples have low-sulphate concentrations and negative δ^{34}S values. These δ^{34}S values are more negative than sedimentary sulphate minerals of any age so that this sulphate must be derived at least in part from oxidized organic sulphur or mineral sulphide.

The Hauensteinbasis 3.135 sample (68), has higher ^{34}S and dissolved sulphate contents than the three samples just discussed. For reasons to be discussed, this sample can be interpreted as a water of origin similar to that of the three more dilute samples (42, 76, 80), with an admixture of sulphate-rich water of other origin (SCHMASSMANN AND OTHERS, 1984, p. 147-150).

The Weissenstein 1.480 sample (56), is remarkably depleted in both ^{34}S and ^{18}O, but it has a rather high dissolved sulphate concentration. The sulphate in this sample is probably derived from the oxidation of ^{34}S-depleted organic sulphur or sedimentary sulphide minerals, as is the sulphate in the samples from Hauensteinbasis 6.785 (76), Schwarzenberg (42), and Durlinsdorf (80). A δ^{34}S value approaching -30 per mil would not be uncommon for such a sulphur (KROUSE, 1980, Figure 11-1).

Oxidation of sulphide minerals could also lead to the relatively depleted ^{18}O contents of the samples from Durlinsdorf (80), Wildegg (30) and Weissenstein 1.480 (56) in Figure 7.1.3. As summarized in INTERNATIONAL ATOMIC ENERGY AGENCY (1983b, p. 82), oxygen in sulphate formed from sulphide oxidation is derived both from the atmosphere and from the water in which the oxidation reaction occurs. Atmospheric oxygen

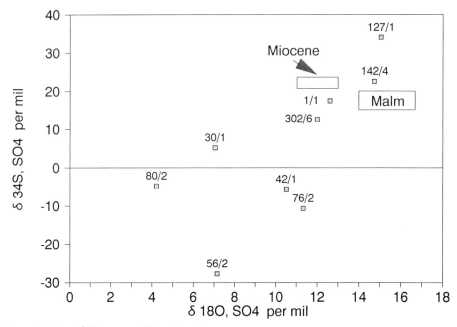

Figure 7.1.2: $\delta^{34}S$ *versus* $\delta^{18}O$ of sulphate dissolved in waters from Tertiary and Jurassic aquifers.

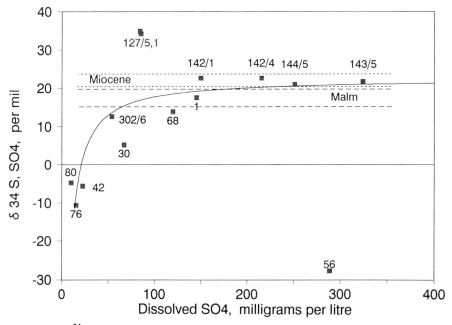

Figure 7.1.3: $\delta^{34}S$ *versus* concentration of sulphate dissolved in waters from Tertiary and Jurassic aquifers.

has a $\delta^{18}O$ value of about +23 per mil. Isotope fractionation accompanying oxidation leads to enrichment of ^{18}O in the product sulphate by 4 to 20 per mil relative to the water in which the oxidation occurs (TAYLOR AND OTHERS, 1984). The differences between the $\delta^{18}O$ values of the sulphate and water of these samples are consistent with this mechanism.

It is certainly possible that the source of oxidized sulphur to the Weissenstein 1.480 sample (56) is sedimentary mineral sulphide. However, this sample was taken from the Weissenstein Railway tunnel which appears to have been completed before 1908 (ISENSCHMID, 1985/86; SCHMASSMANN AND OTHERS, 1984). Presumably, the tunnel was used by steam locomotives before the general electrification of the Swiss railways. Such locomotives would have deposited sulphur from coal smoke on the tunnel walls which would be isotopically light, and because of its fine grain size and exposure to oxygen, easily oxidized. This could account for the association of high sulphate concentrations with low ^{34}S and ^{18}O contents in the Weissenstein 1.480 sample (56). This is also the only sulphate isotope sample from the Lias, and its negative $\delta^{34}S$ value and relatively high sulphate concentration may simply reflect oxidation of reduced sulphur in that formation.

Samples from Eglisau 2, 3, and 4 (142 to 144) and the Weiach 255.0 m sample (302/6) are classified as deep Na-Cl waters, and Aqui (1) and Lottstetten (127) as deep Na-HCO$_3$ waters (SCHMASSMANN AND OTHERS, 1984, Sections 6.1.4 and 6.1.3). The Eglisau samples (142 to 144) are from the Lower Freshwater Molasse, Aqui (1) is from the Upper Marine Molasse, and the Lottstetten (127) and Weiach 255.0 m (302/6) samples are from the Malm. As Figure 7.1.3 shows, Eglisau 2 through 4 (142 to 144) have $\delta^{34}S$ values typical of Miocene sedimentary sulphate, indicating that the dissolved sulphate of these waters is from the solution of mineral sulphate from Miocene formations.

The curved line on Figure 7.1.3 shows the $\delta^{34}S$ values which would result from the dissolution of Miocene sulphate of $\delta^{34}S = +22.5$ per mil in water of the Durlinsdorf (80), Schwarzenberg (42) and Hauensteinbasis 6.785 (76) type. The Weiach 255.0 m (302/6) and Aqui (1) samples are close to this line and may well represent Miocene sulphate dissolved in water also containing sulphate from the oxidation of reduced sulphur.

It is of interest to examine the relationship of waters from the Weiach borehole at 255.0 m (302/6 and 302/9) and those from the Eglisau 2, 3 and 4 (142 to 144) and Aqui (1) boreholes from the standpoint of their water and sulphate isotopic compositions. The water isotopic compositions and residues on evaporation of these samples are displayed in Figures 3.2.1 and 3.2.2. From these figures, it appears that the Eglisau samples could be mixtures of low ^2H and ^{18}O water with a salinity of less than 1 g/l, such as is found in the Upper Marine Molasse at Mainau (128), with a water of at least 5 g/l salinity and enriched in ^{18}O relative to meteoric water, as sampled from the Malm at Weiach 255.0 m (302/6 and 302/9).

As Figures 7.1.2 and 7.1.3 show, however, the sulphate concentration and sulphate isotopic composition of the Weiach 225.0 m sample are intermediate between those of the samples from Eglisau and Aqui and superficial waters such as are found at Schwarzenberg (42), Durlinsdorf (80), and Hauensteinbasis 6.785 (76).

The following hypothesis would qualitatively account for these data. The salinity and ^{18}O content of the Eglisau (141 to 144) and Weiach 255.0 m (302/6 and 302/9) samples result from mixing between superficial water and a water more saline and enriched in ^{18}O than any sampled from the Molasse or Malm as part of the Nagra programme. The superficial waters are young, as shown by the high tritium contents of the Schwarzenberg (42) and Durlinsdorf (80) samples (Table 3.2.1), and could account for at least part of the ^{14}C found in the Eglisau 1 (141) and Weiach 255.0 m (302/6) samples. The sulphate concentration and isotopic composition of the Eglisau and Weiach 255.0 m samples would result from solution of sulphate minerals in the mixed waters.

The Hauensteinbasis 3.135 sample (68), from the Hauptrogenstein is considered a superficial water by SCHMASSMANN AND OTHERS (1984, Section 6.2.2). However, it contains a higher concentration of sulphate relatively enriched in ^{34}S than do other superficial samples from the Tertiary and Jurassic formations. The ^{34}S content of this sample is consistent with sulphate of Keuper age (Table 7.1.1). However, the Hauptrogenstein itself contains no sulphate minerals, so the sulphate dissolved in this sample must have come from other sulphate-bearing horizons through which the water passed before being sampled from the Hauptrogenstein.

The $\delta^{34}S$ value of the Wildegg (30) sample, suggests that it contains less sulphide-derived sulphate than do the superficial samples from Durlinsdorf (80), Schwarzenberg (42) and Hauensteinbasis 6.785 (76). However, the relatively negative ^{18}O value for the Wildegg sample (30), like that for the Weissenstein 1.480 sample (56), would be consistent with a high proportion of sulphide-derived sulphate.

Water from Lottstetten (127) is classified as a deep Na-HCO$_3$ water. As Figure 7.1.2 shows, its ^{34}S content is clearly outside the range of Tertiary evaporites. This water contains H$_2$S (0.21 mg/l), and it is possible that some of the sulphate has been reduced to H$_2$S enriching the residual sulphate in ^{34}S.

It is reported (INTERNATIONAL ATOMIC ENERGY AGENCY, 1983b, p. 79) that during sulphate reduction in solution, residual sulphate is enriched in the heavier isotopes of both sulphur and oxygen, and that the ratio D $\delta^{34}S$/D $\delta^{18}O$ changes by a factor of between two and four. In this ratio:

$$D = \delta(\text{residual sulphate}) - \delta(\text{initial sulphate}).$$

It might be assumed that the sulphate precursor to that found in the Lottstetten sample (127) had a $\delta^{34}S$ value of 22.5 per mil, typical of Miocene evaporites and similar to the values measured in the Eglisau samples (142 to 144). If so, the $\delta^{18}O$ values of the precursor calculated from this relationship would have been between 9.3 and 12.2 per mil. This is more negative than the one $\delta^{18}O$ value measured for Eglisau (142/4, 14.74 per mil), but is within the range of Miocene evaporites as given by CLAYPOOL AND OTHERS (1980), and similar to the $\delta^{18}O$ values of the samples from Aqui (1) and Weiach 255.0 m (302/6).

7.1.3 Samples from the Keuper

Five samples of gypsum and anhydrite from the Gipskeuper of the Böttstein borehole were analysed with the results given in Table 7.1.3. As Table 7.1.1 and Figure 7.1.4 show, the $\delta^{34}S$ and $\delta^{18}O$ values of these samples are within the range of marine evaporites of Keuper age from the literature (Table 7.1.1).

$\delta^{34}S$ values were measured on sulphate dissolved in five waters from the Keuper, but $\delta^{18}O$ values only on two samples. All water samples on which $\delta^{34}S$ measurements were made, except that from Eptingen (23), are saturated with respect to gypsum. As Figure 7.1.5 shows, the $\delta^{34}S$ values of these waters are close to one another, and within the range of values found in mineral samples from the Gipskeuper of the Böttstein core and in Germany.

The $\delta^{18}O$ value of the sulphate dissolved in the Eptingen sample (23) is within the range of the Böttstein mineral sulphates. As shown in Table 2.1.3, this sample consists of modern water. This is consistent with the fact that it is not saturated with respect to gypsum, in spite of the rapid solubility of that mineral. Because of its undersaturation, this water should also be actively dissolving gypsum from the Keuper formation, which would account for the close similarity of both the $\delta^{34}S$ and $\delta^{18}O$ values of its dissolved sulphate to those of gypsum from the Gipskeuper of the Böttstein borehole.

The $\delta^{18}O$ and $\delta^{34}S$ values of the sulphate dissolved in the Riniken 515.7 m sample (303/1) from the Keuper formation are also within the range given by the literature (see Table 7.1.1) for evaporitic sulphate of Keuper age. The Böttstein minerals and all water samples but that of the Riniken 515.7 m sample are from the Gipskeuper. The Riniken sample, however, is from the Gansinger Dolomite and Schilfsandstein.

7.1.4 Samples from the Muschelkalk

The sulphate isotopic compositions of six minerals from the Muschelkalk were analysed with the results given in Table 7.1.3. Four of these samples, B/158, W/825, W/829 and S/1329 are from the upper Muschelkalk, the zone from which all Muschelkalk water samples were taken. The other two, B/241 an W/910, are from the middle Muschelkalk.

As shown in Table 7.1.1, the $\delta^{34}S$ values of the mineral samples from the Muschelkalk are within the range of $\delta^{34}S$ values reported from the Muschelkalk in the literature. The $\delta^{18}O$ values of the samples from the upper Muschelkalk W/829, S/1329, B/158 and more or less W/825 are also within the Muschelkalk range (Table 7.1.1). The two samples B/241 and W/910 from the middle Muschelkalk have more negative $\delta^{18}O$ values.

Table 7.1.5 includes the results of ^{34}S analyses on three samples of sulphide minerals from the lower Muschelkalk of the Böttstein borehole. These sulphides may well be products of the reduction of sulphate of Muschelkalk age. If they were formed at isotopic equilibrium with such sulphate, the $\delta^{34}S$ value of the pyrite would correspond to a formation temperature of about 80°C and those of the galena and sphalerite (Zinkblende) to temperatures of about 200°C (OHMOTO AND LASAGA, 1982). If, as is more likely,

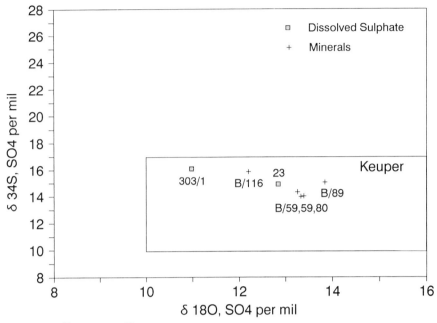

Figure 7.1.4: $\delta^{34}S$ *versus* $\delta^{18}O$ of sulphate minerals and of sulphate dissolved in waters from the Keuper.

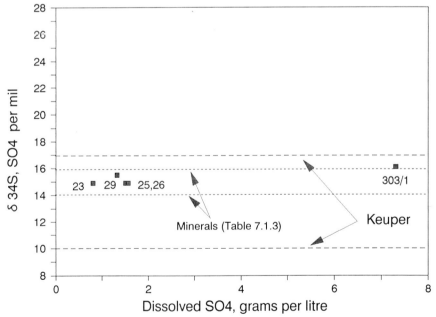

Figure 7.1.5: $\delta^{34}S$ *versus* concentration of sulphate dissolved in waters from the Keuper.

the sulphate reduction was bacterially moderated and not at isotopic equilibrium, no temperature information can be extracted from the $\delta^{34}S$ values of the sulphide minerals.

$\delta^{34}S$ measurements were made on 27 samples of sulphate and on two samples of sulphide dissolved in waters from the upper Muschelkalk. The results are given in Table 7.1.2, and the sulphate results are shown in Figures 7.1.6 and 7.1.7. Muschelkalk samples collected during the Nagra regional programme are discussed by SCHMASSMANN AND OTHERS (1984, Section 6.5).

The $\delta^{34}S$ values of all dissolved sulphates with concentrations above c. 1 gram per litre, except those from Schafisheim (304/3 and 304/14) and Weiach (302/10b), are within the range of Muschelkalk sulphate given in Table 7.1.1. While the $\delta^{34}S$ values of many of the lower-sulphate samples are also within this range, there are a number with lower ^{34}S-contents as well. The $\delta^{18}O$ values of all dissolved sulphates are less positive than the range of values given in the literature.

Samples from the Hauensteinbasis and Alter Hauenstein tunnels (65, 66, 70, 71, 72 and 74) are considered as superficial waters from the upper Muschelkalk by SCHMASSMANN AND OTHERS (1984, Section 6.4.1). They are depleted in ^{34}S by as much as five per mil relative to Muschelkalk minerals, and may include sulphate of Keuper origin. As Table 7.1.1 shows, the $\delta^{34}S$ values of Keuper sulphate should be in the range of 10 to 17 per mil, which encompasses virtually all the $\delta^{34}S$ values of this group of samples.

The Leuggern 74.9 m sample (306/1) also contains less ^{34}S than typical Muschelkalk sulphate minerals. The local hydrogeological situation is such that water in the Muschelkalk is in contact with shallow groundwater from the Quaternary deposits which could contain sulphates of Keuper origin. The radioisotopes data from this sample are also consistent with a mixed origin, as discussed in Chapter 4.

The Weiach 859.1 m sample (302/10b) also has less ^{34}S than typical Muschelkalk minerals. This water is saturated with respect to gypsum and has ^{14}C and ^{39}Ar contents consistent with a water age of < 1 ka (Tables 4.4.2 and 5.2.1). As discussed in Section 3.2.4, the ^{2}H and ^{18}O contents of this sample suggest recharge from the Rhine. Such recharge taking place through the Keuper would account for the $\delta^{34}S$ of the sulphate dissolved in the sample.

The Schafisheim 1260.4 m (304/3) and 1251.1 m (304/14) samples, and perhaps the sample from Schinznach Bad (120) are enriched in ^{34}S relative to typical Muschelkalk sulphate. These samples also contain dissolved sulphide at concentrations of hundreds and tens of mg/l, as H_2S, respectively. Isotope fractionation accompanying sulphate reduction produces sulphide depleted in ^{34}S, and so results in ^{34}S enrichment in the residual sulphate. The enrichment of ^{34}S in the Schafisheim samples is greater than in the Schinznach Bad sample. This could be explained if the former samples had undergone more sulphate reduction than the latter. The high sulphide concentrations of the Schafisheim samples are consistent with this explanation.

Unfortunately, the sulphate reduction process cannot be modelled quantitatively, because it is so closely influenced by bacteria that the SO_4^{-2}/H_2S fractionation factor cannot be

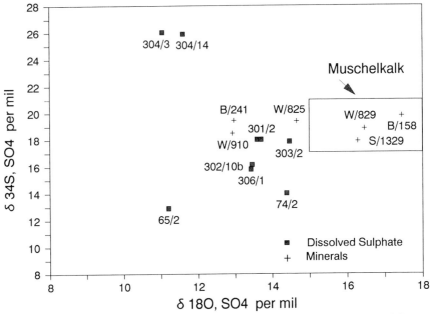

Figure 7.1.6: $\delta^{34}S$ *versus* $\delta^{18}O$ of sulphate minerals and of sulphate dissolved in waters from the Muschelkalk.

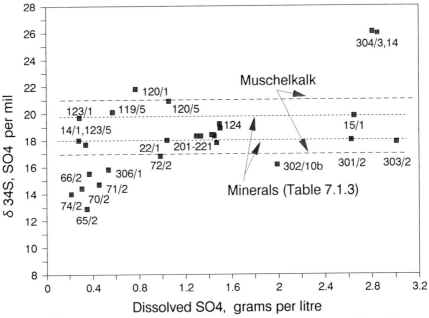

Figure 7.1.7: $\delta^{34}S$ *versus* concentration of sulphate dissolved in waters from the Muschelkalk.

predicted. Even the use of a simple mass balance, as suggested by the INTERNATIONAL ATOMIC ENERGY AGENCY (1983b, p. 81), is dubious because the system is not closed (sulphate minerals may dissolve as sulphate is lost from solution or metallic sulphides may precipitate as the sulphide concentration increases), nor can it be confidently assumed that the isotopic difference between sulphate and sulphide observed in the samples had been maintained throughout the entire sulphate reduction process.

The $\delta^{18}O$ values of the samples from the Weiach (302/10b) and Leuggern (306/1) boreholes, like their $\delta^{34}S$ values, are similar to those of sulphate minerals in the Keuper. ^{18}O in the Böttstein sample (301/2) is depleted in comparison to the Muschelkalk values from the literature. However, this sample has similar $\delta^{34}S$ and $\delta^{18}O$ values to those of the anhydrite from 241 m in the same borehole (designated B/241 in Figure 7.1.6). The $\delta^{34}S$ value of the Riniken sample (303/2) is also typical of the Muschelkalk as well as of the Muschelkalk minerals from the Nagra boreholes. The samples from Schafisheim (304/3 and 304/14) are depleted in ^{18}O relative to the Muschelkalk range. The $\delta^{34}S$ values of the Schafisheim samples can be explained by sulphate reduction, but the origin of their ^{18}O values is uncertain.

Oxygen isotope exchange between dissolved sulphate and water does take place, but only very slowly. WEXSTEEN AND MATTER (1989) have examined the possibility of this exchange influencing samples from the Muschelkalk of the Nagra boreholes. They find that while the ^{18}O in the sulphate and water from the Schafisheim (304/3 and 304/14), Weiach (302/10b), and Riniken (303/2), samples could be in isotopic equilibrium, those of the Böttstein sample (301/2) are not. Oxygen isotope exchange could be an explanation for the distinctive $\delta^{18}O$ value of the sulphate from the Schafisheim samples only if the residence times of that water was significantly longer than that of the Böttstein sample, in which possible isotopic equilibrium does not exist.

7.1.5 Samples from the Buntsandstein and Permian

As discussed in detail by PILOT AND OTHERS (1972), NIELSEN (1979) and CLAYPOOL AND OTHERS (1980), a major shift in the ^{34}S content of evaporite minerals took place during the Early Triassic. Late Permian evaporites of the Rotliegend and Zechstein of Germany have $\delta^{34}S$ values between 5 and 14 per mil. The sulphate minerals of lower Buntsandstein have $\delta^{34}S$ values ranging from 9 to 15 per mil. The $\delta^{34}S$ values of younger evaporites increase markedly to as high as +29 per mil in the upper Lower Triassic Röt evaporites of Germany, and then decrease significantly to the Keuper. These changing ranges, which are summarized in Table 7.1.1., provide the framework for the interpretation of water and mineral samples from the Buntsandstein and Permian.

Two sulphate mineral samples were analysed from the Permian of the Weiach borehole. One is an anhydrite nodule from the Upper Rotliegend at a depth of 1022 m. The other is anhydrite cementing sandstone of the Lower Rotliegend at 1164 m. These analyses are given in Table 7.1.3, and compared with other sulphate isotope measurements in Table 7.1.1 and Figure 7.1.8. The $\delta^{18}O$ values of these mineral samples are nearly the same. Both are within the Rotliegend range reported in the literature. The 1022 m sample has a $\delta^{34}S$ value of 10.6 per mil, close to the upper range of $\delta^{34}S$ values given for Rotliegend

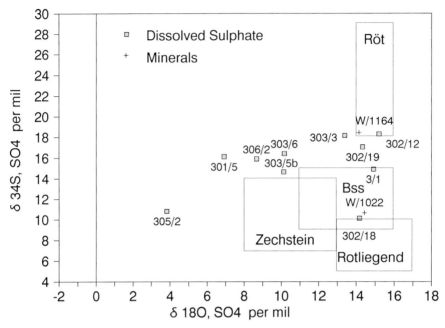

Figure 7.1.8: $\delta^{34}S$ *versus* $\delta^{18}O$ of sulphate minerals and of sulphate dissolved in waters from the Buntsandstein and Permian.

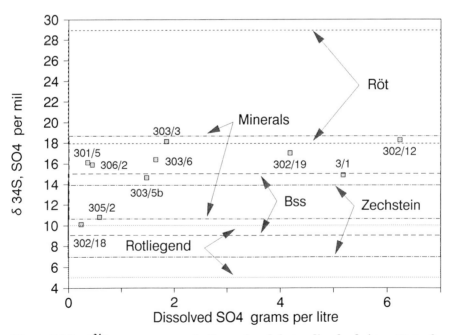

Figure 7.1.9: $\delta^{34}S$ *versus* concentration of sulphate dissolved in waters from the Buntsandstein and Permian.

evaporites (Table 7.1.1). The sample from 1164 m is more enriched in ^{34}S than typical Permian evaporites, but it is within the range of values found in the Röt evaporites of Germany (PILOT AND OTHERS, 1972, and NIELSEN, 1979), those measured on samples from the Muschelkalk of the Böttstein, Weiach, and Schafisheim boreholes, and those reported in the literature (Table 7.1.1).

The isotopic composition of sulphate from a number of water samples from the Buntsandstein, Buntsandstein/weathered crystalline contact zone, and Permian sediments were analysed with results given in Table 7.1.2 and shown in Figures 7.1.8 and 7.1.9. All of these samples are from Nagra test boreholes, except one, from the Kaiseraugst borehole (3/1). Waters from the Buntsandstein and Permian are discussed by SCHMASSMANN AND OTHERS (1984, Section 6.5).

As Figure 7.1.8 shows, the dissolved sulphates in five water samples have ^{18}O contents similar to those of the minerals analysed. These samples are from the Weiach borehole from the Buntsandstein at 985.3 m (302/12), the Upper Rotliegend at 1116.5 m (302/19), and the Lower Rotliegend at 1408.3 m (302/18), and from the Buntsandstein at Riniken 806.6 m (303/3) and at Kaiseraugst (3/1). Sulphate from samples 302/12, 303/3, and 302/19 is enriched in ^{34}S relative to typical Permian or lower Buntsandstein sulphate, but is similar to that of upper Lower Triassic Röt or Middle Triassic Muschelkalk evaporites and to the anhydrite cement analysed from Weiach 1164 m. Sulphate from the 1408.3 m Weiach sample (302/18) is isotopically similar to the anhydrite nodule from 1022 m at Weiach, and to Rotliegend sulphate from Germany (Figure 7.1.8).

Diagenesis of the Buntsandstein in northern Switzerland has been discussed by RAMSEYER (1987). He describes three generations of sulphate minerals: a gypsum/-anhydrite generation formed during eogenesis and two generations of anhydrites formed during mesogenesis. Based on its sulphate isotopic composition, which is equivalent to that of the lowest Buntsandstein in Germany, the nodule from 1022 m at Weiach probably represents sulphate precipitated during eogenesis.

The water from Weiach 1408.3 m (302/18), is the most saline of all samples from either the Buntsandstein or the Permian and the most highly enriched in ^{2}H and ^{18}O as well (Section 3.2.5). This water has a relatively low sulphate content, however (Table 7.1.2 and Figure 7.1.9), and is undersaturated with respect to anhydrite (SI = -0.6). The aggressiveness of this water toward mineral sulphates is a reason for the dissolved sulphate to have an isotopic composition similar to whatever sulphate may be present in the formation. This sulphate is apparently of Permian or of lowest Triassic origin.

The anhydrite cement at 1164 m at Weiach, though sampled from the Upper Rotliegend, is similar isotopically to upper Lower Triassic (Röt) or Middle Triassic (Muschelkalk) sulphate. It may well have formed during an episode of mesogenesis of the overlying Buntsandstein which took place during the late Early or Middle Triassic. The sulphate in the Weiach and Riniken Buntsandstein waters (302/12 and 303/3) and the Weiach Upper Rotliegend water (302/19) could have dissolved from mesogenetic cements like that sampled at 1164 m.

The isotopic composition of the sulphate of these waters could also have developed from the dissolution of Muschelkalk sulphate minerals. The solubility of calcium-sulphate minerals decreases with increasing temperature. Thus, waters saturated with Muschelkalk anhydrite descending into the Buntsandstein and Permian would be warmed and could precipitate anhydrite. If this process has occurred, it would mean that the age of the formation of the anhydrite in the Permian sediments could be much later than the time of deposition of the Muschelkalk.

As discussed in Section 7.2, the $^{87}Sr/^{86}Sr$ ratios from nodular calcite and anhydrite from Weiach Buntsandstein (see Table 7.2.1) are closer to those of Muschelkalk samples than those of crystalline basement samples. This is consistent with the similarity between the isotopic composition of the anhydrite cement at 1164 m in the Permian at Weiach and Muschelkalk sulphate.

The sulphate in the sample from the Riniken Buntsandstein (303/3) is similar isotopically to that from the Weiach Buntsandstein (302/12) and Upper Rotliegend (302/19), suggesting a similar origin for its sulphate. As discussed in Section 3.2.5 and shown in Figure 3.2.8, however, the ^2H and ^{18}O contents of the Riniken sample suggest that it has undergone some oxygen isotopic exchange with rock or evaporation before recharge. The ^2H and ^{18}O values of the Weiach samples show no evidence of these processes.

Sulphate isotopes dissolved in samples from the Upper Rotliegend of Riniken at 965.5. m (303/5b) and 993.5 m (303/6) were also measured. As Figures 7.1.8 and 7.1.9 show, these samples have $\delta^{34}S$ values intermediate between those expected of Rotliegend sulphate and measured in the 1408.3 m Weiach sample (302/18) and the more positive values found in the Riniken (303/3) and Weiach (302/12) Buntsandstein. The water from the Buntsandstein at Kaiseraugst (3/1) appears to contain sulphate of a similar mixed origin.

The Buntsandstein is in direct contact with weathered crystalline basement in the Böttstein and Leuggern boreholes. The isotopic compositions of sulphate from the Buntsandstein/weathered crystalline horizons of Böttstein at 316.6 m (301/5) and 312.5 m (301/25a) and Leuggern at 217.9 m (306/2) are intermediate between the Buntsandstein waters of Weiach (302/12) and Riniken (303/3) and waters typical of the crystalline basement itself.

The ^{34}S content of sulphate from the Upper Rotliegend at Kaisten 284.3 m (305/2) is nearly as low as that of the Rotliegend of Germany (Figures 7.1.8 and 7.1.9). Thus, the origin of the sulphate in this water appears to be the dissolution of mineral sulphate of virtually the age of the formation. This is similar to the origin of the sulphate of the Weiach 1408.3 m sample (302/18).

7.1.6   Samples from the Crystalline

Cores from the crystalline basement from Nagra deep boreholes contain mineral sulphate as barite within fracture fillings. Eight samples of this material were analysed with the results given in Table 7.1.4 and shown in Figure 7.1.10. Their $\delta^{34}S$ values are between 11 and 16 per mil. Two groups of samples can be distinguished based on $\delta^{18}O$ values.

One of the low-^{18}O samples, from 1648.29 m at Leuggern, can be characterized mineralogically as a younger or second generation mineral, while one of the high-^{18}O samples, from 1648.83 m at Leuggern, is an older generation or primary mineral (PETERS AND OTHERS, 1988b).

The δ^{34}S contents of five samples of mineral sulphides from the crystalline were also analysed, with the results given in Table 7.1.5. Two of these samples (Kaisten 664.12 m and Leuggern 1048.98 m) have δ^{34}S values of 11.7 and 15.3 per mil, which are in the range of the barite samples. The remaining samples are considerably depleted, with δ^{34}S values between -2.9 and +1.4 per mil.

The two minerals with relatively high δ^{34}S values may represent original, or "primary", sulphides of the crystalline rock. The mineral sulphides depleted in ^{34}S could result from sulphate reduction, a process which leads to ^{34}S depleted sulphide. For these minerals to have formed at isotopic equilibrium with the sulphate now present in solution would have required unreasonably high temperatures. Bacterial reduction produces considerably less fractionation, and could have formed sulphide with the isotopic composition found in these minerals from sulphates now present in the water at present formation temperatures.

The isotopic composition of dissolved sulphate from waters from the crystalline is given in Table 7.1.2 and shown in Figure 7.1.10, 7.1.11, and 7.1.12. The δ^{34}S values of this sulphate range from 9 to 19 per mil and the δ^{18}O values from -1 to +15 per mil. The ^{34}S contents of these dissolved sulphates are generally within the ranges of values reported in the literature for sedimentary sulphates (Table 7.1.1 and Figure 7.1.10). However, the δ^{18}O values of these sulphates are generally lower (< 8 per mil) than those reported in Table 7.1.1 for sulphates of sedimentary origin.

The dissolved sulphate of waters from the crystalline can be classified using their δ^{18}O values and dissolved sulphate concentrations (Figure 7.1.12).

A first group consists of samples from the crystalline of the Böttstein (301), Weiach (302), Kaisten (305), and Leuggern (306) boreholes, as well as from Zurzach 2 (132) and Säckingen Margarethenquelle (126). These waters are characterized by relatively low sulphate concentrations (< 0.6 g/l SO$_4$) and sulphate δ^{18}O values less positive than 7 per mil. If the same criteria are applied to Buntsandstein and Permian waters (Figures 7.1.8 and 7.1.9), samples 301/5 and 301/25a from the Buntsandstein/weathered crystalline horizon at Böttstein, as well as sample 305/2 from the Upper Rotliegend of Kaisten are also within this group and are, therefore, included in the crystalline figures. The δ^{18}O values of this group of samples are lower than the literature range of sedimentary sulphates, and lower than those of the "first" or "older" generation of barite in the crystalline. On the other hand, the samples of barite characterized as "second" or "younger" generation (L/1648 y, B/619; Figure 7.1.10) could represent a precipitation product of some of the dissolved sulphates in this first group of waters.

A second group is characterized by more positive δ^{18}O values (+8.7 to +10.1 per mil), and by sulphate concentrations of 0.1 to 0.8 g/l SO$_4$ similar to those of the first group. This second group includes waters from the crystalline of Säckingen Badquelle (125) and

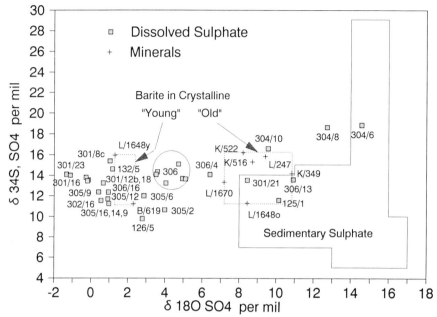

Figure 7.1.10: $\delta^{34}S$ *versus* $\delta^{18}O$ of sulphate minerals and of sulphate dissolved in waters from the Crystalline.

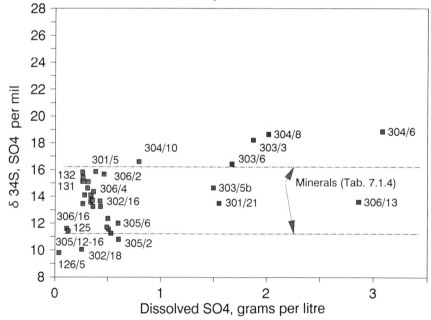

Figure 7.1.11: $\delta^{34}S$ *versus* concentration of sulphate dissolved in waters from the Crystalline and low sulphate water from the Buntsandstein and Rotliegend.

Schafisheim (304/10), as well as from the Buntsandstein/weathered crystalline of Leuggern (306/2). The isotopic composition of these samples is in the range of values which includes the "older" generation of barites, and also within the range of sedimentary sulphate reported in the literature. Samples 303/5b and 303/6 from the Upper Rotliegend of Riniken, and sample 301/21 ($\delta^{18}O$ = 4.88 per mil) from the lower crystalline at Böttstein have $\delta^{18}O$ values within the range of this group. Thus, although they have higher sulphate concentrations (1.5 to 1.7 g/l SO_4), they are included with this second group.

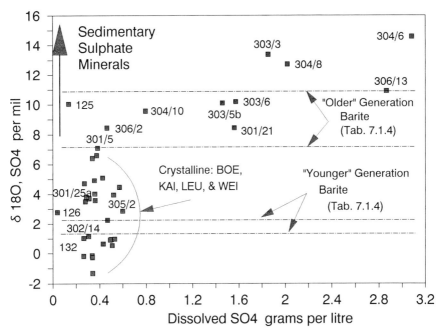

Figure 7.1.12: $\delta^{18}O$ *versus* concentration of sulphate dissolved in waters from samples from the Crystalline, Buntsandstein and Rotliegend waters.

A third group consists of dissolved sulphates from samples from the Buntsandstein/weathered crystalline of Schafisheim (304/6) and the crystalline of Schafisheim (304/8), and Leuggern (306/13). This group has higher sulphate concentrations (1.8 to 3.1 g/l SO_4) and $\delta^{18}O$ values between 11 and 15 per mil. These $\delta^{18}O$ values are still within the upper range of sedimentary sulphates, but are above the range of the "older" generation of barite minerals. Comparison of Figures 7.1.10 and 7.1.12 with Figure 7.1.8 shows that the dissolved sulphates of the waters of this third group have almost identical ranges of $\delta^{18}O$ and $\delta^{34}S$ values to the highly mineralized groundwaters of the Buntsandstein (302/12) and Upper Rotliegend (302/19) of the Weiach borehole, the mineral sulphate sample from the Lower Rotliegend of Weiach (W/1164), as well as the relatively less mineralized water from the Buntsandstein of Riniken (303/3).

The origin of these groups can be explained as follows:

- The <u>first group</u> appears to be of crystalline origin derived from the oxidation of sulphide minerals in the crystalline rock itself, such as those at Kaisten 644.12 m and Leuggern 1048.98 m (Table 7.1.5). This origin was proposed by BALDERER AND OTHERS (1987b) for these Nagra samples and by FONTES AND OTHERS (1989) for similar waters from the Stripa mine;

- The <u>third group</u> comprises waters with high sulphate concentrations and sulphate isotopic compositions similar to those of sedimentary sulphate;

- The <u>second group</u> has the same range of $\delta^{34}S$ and $\delta^{18}O$ values as the "older" generation of barite. The dissolved sulphate of these samples is probably a mixture of sulphates of the first and third group, and contain dissolved sulphate of crystalline and sedimentary origin. The dissolved sulphate in some of these waters, especially those of very low sulphate concentrations, could also be explained as a product of the oxidation of primary sulphides in the crystalline of the type found at Kaisten 664.12 and Leuggern 1048.98 m, which are not depleted in ^{34}S, or by the dissolution of "older" generation barites.

7.1.7 Summary of Sulphur and Oxygen Isotope Results

Sulphate in superficial waters from the Tertiary and Jurassic is derived at least in part from the oxidation of organic sulphur or sulphide minerals. Deep Na-Cl and Na-HCO$_3$ waters from Aqui (1), Eglisau (142 to 144), and Weiach 255.0 m (302/6) may contain mixtures of sulphate originating from the oxidation of mineral sulphides of sedimentary origin with mineral sulphate isotopically similar to Miocene evaporite minerals. Sulphate in the Lottstetten (127) sample has been enriched relative to any reasonable precursor sulphate, probably by H$_2$S production.

Sulphate minerals from the Gipskeuper in the Böttstein core are isotopically like Keuper evaporites reported in the literature. Sulphates dissolved in waters from the Keuper are within the same range and appear to result from sulphate mineral solution.

Sulphate minerals from the middle and upper Muschelkalk from the Böttstein, Weiach, and Schafisheim boreholes have ^{34}S contents within the range of minerals from the Muschelkalk cited in the literature. The ^{34}S contents of sulphate dissolved in many waters from the upper and middle Muschelkalk are also within the range of Muschelkalk minerals. All these samples are generally more depleted in ^{18}O than the Muschelkalk range in the literature. A few samples of superficial waters from the upper Muschelkalk from the Hauensteinbasis and Alter Hauensteintunnels (65, 66, 70, 71, 72 and 74) appear to contain sulphate derived from the Keuper. Samples from the upper Muschelkalk at Schafisheim (304/3 and 304/14) are enriched in ^{34}S, probably as a result of H$_2$S production. The oxygen isotopic composition of sulphate in waters from the upper Muschelkalk resembles minerals from the middle Muschelkalk more closely than minerals from the upper Muschelkalk.

The ^{34}S content of evaporitic sulphates changed dramatically during the early Triassic, so the δ^{34}S values of minerals and waters from the Buntsandstein and Permian sediments are sensitive to the age of the sulphate they contain. An anhydrite nodule from 1022 m at Weiach and sulphate dissolved in water from the Lower Rotliegend at Weiach 1408.3 m (302/18) have δ^{34}S values typical of Rotliegend sulphate. Anhydrite cement from 1164 m at Weiach and dissolved sulphate from the Buntsandstein at Weiach 985.3 (302/12), and Riniken 806.6 (303/3), and from the Upper Rotliegend at Weiach 1116.5 m (302/19) have δ^{34}S values typical of evaporites not older than late Early Triassic (equivalent to the Röt evaporites or late Buntsandstein of Germany), or Middle Triassic (Muschelkalk).

Samples from the Upper Rotliegend at Riniken 963.5 and 993.5 m (303/5b and 6) and from the Buntsandstein at Kaiseraugst (3/1) appear to contain mixed Permian and late-early (Röt) or mid-Triassic (Muschelkalk) sulphate. Water from the Buntsandstein/weathered crystalline contact at Böttstein 316.6 m and 312.5 m (301/5 and 25b) and at Leuggern 217.9 m (306/2) appear to contain sulphate typical of that found in groundwaters in the crystalline. It is also evident from the δ^2H and δ^{18}O values and the ^{87}Sr/^{86}Sr ratios that water sampled from the Buntsandstein of the Böttstein and Riniken boreholes is of a crystalline origin.

Most waters from the crystalline with relatively low sulphate concentrations (300 to 600 mg/l), contain dissolved sulphates with δ^{34}S values of about 9 to 16 per mil, lower than typically found in the Röt or Muschelkalk, and δ^{18}O values from about -1.5 to +7. This range of δ^{34}S values is the same as that of barite found in fractures in the crystalline rock and of two of five sulphide minerals analysed. The ^{34}S and ^{18}O contents of dissolved sulphate are consistent with its production by oxidation of mineral sulphides which are not depleted in ^{34}S (KAI 664.12, LEU 1048.98). Waters from the Schafisheim borehole from the Buntsandstein/weathered crystalline horizon at 1488.3 m (304/6) and the crystalline itself at 1571.1 m (304/8) as well as Leuggern 923.0 m (306/13) with low-water yield show a strong influence of sulphate with sedimentary isotopic composition, such as were found in the Weiach and Riniken Buntsandstein samples (302/12 and 303/2) and the Upper Rotliegend (303/19). Samples from the crystalline zones of low-water yield at Böttstein 1326.0 m (301/21), Schafisheim 1887.9 (304/10), the Buntsandstein/weathered crystalline contact at Leuggern (306/2) and the water from the crystalline at Säckingen Badquelle (125) appear to have a component of Permian sulphate.

Table 7.1.2: $\delta^{34}S$ and $\delta^{18}O$ values and concentrations of sulphate and sulphide dissolved in groundwater from northern Switzerland and adjacent regions. (Page 1 of 2)

Location Sample Number	Date Sampled	Source Formation	Total SO4-2 (mg/l)	H2S (mg/l)	Delta 18O H2O (per mil) GSF	Delta 34S SO4 (per mil) Nielsen Anal. Smpl 1	Delta 34S SO4 (per mil) Nielsen Anal. Smpl 2	Delta 34S SO4 (per mil) Orsay Anal. Prep. by: Fres.	Delta 34S SO4 (per mil) Orsay Anal. Prep. by: Orsay	Delta 18O SO4 (per mil) Orsay Anal. Prep. by: Fres.	Delta 18O SO4 (per mil) Orsay Anal. Prep. by: Orsay
1/1	21-Jul-81	tOMM	145.4	n.n.	-11.89	17.5				12.60	
3/1	16-Jul-81	s-r	5183	0.065	-10.35	14.9				14.92	
14/1	14-Jul-81	mo	281	0.072	-10.23	18.0				13.60	
15/1	22-Jul-81	mo	2648	0.01	-11.09	19.8					
22/1	16-Jul-81	mm	1042	n.n.	-10.01	18.0					
23/1	16-Jul-81	km	811.4	n.n.	-10.09	14.9				12.84	
25/1	17-Jul-81	km	1515	n.n.	-9.29	14.9					
26/1	17-Jul-81	km	1569	n.n.	-9.43	14.9					
29/1	23-Jul-81	km	1325	n.n.	-9.73	15.5					
30/1	21-Jul-81	jo	67.3	n.n.	-10.01	5.2				7.05	
42/1	21-Jul-82	tOSM	22.5		-10.16	-5.6	-5.6			10.50	
56/2	21-Jul-82	ju	288.8	n.n.	-10.96	-27.7			-23.81	7.15	
65/2	19-Jul-82	mo	350	n.n.	-10.32	12.9	12.9			11.20	
66/2	19-Jul-82	mo	370	0.023	-10.27	15.5					
68/2	20-Jul-82	jmHR	119.7	n.n.	-10.08		13.8				
70/2	20-Jul-82	mo	306	n.n.	-9.98	14.4					
71/2	20-Jul-82	mo	455.5	n.n.	-9.96	14.7					
72/2	20-Jul-82	mo	982.8	0.089	-9.84	16.8	16.8				
74/2	20-Jul-82	mo	213.6	n.n.	-10.48	14.2	13.8			14.38	
76/2	20-Jul-82	jmHR	15.6	n.n.	-10.86	-10.6	-10.9			11.32	
80/2	22-Jul-82	jo	10.2	n.n.	-9.26	-4.8				4.20	
119/5	17-Aug-82	mo	574.1	31	-10.13	20.1					
120/1	7-Aug-81	mo	770	70		21.8					
120/5	17-Aug-82	mo	1059	63	-10.24	20.9					
121/1	13-Jul-81	mo	287.1	21		19.7					
123/5	16-Aug-82	mo	340.8	0.37	-10.04	17.7					
124/1	5-Aug-81	mo	1500	15		18.9					
124/5	16-Aug-82	mo	1495	8.4	-10.13	19.2					
125/1	22-Jul-81	KRI	112	n.n.	-8.89	11.6				10.14	
125/5	20-Aug-82	KRI	125.4	n.n.	-8.94	11.4					
126/5	20-Aug-82	KRI	37.6	n.n.	-9.44	9.8				2.78	
127/1	24-Jul-81	jo	85.3	0.21	-10.32	34.1		29.83		15.06	
127/5	19-Aug-82	jo	84.3	0.2	-10.24	34.8					
131/5	17-Aug-82	KRI	260.7	n.n.	-10.06	15.1					
132/1	6-Aug-81	KRI	260	0.000		15.8					
132/5	17-Aug-82	KRI	264.3	n.n.	-10.12	15.4		11.58		1.02	
142/1	10-Aug-81	tUSM	150	n.n.		22.6					
142/4	19-Aug-82	tUSM	215.6	n.n.	-9.41	22.6				14.74	
143/5	19-Aug-82	tUSM	324.1	n.n.	-8.20	21.7					
144/5	19-Aug-82	tUSM	251.3	n.n.	-8.86	21.0					
201/1	08-Jul-81	mo	1450	2.26	-8.99	18.3					
201/3	18-Aug-82	mo	1441	1.5	-9.02	18.4					
203/3	18-Aug-82	mo	1470	1.2	-9.03	17.8					
210/3	18-Aug-82	mo	1332	1.2	-9.21	18.5	18.1				
219/5	18-Aug-82	mo	1296	1.6	-9.06	18.3					
221/3	18-Aug-82	mo	1426	1.1	-9.10	18.4					
301/2	01-Nov-82	mo	2624	n.n.	-11.16			18.00		13.70	
301/5	16-Nov-82	s-KRI	359	0.01	-10.15			16.13		6.92	
301/8c	14-Dec-82	KRI	306.8	0.009	-10.05			14.62		1.16	
301/12b	22-Jan-83	KRI	336.1	n.n.	-10.04			13.56		-0.15	
301/16	09-Aug-83	KRI	340.3	n.n.	-10.01			14.09		-1.32	
301/18	16-Aug-83	KRI	339	n.n.	-9.96			13.80	14.45	-0.58	0.03
301/21	10-Oct-83	KRI	1560		-7.90				13.54		8.44

Chemistry and delta 18O H2O extrapolated from samples 301/19-22; See PEARSON, 1985, Sect. 7.1.

301/23	20-Oct-83	KRI			-9.97			14.01		-1.12	
301/25a	20-Jan-84	s-KRI	286.6	0.042	-10.08			15.35		3.83	

Table 7.1.2: $\delta^{34}S$ and $\delta^{18}O$ values and concentrations of sulphate and sulphide dissolved in groundwater from northern Switzerland and adjacent regions. (Page 2 of 2)

Location Sample Number	Date Sampled	Source Formation	Total SO4-2 (mg/l)	H2S (mg/l)	Delta 18O H2O (per mil) GSF	Delta 34S SO4 (per mil) Nielsen Anal. Smpl 1	Delta 34S SO4 (per mil) Nielsen Anal. Smpl 2	Delta 34S SO4 (per mil) Orsay Anal. Prep. by: Fres.	Delta 34S SO4 (per mil) Orsay Anal. Prep. by: Orsay	Delta 18O SO4 (per mil) Orsay Anal. Prep. by: Fres.	Delta 18O SO4 (per mil) Orsay Anal. Prep. by: Orsay
302/6	6-Mar-83	joki	54	0.1	-5.76				12.59		12.00
302/10b	04-Apr-83	mo	1990	4	-11.76	(1)		16.01	16.25	14.22	12.70
302/12	19-Jul-83	s	6235	0.04	-8.12			18.56	18.30	15.12	15.29
302/14	04-Apr-84	mo	468	0.018	-7.65					2.26	
Corrected for 9% deionized water contamination; See Table 3.1.6.											
302/16	27-Apr-84	KRI	430.7	0.014	-8.17				13.25	0.93	0.37
302/18	15-Jun-84	r	233	n.n.	-4.61				10.10		14.17
302/19	28-Jun-84	r	4185	0.04	-5.32				17.05		14.34
303/1	25-Jul-83	km	7314	0.99	-6.43			15.55	16.60	10.55	11.40
303/2	17-Aug-83	mo	3011	0.35	-10.14			17.88	17.80	14.22	14.70
303/3	16-Sep-83	s	1855	0.01	-6.92			17.91	18.42	13.49	13.28
303/5b	04-Oct-83	r	1477	0.008	-6.73				14.65		10.13
Chemistry corrected for 0.85% brine contamination; See WITTWER, 1986, Sect. 5.3.4.											
303/6	02-Nov-83	r	1653	22	-6.65			16.62	16.20	11.05	9.27
304/3	17-Feb-84	mo	2805	743	-11.02				26.05		11.04
304/6	02-Apr-84	s-KRI	3083	0.32	-5.42				18.85	14.45	14.71
304/8	02-May-84	KRI	2018	0.083	-6.66				18.65	13.36	12.08
304/10	17-Jun-84	KRI	794	0.072	-5.55				16.60		9.56
304/14	21-Jan-85	mo	2850	950	-10.81	(2)			25.91		11.60
305/2	01-Mar-84	r	586.5	n.n.	-10.36				10.80		3.83
305/4	15-Mar-84	KRI	570.4	n.n.	-10.29					4.47	
305/6	03-Apr-84	KRI	598.3	n.n.	-10.34				12.00	3.17	2.55
305/9	02-May-84	KRI	501.3	n.n.	-10.32				12.35	1.20	0.68
305/12	05-Jun-84	KRI	507.9	n.n.	-10.61				11.55		0.54
305/14	13-Aug-84	KRI	491	0.4	-10.40				11.70		0.91
305/16	27-Aug-84	KRI	529	0.23	-10.46				11.25		0.97
306/1	18-Jul-84	mo	536.8	n.n.	-9.96				15.80		13.43
306/2	08-Aug-84	s-KRI	445	n.n.	-10.39				15.90		8.68
306/4	14-Aug-84	KRI	338	n.n.	-10.49				14.10		6.41
306/5	14-Sep-84	KRI	355	n.n.	-10.35				13.70		4.93
306/7	27-Sep-84	KRI	362	n.n.	-10.37				14.35		3.59
306/9	17-Oct-84	KRI	426	n.n.	-10.52				13.65		5.10
306/13	07-Dec-84	KRI	2864		-10.45				13.60		10.91
306/16	15-Feb-85	KRI	263	n.n.	-10.25				13.45		-0.19
306/17	25-Mar-85	KRI	357.9	0.04	-10.48				13.25		4.04
306/20	22-Apr-85	KRI	270	0.32	-10.42				15.10		4.72
306/26	14-May-85	KRI	278	0.6	-10.02				14.10		3.51

(1): Delta 34S dissolved H2S = 1.91 per mil

(2): Delta 34S dissolved H2S = 4.40 per mil

Table 7.1.3: $\delta^{34}S$ and $\delta^{18}O$ values of sulphate minerals from sedimentary section of the Nagra deep boreholes.

Borehole	Depth (metres)	Geology	Description of Sample	$\delta^{34}S$ (SO$_4$) per mil	$\delta^{18}O$ (SO$_4$) per mil
Böttstein	59.37	Gipskeuper	Gypsum nodule, with anhydrite; Sabkha deposit	14.31	13.26
	59.00	Gipskeuper	Satin spar	13.98	13.33
	80.00	Gipskeuper	Satin spar	14.00	13.39
	88.90	Gipskeuper	Anhydrite pseudomorphic after satin spar	15.04	13.84
	115.50	Gipskeuper	Subaqueous gypsum with dolomite layers	15.85	12.20
	158.23	Upper Muschelkalk (Plattenkalk)	Anhydrite crystals from druse	19.77	17.45
	241.16	Middle Muschelkalk (Upper Sulphatgruppe)	Nodular anhydrite, Sabkha type	19.46	12.95
Weiach	824.85	Upper Muschelkalk (Trigonodus Dolomit)	Blocky anhydrite from nodule	19.21	14.59
	828.72	Upper Muschelkalk (Trigonodus Dolomit)	Felted anhydrite from nodule	18.70	16.50 / 16.30
	910.15	Middle Muschelkalk (Upper Sulphatgruppe)	Massive anhydrite (subaqueous)	18.47	12.93
	1022.45	Permian (Upper Rotliegend)	Anhydrite nodule	10.64	14.43
	1164.40	Permian (Lower Rotliegend)	Sandstone with approx. 3% anhydrite cement	18.45	14.14
Schafisheim	1329.20	Upper Muschelkalk (Hauptmuschelkalk)	Anhydrite from open vein	17.67	16.18

Table 7.1.4: $\delta^{34}S$ and $\delta^{18}O$ values of barite from the crystalline of the Nagra deep boreholes.

Borehole	Depth (metres)	Description of Sample	$\delta^{34}S$ (SO$_4$) per mil	$\delta^{18}O$ (SO$_4$) per mil
Böttstein	618.72	Fracture in pegmatite in granite	11.25	2.29
Kaisten	348.61	Large fracture along an old cataclastic zone with coarse sparry barite	14.20	10.84
	516.50	Open fracture with fluorite, sphalerite, and barite	15.30	8.71
	522.00	Small open fracture in old cataclastic zone with fluorite and barite	16.20	8.21
Leuggern	247.15	Fracture at the lamprophyre-granite boundary, with barite, calcite, hematite and chalcopyrite	15.85	9.41
	1648.29 (Young)	Compact clear rhombohedral barite from the hanging fracture face of a water-bearing zone	15.95	1.29
	1648.83 (Old)	Fracture along a hydrothermal vein filling with coarsely sparry barite	11.30	8.43
	1670.09	Fracture along a hydrothermal quartz-illite vein with barite	13.35	7.18

Table 7.1.5: $\delta^{34}S$ values of sulphide minerals from the sediments and the crystalline of the Nagra deep boreholes.

Borehole	Depth (metres)	Description of Sample	Mineral	(SO_4) per mil
Böttstein	301.00	Lower Muschelkalk/Wellendolomite (galena bed: Biopelsparite with dissolution pores and galena	Pyrite Galena Sphalerite	-33.05 -8.27 -7.73
Kaisten	664.12	Biotite-plagioclase gneiss with fine open fractures with quartz, pyrite (albite)	Pyrite	11.69
	817.56	Pyrite-clay mineral incrustation in water-bearing fracture	Pyrite	-2.91
Leuggern	919.20	Open fracture in vergrüntem (?) gneiss with native arsenic, sphalerite and synchisite	Sphalerite	1.09
	1048.98	Biotite plagioclase gneiss with partly open fractures with pyrite, barite, calcite, chlorite, galena and clay minerals	Pyrite	15.29
	1298.18	White aplite with open fractures with calcite, galena, sphalerite, native arsenic and clay minerals	Galena	1.42

7.2 Strontium Isotopes in Groundwaters and Minerals

A. Matter, Tj. Peters, and K. Ramseyer

7.2.1 Introduction

$^{87}Sr/^{86}Sr$ isotope studies can assist the evaluation of the extent of groundwater-rock interaction in a variety of subsurface settings (*e.g.*, FRITZ AND OTHERS, 1986, groundwaters from Stripa, Sweden; McNUTT AND OTHERS, 1984, groundwaters from the Precambrian Shield of Canada; LAND, 1980, pore waters in Cretaceous sediments from Texas).

Groundwaters, host rocks and minerals in Nagra's deep boreholes were analysed for Sr content and $^{87}Sr/^{86}Sr$ with the aim of ascertaining whether:

- the groundwaters in the crystalline basement, the Buntsandstein and the Muschelkalk aquifers could be characterized on the basis of these parameters;

- the groundwaters occurring today in fractures and pores were in equilibrium with the host rock, or at least with specific primary or authigenic minerals therein;

- the Sr content and isotope ratio of the groundwaters had changed since the deposition of the host sediments, or since growth of crystals in fissures and fractures.

Sr present in authigenic minerals inherits its isotopic composition directly from the solution out of which precipitation takes place. No isotopic fractionation occurs during precipitation, and, moreover, (in contrast to oxygen and carbon isotopes) the Sr isotope ratio is not temperature-dependent.

The initial Sr isotope ratio of a mineral remains unchanged, providing that no recrystallization has taken place, except in Rb-bearing minerals (*e.g.*, micas, illites and feldspars). Radiogenic ^{87}Sr is continually being formed in these minerals by the β-decay of ^{87}Rb. This effect is significant where Rb-bearing minerals are present in rock-forming proportions, for example in igneous and metamorphic rocks, and in immature siliciclastic sediments derived from them. No measurable Rb (detection limit 4 ppm) is present in carbonate and sulphate minerals. A change in the $^{87}Sr/^{86}Sr$ ratio of rocks made up of these minerals or in associated groundwaters is only possible where the system is not closed and there is an external source of Sr.

However, given equilibrium precipitation in an open system, the amount of Sr incorporated into calcite, dolomite, gypsum or anhydrite is dependent upon:

a) the relative Sr and Ca concentrations in the solution;

b) the temperature;

c) the mineral species being precipitated (different minerals have different Sr/Ca distribution coefficients); and

d) the rate of crystal growth.

If these variables are defined, then the Sr content of a precipitating mineral can be calculated according to VEIZER (1983), providing that the amount of precipitated mineral is minor relative to the amount of solution (*i.e.,* as long as the composition of the solution does not change significantly during precipitation). In addition, given the Sr content of a particular carbonate or sulphate mineral, it is possible to determine whether it was precipitated from groundwaters currently in contact with that mineral.

7.2.2   Analytical Procedures

The sample preparation was carried out following the technique described by JÄGER (1962, 1979). Rb and Sr isotopes were measured at the Isotope Geology Division of the Mineralogical Institute, University of Berne, using an Ion Instrument solid-state mass spectrometer with 35 cm radius, 90° deflection and with a three-filament ion source (BRUNNER, 1973). The isotope ratios and radiometric ages were calculated using the approach of YORK (1969) and the programmes published by FAURE (1977). Decay constants applied were as recommended by the IUGS Subcommission on Geochronology (STEIGER AND JÄGER, 1977). The Rb content was determined by mass spectrometry for all minerals except barite, which was analysed using an ARL-SEMQ electron microprobe. The Rb content of rocks from the crystalline basement was analysed using X-ray fluorescence.

7.2.3   Samples Analysed

Sr contents and isotope ratios were measured for groundwaters present in the crystalline basement and in the Buntsandstein and Muschelkalk aquifers. Sr analyses were also performed for the host rocks themselves, as well as for vein minerals and cements. This work was carried out to assist determination of the origin of the groundwaters and to assess the extent of their interaction with their host rocks. The results are given in Table 7.2.1, at the end of Section 7.2.

The crystalline host rocks comprise high-grade gneissic metasediments, diorite and syenite intrusives and late Variscan granites. These rocks possess fracture zones showing cataclastic deformation (MEYER, 1987) and hydrothermal alteration (PETERS, 1987). Interpretation of radiometric data (HUNZIKER AND OTHERS, 1987) and of supporting stratigraphic evidence, as well as studies of the thermal maturation of organic matter in Permo-Carboniferous sediments from Weiach (KEMPTER, 1987), all indicate that a hydrothermal event took place during the Permian. However, further evidence from radiometric data and from fluid inclusion investigations also suggests subsequent water circulation along these fractures.

Almost all fractures in the crystalline basement are lined by clay minerals, calcite and quartz, and, at shallow depths, by hematite. Calcite, quartz, fluorite, barite and celestite crystals were observed frequently in the open fissures. Sr isotopic ratios were performed on whole rock samples (Table 7.2.1) from apparently fresh granites (Böttstein 1490.7 to 1491.1 m, Leuggern 1525.97 m and 1636.90 m), fresh syenite (Schafisheim 1879.78 to 1883.76 m), fresh gneisses (Kaisten 1038.83 m, Leuggern 1636.90 m), hydrothermally altered gneiss (Weiach 2220.12 m, Kaisten 1046.36 m) and granite (Leuggern 1648.29 m). Table 7.2.1 also includes analyses on rock forming minerals which had previously been investigated for the purposes of radiometric age determination (PETERS AND OTHERS, 1986, 1988a,b; MATTER AND OTHERS, 1988b). The analysed calcites and the first generation barite (I) occur as massive vein fillings. The second generation barite (II) was encountered in one drusy cavity.

The siliciclastic sediments of the Buntsandstein lie unconformably either directly upon the crystalline basement (Böttstein, Leuggern and Schafisheim) or upon the Permian sediments (MÜLLER AND OTHERS, 1984). The only detrital minerals present in the Buntsandstein which may contain Rb as well as Sr are illite and K-feldspar. Both of these minerals occur in variable amounts. Radiogenic ^{87}Sr is continuously being produced from ^{87}Rb in these minerals.

The K-feldspar frequently shows dissolution features. Dissolution may have released Sr, which was incorporated in the feldspar lattice, out into the groundwaters (RAMSEYER, 1987).

Authigenic minerals observed in the Buntsandstein include anhydrite, gypsum, barite, celestite, calcite, dolomite and illite (Table 7.2.1). The Sr isotopic ratios of the eogenetic, mesogenetic and telogenetic minerals record the change in Buntsandstein porewater composition during burial and subsequent uplift (RAMSEYER, 1987).

Rock-forming dolomite, anhydrite and halite samples were also analysed, chiefly from the upper and middle Muschelkalk. Analyses from the middle Muschelkalk anhydrite marker horizon, from a Keuper anhydrite nodule and its gypsum rim, and from satin spar were performed in an investigation into the likely extent of diagenetic modification of ^{87}Sr/^{86}Sr ratios and Sr content.

7.2.4   Results

7.2.4.1   Groundwater Data

Twelve deep groundwater samples taken from the crystalline basement were analysed for Sr content and Sr isotopic composition. The ^{87}Sr/^{86}Sr analyses from the Böttstein, Leuggern, Kaisten and Weiach boreholes showed a narrow range of values between 0.7166 and 0.7178. The water sample taken from the crystalline basement in the Schafisheim borehole had a distinctly lower isotope ratio (0.7144; see Table 7.2.1).

One sample of the groundwaters flowing through the Buntsandstein siliciclastic sediments was also analysed for each well. The Sr isotope values of these waters were generally lower than those of the groundwaters from the crystalline basement. Moreover, the Sr content of the waters present in the Buntsandstein was 1.4 to 9 times that measured for the crystalline basement waters (Figure 7.2.1, Table 7.1.1).

However, $^{87}Sr/^{86}Sr$ ratios of groundwaters from the Muschelkalk were significantly lower (0.7087 to 0.7097) than the values obtained from groundwaters in the crystalline basement and in the Buntsandstein. The Sr content of groundwaters present in the Muschelkalk aquifer was similar to that of groundwaters in the Buntsandstein, but mostly higher than that of groundwaters sampled from the crystalline basement.

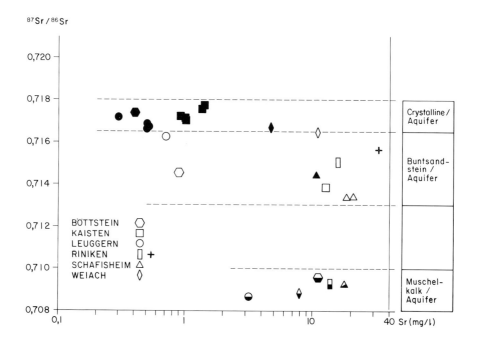

Figure 7.2.1: Plot showing $^{87}Sr/^{86}Sr$ ratio *versus* Sr contents of groundwaters.

7.2.4.2   Crystalline Basement

$^{87}Sr/^{86}Sr$ whole rock analyses of the granitic bodies (Böttstein 1490.7 to 1491.1 m; Leuggern 1525.97 m, 1636.90 m, 1648.83 m; Kaisten 1038.83 m) gave values between 0.7239 and 0.8239, much higher than the values obtained from the associated groundwaters. A single groundwater sample (2211.6 to 2224.6 m), from a fracture in the crystalline basement at Weiach, had an $^{87}Sr/^{86}Sr$ ratio significantly higher than that of the adjacent host rock (2220.12 m), which itself gave an unusually low $^{87}Sr/^{86}Sr$ ratio of 0.7144.

In two locations in the crystalline basement, the Sr isotopic composition of fillings (0.7175 and 0.7170) matched the $^{87}Sr/^{86}Sr$ ratios of the groundwaters in adjacent fractures. These were barite crystals taken from an open fissure from Leuggern (1648.29 m), and vein-filling calcite from Kaisten (1046.36 m). Conversely, illites from hydrothermal vein-fillings (Leuggern 1648.83 m, Kaisten 1046.36 m) showed significantly higher $^{87}Sr/^{86}Sr$ values (0.8014 and 0.8097, respectively). Illites taken from a mixed quartz-illite fissure-fill in the Leuggern borehole (1648.36 m) also showed high $^{87}Sr/^{86}Sr$ ratios. In contrast, calcites taken from calcite-albite and calcite vein-fillings had lower $^{87}Sr/^{86}Sr$ ratios (0.7115 and 0.7110 for each sample suite) than the waters now present in the crystalline basement (Table 7.2.1).

7.2.4.3   Sedimentary Rocks

The $^{87}Sr/^{86}Sr$ ratios of mesogenetic and telogenetic calcite, anhydrite and gypsum cements taken from the Buntsandstein generally ranged from 0.7122 to 0.7166, whereas two eogenetic cements gave values of 0.7104 and 0.7102 (Table 7.2.1).

Table 7.2.1 also shows that calcite, dolomite, anhydrite and halite from the Muschelkalk and Keuper were found to have rather low $^{87}Sr/^{86}Sr$ ratios (0.7078 to 0.7105). Three samples taken from a middle Muschelkalk anhydrite marker horizon in the Leuggern, Schafisheim and Weiach boreholes displayed identical $^{87}Sr/^{86}Sr$ ratios and very similar Sr contents.

An investigation into the effects of dehydration and hydration was carried out on a number of samples with the aim of ascertaining whether the primary selenite (transparent gypsum of submarine origin) formerly present in the marker horizon had retained its original Sr isotopic composition and content during conversion to anhydrite. Comparison between the core of an anhydrite nodule and its gypsum rim revealed no change in $^{87}Sr/^{86}Sr$ ratio during hydration. However, the Sr content decreased from 2150 ppm (1700 ppm if recalculated in terms of gypsum composition) in the anhydrite core to 1576 ppm in the gypsum rim (Table 7.2.1).

Also, anhydrite replacing satin spar gypsum veins and a younger phase of gypsum satin spar showed the same Sr isotopic compositions as the anhydrite cores and the gypsum rim of the nodules. However, a lower Sr content was measured from the younger satin spar vein filling. The lowest Sr contents (58 to 233 ppm) were recorded in authigenic calcites from drusy cavities and fractures in the upper Muschelkalk (see Table 7.2.1).

7.2.5 Discussion

Deep groundwaters from the crystalline basement, the Buntsandstein and the Muschelkalk can be distinguished from each other on a cross-plot of $^{87}Sr/^{86}Sr$ and Sr content (Figure 7.2.1). In particular, waters from the Muschelkalk aquifer have a distinctly lower $^{87}Sr/^{86}Sr$ ratio than waters taken from the crystalline basement. This difference indicates that there may have been little or no interaction between water from the Muschelkalk and lower aquifers (DIEBOLD, 1986).

Neither the crystalline host rocks nor hydrothermal vein-fillings display $^{87}Sr/^{86}Sr$ compositions identical to those of the groundwaters currently occurring in fractures or in veins. This implies that the Sr present in the groundwaters was at least partly derived from an external source, that it originated from dissolution of Sr-bearing minerals (feldspars, micas), or that a combination of external and solution-related sources were important.

A plot of $^{87}Sr/^{86}Sr$ as a function of $^{87}Rb/^{86}Sr$ ratios for whole rock samples from the crystalline basement gives an isochron with an age of 279 ± 5 million years (Figure 7.2.2). This Permian age corresponds to a hydrothermal event dated using a large number of K-Ar illite ages (HUNZIKER AND OTHERS, 1987). This hydrothermal event was apparently so pervasive that it modified the Sr composition of all the granites (Table 7.2.1) including those which appear "fresh" today, such as the granite encountered in the lower part of the Böttstein borehole. The intercept of the isochron plot reveals an initial $^{87}Sr/^{86}Sr$ ratio of 0.714 ± 0.001 (Figure 7.2.2).

If the Sr-bearing minerals, such as barite, calcite, or fluorite, had formed during this Rotliegend event, they would have inherited their $^{87}Sr/^{86}Sr$ ratios from their parent solutions. The $^{87}Sr/^{86}Sr$ values would have then remained unchanged with time, because these minerals contain no Rb, and hence, no additional radiogenic ^{87}Sr would have been produced. If the Sr present in modern groundwaters in the crystalline basement had originated from dissolution of the calcite crystals, for example, which generally replace plagioclase, then the groundwater $^{87}Sr/^{86}Sr$ ratio would have been 0.714. This value was recorded in groundwaters from the crystalline basement only in the case of the Schafisheim borehole. In order to explain the values of 0.716 to 0.718 characteristically occurring in groundwaters taken from the crystalline basement in all the other boreholes, it is necessary to invoke a Sr source with a higher $^{87}Sr/^{86}Sr$ ratio. One possible source fulfilling this requirement is the whole rock itself. Radiogenic ^{87}Sr occurring in Rb sites in a mineral lattice is more weakly bonded than Sr occurring in Ca sites. Thus, movement of radiogenic ^{87}Sr into solution is energetically favoured. An alternative possibility is that younger minerals of neomorphic origin were produced less than 280 million years ago during a later stage of the hydrothermal event. Such minerals would have a higher $^{87}Sr/^{86}Sr$ ratio, as a result of the incorporation of additional, radiogenically-derived, ^{87}Sr. The resultant isotopic composition of these minerals would, most probably, lie within the present range (0.716 to 0.718) of $^{87}Sr/^{86}Sr$ compositions for groundwaters encountered in basement.

The Sr isotopic composition of the present groundwaters is not compatible with Sr derivation either from congruent dissolution of the whole rock or from dissolution of Rb-

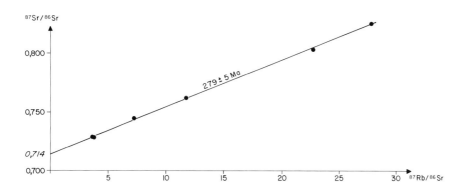

Figure 7.2.2: Isochron diagram for values from crystalline basement rocks at Böttstein and Leuggern, and from an illite vein fill (Leuggern, 1648.83 m).

rich minerals, such as biotite or illite (Table 7.2.1). If these possibilities are ruled out, then the only remaining potential Sr source is from Rb-poor minerals with low $^{87}Sr/^{86}Sr$ ratios, such as plagioclase, or, alternatively, calcite replacements of the anorthite component in plagioclase.

KAY AND DARBYSHIRE (1986) demonstrated that deep groundwaters in the Carnmenellis granite of Cornwall had $^{87}Sr/^{86}Sr$ ratios nearly identical to those of fluorites and calcites (formed during a hydrothermal event at 290 million years) and of hydrothermally altered plagioclases. KAY AND DARBYSHIRE (1986) concluded that plagioclase was the likely major source of Sr. $^{87}Sr/^{86}Sr$ ratios obtained from two calcite samples taken from closed, calcite-illite filled fissures in the Leuggern borehole (1511.32 m, 1525.97 m) are lower (0.711) than those recorded from the adjacent waters in the crystalline basement (Table 7.2.1). These calcites must, therefore, have precipitated from waters of external origin (for example from waters responsible for the dissolution of Buntsandstein caliche or anhydrite), or else they must pre-date the Permian hydrothermal event. However, this latter possibility is thought unlikely in view of the intimate intergrowth between the calcite and hydrothermally-formed illite.

The comparability in the $^{87}Sr/^{86}Sr$ values of groundwaters in the crystalline basement, in the Buntsandstein and in the Rotliegend suggests that the Sr isotopic composition of the groundwaters is controlled by the similarity in mineralogical composition of the host rocks. It appears to be unimportant in terms of $^{87}Sr/^{86}Sr$ values whether the Buntsandstein lies upon the Rotliegend or directly upon the crystalline basement. The Buntsandstein consists in some boreholes of first-cycle detrital sediments derived from the erosion of basement massifs, and in others of sediments derived from the reworking of Rotliegend sandstones. In each case, the Buntsandstein contains the same mineral assemblage as the crystalline basement or as the Rotliegend, respectively. A notable exception occurs in the case of plagioclase, which is missing in the Buntsandstein. The absence of plagioclase was attributed by RAMSEYER (1987) to breakdown during synsedimentary surface weathering. This interpretation is supported by similarity of the $^{87}Sr/^{86}Sr$ composition of groundwaters present in the crystalline basement and in the Buntsandstein at Schafisheim. The heavy mineral data suggests that the Buntsandstein is of local derivation here. The low $^{87}Sr/^{86}Sr$ ratio appears to have been inherited from the local basement rocks (Table 7.2.1).

However, the groundwaters present in the Buntsandstein are characterized by slightly lower Sr-isotope ratios and higher Sr contents than groundwaters in the basement. It is thought that the waters in the Buntsandstein at Böttstein and Leuggern evolved chiefly from waters originally present in the basement, with additional compositional influence coming from the dissolution of calcite and sulphate cements present in the Buntsandstein.

The waters currently present in the Buntsandstein, at Schafisheim, have a measured Sr content of 19 ppm and an $^{87}Sr/^{86}Sr$ ratio of 0.71343. How could these values be achieved? If the Sr were derived only from waters present in the crystalline basement, the Sr content (11 ppm, $^{87}Sr/^{86}Sr$ 0.71442) would be too low, so that an additional source of Sr must be implied. Calculations suggest that dissolution of vein calcites present in Buntsandstein strata could yield a further 8 ppm of Sr giving a theoretical resultant $^{87}Sr/^{86}Sr$ of 0.71360. This is sufficiently close to the measured value to indicate that some Sr may indeed have been derived by this process.

Exceptionally, in the case of the waters present in the Buntsandstein of the Weiach borehole, the elevation of Sr contents above basement groundwater values cannot be attributed to dissolution of calcite or sulphate cements, because thermodynamic calculations (PEARSON AND OTHERS, 1989) show that the Buntsandstein waters are saturated with respect to these phases. Moreover, dissolution of these cements (with $^{87}Sr/^{86}Sr$ ratios lower than those of the Buntsandstein waters) in crystalline basement groundwaters would give far too low a resultant $^{87}Sr/^{86}Sr$ composition. It is concluded from this evidence that additional calcite- and sulphate-saturated waters with higher $^{87}Sr/^{86}Sr$ ratio and Sr content infiltrated into the Buntsandstein at Weiach.

Waters of potentially suitable composition could include either those present in the continental Rotliegend, or evolved waters originally derived from the crystalline basement but already displaying compositional evolution as a result of contamination from sedimentary sources.

It was clearly of interest to ascertain whether the host rocks or vein minerals had the same Sr isotope ratios as their associated groundwaters, because, if this were the case, then these rocks or minerals could have acted as sources or sinks for Sr during dissolution or precipitation.

The Sr-isotope ratios of calcite crystals found in drusy cavities in the Buntsandstein of the Riniken (805.0 m) and Leuggern (218.61 m) boreholes are very similar to those of their respective associated groundwaters. These crystals, then, might be in thermodynamic equilibrium with the groundwaters.

However, calculation of the Sr content of calcite precipitated from present groundwaters (using a distribution coefficient of 0.13, precipitation temperatures of 25°C and 100°C, respectively, and assuming chemical equilibrium) gives theoretical Sr contents which are high in comparison with the actual recorded values in calcites present in the fissures (BLÄSI AND OTHERS, 1989).

In addition, these calcites are not in oxygen isotopic equilibrium with the groundwaters at currently prevailing temperatures. The minimum precipitation temperatures suggested for these calcites by fluid inclusion measurements (65° to 70°C) also preclude precipitation from waters currently present, which are significantly cooler (Leuggern 18°C; Riniken 51°C).

The effects of mineral reactions on $^{87}Sr/^{86}Sr$ groundwater values are clearly demonstrated by the gypsum and anhydrite samples from the Keuper at Böttstein (Table 7.2.1). Neither rehydration of anhydrite to gypsum, nor anhydrite dissolution and satin spar precipitation resulted in modification of the primary Sr-isotopic ratio. The lack of change in $^{87}Sr/^{86}Sr$ during these processes mainly reflects the high Sr (900 to 2700 ppm) and low Rb contents (below detection limit of 4 ppm) of these samples, which meant that the contribution of additional ^{87}Sr from radiogenic sources was insignificant. It is thus not surprising that the anhydrite marker horizon in the middle Muschelkalk retained a $^{87}Sr/^{86}Sr$ ratio of 0.70784 characteristic for Muschelkalk sea water (0.70780, according to BURKE AND OTHERS, 1982). Conversely, the higher $^{87}Sr/^{86}Sr$ ratios recorded for halite from Riniken may reflect admixture by influx of some continental water into the salt ponds.

The Sr-isotopic ratios of all measured phases and host rocks from the upper Muschelkalk exceed the relevant values both for Muschelkalk sea water and for the anhydrite marker horizon. This implies that the Sr isotopic composition of groundwaters in the Muschelkalk aquifer (mean c. 0.709) cannot simply be the result of the dissolution of marine sulphates. The $^{87}Sr/^{6}Sr$ ratios rather suggest progressive evolution of the Muschelkalk waters as a result of the dissolution either of carbonates (mainly dolomite; $^{87}Sr/^{86}Sr$ 0.7081 to 0.7105) and continental sulphates from the overlying Keuper ($^{87}Sr/^{86}Sr$ = 0.7081), or of diagenetic sulphates in the upper Muschelkalk ($^{87}Sr/^{86}Sr$ 0.7087 to 0.7095).

However, $\delta^{34}S$ and $\delta^{18}O$ data from the sulphates would seem to rule out the continental sulphates (Keuper) as a contributor. Alternatively, if middle Muschelkalk halite (which is a possible source of chloride ions) were dissolved, then the $^{87}Sr/^{86}Sr$ values would not have been affected because the Sr-isotope ratios of halite and of Muschelkalk groundwaters are similar.

Sr-isotope analyses of a calcite from Leuggern pre-dating Jura folding, and of syntectonic calcite crystals from Schafisheim which grew into an open fissure during folding (BLÄSI AND OTHERS, 1989), gave values of 0.7090 and 0.7096, respectively. In addition, available evidence (from early diagenetic, or pedogenic, calcite/caliche and anhydrite) suggests that the syndepositional porewaters present during Buntsandstein times probably had $^{87}Sr/^{86}Sr$ ratios of c. 0.710. These are significantly lower than the values of present day waters and minerals in the Buntsandstein strata.

This seems to indicate that the Sr isotopic composition of groundwaters fluctuated through geological time, a hypothesis borne out by comparison of these values with current groundwater ratios. In terms of Sr-isotopic composition and Sr content, neither of these calcite suites are in equilibrium with the groundwaters now occurring in the Muschelkalk and the Buntsandstein.

Table 7.2.1: Results of strontium and rubidium analyses. (Page 1 of 3)

Name	Depth m	Sample Material	Water Sample Number	Occurrence	Sr ppm	87Sr/86Sr	87Rb/86Sr
BÖTTSTEIN							
Keuper	53.37	Gypsum		Marginal zone of anhydrite nodule	1576	0.70811 ± 0.00008	
	53.37	Anhydrite		Nodule	2150	0.70814 ± 0.00009	
	59	Gypsum		Satin spar	886	0.7081 ± 0.00007	
	88.9	Anhydrite		Pseudomorph after fibrous gypsum	2701	0.70807 ± 0.00004	
Muschelkalk	122.65	Calcite		Drusy cavity	100	0.70936 ± 0.00006	
	123-202.5	Water	301/2		11.0	0.70968 ± 0.00006	
	137.2	Dolomite		Whole rock	586	0.71053 ± 0.00006	
Buntsandstein	305.2-319.8	Water	301/5		0.9	0.71455 ± 0.00006	
Crystalline Basement	393.9-405.1	Water	301/8c		0.4	0.7173 ± 0.0001	
	618.5-624.1	Water	301/12b		0.6	0.71752 ± 0.00006	
	1490.7-1491.1	Biotite			22.4	1.10808 ± 0.00007	113.738
	1490.7-1491.1	Whole rock			207	0.72716 ± 0.00006	3.58725
WEIACH							
Keuper	749.6	Anhydrite		Nodule	1893	0.70821 ± 0.00009	
Muschelkalk	822.0-896.1	Water	302/10b		7.9	0.70881 ± 0.00006	
	824.85	Anhydrite		Nodule	2437	0.70953 ± 0.00003	
	828.72	Anhydrite		Nodule	1222	0.70874 ± 0.00003	
	910.15	Anhydrite		Marker horizon	1105	0.70783 ± 0.00006	
Buntsandstein	981.0-989.6	Water	302/12,13		10.9/11.4	0.71652 ± 0.00008	
	985.7	Anhydrite		Sandstone cement	2614	0.71024 ± 0.00009	
	988.56	Calcite		Caliche nodule	1092	0.71042 ± 0.00007	
Crystalline Basement	2211.6-2224.6	Water	302/16		4.7	0.71671 ± 0.00006	
	2220.12	Whole rock		Chloritised biotite-plag. gneiss	244 *	0.71441 ± 0.00005	
	2224.95	Calcite		Calcite-prehnite vein	584	0.70992 ± 0.00006	
RINIKEN							
Muschelkalk	617.3-696.0	Water	303/2		13.6	0.70924 ± 0.00004	
	742.29	Halite			2.4	0.70915 ± 0.00025	
Buntsandstein	793.0-820.2	Water	303/3,4		5.7/16.1	0.71506 ± 0.00007	
	805	Calcite		Drusy cavity	290	0.71504 ± 0.00009	

Table 7.2.1: Results of strontium and rubidium analyses (Page 2 of 3)

Name	Depth m	Sample Material	Water Sample Number	Occurrence	Sr ppm	87Sr/86Sr		87Rb/86Sr
RINIKEN	(continued)							
Upper Rotliegendes	958.4-972.5	Water	303/5		32.4	0.7157	± 0.00006	
	1045.45	Calcite		Open fracture	286	0.71618	± 0.00008	
KAISTEN								
Buntsandstein	95.17	Gypsum		Satin spar	199	0.71224	± 0.00006	
	97.0-129.9	Water	305/1		12.6	0.71386	± 0.00012	
Crystalline	475.5-489.8	Water	305/6,7		0.9/ 1.1	0.71722	± 0.00008	
Basement	816.0-822.9	Water	305/9,10		1.3/ 1.4	0.71767	± 0.00006	
	1021.0-1040.9	Water	305/12		1.4	0.71778	± 0.00003	
	1038.83	Whole rock		Migmatitic metapelite, fresh	-	0.72392	± 0.00004	
	1046.36	Calcite		Xenomorphic fissure lining	148	0.71701	± 0.00011	
	1046.36	Illite		Fissure lining	49.7	0.80971	± 0.00005	50.777
	1046.36	Whole rock		Chloritised metapelite	-	0.72305	± 0.00013	
	1238.0-1305.8	Water	305/16		1.0	0.71711	± 0.00008	
LEUGGERN								
Muschelkalk	53.5-96.4	Water	306/1		3.1	0.70867	± 0.00004	
	54.08	Calcite		Drusy cavity	58	0.70905	± 0.00003	
	116.43	Anhydrite		Marker horizon	1196	0.70784	± 0.00007	
Buntsandstein	208.2-227.5	Water	306/2		0.7	0.71629	± 0.00022	
	218.61	Calcite		Drusy cavity	107	0.71659	± 0.00014	
Crystalline	507.4-568.6	Water	306/7		0.5	0.71672	± 0.0001	
Basement	702.0-709.5	Water	306/9		0.5	0.71656	± 0.00008	
	1179.3-1227.2	Water	306/20		0.3	0.71729	± 0.00009	
	1427.4-1439.4	Water	306/26		0.7	0.71741	± 0.00006	
	1511.32	Calcite		Zoned calcite-albite vein filling	-	0.71156	± 0.00008	
	1525.97	Whole rock		Biotite granite, fresh	206 *	0.72859	± 0.00008	3.8771
	1525.97	Calcite		Calcite vein fill	-	0.71105	± 0.00008	
	1636.9	Whole rock		Cordierite-bearing two-mica, granite, fresh	290 *	0.76098	± 0.00005	11.7798
	1642.2-1688.9	Water	306/23		0.4	0.71747	± 0.00009	
	1648.29	Barite II		Transparent blocky rhombohedral crystal	1090 +	0.71755	± 0.00017	

Table 7.2.1: Results of strontium and rubidium analyses. (Page 3 of 3)

Name	Depth m	Sample Material	Water Sample Number	Occurrence	Sr ppm	87Sr/86Sr		87Rb/86Sr
LEUGGERN								
Crystalline Basement	(continued)							
	1648.29	Whole rock		Cordierite-bearing two-mica granite, altered		0.8239		27.7169
	1648.83	Illite		Hydrothermal vein filling	60.6	0.80142	± 0.00006	22.6114
	1648.83	Barite I		Secondary vein filling	16500 +	0.71575	± 0.00006	
	1648.83	Whole rock		Hydrothermal vein filling	38 *	0.74421	± 0.00006	7.2449
SCHAFISHEIM								
Muschelkalk	1227.8-1293.0	Water	304/3		17.2	0.70919	± 0.00004	
	1233.93	Dolomite		Whole rock (includes 2% anhydrite)	85	0.70812	± 0.00009	
	1329.2	Calcite		Fracture filling	233	0.70966	± 0.00005	
	1374.25	Anhydrite		Marker horizon	1142	0.70785	± 0.00003	
Buntsandstein	1476.0-1500.4	Water	304/6		8.0/ 20.9	0.71343	± 0.00009	
	1485.75	Calcite		Fissure	3566 0.5/	0.713	± 0.00023	
	1564.5-1577.7	Water	304/8,9		11.6	0.71442	± 0.00002	
Crystalline Basement	1879.78-1883.76	Apatite			1581	0.70872	± 0.00009	0.002415
	1879.78-1883.76	Perthite			2978	0.70958	± 0.00009	0.2741
	1879.78-1883.76	Albite + Quartz			928	0.70964	± 0.00004	0.2662
	1879.78-1883.76	Hornblende			261	0.70988	± 0.00006	0.2192
	1879.79-1883.76	Biotite			16.8	1.14886	± 0.00005	98.0801
	1879.79-1883.76	Whole rock		Syenite, fresh	494	0.71405	± 0.00005	1.2351

* Data obtained by x-ray fluorescence

+ Electron microprobe analyses

Data from: PETERS AND OTHERS (1986, 1988a,b), MATTER AND OTHERS (1987, 1988a,b) and BLÄSI AND OTHERS (1989)

335

7.3 Uranium and Thorium-Series Nuclides

P. Baertschi, W. Balderer, and M. Ivanovich

The concentration and distribution of uranium and thorium nuclides and their daughter products in rock-water systems are important indicators of processes of mobilization, transport and fixation which may also be relevant to highly active waste repository sites.

7.3.1 Origin and Significance of Decay Series Disequilibria

In a rock mass which has remained undisturbed for some 10^7 years and may, therefore, be considered as closed system, the decay series nuclides of the long-lived parents ^{238}U, ^{235}U and ^{232}Th are in secular radioactive equilibrium. Secular equilibrium means that the parent nuclide and all successive members of its decay chain, shown in Figure 7.3.1, are present with the same activity so that all activity ratios (AR) are one.

Opening the system by a geologic disturbance such as interaction of the rock with flowing groundwater will, in general, lead to radioactive disequilibria (AR $\neq$ 1) by two processes:

a) Fractionation induced by α recoil.

CHERDYNTSEV AND OTHERS (1955) found activity ratio ^{234}U/^{238}U differed greatly from one, mainly in groundwaters. This unexpected isotopic fractionation in rock-water systems has since been investigated extensively and may be described as follows:

As shown in Figure 7.3.1, α decay of ^{234}U (4.2 MeV) leads to a "hot" ^{234}Th atom with an α recoil energy of 72 KeV. The short-lived ^{234}Th may thus be displaced by 20 nm or more in the rock lattice, producing severe damage along the track, or it may even be expelled into the groundwater if the decay happened close enough to the rock-water interface. Because ^{234}Th and ^{234}Pa are short-lived intermediates, they will have little influence on the fate of the ^{234}U formed. Thus, rock-water interaction will generally lead to an enrichment of ^{234}U in the groundwater either by direct recoil expulsion of ^{234}Th into the water phase, or by preferential dissolution from recoil-damaged lattice sites.

b) Fractionation caused by differences in the physical and chemical behaviour of the various elements in a decay series.

As an example, interaction of rocks with oxidizing, carbonate-bearing groundwater will result in preferred dissolution of uranium, leaving the insoluble thorium and protactinium behind. Therefore, the ^{230}Th/^{234}U and ^{231}Pa/^{235}U activity ratios will rise above one in the rock, and at the same time, drop to very low values (AR << 1) in the interacting groundwater.

Such chemical fractionation will always occur between different elements in a decay series, independent of any preceding α recoil fractionation. It must also

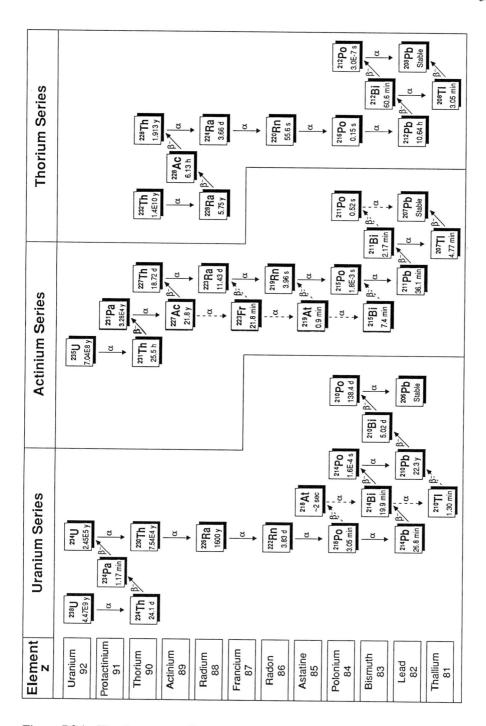

Figure 7.3.1: The three naturally occurring radioactive decay series.

be considered for activity ratios between isotopes such as ^{228}Th/^{232}Th, which have an intermediate nuclide of a different element. In this case, the intermediate is ^{228}Ra with a half-life comparable to that of the daughter nuclide ^{228}Th.

If a system which is in disequilibrium becomes closed again at a certain moment (t = 0), e.g., by a stop of water flow through the rock, then after a time t, secular equilibrium will be approached again according to

$$(AR_t - 1) / (AR_0 - 1) = \exp(-\lambda t) = 2^{-t/T_{1/2}} \qquad (7.3.1)$$

(λ = decay constant, $T_{1/2}$ half-life of daughter nuclide)

Equation 7.3.1 applies when the half-life of the daughter nuclide is much shorter than that of the parent, as is the case with ^{234}U/^{238}U and ^{226}Ra/^{230}Th. The useful time scale comprises a few half-lives of the daughter nuclide. That is, the approach to equilibrium, as given by the left hand side of equation 7.3.1, will be ≈ 97 per cent after five half-lives of the daughter nuclide (t = 5 $T_{1/2}$). The five half-lives may be used to estimate the minimum time of undisturbed rock history if equilibrium has been measured with a limit of error of a few per cent for the activity ratio in question.

The various aspects and hydrogeochemical applications of nuclear series disequilibria were treated by several authors in IVANOVICH AND HARMON (1982) and INTERNATIONAL ATOMIC ENERGY AGENCY (1983a and b), and by GASCOYNE (1987) and FLEISCHER (1988).

7.3.2 Analysis of Concentrations and Activity Ratios

Measurements were performed by AERE, Harwell, by PSI and by Fresenius. Additional data on uranium and thorium contents of rocks of the deep boreholes were provided by the gamma-logs and by radiometric analyses of selected rocks (references given in Tables 7.3.9 through 7.3.13). Methods applied at AERE are described by LALLY (1982) and by CUTTELL (1983), those used at PSI and by Fresenius are summarized by KUSSMAUL AND ANTONSEN (1985).

For uranium concentrations in groundwater, the AERE results (α spectroscopy) and PSI results (photometric analysis) are of comparable quality, whereas the Fresenius data (fluorometric analysis) show larger scatter and less sensitivity (see Figure 2.3.3). Therefore, only the Harwell and the PSI results are considered further. According to KEIL (1979), the PSI method has a sensitivity of 0.05 ppb leading to standard deviations (for multiple samples) of about ±10 per cent at the 1 ppb level and of ±1 per cent at levels of 10 ppb or higher. The Harwell method apparently has a higher sensitivity, but the 1 σ errors given in Tables 7.3.1 through 7.3.7 apply to counting statistics only, and may in fact be much higher in the analysis of multiple water samples.

The water samples for analysis by Harwell and by PSI were taken separately. If possible, 20 litre samples were shipped to Harwell where, after two weeks or more, they were filtered and acidified and used for analyses of the whole decay series. PSI samples for chemical U analysis were collected in 0.5 litre polyethylene bottles. In the laboratory, usually only about one day after sampling, any coarse particulate matter collecting at the bottom was removed and the water was then acidified in the sample bottle without being filtered.

This procedure minimized any loss by sorption processes but did not exclude desorption from, or dissolution of, colloidal material. Due to these differences in sampling and pretreatment, it is clear that best agreement between the two laboratories was obtained for samples of abundant and pure groundwaters as mainly found in the regional hydrochemical programme (Table 7.3.7).

Pronounced deviations between Harwell and PSI data are mainly found in low-uranium waters. Much lower U contents were measured at Harwell for Böttstein (301/18), Weiach (302/12, 302/19 and 302/17), Schafisheim (304/2, 304/10) and Leuggern (306/12, 306/13). The Harwell value is much higher only for Schafisheim (304/3). The differences may either be due to changes of U concentration during sampling or--more probably--to different sorption/desorption processes in the pre-analytical procedures as indicated above. However, the deviations cannot be attributed to analytical inaccuracies as shown by the multiple analyses of Leuggern (306/11b, 12, 13) (Table 7.3.6), which show acceptable reproducibility within the data sets of the two laboratories, even for samples taken at different dates and in spite of the large difference between the Harwell and PSI data.

It is also possible to develop uranium concentrations from the PSI determination of total uranium α activity ($^{234}U + ^{238}U$) and the $^{234}U/^{238}U$ activity ratio as reported by WITTWER (1986). These determinations were done independently of the chemical U analyses on separate water samples, and by another PSI laboratory. For U contents > 1 ppb, the data were found to agree quite well with the results of the spectrophotometric U analysis, but for lower U contents the error of ±0.3 ppb (for AR = 1) may become too large. Unfortunately, no data are available for Leuggern (306/12, 13). For other samples showing large deviations, such as Schafisheim (304/2), and Weiach (302/17 and 302/12), the results and their large errors do not allow a decision for either Harwell or PSI data. For the sample Schafisheim (304/3) with 0.9 ppb U (Harwell) and 0.06 ppb (PSI), the activity measurement leads to a U content of 0.06 ± 0.06 ppb, supporting the lower PSI value. Except by the use of this technique, it is not possible to decide whether the Harwell or the PSI data set is the more realistic.

For the $^{234}U/^{238}U$ activity ratio, the α-spectroscopic methods of Harwell and PSI are similar. The Harwell technique, using ^{236}U spike, however, is more precise and sensitive than the PSI method which is based on external calibration. There is generally satisfactory agreement for U contents > 0.5 ppb. For lower concentrations, the PSI values apparently are not reliable. They are reported by WITTWER (1986) and by SCHMASSMANN AND OTHERS (1984) but are not considered further in this report except for some groundwater samples from the regional hydrochemical programme (Table 7.3.7) for which no Harwell data exist.

Other activity ratios, including $^{228}Ra/^{226}Ra$ in groundwater, $^{238}U/^{232}Th$, $^{231}Pa/^{235}U$, $^{226}Ra/^{230}Th$ in rocks and $^{230}Th/^{234}U$, $^{228}Th/^{232}Th$, $^{230}Th/^{232}Th$ in both groundwaters and rocks, were determined only by Harwell.

The 1-σ error given for most Harwell data applies to counting statistics only, and does not include any errors in connection with counting sample preparation. Thus, measured activities and concentrations as given for uranium and radium and activity ratios involving different elements, such as $^{230}Th/^{234}U$, may deviate more from reality than indicated by the limits of error. For activity ratios of isotopes of the same element as, *e.g.*, $^{234}U/^{238}U$, the limits of error may, however, be quite realistic since no fractionation effects are to be expected in the analytical procedures.

The radon activities in groundwaters were measured by PSI and Fresenius using methods sketched by KUSSMAUL AND ANTONSEN (1985). The PSI method uses gamma spectrometric determination of the short-lived daughters ^{214}Pb and ^{214}Bi in five one-litre water samples with a detection limit around 1 Bq/l. The reproducibility depends on the activity level, and is around ±10 per cent for activities > 30 Bq/l, but never better than ±2 Bq/l. Fresenius uses the classical method of electrometer discharge either at the sampling place or in the laboratory. A detection limit of 0.74 Bq/l (20 pCi/l) and a reproducibility of ±0.74 Bq/l (±20 pCi/l) is quoted.

The PSI and the Fresenius data are well correlated (r = 0.96), but the PSI values are systematically higher, up to 30 per cent for radon activities above 50 Bq/l. These discrepancies are probably due to different losses in the sampling procedures, and may also involve calibration errors in one or both of the analytical methods.

7.3.3 Interpretation of Groundwater Analyses

Uranium content and decay-series disequilibria in groundwaters were discussed by OSMOND AND COWART (1976), OSMOND AND OTHERS (1983) and, in a short summary, by BALDERER (1983). Results of other relevant hydrochemical investigations and thermodynamic model calculations on the groundwaters considered here are given by WITTWER (1986), PEARSON (1985) and PEARSON AND OTHERS (1989).

Uranium and decay-series data on groundwater samples from six Nagra deep boreholes and from selected boreholes and wells of the Nagra regional hydrochemical programme are compiled in Tables 7.3.1 through 7.3.7 at the end of Section 7.2. The sampling locations are shown in Plate 1 and in Figure 7.3.6. The data for the Böttstein borehole have already been published and partly evaluated by BALDERER (1985). Samples chosen from the regional hydrochemical programme were mainly those for which Harwell measurements are available. There are, however, many other sample locations for which only PSI and Fresenius data exist and which are not considered here. Many of these have been compiled and discussed by SCHMASSMANN AND OTHERS (1984).

7.3.3.1   Uranium Content

The uranium content of groundwaters from strongly reducing aquifers is generally very low, usually below 0.1 ppb, whereas under oxidizing conditions, in the stability range of U(VI) (as UO_2^{+2}) it may reach values as high as some mg/l. Thermodynamic model calculations by LANGMUIR (1978) and SCHWEINGRUBER (1981) confirmed this strong dependence of uranium solubility on the oxidation potential. The solubility in a typical crystalline groundwater rises by some five orders of magnitude between -0.15 V and +0.05 V (or pE -2.5 and +0.85) at near neutral pH values. However, for the groundwater samples considered here, it was found that there is only a very poor correlation between the oxidation potential measured with a platinum electrode and the uranium content. The main reason for this is probably oxygen contamination in the sampling procedure. Therefore, PEARSON (1985) and PEARSON AND OTHERS (1989) calculated pE values with the PHREEQE computer code using various redox couples as analysed in the groundwaters. Some of these redox couples, particularly As(V)/As(III), led to pE values which correlate well with the U content. Because of the strong dependence of uranium solubility on oxidation potential, the U content itself may also be used as an indicator of the redox state of a groundwater.

A related possibility for judging the redox state of a groundwater is its $^{230}Th/^{234}U$ activity ratio as suggested by CUTTELL AND OTHERS (1986). Uranium in its reduced 4-valent state has a very low solubility, similar to that of thorium. The $^{230}Th/^{234}U$ activity ratio may approach one in extremely reducing groundwaters, whereas under more oxidizing conditions, it usually will be well below one because of the higher abundance of the soluble hexavalent uranium species. There is, in fact, a reasonable correlation (r = 0.86) between the equivalent ^{234}U content (ppb U · $^{234}U/^{238}U$ activity ratio) and the $^{234}U/^{230}Th$ activity ratio for 65 groundwaters of Tables 7.3.1 through 7.3.7, indicating a rather constant ^{230}Th content.

In addition to the redox state and the chemical composition of the groundwater, other parameters may strongly influence its uranium content. These include the availability of uranium in the rock-matrix and the presence of strongly sorbing phases such as iron hydroxides. Therefore, oxidizing groundwaters with pE values > -2.5 may show a wide range of uranium concentrations.

7.3.3.2   Activity Ratios

The most useful activity ratio for groundwater is $^{234}U/^{238}U$ because it will not be influenced by fractionation during chemical reactions or sorption. Other activity ratios such as $^{230}Th/^{234}U$ or $^{226}Ra/^{230}Th$ are sensitive to such processes. They are more difficult to interpret, but may be useful as additional indicators. Most groundwaters have $^{234}U/^{238}U$ activity ratios greater than one, and up to about ten or even higher. The ^{234}U excess depends on the time of water-rock interaction, the concentration, chemical state and distribution of uranium in the aquifer rock, and on the contact surface and chemical properties of the reacting groundwater.

Weathering and fast extraction of uranium-containing rocks under oxidizing conditions, as found at some places in the Black Forest crystalline, will lead to both high uranium contents and to activity ratios usually around one, which are close to those of the solid phases in the fractures of the rock (HOFMANN, 1989).

Slowly moving (quasi-stagnant) porewater in a uranium-containing aquifer rock may develop high activity ratios. The time t needed to reach an activity ratio AR, starting from an initial activity ratio AR_i may be estimated from an equation given by ANDREWS AND OTHERS (1982):

$$AR = 1 + (AR_i - 1) \exp(-\lambda_4 t) + 0.235\, \rho SR\, [(1 - \exp(-\lambda_4 t)](U_r)/(U_s) \quad (7.3.2)$$

where:

AR_i	initial state of disequilibrium
λ_4	decay constant of ^{234}U ($2.77 \cdot 10^{-6} a^{-1}$)
t	residence time (a)
0.235	recoil escape probability
ρ	dry density of aquifer material (g/cm^{-3}) (2.7 for crystalline rocks)
S	contact surface of rock with 1 cm^3 of porewater (cm^2/cm^3)(typically $\approx 10^4$ cm^{-1})
R	recoil distance of ^{234}Th in light materials ($\approx 3 \cdot 10^{-6}$ cm)
$(U_r),(U_s)$	uranium contents of rock available to exchange and of solution, respectively (typically $U_r/U_s \approx 1000$)

Equation 7.3.2 applies if ^{234}U enrichment is caused solely by α recoil of ^{234}Th into the porewater. Inserting $AR_i = 1$ and the typical data for crystalline rocks as given above into equation 7.3.2 leads to the following "ages":

AR:	1.5	2	3	4	5.6
t(a):	10^4	$2 \cdot 10^4$	$4 \cdot 10^4$	$6 \cdot 10^4$	10^5

The critical variable in equation 7.3.2 is $S \cdot (U_r/U_s)$ which equals 10^7 cm^{-1} for the data given above. This product varies greatly according to the crystalline aquifer considered. A higher value, e.g., 10^8 cm^{-1}, would lead to ages which are smaller by about a factor of ten. If preferential leaching of ^{234}U is involved, as is usual, the activity ratio may rise faster, but not to such high final values. An example of such an augmenting system is groundwater slowly moving through fractured granite at intermediate redox conditions. On the other hand, if reducing groundwater moves into an aquifer rock with low uranium content, the excess ^{234}U will decay according to its half-life of $2.5 \cdot 10^5$a and the activity ratio AR according to equation 7.3.1 or in this case

$$AR = 1 + (Ar_i - 1) \exp(-\lambda_4 t) \qquad (7.3.1a)$$

Decaying systems of this type are found in deep aquifers exhibiting reducing conditions and very low solubility of uranium (OSMOND AND COWART, 1982).

Modelling uranium content and activity ratios in groundwater has been attempted for various situations, most successfully for roll-front uranium ore deposits in large sandstone aquifers (*e.g.,* OSMOND AND OTHERS, 1983; ANDREWS AND PEARSON, 1984). Here, the relative clear situation of a well-defined uranium source in a large and rather homogenous aquifer allows the investigation of groundwater movement and groundwater ages by following the decay of ^{234}U along the flow path. In more complicated situations, without a known localized uranium input, such as in the formations of northern Switzerland, one might expect at best some qualitative information on the prevailing groundwater situation. Thus, high activity ratios, especially in connection with high uranium contents, indicate long-lasting and intimate contact of groundwater with uranium-bearing rock formations. Low activity ratios (close to one) in high-uranium waters are typical of young groundwaters in recharge zones, and low activity ratios in connection with very low uranium content may represent old, reducing groundwaters in decaying systems.

In reducing systems, no correlation apparently exists between the uranium content of groundwater and that of aquifer rock. In oxidizing systems, however, such a correlation can be found under certain conditions.

The ^{222}Rn activity is useful as a qualitative clue to the short-term history of a groundwater in the vicinity of the sampling point. Since radon is highly soluble in water and is not adsorbed by any solid, its activity is usually greater than that of uranium or ^{226}Ra by a factor of typically 10^3 to 10^5. Thus, high ^{222}Rn contents are typical for water in porous aquifers with high contents of finely dispersed uranium (or ^{226}Ra) in contact with the pore water, and also for a fast transfer of such pore water to the groundwater at the sampling point. The behaviour and significance of radon in groundwater is discussed by ANDREWS AND WOOD (1972), ANDREWS AND LEE (1979) and by CUTTELL AND OTHERS (1986).

Finally, the uranium content and the ^{234}U/^{238}U activity ratio may also be considered as natural tracers in groundwater mixing problems. However, this application is restricted to those cases in which large anomalies exist, and fast flow through fractures can be assumed.

Figure 7.3.2 through 7.3.5 are graphs of U content against ^{234}U/^{238}U activity ratio for the various aquifers. When Harwell (h) and PSI (e) reported very different values for the U

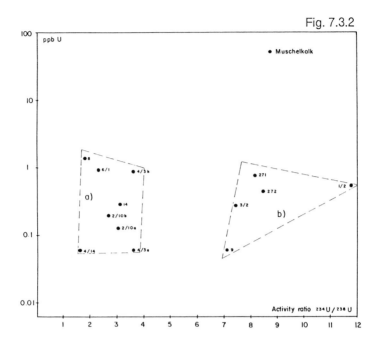

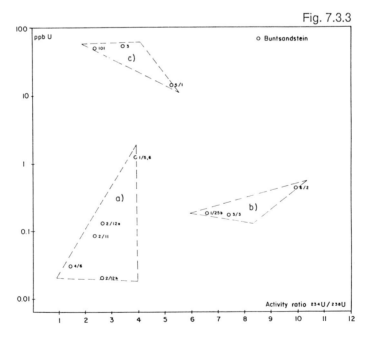

Figure 7.3.2 to 7.3.5: Graphs of uranium content and $^{234}U/^{238}U$ activity ratio of groundwater samples from Nagra boreholes. (Page 1 of 2)

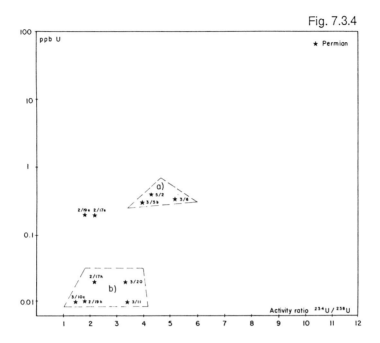

Fig. 7.3.4

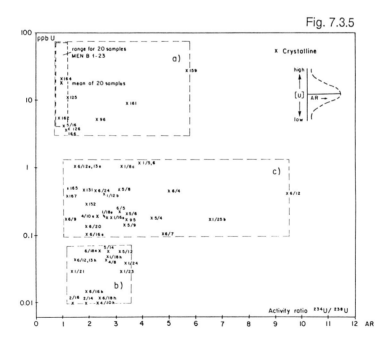

Fig. 7.3.5

Figure 7.3.2 to 7.3.5: Graphs of uranium content and $^{234}U/^{238}U$ activity ratio of groundwater samples from Nagra boreholes. (Page 2 of 2)

content of samples from the same source, both analyses are given. In all other cases the average of both laboratories was used based on the data in Tables 7.3.1 through 7.3.7.

7.3.3.3 Water from the Tertiary, Malm and Keuper

There are five samples from four locations: Weiach (302/6 and 8), Riniken (303/1), Schafisheim (304/2), and Meltingen (19).

Uranium contents range from 0.02 to 1.2 ppb, activity ratios are rather low (2 to 2.6) except sample 303/1 with a ratio of 5.7 indicating a different age, either higher or lower. However, there is hardly any connection between these few samples from the various upper aquifers, and so they are not considered in the figures. Together with many other samples from the regional hydrochemical programme, they are discussed by SCHMASSMANN (in preparation).

7.3.3.4 Water from the Muschelkalk

Twelve samples from ten locations are shown in Figure 7.3.2. They have low to medium uranium contents (0.06 to 1.4 ppb) and, as in the overlying aquifers, no spatial distribution pattern. Based on activity ratios, two types of groundwaters may be distinguished:

a) A type with low to medium AR's (1.6 to 3.6), including Weiach (302/10a and 10b), Schafisheim (304/3 and 14), Leuggern (306/1), Pratteln (8), and Densbüren (14);

b) A type with high AR's (7.1 to 11.8), including Böttstein (301/2), Riniken (303/2), Frenkendorf (9), and Magden Weiere (271) and Magden Falke (272).

The low AR's suggest younger groundwaters--with the possible exception of the Schafisheim samples with their low uranium contents--because the moderate uranium contents of these samples indicate some uranium exchange during rock-water interaction. Therefore, they represent augmenting rather than decaying activity ratios.

7.3.3.5 Water from the Buntsandstein

Three types of groundwater may be distinguished as shown in Figure 7.3.3:

The seven waters of low to moderate uranium contents (0.03 to 1.23 ppb) show an activity ratio pattern similar to that in the Muschelkalk.

a) Waters with low to moderate AR's (1.4 to 3.9), including Böttstein (301/5 and 6), Weiach (302/11 and 12), and Schafisheim (304/6);

b) Waters with high AR's (5.3 to 9.5) from Böttstein (301/25b), Riniken (303/3), and Leuggern (306/2).

Of the low AR waters, Böttstein (301/5 and 6) with 1.5 ppb U may also belong to an augmenting system but from another flow path than the high AR sample 301/25b. The very low uranium contents of the samples from Weiach and Schafisheim, however, rather suggest older groundwaters of the decaying AR type;

c) In a third type of Buntsandstein groundwaters, high uranium contents of 14.5 to 57 ppb indicate oxidizing conditions. The three samples involved are from Kaisten (305/1), Grenzach (101) and Kaiseraugst (3). They are located north of the low-uranium waters (see Figure 7.3.6) and seem to be connected to the uranium-rich Permian and Black Forest crystalline as recharge aquifers. The observed activity ratios of 2.4 to 5.3 probably correspond to increasing travel and residence times. According to equation 7.3.2, the time needed to reach a certain AR value increases with increasing uranium content of the groundwater $[U_s]$, thus indicating high ages for these high uranium waters.

7.3.3.6 Water from the Permian

The eight permian groundwaters from three deep boreholes shown in Figure 7.3.4 may be separated into two groups:

a) The three samples from the uppermost 200 m of the Permian, Riniken (303/5b and 6); Kaisten (305/2) have intermediate uranium contents (0.31 to 0.4 ppb) and activity ratios (4 to 5.2). They are in the range of the intermediate crystalline groundwaters. In fact, the Kaisten sample (305/2) shows characteristics typical of these crystalline groundwaters. The moderate U content and the very low $^{230}Th/^{234}U$ AR indicate intermediate redox conditions;

b) The second group comprises five samples from the Permo-Carboniferous Trough from Riniken (303/20, 10c and 11), and from Weiach (302/19 and 17). Their very low U contents of 0.01 to 0.02 ppb (neglecting the higher PSI values) and the high $^{230}Th/^{234}U$ activity ratio (0.14 to 0.87) suggest strongly reducing conditions comparable to the deepest crystalline groundwaters. The low activity ratios (1.4 to 3.4) indicate high ages in decaying systems which must be assumed under these reducing conditions.

7.3.3.7 Water from the Crystalline

The 48 crystalline groundwaters in Figure 7.3.5 show the typical distribution described by OSMOND AND COWART (1976), with low AR's at very low and very high U concentrations and a peak in the AR values at uranium concentrations between 0.1 and 1 ppb. In general, moving from low to high AR's on the high uranium side of the peak means longer water-rock interaction time (augmenting systems).

On the low uranium side of the peak, which represents reducing conditions in the deep crystalline, activity ratios usually are found to decrease with time (decaying systems).

For intermediate U concentrations in the range of about 0.1 to 1 ppb, the meaning of low activity ratios is ambiguous. Whether they represent young or rather old groundwaters must be decided from other considerations.

a) High-uranium waters:

They are found only in the crystalline of the Black Forest and somewhat to the south and along the Rhine as indicated in Figure 7.3.6.

The situation in a uranium-rich recharge zone of the Black Forest crystalline has been investigated by HOFMANN (1989) and DEARLOVE (1989) at the Krunkelbach pilot uranium mine at Menzenschwand (162, 164, and MENB, Table 7.3.7). Groundwaters to a depth of about 300 m have high U contents (4 to 82 ppb and up to 3000 ppb in the ore zone) and low activity ratios (0.79 to 1.17). Ratios below one arise from fast dissolution of U minerals, which were already depleted in ^{234}U by former leaching processes. Varying high tritium and dissolved oxygen contents of all mine waters indicate mixtures of recent precipitation and older waters of the time before 1953. Age investigations of secondary uranyl-minerals along the water paths in the mine show that oxidizing conditions and uranium dissolution have persisted for some 10^7 years. The total loss of uranium during this time and under these very unfavourable conditions is estimated to be below ten per cent (HOFMANN, 1989). In this connection, it is interesting to note that practically no uranium enrichment was found in well, surface and drinking water from the vicinity of the mine for the period before 1972, when excavation of the mine started and the hydrological situation was not yet disturbed. According to FAUTH AND OTHERS (1985), the uranium contents of surface waters in the Black Forest are generally low, and do not show signs of near-surface uranium enrichments in the crystalline. This may indicate that uranium-containing groundwater from the mine, and probably also from other recharge areas in the southern Black Forest, is flowing rather quickly to deeper regions of the crystalline and then southward in direction of the Rhine valley. In fact, several high uranium crystalline groundwaters with 2 to 29 ppb were found along the Rhine, including Rheinfelden (159), Säckingen (D) (125, 126 and 161), Kaisten (305/16) and possibly Leuggern (306/11b, 12 and 13), if the high PSI values for the U contents are accepted. The three high uranium samples from the Buntsandstein mentioned in Section 7.3.3.5 may also be of Black Forest origin.

Activity ratios of these high uranium crystalline waters are around one for the recharge zone and may increase during flow in the high uranium granite to values of 3.5, as observed at Säckingen (161), and up to 5.7, at Rheinfelden (159). The wide range of activity ratios (1.2 to 5.7) found in these high-uranium groundwaters from the crystalline along the Rhine valley indicate a complex local pattern of recharge areas and ages.

At present, there is no explanation for the high U content of the deep sample from the crystalline at Kaisten (305/16) (Table 7.3.5). The high uranium content of this sample suggests at least moderately oxidizing conditions, while its high

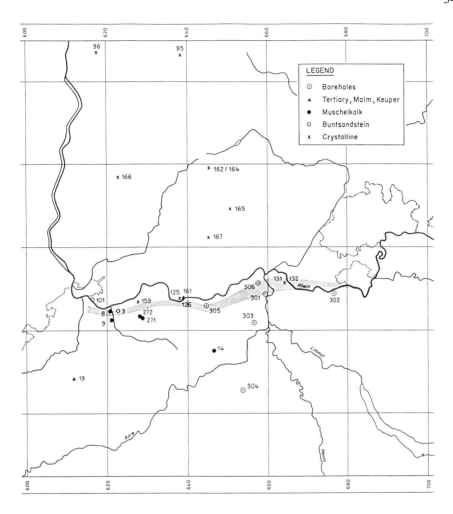

Figure 7.3.6: Location of uranium samples and of transition zone (shaded) between high-uranium, oxidising waters and low-uranium, reducing waters in the Buntsandstein and Crystalline.

^{230}Th/^{234}U AR of 0.41 either indicates strongly reducing conditions or an anomalous high ^{230}Th content. The uranium AR of 1.03 could signify either recharge conditions, or the final state in a decaying system. For the latter, however, the U content of ≈ 5 ppb seems to be much too high. The gneisses of this horizon have a normal U content around 5 ppm and some indication of recent uranium deposition from the groundwater (see Table 7.3.12). The ambiguous groundwater characteristics could possibly be explained by a mixture of very young and very old water.

b) Low-uranium waters:

In Figure 7.3.5 there are 13 groundwaters from five boreholes with uranium contents below 0.1 ppb (0.01 to 0.06 ppb), if one accepts the lower Harwell values. These are from Böttstein (301/18), Schafisheim (304/10) and Leuggern (306/12, 13 and 16). They belong to the deep crystalline and show low U contents and--with a few exceptions--high $^{230}Th/^{234}U$ activity ratios (average 0.34), indicating strongly reducing conditions. The low to medium activity ratios (1.4 to 3.4) probably correspond to different stages in decaying systems according to equation 7.3.1a.

c) Intermediate uranium waters:

Most of the 24 intermediate waters in Figure 7.3.5 with uranium contents between 0.1 and 1.5 ppb belong to the upper crystalline, and show activity ratios between 1.1 and 9.5. AR values above three are generally found in the uppermost 500 m of the crystalline. The highest values, 6.6 and 9.5, are in samples from the contact with the Buntsandstein from Böttstein (301/25b), and Leuggern (306/2).

These medium to high AR waters of the uppermost crystalline are probably connected to the augmenting groundwater system of the Black Forest, which also leads to the high-uranium waters along the Rhine valley. The low-AR, intermediate waters (AR < 3) are related more to the low-uranium waters of the deep crystalline, however, because they are somewhat less reducing (AR $^{230}Th/^{234}U$ << 1). The data will probably not allow the extraction of more information on the origin and the history of these waters.

7.3.3.8 Conclusions

The deeper aquifers, particularly the Buntsandstein and crystalline, show zones of high- and low-uranium groundwaters. High-uranium groundwaters are typical of the oxidizing recharge zone in the southern Black Forest crystalline and flow south. As they do so--in common with most groundwaters--they become reducing. This leads to the lower uranium contents observed south of a transition zone along the Rhine, shown as a shaded region in Figure 7.3.6.

The uranium activity ratio in the oxidizing zone increases with the age of the groundwater (augmenting system), and high AR's are found along the transition zone in the Rhine valley. In the low uranium reducing zone, the AR may decrease again by the decay of the surplus ^{234}U.

The groundwaters from the upper aquifers, the Muschelkalk, Keuper, and Tertiary, don't show this north/south decrease of uranium content.

7.3.4　　　Interpretation of Rock Sample Analyses

In the search for suitable locations for a final radioactive waste repository in geologic formations, it is important to know the extent to which the potential host rock has been exposed to circulating groundwater within the last 10^6 to 10^7 years. The study of natural decay-series disequilibria in the rock may give an answer to this question, as first proposed by SCHWARCZ AND OTHERS (1982). Thus, any disequilibrium (activity ratio AR ≠ 1) would indicate interaction of the rock with flowing water, with the maximum time elapsed since such an event roughly being given by about five half-lives of the daughter nuclide.

In general, such disequilibria indicate flow of groundwater in nearby water conducting structures from where the solutes diffuse into the surrounding rock matrix. Besides indicating some nearby water flow, this matrix diffusion may, however, also have a favourable effect of radionuclide retardation along the main flow channel (NERETNIEKS, 1980).

Usually only isotopes of uranium, protactinium, thorium and radium with half-lives ranging from $2.5 \cdot 10^5$ (^{234}U) to 1.9 years (^{228}Th) are considered. In principle, these natural clocks would allow the approximate dating of the geochemical disturbance. Table 7.3.8 summarizes the principal processes which will lead to the observed disequilibria.

Table 7.3.8:　Natural decay-series disequilibria in rocks useful for single event dating in rock-water interaction.

Activity ratio AR	Half-life of daughter, years	Single Event AR > 1	Single Event AR < 1	Maximum elapsed time ≈ 5 half-lives, years
^{234}U/^{238}U	245,000	accumulation of U U from solutions with excess ^{234}U	slow loss of U, interaction with large amounts of water	1,200,000
^{230}Th/^{234}U	75,400	loss of U	accumulation of U	380,000
^{231}Pa/^{235}U	32,800	loss of U	accumulation of U	160,000
^{228}Th/^{232}Th	1.91	accumulation of ^{228}Ra	loss of ^{228}Ra	30
(^{228}Ra	5.75)			
^{226}Ra/^{234}U	1,600	loss of U, accumulation of Ra	accumulation of U, loss of Ra	8,000

Under closed-system equilibrium conditions there may also exist a large elemental or isotopic fractionation between the rock matrix and the stagnant pore water. However, since the disequilibria in the water and the rock phase must compensate each other, and the inventory of decay series nuclides is much higher in the rock (typically by a factor of 10^5), the resulting disequilibrium in the rock usually remains far below experimental detectability. Measurable disequilibria in rocks thus clearly indicate interaction with flowing groundwater. Such disequilibria is usually much smaller than that found in groundwater, primarily because whole-rock analysis is applied, which always involves varying percentages of uranium and thorium that did not interact with the pore water.

The measurements of uranium and thorium concentrations and of the various activity ratios in rock samples from the six deep boreholes are given in Tables 7.3.9 through 7.3.13 at the end of Section 7.3. They were performed at Harwell using the methods of IVANOVICH AND WILKINS (1984), and IVANOVICH AND OTHERS (1986). Short descriptions of the samples by PETERS, MATTER AND GAUTSCHI (private communication) are included. The Nagra Technical Reports, referred to in Tables 7.3.9 through 7.3.13, give more details on the geology and mineralogy of the boreholes, and additional data on U and Th contents from borehole gamma-logs and gamma spectrometry of selected rock samples.

The following activity ratios were determined: $^{234}U/^{238}U$, $^{230}Th/^{234}U$, $^{230}Th/^{232}Th$, $^{238}U/^{232}Th$, $^{228}Th/^{232}Th$, $^{231}Pa/^{235}U$ and, for a few Böttstein samples, also $^{226}Ra/^{230}Th$. In the last column, the groundwater samples as given in Tables 7.3.1 through 7.3.6, are indicated, to which the rock samples could possibly be related.

From the activity ratios in Tables 7.3.9 through 7.3.13, absolute activities can be calculated by using specific activities of ^{238}U and ^{232}Th and their measured concentration (ppm):

$$1 \text{ ppm } ^{238}U = 12.44 \text{ Bq/kg} = 0.746 \text{ dpm/g}$$

$$1 \text{ ppm } ^{232}Th = 4.07 \text{ Bq/kg} = 0.244 \text{ dpm/g}$$

The activity ratios in Tables 7.3.9 through 7.3.13 provide site-specific information on geochemical events involving mobilization and deposition of decay-series elements during the last $\approx 10^6$ years as sketched in Table 7.3.8.

For example, ^{230}Th is much less soluble in normal oxidizing groundwater than its parent ^{234}U, leading to a $^{230}Th/^{234}U$ activity ratio (AR) above one in the rock. Assuming an end of this geochemical disturbance at t = 0, the activity ratio AR will decrease to one as a function of time t according to equation 7.3.1, or in this special case, according to

$$AR_t = 1 + (AR_o - 1) \exp(-\lambda_{230} t) \tag{7.3.1b}$$

The half-life of ^{230}Th is $7.52 \cdot 10^4$ a, so $\lambda_{230} = 9.22 \cdot 10^{-6}$ a^{-1}.

Equation 7.3.1b is valid only if the ^{234}U is in equilibrium with the parent ^{238}U. For the rather rare cases of considerable $^{234}U/^{238}U$ disequilibrium in the rock, more complicated relations such as given by SCHWARCZ AND OTHERS (1982) should be used. In any case, after some five half-lives of undisturbed rock history ($\approx$ 380,000 years for ^{230}Th), secular equilibrium will be restored to 97 per cent according to equation 7.3.1, and thus, within the limits of usual experimental accuracy.

Thus, activity ratios, as given in Tables 7.3.9 through 7.3.13, may allow a crude estimate of the periods of time between which certain processes of nuclide mobilization or accumulation did occur. More quantitative interpretations of this kind, however, must be made with caution because the simple assumption of a single fractionating event may be wrong, and because it will usually be difficult to distinguish between old but intense, and young but only mild rock-water interaction processes. The degree of rock alteration may give some additional information. Thus, one can be sure that in a strongly altered rock showing no radioactive disequilibrium, the transition took place more than a million years ago. Disequilibria can be interpreted in terms of loss or accumulation of the nuclides in question, and the maximum age of such processes can be estimated as shown in Table 7.3.8. Quantitative relations and model calculations based on disequilibria in rocks are given by THIEL AND OTHERS (1983).

7.3.4.1 Lias

The only sample, WEI 679.57 m, is from a belemnite layer also showing a pronounced anomaly in the gamma-log. The high U content (29.8 ppm) and the high activity ratios ^{234}U/^{238}U (1.28) and ^{230}Th/^{234}U (1.40) indicate uranium uptake and loss during the last $\approx 5 \cdot 10^5$a with a complex multi-event geochemical history. The uranium enrichment is probably associated with phosphorite and glauconite nodules and is limited to the narrow belemnite band. The uranium uptake from groundwater with excess ^{234}U was probably terminated some 10^5 years ago according to the equilibrium activity ratio of ^{231}Pa/^{235}U (AR = 1.0). The apparent excess of ^{228}Th with respect to its grandparent ^{232}Th indicates current ^{228}Ra (or ^{228}Th) uptake from local groundwater.

Testing for groundwater was done between 692.0 to 706.0 m with the large volume DST apparatus but without success. This indicates only limited groundwater circulation.

7.3.4.2 Muschelkalk

BOE 138.15, 187.60 and 191.60 m; WEI 848.60 m; RIN 626.77 m; SHA 1227.90 m; LEU 52.10 m

The seven samples contain moderate amounts of uranium (0.5 to 3 ppm) and low amounts of thorium (0.1 to 0.6 ppm). The four shallow samples, BOE 138.15 m, 187.60 m and 191.60 m, and LEU 52.10 m show ^{234}U/^{238}U ratios > 1 and ^{230}Th/^{234}U ratios of 1.0 indicating an old uranium accumulation some $3 \cdot 10^5$ to 10^6 years ago. This is supported by the equilibrium AR of ^{231}Pa/^{235}U for the three Böttstein samples. Current interaction with water and loss of radium is evident from ^{228}Th/^{232}Th AR (= 0.57) for the shallow sample, BOE 138.15 m.

The three deep samples, WEI 848.60 m, RIN 626.77 m, and SHA 1227.90 m show all activity ratios to be in equilibrium, indicating long-term geochemical stability (= 10^6a) of these Muschelkalk rocks.

7.3.4.3   Buntsandstein

BOE 311.87 m; WEI 982.28, 984.10 and 987.00 m; RIN 797.14 m; SHA 1481.83 m; KAI 107.40 m; LEU 213.95 m

Concentration ranges for the eight samples are 0.8 to 5 ppm U and 0.1 to 18 ppm Th. The activity ratios $^{234}U/^{238}U$, $^{230}Th/^{234}U$, $^{231}Pa/^{235}U$, $^{238}Th/^{232}Th$ for all samples except BOE 311.87 m and SHA 1481.83 m equal one within 2 σ, suggesting that the formation remained undisturbed for at least $30 \cdot 10^4$a. BOE 311.87 m shows an AR $^{234}U/^{238}U$ of 1.13 ± 0.06 very similar to the Muschelkalk samples and also interpretable as old uranium accumulation. The SHA 1481.83 m sample has yielded diverse results from two analyses which may be due to a high inhomogeneity in this porous sandstone. Such local inhomogeneities even over a few millimetres are known to exist, in reduction haloes of Permian redbeds from the Kaisten and Riniken deep borehole, for example, (HOFMANN AND OTHERS, 1987). The groundwaters from Böttstein, Leuggern, Riniken and Kaisten, sampled at or near these horizons, all have high $^{234}U/^{238}U$ activity ratios, which probably have been acquired in the Black Forest crystalline.

7.3.4.4   Permian and Carboniferous

WEI 1116.0, 1404.79 and 1590.00 m; RIN 1710.20 m; KAI 285.78 m

The concentration ranges for the five samples are U 1.2 to 15.9 ppm, Th 2.2 to 22.7 ppm. All $^{234}U/^{238}U$ and $^{231}Pa/^{235}U$ activity ratios equal one within 2 σ. The KAI 285.78 m sample exhibits equilibrium (AR = 1 within 2 σ) for all relevant activity ratios, indicating long-term geochemical stability even in groundwater of medium U content and thus, only moderate reducing properties (sample 305/2).

A very small loss of ^{228}Ra may exist in WEI 1404.79 m. The WEI 1116.10 and 1590.00 m (coal) and RIN 1710.20 m samples from the Permo-Carboniferous Trough show $^{230}Th/^{234}U$ activity ratios below one, indicating accumulation of uranium under the prevailing strongly reducing conditions. The equilibrium found for $^{231}Pa/^{235}U$ suggests an age of $(15 \text{ to } 20) \cdot 10^4$a for this uranium accumulation. Since the $^{234}U/^{238}U$ ratios in the three rock samples are one, the uranium assimilated from the groundwater could not have had an appreciable excess of ^{234}U. In fact, the $^{234}U/^{238}U$ activity ratios in the present groundwaters 302/19 and 17 and 303/10c from these horizons are low, between 1.4 and 2.2 (see Section 7.3.3.4).

7.3.4.5 Crystalline

BOE 316.30, 481.92, 492.00, 671.30 and 696.14 m; WEI 2220.12 m; SHA 1566.24 m; KAI 478.47, 488.88, 819.67, 1038.83, 1046.36 and 1273.94 m; LEU 539.76, 701.68, 916.87, 1197.62, 1525.97, 1636.90 and 1648.27 m

The thorium content of the 20 samples of crystalline basement rock is relatively constant, and of the order of 10 to 14 ppm. Exceptions with very low thorium contents are from LEU 539.76 m (0.69 ppm) and 701.68 m (2.29 ppm), while granites with higher contents are from LEU 1525.97 m (21.74 ppm) and SHA 1566.24 m (18.8 ppm). The uranium content is less uniform, reflecting the higher mobility of uranium. On average, the uranium content is of the order of 2 to 3 ppm. Samples depleted in uranium and thorium are from LEU 539.76 m (1.85 ppm) and 701.68 m (1.34 ppm). Rather high uranium and thorium contents occur in the fresh granites LEU 1525.97 m (10.37 ppm) and 1636.90 m (8.01 ppm). An interesting contrast between uranium and thorium mobilities can be observed in the samples from LEU 1636.90 m (fresh granite) and 1648.27 m (altered granite). The thorium content of both samples is about 14 ppm, but the uranium content of 8 ppm in the fresh granite decreases to 3.3 ppm in the altered sample. The latter rock contains argillaceous material and is located near a water-bearing zone. The details on the geology and mineralogy of the Leuggern drill core are described by PETERS AND OTHERS (1988b).

Most unaltered crystalline rock samples have $^{234}U/^{238}U$ and $^{230}Th/^{234}U$ activity ratios of unity within a 2 σ uncertainty. In contrast, most of the samples showing disequilibrium activity ratios show some kind of alteration such as the calcite crystals at SHA 1566.24 m, KAI 819.67 m, KAI 1273.94 m, and LEU 1197.62 m, argillaceous alterations near a water-bearing zone at LEU 1648.27 m, or other kinds of mineral alteration at BOE 316.30 m, BOE 481.92 m, BOE 492.00 m and BOE 696.14 m. All these samples have $^{230}Th/^{234}U$ activity ratios less than one, and most also have low $^{231}Pa/^{235}U$ ratios, indicating uranium accumulation within the last $\approx 20 \cdot 10^4$a. Further support is also given by some $^{234}U/^{238}U$ ratios being larger than one, reflecting the excess of ^{234}U usually present in the groundwaters. Most of the samples also show current uptake of ^{228}Ra because their $^{228}Th/^{232}Th$ activity ratio is larger than one. For the BOE 481.92 m sample, the $^{226}Ra/^{230}Th$ activity ratio was also measured. Its value, 1.97, indicates that radium accumulation has probably proceeded for many thousands of years. On the other hand, the weathered granites at BOE 316.30 m and LEU 539.76 m with $^{228}Th/^{232}Th$ activity ratios below one, seem to be losing some of their ^{228}Ra currently. With reference to the groundwaters circulating in the crystalline regions investigated, the following relations may be interesting:

- Crystalline rocks associated with reducing groundwaters with uranium contents below 0.1 ppb exhibit equilibrium activity ratios indicating minimal geochemical interaction between groundwater and aquifer rocks. This applies for the samples WEI 2220.12 m, KAI 1038.83 m and LEU 916.87 m, for example.

- For more oxidizing groundwaters with uranium contents > 0.1 ppb, two different situations may prevail:

a) If there is no visible calcite present in the crystalline rock, then radioactive equilibrium is generally maintained, as in the case of low uranium reducing groundwater;

b) When appreciable amounts of calcite are present, radioactive disequilibrium will be observed in those samples.

An interesting example for case b) is the sample from KAI 1273.94 m containing vugs filled with two generations of calcite. (This calcite will be discussed further).

7.3.4.6 Fracture Infillings and Minerals

BOE 501.90, 542.16, 618.70, 619.10, 654.12 and 902.69 m; SHA 1329.20 m; KAI 348.61 and 1273.94 m; LEU 704.44, 1648.83 (qz-ill) and 1648.83 m (barite)

There are 11 samples from the crystalline and one (SHA 1329.20 m) from the Muschelkalk.

Of the six fracture fillings from the Böttstein borehole, three show current ^{228}Ra accumulation, namely BOE 501.90, 618.70 and 619.10 m. Sample BOE 501.90 m also shows a pronounced loss of uranium in the last $1.5 \cdot 10^5$ years, although its equilibrium activity ratio for ^{234}U/^{238}U indicates some compensating accumulation of ^{234}U-enriched uranium within the last 10^6 years. The other three Böttstein samples indicate stable conditions for the last million years.

The pegmatite and granite of and near the fracture at BOE 618.70 m has been studied in detail by SMELLIE AND OTHERS (1986) and by ALEXANDER AND OTHERS (1988) with respect to element migration and matrix diffusion. Mobilization of uranium within the last $\approx 3 \cdot 10^5$ years and more recent uptake of ^{234}U and ^{230}Th have been found.

Of the six fracture fillings from three other boreholes, the calcite and barite crystals could, in principle, lead to more quantitative information (IVANOVICH AND OTHERS, 1986). Assuming the crystal growth to be a one-step irreversible process resulting in a geochemically closed solid, then the precipitation of these crystals can be dated by the ^{230}Th/^{234}U-dating method (IVANOVICH, 1982; SZABO AND KYSER, 1985; MILTON, 1986). Since the uranium in these samples is derived from the local groundwater which, under moderate redox conditions contains virtually no ^{230}Th relative to uranium, it can be assumed that the initial ^{230}Th content of the calcite is zero. With these presuppositions, the average growth age t of a crystal with a ^{230}Th/^{234}U activity ratio AR_{04} and a ^{234}U/^{238}U activity ratio $AR_{48} \approx 1$ is approximately given by

$$t \approx (1/\lambda_0) \ln 1/(1 - AR_{04}) = 1.09 \cdot 10^5 \ln 1/(1 - AR_{04}) \qquad (7.3.3)$$

$$(\lambda_0 = 9.22 \cdot 10^{-6} \text{ a}^{-1}: \text{decay constant of } ^{230}\text{Th})$$

The resulting ages for the calcite crystals at SHA 1329.20 m, with $AR_{04} = 0.90 \pm 0.09$, and at LEU 704.44 m, with $AR_{04} = 0.93 \pm 0.09$, are $2.5 \cdot 10^5$ and $2.9 \cdot 10^5$a, respectively. The near equilibrium values of AR_{04} and their errors, however, indicate that the calculated ages are very uncertain and should rather be interpreted as minimum ages. In fact, based on geologic reasoning (MATTER AND OTHERS, 1988b, PETERS AND OTHERS, 1988b) a much older age was inferred for these fillings. Their near equilibrium $^{234}U/^{238}U$ activity ratio would also be compatible with a higher age.

For the sample at KAI 348.60 m, consisting of coarse and almost perfect barite crystals with $AR_{04} = 0.75 \pm 0.09$ but showing a rather high $^{234}U/^{238}U$ activity ratio of $AR_{48} = 1.89 \pm 0.27$, a more complicated time equation must be applied:

$$e^{-\lambda_0 t} = (1 - AR_{04}) / (e^{\lambda_4 t} [1 - 1/AR_4] + 1/AR_{48}) \qquad (7.3.4)$$

($\lambda_4 = 2.83 \cdot 10^{-6}$ a^{-1}: decay constant of ^{234}U)

Equation 7.3.4 leads to an age of $(18.0 \pm 4.5) \cdot 10^4$a which is also compatible with geologic considerations (PETERS AND OTHERS, 1988a). Moreover, the pronounced uranium disequilibrium itself, with $AR_{48} = 1.89$, would limit the age of the barite crystals to some $3.0 \cdot 10^5$a according to equation 7.3.1a. If the initial ^{230}Th content was not zero, as might be indicated by the relatively high Th content of 0.22 ppm, then the age would be less than the calculated $18.0 \cdot 10^4$a.

The method is probably not applicable to the calcite at KAI 1273.94 m and the barite at LEU 1648.83 m for which the associated groundwaters have high $^{230}Th/^{234}U$ activity ratios, suggesting a non-zero initial ^{230}Th content in the crystals. Also, the AR_{04} values are too close to equilibrium.

If the assumptions of zero initial ^{230}Th and of closed geochemical system are not valid, only qualitative conclusions according to Table 7.3.8 are possible.

Efforts to date calcites and opals formed in fractured crystalline rocks have been reported by ZIELINSKI AND OTHERS (1986), SZABO AND KYSER (1985) and MILTON (1986).

7.3.4.7 Conclusions

The investigation of isotopic and elemental fractionation between various longer-lived daughter/parent pairs from the three natural radioactive decay series in whole-rock samples has proved to be useful in the chronology of geochemical events of the last million years. Fifty-three solid samples were selected from the six deep boreholes including 21 sedimentary rocks, 20 crystalline rocks and 12 secondary fracture fillings. About half of the samples of each category show some kind of radioactive disequilibria between ^{238}U, ^{234}U and ^{230}Th and between ^{235}U and ^{231}Pa, indicating transport processes from and to the rocks by groundwater in the last 10^6 years. In crystalline rocks, disequilibria are concentrated in strongly altered granites or those containing calcite. About one-third of all samples show disequilibrium between ^{232}Th and ^{228}Th, indicating very recent ongoing rock-water interaction with high mobility of ^{228}Ra. These samples are about evenly distributed between those showing or not showing long-term disequilibria. The interpretation of specific results may be severely disturbed by strong local inhomogeneities of uranium distribution. On the other hand, analytical results on natural uranium- and thorium-series nuclides in groundwater from corresponding horizons are helpful. In general, the information on the fractionating events, their age and duration remains qualitative. An exception may come from radiometric results on calcite, barite and possibly low-temperature fracture filling materials precipitated from the local groundwater. These materials have the potential for setting up time scales for rock-water interaction in fracture flow, and thus, indirectly the time scales relevant to the understanding of the geochemical stability of the surrounding rock. It is this surrounding rock that is the possible host rock for a radioactive waste repository.

Table 7.3.1: Groundwater Samples: Uranium- and thorium-natural series nuclides abundance data, Böttstein borehole.

Sample Nr. 301/	Depth av. m	For-ma-tion	Sampling method	Date of Sampling	Qual-ity Block	[U] µg/l PSI	[U] µg/l Harwell	234U/238U	230Th/234U AR Harwell	230Th/232Th	228Th/232Th	228Ra/226Ra	226Ra dpm/l Fres	228Ra dpm/l Har-well +Fres	222Rn Bq/l PSI	222Rn Bq/l Fres
2	162.9	mo	p.t	1.11.82(1)	1	0.55	0.5 ±0.01	11.8 ±0.22	0.014±0.001	1.6±0.3	- - -	0.38 ±0.01/ 0.30±0.02	13.3	4.4	18.5	15.5
25b	312.5	s/KRI	p.t	20.1.84(2)	2	0.2	0.1 ±0.008	6.59 ±0.27	0.00 ±0.002	>1000	- - -	- - -			25.9	18.1
5	316.6	s/KRI	p.t DST	16.11.82(1) 16.11.82(1)	2 2	0.9 1	1.5 ±0.04 1.4 ±0.04	3.92 ±0.08 3.89 ±0.09	0.11 ± 0.01 0.11 ± 0.01	1.6±0.1 1.2±0.1	- - -	0.3 ±0.01	2	0.7	29.6	21.5
6		(200)											2		29.6	
8c	399.5	KRI	art.	14.12.82(1)	1	1.1	1.1 ±0.02	3.26 ±0.05	0.10 ± 0.004	2.2±0.1	- - -		0.7		81.4	78.8
12b	621.3	KRI	art.	22.1.83(1)	1	0.45	0.4 ±0.01	2.59 ±0.07	0.06 ± 0.003	1.7±0.1	- - -		1.6	1.1	49.2	34.4
16	618.4	KRI	p.t.	9.8.83(2)	2	0.23	0.1 ±0.008	2.85 ±0.15	0.01 ± 0.003	3.0±0.2	- - -	0.74±0.01/ 0.75±0.06	0.9	0.7	31.1	27.8
23	649	KRI	p.t.	20.10.83(2)	2		0.0 ±0.003	3.21 ±0.34	0.03 ± 0.01	2.1±0.1		0.73±0.01				
24	649.1	KRI	p.t.	4.11.83(2)	2		0.0 ±0.004	3.39 ±0.28	0.02 ± 0.01	0.8±0.4		0.75±0.01				
18	792.4	KRI	p.t.	16.8.83(2)	2	0.19	0.0 ±0.005	2.69 ±0.28	0.03 ± 0.01	3.8±0.3		0.73±0.03	0.7	0.4	42.9	39.6
21	1326.2	KRI	Preus-sag p.v. (1300)	10.10.83(2)	2	0.05	0.0 ±0.006	1.36 ±0.51	1.33 ± 0.41	1.4±0.3						
df			tap water	14.12.82			0.6 ±0.01	1.47 ±0.04	0.33 ± 0.01	50.6±9.2	- - -	1.06 ±0.04				

art. artesian discharge
p.t. pumping test discharge
p.v. pressure vessel
df drilling fluid
(1) drilling phase
(2) test phase

AR = activity ratio
1 Bq/l = 60 dpm/l
1 dpm/l 238U ---> 1.34 µg/l
1 dpm/l 232Th ---> 4.09 µg/l
df drilling fluid

1 dpm/l 226Ra ---> 4.55 · 10⁻⁷ µg/l
1 dpm/l 228Ra ---> 1.65 · 10⁻⁹ µg/l

Table 7.3.2: Groundwater Samples: Uranium- and thorium-natural series nuclides abundance data, Welach borehole. (Page 1 of 2)

Sample Nr. 302/	Depth av. m	Formation	Sampling method (depth, m)	Date of Sampling	Quality Block	[U] µg/l PSI	[U] µg/l Harwell	234U/ 238U	230Th/ 234U	230Th/ 232Th AR Harwell	228Th/ 232Th	228Ra/ 226Ra	226Ra dpm/l Fres	228Ra dpm/l Harwell +Fres	222Rn Bq/l PSI	222Rn Bq/l Fres
5	255	jo	GTC p.v. (110)	05.03.83	2	0.12									<3.7	7.8
6			DST u.c. (220)	06.03.83	2	0.06	0.08±0.01	2.57±0.16	0.07±0.01	1.9±0.5						
8			Pengo-bailer (110)	06.03.83	2		0.08±0.01	1.96±0.26	0.06±0.01	1.6±0.4						1.5
10a	859.1	mo	p.t.	03.04.83	1		0.13±0.01	3.02±0.12	0.12±0.01	1.2±0.1		0.15±0.01				
10b			p.t.	04.04.83	1	0.17	0.24±0.01	2.68±0.09	0.03±0.003	2.2±0.4		0.12±0.01				
11	984.2	so	p.t.	01.05.83	2	0.11	0.06±0.00	2.37±0.19				0.15±0.01	1.6	0.2*	13.0	14.4
12	985.3	s	p.t.	18.- 19.07.83	1	0.13	0.02±0.00	2.64±0.44	0.11±0.02	2.7±1.1		0.56±0.01			14.8	13.0
13			GTC p.v. (600)	19.07.83	1	0.24						0.57±0.01				
19	1116.5	r	p.t.	28.06.84	1	0.20	0.01±0.003	1.83±0.66	0.68±0.19	6.9	>1000	3.68±0.01	4.64±0.08	17.1±0.6	7.4	
20			GTC p.v. (360)	29.06.84	1	0.20										
17	1408.3	r	Preus-sag p.v. (1385)	05.06.84	2	0.20	0.02±0.006	2.18±0.77	0.14±0.05	2.3	48				23.7	
18			Preus-sag p.v. (1385)	15.06.84	2										29.6	18.9

Table 7.3.2: Groundwater Samples: Uranium- and thorium-natural series nuclides abundance data, Welach borehole. (Page 2 of 2)

Sam-ple Nr. 302/	Depth av. m	For-ma-tion	Sam-pling method (depth, m)	Date of Sam-pling	Qual-ity Block	[U] µg/l PSI	[U] µg/l Harwell	$234U/238U$ AR Harwell	$230Th/234U$	$230Th/232Th$	$228Th/232Th$	$228Ra/226Ra$	$226Ra$ dpm/l Fres	$228Ra$ dpm/l Har-well +Fres	$222Rn$ Bq/l PSI	$222Rn$ Bq/l Fres
16	2218.1	KRI	Preus-sag p.v. (2190)	27.04.84	2	<0.05	0.01±0.002	1.36±0.49	0.50±0.21	1.2	318	– – – – –			22.2	19.2
14	2267	KRI	Preus-sag p.v. (2245)	04.04.84	2	0.05	0.01±0.004	1.90±0.69	0.46±0.18	1.2	247	– – – – –			81.4	66.6
15			schlum. p.v. (2245)	06.04.84	2	0.20										
df			tap water	06.03.83			1.07±0.02	1.16±0.02	0.01±0.002	5.5±2.2						

p.t. pumping test discharge
p.v. pressure vessel
u.c. upper chamber
l.c. lower chamber
df drilling fluid

AR = activity ratio
1 Bq/l = 60 dpm/l
1 dpm/l 238U ---> 1.34 µg/l
1 dpm/l 232Th ---> 4.09 µg/l
1 dpm/l 226Ra ---> 4.55 · 10−7 µg/l
1 dpm/l 228Ra ---> 1.65 · 10−9 µg/l
* Fresenius

Table 7.3.3: Groundwater Samples: Uranium- and thorium-natural series nuclides abundance data, Riniken borehole.

Sample Nr. 303/	Depth av. m	For-ma-tion	Sampling method (depth, m)	Qual-ity Sampling Block	[U] µg/l PSI	[U] µg/l Harwell	Date of Sampling	234U/238U	230Th/234U	230Th/232Th AR Harwell	228Th/232Th	228Ra/226Ra	226Ra dpm/l Harwell	228Ra dpm/l Harwell	222Rn Bq/l PSI	222Rn Bq/l Fres
1	515.7	km	p.t.	1	0.13	0.02±0.004	25.07.83	5.70±1.13	0.08±0.016	1.4±0.3	---	0.40±0.01			24.4	21.8
2	656.7	mo	p.t.	1	0.32	0.23±0.01	17.08.83	7.46±0.28	0.003±0.001	12.5±1.3	---	0.11±0.03			10.8	10.0
3	806.6	s-r	p.t.	1	0.2	0.13±0.01	16.09.83	7.41±0.35	0.01±0.001	7.2±0.8	---	0.61±0.03			10	10.7
4			GTC p.v. (300)	1	0.2		17.09.83									
5b	965.5	r	Nagra bailer (930)	1	---	0.31±0.01	04.10.83	3.95±0.12	0.02±0.002	1.2±0.1						23.3
6	993.5	r	p.t.	2	0.5	0.20±0.01	02.11.83	5.21±0.23	0.04±0.003	1.3±0.1		3.77±0.10	41.7±0.2	157.1±4.2	63.3	51.8
7			Preus-sag p.v. (900)	2	0.3		03.11.83									
8/9			GTC p.v. (500)	2	0.6		03.11.83									
20	1361.5	r	Preus-sag p.v. (1340)	2	---	0.02±0.003	15.07.85	3.30±0.68	0.87±0.14	3.1	410					1.1
10c	1709.3	r	Preus-sag p.v. (1625)	3	<0.5	0.01±0.004	27.07.84	1.44±0.61	0.59±0.21	1.5	369	28.3±0.3				5.0
11			(1630)	3		0.01±0.002	29.11.84	3.38±0.81	0.38±0.12	2.5	>1000					

art. = artesian discharge
p.t. = pumping test discharge
p.v. = pressure vessel

AR = activity ratio
1 Bq/l = 60 dpm/l
1 dpm/l 238U ---> 1.34 µg/l
1 dpm/l 232Th ---> 4.09 µg/l
1 dpm/l 226Ra ---> 4.55 · 10⁻⁷ µg/l
1 dpm/l 228Ra ---> 1.65 · 10⁻⁹ µg/l

Table 7.3.4: Groundwater Samples: Uranium- and thorium-natural series nuclides abundance data, Schafisheim borehole.

Sample Nr. 304	Depth av. / m	Formation	Sampling method depth (m)	Date of sampling	Quality Block	[U] µg/l PSI	[U] µg/l Harwell	$234U/238U$	$230Th/234U$ AR Harwell	$230Th/232Th$	$228Th/232Th$	$228Ra/226Ra$	$226Ra$ dpm/l Harwell	$228Ra$ dpm/l Harwell	$228Rn$ Bq/l PSI	$222Rn$ Bq/l Fres
1	558.0	tUSM	p.t.	20.12.83	1	0.40									12.6	14.8
2			GTC	20.12.83	1	0.70	0.05±0.004	2.67±0.23	0.07±0.02	3.0			24.5±0.2			
14	1251.1	mo	p.v. (420) p.t.	21.1.85	2		0.03±0.004	1.62±0.27	0.79±0.11	5.1	59	0.05±0.01	60.8±0.2	3.2±0.4	11.1	11.1
3	1260.4	mo	p.t.	17.2.84	1	0.06	0.90±0.06	3.61±0.11	0.01±0.004	3.0	54		70.0		11.1	7
5			GTC	1.4.84	2	0.06										
6			p.v. (410) Preus-sag	2.4.84	2	traces	0.03±0.004	1.43±0.25	0.14±0.05	2.0	89				99.9	113.6
7			p.v. (1465) GTC	2.4.84	2	0.20										
8	1571.1	KRI	p.v. (410) p.t.	2.5.84	2	0.05	0.04±0.004	2.64±0.34	0.25±0.04	1.6	550	4.13±0.07	8.99±0.07	37.1±0.7	103.6	63.3
9			GTC	2.5.84	2	traces										
10	1887.9	KRI	p.v. (430) p.t.	17.6.84	1	0.20	0.01±0.002	2.33±0.75	0.50±0.15	1.4	760	8.66±0.22	2.14±0.06	18.5±0.7	37.0	23.3
11			GTC	18.6.84	1	<0.05										
			p.v. (340)													

AR = activity ratio
p.t. pumping test discharge
p.v. pressure vessel

1 Bq/l = 60 cpm/l
1 dpm/l 238U ---> 1.34 µg/l
1 dpm/l 232Th ---> 4.09 µg/l
1 dpm/l 226Ra ---> 4.55 · 10⁻⁷ µg/l
1 dpm/l 228Ra ---> 1.65 · 10⁻⁹ µg/l

Table 7.3.5: Groundwater Samples: Uranium- and thorium-natural series nuclides abundance data, Kaisten borehole.

Sample Nr. 305/	Depth av. m	Sampling or-method	Date of Sampling	Quality Block	[U] µg/l PSI	[U] µg/l Harwell	$234U/238U$ AR	$230Th/234U$ AR	$230Th/232Th$ Harwell	$228Th/232Th$	$228Ra/226Ra$	$226Ra$ dpm/l Harwell	$228Ra$ dpm/l Harwell	$222Rn$ Bq/l PSI	$222Rn$ Bq/l Fres
1	113.5	s	22.2.84	1	15.1	13.94±0.56	5.29±0.14	0.001±0.0003	7.6	4.3		4.21±0.03		25.9	22.2
2	284.3	r	1.3.84	1		0.50±0.02	4.31±0.08	0.002±0.001	0.8	22	0.70±0.01	0.72±0.02	0.50±0.01	7.4	5.6
4	310.4	KRI	p.t. 15.3.84	1	0.20	0.18±0.02	4.39±0.32	0.01±0.003	>1000	>1000	1.55±0.03	0.78±0.02	1.21±0.03	18.5	14.8
5			GTC 16.3.84 p.v. (200)	1	0.20										
6	482.6	KRI	p.t. 3.4.84	1	0.30	0.14±0.01	3.43±0.19	0.01±0.005	>1000	>1000	1.56±0.03	0.84±0.01	1.31±0.01	33.3	23.7
7			GTC 4.4.84 p.v. (210)	1	0.10										
8	656.4	KRI	p.t. 23.4.84	3	0.60	0.41±0.02	3.15±0.08	0.01±0.001	11.8	245	0.82±0.02	2.42±0.04	1.99±0.06	214.6	149.1
9	819.4	KRI	p.t. 2.5.84	1	0.20	0.11±0.01	3.37±0.17	0.01±0.002	3.4	>1000	0.07±0.004	23.0±0.2	1.70±0.09		
11			GTC 3.5.84 p.v. (270)	1	0.10										
12	1031.0	KRI	art. 5.6.84	1	0.10	0.03±0.004	3.19±0.48	0.09±0.02	11.2	140	0.30±0.01	2.31±0.04	0.68±0.03	122	81.8
13			GTC 6.6.84 p.v. (300)	1	0.20										
14	1153.3	KRI	Preus- 13.8.84 sag	1		0.06±0.01	2.74±0.30	0.07±0.01	2.7	58		13.9±0.1			116
15			p.v. (1120) GTC 14.8.84 p.v. (410)	1	<0.05										
16	1271.9	KRI	p.t. 27.8.84	1	6.57	4.29±0.26	1.03±0.01	0.41±0.03	>1000	268		5.49±0.05		251.6	138
17			GTC 28.8.84 p.v. (410)	1	2.26							3.70±0.04			

art. = artesian discharge
p.t. = pumping test discharge
p.v. = pressure vessel

AR = activity ratio
1 Bq/l = 60 dm/l
1 dpm/l 238U ---> 1.34 µg/l
1 dpm/l 232Th ---> 4.09 µg/l
1 dpm/l 226Ra ---> 4.55 · 10^{-7} µg/l
1 dpm/l 228Ra ---> 1.55 · 10^{-9} µg/l

Table 7.3.6: Groundwater Samples: Uranium- and thorium-natural series nuclides abundance data, Leuggern borehole. (Page 1 of 2)

Sample Nr. 306/	Depth av. m	For-ma-tion	Sampling method (depth, m)	Date of Sampling Block	Qual-ity	[U] µg/l PSI	[U] µg/l Harwell	234U/ 238U	230Th/ 234U -- AR Harwell	230Th/ 232Th	228Th/ 232Th	228Ra/ 226Ra	226Ra dpm/l Harwell	228Ra dpm/l Harwell	222Rn Bq/l PSI	222Rn Bq/l Fres
1	74.9	mo	p.t.	18.7.84	1	1.00	0.93±0.03	2.32±0.04	0.004±0.001	2.5	75	---	10.4		51.8	41.1
2	217.9	s/	p.t.	8.8.84	2	0.50	0.37±0.02	9.47±0.28	0.002±0.0003	14.1	>1000	---	1.57±0.03		40.7	28.5
3		KRI	GTC p.v. (190)	8.8.84	2	0.50										
4	251.2	KRI	p.t.	14.8.84	1	0.50	0.43±0.01	5.05±0.11	0.02±0.001	3	1.6	0.88±0.04	0.80±0.03	0.70±0.05	66.6	41.8
5	444.2	KRI	p.t.	14.9.84	1	0.29	0.20±0.01	3.18±0.14	0.01±0.002	5.3	31	---	0.21±0.02		37.0	19.2
6			GTC a.v. (340)	15.9.84	1	0.15										
7	538.0	KRI	p.t.	27.9.84	1	0.11	0.11±0.01	4.80.±0.20	0.02±0.003	2.9	22	0.57±0.02	0.33±0.02	0.58±0.03	51.8	32.9
8			GTC p.v. (300)	28.9.84	1	0.07										
9	705.7	KRI	p.t	17.10.84	1	0.17	0.20±0.01	1.12±0.08	0.02±0.004	1.4	27		0.40±0.02		66.6	49
10			GTC p.v. (400)	18.10.84	1	0.07										
17	847.0	KRI	p.t.	25.3.85	2	0.05	0.012±0.002	2.42±0.45	0.26±0.06	5.9	29		0.48±0.01		40.7	24.4
18			p.t.	26.3.86	2	0.06										40.7	27
11b	923.0	KRI	Preus-sag p.c. (890)	28.11.84	2	2.10										
12/1			do (890)	30.11.84	2	2.03	0.04±0.01	1.30±0.23	0.32±0.07	0.9	2.8					
12/2*)			do (890)	30.11.84	2		0.05±0.01	1.63±0.25	0.50±0.08	1.1	2.1					

Table 7.3.6: Groundwater Samples: Uranium- and thorium-natural series nuclides abundance data, Leuggern borehole. (Page 2 of 2)

Sample Nr. 306/	Depth av. m	Formation	Sampling method (depth, m)	Date of Sampling	Quality Block	[U] µg/l PSI	[U] µg/l Harwell	234U/238U - - - AR Harwell	230Th/234U - - - AR Harwell	230Th/232Th - - -	228Th/232Th - - -	228Ra/226Ra - - -	226Ra dpm/l Harwell	228Ra dpm/l Harwell	222Rn Bq/l PSI	222Rn Bq/l Fres
13/1			do (885)	7.12.84	2	2.13	0.02±0.003	1.94±0.34	0.24±0.06	1.1	21					
13/2*)			do (885)	7.12.84	2		0.07±0.01	1.10±0.15	0.26±0.05	1.0	46					
20	1203.2	KRI	Preus-sag p.v. (1150)	22.4.85	2		0.14±0.01	1.90±0.15	0.031±0.007	3.2	---		0.14±0.01			9.3
21			GTC p.v. (470)	23.4.85	2	0.20										
24	1433.4	KRI	Nagra bailer (1405)	9.5.85	2		0.48±0.02	2.26±0.10	0.09±0.01	13.2	21	---	0.52±0.02			
26			Preus-sag p.v. (1405)	14.5.85	2										74.0	51.8
16	1643.4	KRI	p.t.	15.2.85	1	0.11	0.015±0.002	1.91±0.33	0.39±0.08	3.6	33	0.42±0.01	0.80±0.30	0.34±0.02	59.2	57.7

p.t. pumping test discharge
p.v. pressure vessel
*) duplicate U/Th series data

AR = activity ratio
1 Bq/l = 60 dpm/l
1 dpm/l 238U ---> 1.34 µg/l
1 dpm/l 232Th ---> 4.09 µg/l
1 dpm/l 226Ra ---> 4.55 · 10-7 µg/l
1 dpm/l 228Ra ---> 1.65 · 10-9 µg/l

Table. 7.3.7: Groundwater Samples: Uranium- and thorium-natural series nuclides abundance data from the Nagra regional hydrochemical programme. (Page 1 of 2)

Sample Nr.	Location	Formation	Date of Sampling	Quality Block	[U] µg/l PSI	[U] µg/l Harwell	234U/238U	230Th/234U AR Harwell	230Th/232Th	228Ra/226Ra	222Rn Bq/l PSI	222Rn Bq/l Fres
19	Meltingen 2	km	8.9.81	1	1.2		2.0 (PSI)				13.0	10.7
8/2	Pratteln	mo	7.2.84	1	1.4	1.45±0.05	1.83	0.001±0.0002	2		11.1	10.7
9/2	Frenkendorf	mo	8.2.84	1	<0.05	0.06±0.004	7.11±0.52	0.008±0.003	1.7		14.8	11.1
14/3	Densbüren Felsbohrung	mo	10.2.84	1	0.3	0.28±0.01	3.11±0.09	0.13±0.002	1.3		18.5	13.7
271/101	Magden Weiere	mo	14.9.83		0.5	0.96±0.03	8.16±0.12	0.00001	1.8		9.3	
272/102	Magden Falke	mo	25.11.85		0.5	0.40±0.02	8.48±0.27	0.008	26	0.092±0.004	7.4	
101/1	Grenzach 1	mu-so	25.11.81	1	51.2		2.4 (PSI)				26.3	21.5
3/3	Kaiseraugst	s-r	6.2.84	1	57.3	55.8±1.6	3.44±0.07	0.00001	3.1		88.8	73.3
125/11	Säckingen Badquelle	KRI	22.2.84	1	11.9		1.2 (PSI)				179	155.4
126/11	Säckingen Margarethen	KRI	22.2.84	1	4.0		1.2 (PSI)				55.5	41.4
161/1	Säckingen Stammelhof	KRI	22.2.84	1	9.4	9.55±0.42	3.44±0.11	0.002±0.0000	1000		44.4	39.2
131/5	Zurzach 1	KRI	17.8.82	1	0.50		1.8 (PSI)				18.5	18.1
132/5	Zurzach 2	KRI	17.8.82	1	0.30		1.9 (PSI)				18.5	15.9
159/4	Rheinfelden	KRI	19.12.83	1	29.2	28.43±1.06	5.68±0.1	0.00001	1000	0.63±0.01	44.6	---
162/1	Menzenschwand 1	KRI	19.9.85	1	5.7	5.81±0.09	0.88±0.01	0.144±0.005	16		547.6	347
164/1	Menzenschwand 2b	KRI	19.9.85	1		21.9±0.03	0.97±0.01	0.04±0.01	55.8		---	
MENB	Menzenschwand, U-pilot mine 1)	Average of 20 samples	25.-27.8.86		18.8	18.1	0.96	0.11	48		990	
	Ranges of 20 samples				4.4-78.4	4.4-82	0.79-1.17	0.003-0.54	5.5-170		662-1589	

Table. 7.3.7: Groundwater Samples: Uranium- and thorium-natural series nuclides abundance data from the Nagra regional hydrochemical programme. (Page 2 of 2)

Sample Nr.	Location	Formation	Date of Sampling	Quality Block	[U] µg/l PSI	[U] µg/l Harwell	234U/ 238U	230Th/ 234U AR Harwell	230Th/ 232Th	228Ra/ 226Ra	222Rn Bq/l PSI	222Rn Bq/l Fres
165/1	St. Blasien Erzgruben	KRI	19.9.85	1	0.5	0.53±0.02	1.19±0.05	0.05±0.001	3.4	- - -	118.4	100
166/1	Sulzburg Waldhotel	KRI	20.9.85	1	3.8	3.75±0.06	1.16±0.02	0.005±0.001	4.2		74	61.8
167/1	Görwihl Bohrloch 30	KRI	20.9.85	1	0.4	0.4±0.01	1.17±0.06	0.053±0.009	3.9		462.5	311
95/2	Waldkirch	KRI	11.4.84	1	0.2	0.16±0.01	3.43±0.2	0.012±0.003	3.6			2.5
96/2	Oberbergen	tvu	11.4.84	1	5.6	5.45±0.35	2.32±0.02	0.00001	1000		144.3	124
231/1	Amsteg	KRI	11.9.85	1	3.6	3.84±0.15	1.03±0.01	0.002±0.0001	1.5		74	46.6
232/1	Gotthardstrassent.	KRI	11.9.85	1	32.9	19.2±1.0	0.99±0.01	0.002±0.0002	1000		129.5	107
233/1	Grimsel Transitgas	KRI	12.9.85	1	72.8	224.9±10	1.043±0.004	0	1000		129.5	110

1) Hofmann (1989)
AR = activity ratio

1 Bq/l = 60 dpm/l
1 dpm/l ^{238}U ---> 1.34 µg/l
1 dpm/l ^{232}Th ---> 4.09 µg/l
1 dpm/l ^{226}Ra ---> 4.55 · 10^{-7} µg/l
1 dpm/l ^{228}Ra ---> 1.65 · 10^{-9} µg/l

Table 7.3.9: Rock Samples: Uranium- and thorium-natural series nuclides abundance data, Böttstein borehole.

Sample Depth m BOE	Formation	Description PETERS AND OTHERS (1986)	Corresp. groundwater 301/	[U] ppm	[Th] ppm	234U/ 238U	230Th/ 234U	230Th/ 232Th	228Th/ 232Th	231Pa/ 235U	226Ra/ 230Th
								AR Harwell*			
138.15	mo 3	Trigonodus-Dolomit, open fractures	2	1.54±0.05	0.25±0.20	1.12±0.05	0.95±0.05	20.05±15	0.57±0.17	0.98±0.1	--
187.60	mo 1	Trochitenkalk, open fractures	2	0.93±0.03	0.06±0.01	1.12±0.05	0.96±0.04	49.3±11	0.87±0.22	0.97±0.1	--
191.60	mm	Muschelkalk, pores from dissolved gypsum crystals	2	1.29±0.05	0.09±0.01	1.11±0.05	1.01±0.04	50.9±7.3	1.14±0.21	1.09±0.1	--
311.87	s	Buntsandstein	25b,5	0.84±0.04	0.07±0.01	1.13±0.06	1.04±0.06	46.3±8.6	1.94±0.48	1.03±0.1	--
316.30	KRI	weathered granite (1 m below Buntsandstein)	5	2.51±0.08	8.34±0.24	1.03±0.03	0.77±0.03	0.73±0.04	0.73±0.03	0.80±0.08	--
481.92	KRI	strongly altered granite		1.86±0.04	10.10±0.24	1.01±0.02	0.77±0.03	0.44±0.02	1.81±0.06	0.80±0.08	1.97±0.2
492.00	KRI	medium altered granite (Qz 33%, K-spar 28%, calcite 1%, clay min., kaol.)		3.83±0.07	13.05±0.28	1.01±0.02	0.92±0.03	0.82±0.04	1.70±0.05	0.96±0.1	1.15±0.12
671.30	KRI	very slightly altered granite (slight seritization of plagioclase)		6.76±0.10	8.34±0.32	0.98±0.01	1.01±0.02	1.24±0.03	1.58±0.04	1.04±0.1	1.22±0.12
696.14	KRI	Kakirite (Qz 50% K-spar 8%, clay min. 40% (illite + mixed layers)		11.9±0.18	15.66±0.32	1.00±0.01	0.42±0.01	0.98±0.03	1.59±0.05	0.44±0.05	1.31±0.13
501.90	KRI	fracture mat. chlorite + mixed layers		4.15±0.07	5.69±0.12	1.02±0.01	2.02±0.05	4.58±0.14	1.57±0.05	1.71±0.17	--
542.16	KRI	fracture mat. calcite + traces of ore minerals		11.6±0.25	15.46±0.45	1.00±0.02	1.00±0.03	2.30±0.10	0.98±0.04	1.00±0.1	--
618.70	KRI	fracture mat. open fractures in pegmatite calcite, quartz, clay minerals, apatite	12b,16	2.39±0.08	0.98±0.04	1.04±0.04	0.97±0.04	7.42±0.30	1.60±0.09	0.98±0.1	--
619.10	KRI	apatite with open fracture	12b,16	2.63±0.10	4.25±0.12	1.06±0.04	0.90±0.04	1.80±0.06	1.64±0.06	0.92±0.1	--
654.12	KRI	fract. mat. illite + serizite + Qz + calcite	23, 24	2.36±0.05	7.53±0.40	0.99±0.03	0.98±0.05	0.93±0.04	0.96±0.07	0.99±0.1	--
902.69	KRI	fract. mat. illite + serizite		2.45±0.10	9.86±0.25	1.08±0.05	0.90±0.05	0.74±0.02	1.03±0.04	0.93±0.1	--

* AR = activity ratio

369

Table 7.3.10: Rock Samples: Uranium- and thorium-natural series nuclides abundance data, Weiach borehole.

Sample Depth m WEI	For- ma- tion	Description of Rock MATTER AND OTHERS (1988a)	Corres. ground- water 302/	[U] ppm	[Th] ppm	$^{234}U/^{238}U$	$^{230}Th/^{234}U$	$^{230}Th/^{232}Th$ AR Harwell	$^{238}U/^{232}Th$	$^{228}Th/^{232}Th$	$^{231}Pa/^{235}U$
679.57	ju	belemnite limestone with phosphorite and glauco- nite nodules		29.81±0.90	4.30±0.20	1.28±0.02	1.40±0.06	37.7±1.6	21.0±1.2	1.49±0.09	0.96±0.15
848.60	mm	dolomite with solution pores	10a,10b	0.48±0.02	0.19±0.02	0.98±0.05	1.12±0.06	8.7±0.8	7.9±0.7		
982.28	s	laminated, argillaceous sandstone	12,13	5.36±0.14	17.99±0.65	1.05±0.03	1.00±0.04	0.95±0.02	0.90±0.04		1.09±0.07
984.10	s	sandstone with anhydrite	12,13	1.72±0.06	5.25±0.11	0.98±0.04	0.94±0.04	0.92±0.02	0.99±0.04		1.28±0.16
987.00	s	compact sandstone	12,13	0.91±0.04	1.96±0.07	1.10±0.07	0.95±0.05	1.46±0.07	1.40±0.08	0.93±0.07	
1116.10	r	porous sandstone	19,20	15.88±0.40	22.42±0.51	0.96±0.02	0.87±0.03	1.79±0.02	2.1±0.1		0.95±0.06
1404.79	r	sandstone	17,18	1.44±0.11	4.31±0.11	0.92±0.09	0.94±0.08	0.87±0.03	1.01±0.08	0.96±0.01	
1590.00	r	coal, above fractured sandstone		1.22±0.05	2.25±0.07	1.00±0.05	0.72±0.04	1.18±0.05	1.65±0.08	0.98±0.04	
2220.12	KRI	chloritized plag gneiss	16	4.03±0.12	12.11±0.20	1.06±0.03	0.99±0.04	1.07±0.02	1.01±0.03	0.96±0.01	1.17±0.18

AR = activity ratio

Table 7.3.11: Rock Samples: Uranium- and thorium-natural series nuclides abundance data, Riniken and Schafisheim boreholes.

Sample Depth m	Formation	Description of Rock MATTER AND OTHERS (1987) MATTER AND OTHERS (1988b)	Corresp. ground-water 303/	[U] ppm	[Th] ppm	234U/ 238U	230Th/ 234U	230Th/ 232Th	238U/ 232Th	228Th/ 232Th	231Pa/ 235U
RIN								-- AR Harwell --		-- -- --	-- -- --
626.77	m	dolomite, pitted	2	2.91±0.07	0.61±0.04	0.96±0.02	1.04±0.04	14.45±0.96	14.4±1.0	1.44±0.05	1.20±0.15
797.14	s	sandstone, porous, argillaceous with hematite	3,4	4.28±0.12	1.04±0.08	1.12±0.10	1.06±0.08	0.88±0.03	0.74±0.06	0.96±0.04	1.04±0.15
1710.20	r	sandstone with hematite	10c,11	4.14±0.08	22.73±0.65	1.04±0.02	0.82±0.04	0.47±0.02	0.55±0.02	0.96±0.04	1.14±0.12
SHA			304/								
1227.90	mo	dolomite, porous bonebed (1) (2)	3	1.91±0.07 1.78±0.06	0.19±0.02 0.16±0.02	1.02±0.04 1.01±0.04	0.98±0.04 1.04±0.04	30.4±2.7 38.3±4.7	30.5±2.9 36.5±4.7	--- 0.16±0.16	--- 1.15±0.15
1329.20	mo	calcite crystals, secondary		0.19±0.01	0.004±0.002	1.15±0.11	0.90±0.09	98.0±0.1	194.7±4.5	---	---
1481.83	s	porous sandstone (1) (2)	4,5,6,7	1.35±0.07 2.69±0.09	10.8±0.5 7.9±0.2	1.11±0.07 1.20±0.04	2.61±0.18 0.70±0.03	1.1±0.01 0.89±0.02	0.38±0.03 1.03±0.02	0.99±0.05 1.02±0.03	--- ---
1566.24	KRI	granite, with cataclastic zones and calcite filled fractures	8,9	2.16±0.07	18.8±0.4	1.12±0.04	0.91±0.04	0.35±0.01	0.35±0.01	1.25±0.02	---

* AR = activity ratio

Table 7.3.12: Rock Samples: Uranium- and thorium-natural series nuclides abundance data, Kaisten borehole.

Sample Depth m KAI	Formation	Description of Rock PETERS AND OTHERS (1988a)	Corresp. groundwater 305/	[U] ppm	[Th] ppm	234U/ 238U AR	230Th/ 234U AR	230Th/ 232Th AR	238U/ 232Th Harwell	228Th/ 232Th	231Pa/ 235U
107.40	s	porous sandstone	1	1.78±0.08	2.3±0.1	0.95±0.06	1.04±0.06	2.28±0.09	2.31±0.13	0.95±0.02	1.10±0.42
285.78	r	sandstone with hematite	2,3	1.42±0.06	5.6±0.1	1.09±0.05	0.93±0.04	0.77±0.02	0.76±0.03	0.97±0.05	1.09±0.25
478.47	KRI	cataclastic pelitic gneiss	6,7	2.80±0.07	10.1±0.2	1.04±0.03	0.95±0.03	0.83±0.02	0.84±0.03	0.96±0.02	1.31±0.27
488.88	KRI	fresh pelitic gneiss	6,7	2.67±0.06	13.21±0.40	0.96±0.03	1.04±0.05	0.61±0.02	0.61±0.02	0.99±0.10	1.01±0.28
819.67	KRI	fresh pelitic gneiss with calcite filled vugs	9,10,11	4.15±0.25	12.46±0.51	1.18±0.07	0.75±0.05	0.89±0.04	1.01±0.07	0.93±0.06	0.86±0.20
1038.83	KRI	fresh pelitic gneiss (Leucosome-rich) directly besides water bearing joint	12,13	2.97±0.07	12.50±0.22	1.00±0.03	0.93±0.03	0.67±0.01	0.72±0.02	0.97±0.06	1.05±0.25
1046.36	KRI	chloritized pelitic gneiss		2.36±0.05	10.12±0.24	1.00±0.02	0.85±0.03	0.60±0.02	0.71±0.02	1.00±0.02	1.01±0.21
1273.94	KRI	chloritized bi-plag gneiss with vugs filled by 2 generations of calcite	16,17	4.42±0.08	13.84±0.30	1.09±0.02	0.92±0.03	0.96±0.02	0.97±0.03	0.95±0.09	1.15±0.25
348.61	KRI	baryte, coarse grained crystals from vug		0.04±0.005	0.22±0.02	1.89±0.27	0.75±0.09	0.79±0.10	0.56±0.08	0.41±0.21	
1273.94/1	KRI	first generation calcite from vug of sample 1273.94 (1)	16,17	0.017±0.004	0.57±0.04	3.09±0.77	0.89±0.16	0.25±0.03	0.09±0.02	0.69±0.10	
1273.94/2	KRI	(2)		0.017±0.003	0.54±0.03	2.69±0.64	0.87±0.15	0.22±0.03	0.09±0.02	0.79±0.07	

AR = activity ratio

Table 7.3.13: Rock Samples: Uranium- and thorium-natural series nuclides abundance data, Leuggern borehole.

Sample Depth m LEU	For-ma-tion	Description of Rock PETERS AND OTHERS (1988b)	Corresp. ground-water 306/	[U] ppm	[Th] ppm	$^{234}U/^{238}U$	$^{230}Th/^{234}U$	$^{230}Th/^{232}Th$	$^{238}U/^{232}Th$	$^{228}Th/^{232}Th$	$^{231}Pa/^{235}U$
								AR Harwell			
52.10	mm	dolomite with solution pores, dark components (bones?) and small crystals	(1)	0.57±0.03	0.23±0.02	1.22±0.08	1.00±0.06	9.25±1.04	7.54±0.91	0.59±0.02	-
213.95	s	porous sandstone	2,3	1.87±0.08	4.91±0.13	0.90±0.05	0.99±0.05	1.03±0.03	1.15±0.06	0.94±0.04	-
539.76	KRI	aplitic part in gneiss with vug	7,8	1.85±0.08	0.69±0.04	0.98±0.05	0.91±0.05	7.30±0.43	8.15±0.57	0.75±0.08	-
701.68	KRI	chloritized hbl-bi-plag gneiss	9,10	1.34±0.03	2.29±0.19	1.04±0.03	0.94±0.07	1.74±0.17	1.77±0.15	0.91±0.11	1.07±0.24
916.87	KRI	chloritized bi-plag gneiss close by joint with free arsenic	11b, 12	2.46±0.09	10.4±0.3	0.96±0.04	0.97±0.05	0.67±0.02	0.72±0.03	0.93±0.04	1.15±0.28
1197.62	KRI	chloritized pelitic gneiss, part with argillaceous alteration and with calcite filled vugs	20,21	2.92±0.07	11.9±0.3	1.05±0.03	0.89±0.04	0.70±0.02	0.74±0.03	0.93±0.03	1.36±0.34
1525.97	KRI	fresh bi-granite		10.37±0.17	21.74±0.60	0.99±0.01	0.95±0.03	1.36±0.01	1.45±0.05	1.00±0.03	1.20±0.34
1636.90	KRI	fresh cord-bearing, two-mica granite above water bearing zone	(16)	8.01±0.12	14.08±0.28	1.00±0.01	0.93±0.03	1.61±0.03	1.72±0.04	0.97±0.02	1.17±0.18
1648.27	KRI	cord-bearing, two-mica granite with argillaceous alteration of plag close by the hanging hanging wall of the wate bearing zone	16	3.31±0.12	14.31±0.28	1.06±0.04	0.91±0.04	0.67±0.01	0.70±0.03	0.97±0.03	0.99±0.15
704.44	KRI	coarse grained calcite (vein filling)	9,10	0.16±0.01	0.06±0.01	1.09±0.10	0.93±0.09	7.70±1.63	7.60±1.61		
1648.83	KRI	hydrothermal vein filling (quartz and illite)	16	0.30±0.04	0.10±0.01	1.18±0.22	0.75±0.10	8.81±1.12	9.18±1.61		
1648.83	KRI	coarse grained baryte from joint along qz-ill-vein	16	0.17±0.01	0.09±0.01	1.08±0.08	0.95±0.07	5.88±0.82	5.78±0.82		

Ar = activity ratio

8. SYNTHESIS OF ISOTOPE RESULTS

This chapter provides a synthesis by aquifer and region of the isotope evidence on recharge conditions and source areas, residence times and flow paths, and the chemical and isotopic evolution of groundwater in the bedrock of northern Switzerland. The waters discussed are of deep origin, even though they may be present at shallow depths. The chapter is arranged by water-bearing unit, with sections on aquifers in the Tertiary and Malm, the Dogger, Lias and Keuper, the Muschelkalk, and the Buntsandstein, Permian and crystalline. Each section summarizes the results of the various isotope studies described in the preceding chapters, within the framework of the chemical character of the water in each unit, as described in Section 1.5. Each section concludes with a synthesis of these results.

Primary evidence about recharge conditions is given by the ^{2}H and ^{18}O contents of waters, and by temperatures calculated from noble gas concentrations, as discussed in Chapter 3. Information about residence times can be extracted from ^{3}H, ^{85}Kr, and ^{14}C data, as described in Chapters 4 and 5. ^{39}Ar in some samples is related to water residence time, but in others it appears to have been produced underground and reflects water-rock interaction processes, as pointed out in Chapters 4 and 6. ^{40}Ar and helium concentrations are also determined by underground processes, as discussed in Chapter 6, but they may also have water age implications.

Other isotopes provide information on the origin of substances dissolved in groundwaters and on geochemical reactions between the waters and the rock through which they have passed. The sources of dissolved sulphate and strontium are discussed in Sections 7.1 and 7.2, respectively, and information about the oxidation state of the waters provided by the uranium series isotopes is discussed in Section 7.3. Reactions between dissolved and mineral carbonate are covered in Chapter 5, and oxygen isotope exchange as an indicator of other water-rock reactions in Section 3.1. Finally, the concentration of ^{36}Cl and ^{4}He produced underground is correlated with the general salinity of the water, as measured by its chloride content, as discussed in Chapter 6.

Even a study as comprehensive as this one will raise questions, in addition to providing conclusions. Thus, this chapter ends with suggestions about additional isotope studies which would address both new and old unanswered questions about the hydrology and geochemistry of deep groundwaters in northern Switzerland.

8.1 Tertiary and Malm

Groundwaters in these formations can be described as members of three water types with distinguishing chemical and isotopic characteristics (Section 1.5.3, Table 1.5.1). These are a Ca-Mg-HCO$_3$, a Na-HCO$_3$, and a Na-Cl water type. In the Molasse Basin, these three types are found one above another but in discordance to the stratigraphic formation boundaries (Figure 1.5.1). Samples from these formations are listed in Table 8.1.1, with selected isotopic data on them.

Chemical characteristics have been discussed in SCHMASSMANN AND OTHERS (1984) and in NAGRA (1988, Chapter 4.8.2). A detailed hydrochemical synthesis of Tertiary and Malm aquifers is in preparation (SCHMASSMANN, *in preparation*).

8.1.1　Calcium-Magnesium-Bicarbonate Groundwaters

This water type is represented in the Nagra regional programme by samples from the Molasse Basin, from the Jura mountains and from the Upper Rhine Graben. They can be grouped as follows:

- Young, shallow groundwaters with high-tritium contents (generally ≥ 20 TU). In this group are waters of Schwarzenberg (42), Gränichen (130), Fulenbach (33), Attisholz (32), Malleray (7), Welschingen (99) and Durlinsdorf (80). The waters of Wildegg (30), and Steinenstadt (85) have tritium contents exceeding 20 TU, but are mixtures of young and old waters. They are included in this group;

- Deep groundwaters with low tritium contents or mixtures between an old deep groundwater and a young water component. Only groundwaters with tritium contents less than 20 TU are included in this group.

Except for the mention of ^{2}H, ^{18}O and ^{34}S, the following points describe only the deep, low-tritium, Ca-Mg-HCO$_3$ group of groundwaters. The isotope characteristics of the Ca-Mg-HCO$_3$ groundwaters include:

- δ^2H and δ^{18}O values centre on the meteoric water line and range from -61 to -77 per mil and from -8.8 to -10.6 per mil, respectively. For waters with tritium contents exceeding 20 TU, the variation is mainly a function of the average recharge altitude (SCHMASSMANN AND OTHERS (1984); Section 3.1). Shallow and deep groundwaters are within the same range;

- δ^{13}C values range from -10.5 to -14.2 per mil. These depleted δ^{13}C values clearly differ from those of the other groundwater types from the Tertiary and Malm aquifers;

- Measured ^{14}C values are between 40 and 70 pmc with different tritium contents indicating young groundwaters (Beuren, 100 and 149) or groundwater recharge just prior to the start of the nuclear era (Schönenbuch, 5), and, in two cases, probably mixtures of young and pre-nuclear groundwaters (Oberbergen, 96; Sauldorf, 502) (Section 5.2.2.1);

- Depleted ^{34}S values at low sulphate concentrations are characteristic of the shallow groundwaters of Schwarzenberg (42), Durlinsdorf (80) and Wildegg (30). Compared to the deep groundwaters of the Tertiary and Malm aquifers, both the δ^{34}S values (-5.6 to +5.2 per mil) and the δ^{18}O values (+4.2 to 10.5 per mil) in the sulphates of the shallow Ca-HCO$_3$ groundwaters are more negative than sedimentary sulphate minerals of any age. Thus, the sulphate must be derived at least in part from the oxidation of ^{34}S-depleted organic sulphur or sulphide minerals.

The isotope results indicate recharge under present climatic conditions for the Ca-HCO$_3$ deep groundwaters investigated. The shallow groundwaters have characteristics of recharge during the nuclear era.

8.1.2 Sodium-Bicarbonate Groundwaters

This water type is represented by samples from the Nagra regional programme as well as by data from BERTLEFF (1986). This group of Tertiary and Malm waters are listed in Tables 8.1 and 1.5.1. The sample from Neuwiller (77) from the Upper Rhine Graben is included in this group as is the high-^{3}H sample of mixed old and young water from Reichenau (501). Although the Neuwiller borehole reaches the Dogger, it yields water from the Malm.

The Na-HCO$_3$ waters have the following isotope characteristics:

- $\delta^{18}O$ and δ^2H plot close to the meteoric water line. However, they typically have a lighter isotopic composition than the Ca-HCO$_3$ waters. Most Na-HCO$_3$ waters have $\delta^{18}O$ values ranging from -11.6 to -12.7 per mil and δ^2H values from -83 to -92 per mil. The waters of Lottstetten (127) and Neuwiller (77) are exceptions, with δ-values falling into the group of Ca-HCO$_3$ waters. Except for these two samples, this group is clearly distinct from all other waters in the Tertiary and Malm aquifers;

- Noble gas temperatures of Aqui (1), Reichenau (501), Birnau (508), Mainau (128) and Singen (97) plot close to the recent δ^2H-temperature line of northern Switzerland, but are clearly shifted to lower temperatures compared to recent climatic conditions. Including their 1-σ errors, the noble gas temperatures of these samples range between about 0°C and 4°C;

- A ^{39}Ar measurement is available only for the water of Aqui (1) and suggests an age greater than 1100 a. The moderately increased He concentration of this sample is consistent with an elevated age;

- From ^{14}C analyses, younger limiting ages for the waters of Aqui (1), Neuwiller (77), Singen (97), Konstanz (98), Lottstetten (127), Mainau (128), Birnau (508) and Ravensburg (509) were obtained, ranging from > 11 ka to > 31 ka. The limiting age of the old component of the high-^{3}H mixed sample from Reichenau (501) is > 5.3 ka;

- δ^{13}C values range between -6.4 and -3.0 per mil, clearly different from those of the Ca-HCO$_3$ waters of the Tertiary and Malm aquifers. This indicates considerable carbonate chemical evolution brought about by sodium-for-calcium exchange, which is reflected in the higher uncertainties in the ^{14}C model ages;

- The water of Aqui (1) is presumed to contain a mixture of sulphate dissolved during recharge, with mineral sulphate isotopically similar to Miocene evaporite minerals. ^{34}S has been enriched in the Lottstetten sample relative any reasonable precursor sulphate, probably by H$_2$S production.

The $\delta^{18}O$ and δ^2H values from sample locations near the Bodensee and Zürichsee are significantly more negative than those of recent high tritium water. These locations are Aqui (1), Singen (97), Konstanz (98), Mainau (128), Birnau (508), and Ravensburg (509). The isotopic compositions of these samples are similar to those of the lakes, and from the ^{18}O and ^2H isotope data alone, one might conclude that the groundwaters contain some water derived from the lakes themselves. This explanation is not possible for the sample from Singen (97) because the hydraulic potential there is higher than that of the Bodensee or the Rhine River (SCHMASSMANN AND OTHERS, 1984, p. 120). This is also true for Ravensburg, which has an artesian pressure corresponding to an elevation which is about 110 m higher than the level of the Bodensee (NAGRA, 1988, Beilage 4.5; BERTLEFF AND OTHERS, 1987, Tab. 1 and Abb. 2; BERTLEFF AND OTHERS, 1988, Abb. 6). Therefore, the low noble-gas infiltration temperatures of these samples indicate a recharge of the Na-HCO$_3$ waters during a colder climatic period than today.

The ^{39}Ar age of Aqui (1) and the younger limiting ^{14}C model ages of the low-^3H samples (> 11 ka) also demonstrate a greater age for the Na-HCO$_3$ waters than for the Ca-Mg-HCO$_3$ waters, and are consistent with an interpretation that the Na-HCO$_3$ waters possibly infiltrated during the last or an earlier glaciation. No noble gas analyses are available for the waters of Neuwiller (77) and Lottstetten (127), but their stable isotope values are not depleted as are the other Na-HCO$_3$ waters. Therefore, the interpretation of an infiltration during a colder period seems inadequate for these two samples. These two waters seem to be old waters, but probably did infiltrate under climatic conditions like those of the present.

8.1.3 Sodium-Chloride Groundwaters

Groundwaters that belong to this group are listed in Tables 8.1.1 and 1.5.1. The isotope results on these waters are mainly based on the Nagra regional programme. Only the Weiach (302/6) and Schafisheim (304/1, 2) samples are from the Nagra boreholes. The isotopic features of the Na-Cl waters, which are not as uniform as those of the Ca-Mg-HCO$_3$ and Na-HCO$_3$ waters, can be summarized as follows:

- $\delta^{18}O$ and δ^2H values of Na-Cl waters all fall below the global meteoric water line. The δ^2H values are within about the same range as the δ^2H values of the Ca-HCO$_3$ waters, but the $\delta^{18}O$ values tend to be shifted to more positive values, especially for the more saline waters. This deviation from the global meteoric water line reflects oxygen isotope exchange. All waters are members of a mixing sequence between two end-members, one an isotopically light Na-HCO$_3$ water and the other a saline, isotopically heavier water which probably is a brackish or even a marine water of the Molasse period (SCHMASSMANN AND OTHERS, 1984; NAGRA, 1988; SCHMASSMANN, *in preparation*);

- Noble gas analyses were made only for the OMM sample from Tiefenbrunnen (2). The large error of this measurement (9.7 ± 5.4°C) precludes an interpretation;

- The ^{39}Ar and ^{14}C contents of the water from Tiefenbrunnen (2) lead to ages of > 700 a and > 13 ka, respectively. The ^{40}Ar/^{36}Ar ratio and helium concentration of this water are consistent with a long residence time. The tUSM sample from the Eglisau 1 borehole (141) is a mixed water. The ^{14}C model age of its old component is > 15 ka. The Malm water of Weiach (302/6), probably contains some ^{14}C from drilling fluid and (or) DOC, and thus its old component has a relatively low limiting age (> 1 ka);

- The δ^{13}C values of the Na-Cl waters range from -8.7 to -3.1 per mil. The water of Tiefenbrunnen (2) is within the δ^{13}C range of the Na-HCO$_3$ waters, whereas the waters of Eglisau (141) and Schafisheim (304/1, 2), both situated at the northern border of the Molasse Basin, have δ^{13}C values between the water of Tiefenbrunnen (2) and the range of the Ca-Mg-HCO$_3$ waters;

- The water of Schafisheim (304/1, 2) is characterized by an extremely high ^{40}Ar/^{36}Ar ratio of 357.3. The relatively low He content is explained by gas loss during sampling. The high ^{14}C content of this sample is inconsistent with its lack of tritium and high ^{40}Ar/^{36}Ar ratio and may represent ^{14}C from drilling fluid or DOC. The high ^{14}C of this sample and that from the Malm at Weiach (302/6) could also result from natural admixing of high-^{14}C water (BALDERER, 1990);

- The Eglisau samples (142 to 144) show ^{34}S values typical of Miocene sedimentary sulphates. The more negative δ^{34}S value of the Malm water of Weiach (302/6) may have been produced by a mixture of Miocene sulphate from the oxidation of reduced sulphur.

8.1.4   Conclusion

The three main water types distinguished by hydrochemical features can also be confirmed by isotope results. Generally, the ^{3}H and ^{14}C results show that the Na-HCO$_3$ and Na-Cl waters from the regional programme are older than the Ca-Mg-HCO$_3$ waters. δ^2H and δ^{18}O results of the Ca-Mg-HCO$_3$ waters suggest recharge under post-Pleistocene climatic conditions. Most of the Na-HCO$_3$ groundwaters probably represent recharge during an earlier, colder climatic period, as during glaciation. This would account for their high ^{14}C model ages, depletion in ^{2}H and ^{18}O and low noble gas infiltration temperatures. A few Na-HCO$_3$ waters can also be considered as old, based on their ^{14}C model ages, but their undepleted stable isotope values indicate recharge during climatic conditions similar to those of the present. The observed mixing sequence from a low mineralized Na-HCO$_3$ water through the Na-Cl waters of Eglisau to Weiach suggests an end-member even more saline and more enriched in δ^{18}O and δ^2H than the investigated samples, as, for example, a water of marine or perhaps brackish origin (NAGRA, 1988, Figure 4.30). Isotope results are also consistent with, or at least not contradictory to, the model derived from hydrochemistry (NAGRA, 1988) that the Na-HCO$_3$ water layer is wedging out at the northern border of the Molasse Basin, and therefore, the Ca-Mg-HCO$_3$ waters are placed directly upon the Na-Cl waters in this area (Figure 1.5.1).

8.2 Dogger, Lias, and Keuper

Isotope data on low-^{3}H samples from these aquifers are summarized in Table 8.2.1. The chemistry of these waters is summarized in Table 1.5.2 and discussed in Section 1.5.4.

Several samples from the Hauptrogenstein aquifer within the Dogger were analysed. All those with < 20 TU were from the region of the Rhine Graben and are not directly applicable to central northern Switzerland (Section 1.5.4). As Figure 3.2.3 shows, samples with high-^{3}H concentrations plot along the meteoric water line, while samples of low-^{3}H from Neuwiller (77), Bellingen (81 and 83), and Steinenstadt Thermal (84) are slightly enriched in ^{18}O. The only ^{14}C measurement was on the sample from Neuwiller (77) and corresponded to a model age of > 26 ka. Oxygen enrichment by water-rock exchange would be consistent with a high residence time and the fact that the other samples are associated with thermal features.

The isotopic composition of sulphate was measured on the high-^{3}H samples from the Hauensteinbasis Tunnel at 6.785 km (76) and 3.135 km (68). The sulphate concentration in the sample from location 76 is low, and is depleted in ^{34}S (δ^{34}S = -10.9 per mil). Thus, it is probably derived at least in part from the oxidation of organic sulphur or mineral sulphide. The sulphate and ^{34}S contents of the sample from location 68 are higher due to an admixture of a sulphate-rich water of another origin (Section 7.1.2).

Waters from the Keuper were sampled at a number of locations, but all except two had ^{3}H contents above 20 TU. The ^{14}C contents of two high-^{3}H samples from the Keuper, Meltingen 2 (19) and Eptingen (23), were also measured. The adjusted ^{14}C contents of the Meltingen 2 (19) samples is 127 ± 8 pmc, while the measured ^{14}C content of the Eptingen (23) sample is 122 ± 1 pmc. The ^{13}C content of the Eptingen sample is unusually high (δ^{13}C = -3.6 per mil) for such a high ^{14}C water, and cannot presently be explained.

Many of these waters have relatively high mineralization of up to 2.7 g/l, most of which is calcium and sulphate. Sulphate isotope ratios were measured on samples from Eptingen (23), Sissach (25), Wintersingen (26), and Schinznach Dorf (29). All were in the range of Keuper mineral sulphate (Section 7.1.3).

The two Keuper samples with ^{3}H concentrations below 20 TU are from the Beznau borehole at 105 m (171) and from the Riniken borehole between 501 and 530 m (303/1). Both samples have salinities > 14 g/l and ^{14}C model ages of > 12 and > 17 ka, respectively. Thus, these samples are clearly associated with zones of limited circulation. The ^{2}H and ^{18}O content of the Riniken sample (303/1) falls on the meteoric water line, but it is so strongly enriched relative to modern meteoric water that it cannot have been recharged under climatic conditions like those of the present. Only an ^{18}O analyses is available for the Beznau sample (171). This value, -8.8 per mil, is more positive than that of modern waters, but is not as enriched as that of the Riniken sample, -6.4 per mil.

Sulphur and oxygen isotope analyses were made on sulphate from the Riniken water. Its sulphur isotopic composition is consistent with minerals from the Gipskeuper and with dissolved sulphate in high-^{3}H waters from the Keuper. The ^{18}O in the sulphate is depleted relative to that of Gipskeuper minerals from Nagra boreholes or dissolved sulphate in other Keuper groundwater samples, (Section 7.1.3), but it is within the range of $\delta^{18}O$ values found in minerals of Keuper age elsewhere in the world (Table 7.1.1, Figure 7.1.4). Enrichment in ^{18}O could also result from oxygen isotope exchange between sulphate and water during the long residence time of this water, as suggested by WEXSTEEN AND MATTER (1989) for Muschelkalk waters.

The high-^3H samples from the Keuper are from shallow sampling points in the Folded and western Tabular Jura. These samples represent regions of very active groundwater circulation. Many of these waters have dissolved significant quantities of gypsum from the Gipskeuper. The two samples from boreholes in the eastern Tabular Jura, on the other hand, have long residence times and represent regions of restricted or stagnant groundwater circulation.

8.3   Muschelkalk

Most samples from the Muschelkalk were taken from sites in the Folded Jura and the western Tabular Jura. With a few exceptions, these waters are chemically Ca/Mg-HCO$_3$-SO$_4$ and Ca/Mg-SO$_4$-HCO$_3$ types, and have relatively low chloride contents. Samples from the eastern Tabular Jura were taken from deep boreholes and are Ca/Mg-SO$_4$-Cl to Na-Cl-SO$_4$ waters. Muschelkalk waters from boreholes within and along the northern border of the Molasse Basin are characterized chemically by H$_2$S contents higher than in the other groups of waters, and are typically mixtures of older waters of several different types with recent waters. A few Muschelkalk samples were also taken from locations along the flank of the Rhine Graben. The chemistry of samples from the upper Muschelkalk is discussed in Section 1.5.5 and illustrated in Table 1.5.3. Isotope data useful for determining the recharge conditions and residence times of Muschelkalk waters, except ^{36}Cl data, are summarized in Table 8.3.1 for samples on which ^{14}C, noble gas or ^{39}Ar measurements were made.

8.3.1   Recharge Conditions and Residence Times

Muschelkalk waters in northern Switzerland which were recharged under present conditions have δ^2H values from about -63 to -75 per mil and $\delta^{18}O$ values from about -9 to -10.5 per mil. These are the ranges of values from virtually all samples of superficial waters (those with ^3H contents of > 20 TU) shown in Figure 3.2.5. All group closely along the meteoric water line.

Most of the deeper Muschelkalk waters (containing < 20 TU) have δ^2H and $\delta^{18}O$ values in the same range as those of the higher ^3H waters, as shown in Figure 3.2.6. All samples from the Folded and western Tabular Jura are in this group.

With a few exceptions, all samples from the Folded and western Tabular Jura have ^{14}C model ages of 10 ka or less. All have detectable ^{39}Ar concentrations as well, some of which are high enough to provide ^{39}Ar residence times. These are not inconsistent with the ^{14}C model ages of the same samples.

Noble gas measurements yielded consistent temperature data on seven samples from six Muschelkalk locations in the Folded and western Tabular Jura. As shown in Figure 3.3.8b, the relationship between the δ^2H and noble gas temperatures of all samples from this region agree with the relationship between δ^2H and the mean annual temperatures of precipitation and surface water in northern Switzerland. The ^2H and ^{18}O data and noble gas temperatures show that these waters were recharged under conditions like those of the present, consistent with the ^{14}C and ^{39}Ar results.

Noble gas temperatures are also available on five samples from the eastern Tabular Jura and the northern border of the Molasse Basin. The relationship between the δ^2H values and noble gas temperatures of all samples except that from Weiach (302/10) are consistent with the patterns in modern waters.

Certain samples have δ^2H and (or) $\delta^{18}O$ values outside the range of high-^3H waters. These include:

- Numbers 201 through 221, the thermal waters of Baden and Ennetbaden. These are waters of mixed deep and shallow origin from the northern border of the Molasse Basin, with generally higher H_2S contents than samples from the other groups. These samples have ^2H contents within the range of high-^3H, recent recharge waters, but they are enriched in ^{18}O relative to meteoric water. This provides information about the geochemical evolution of these waters, as discussed in the next section;

- Samples from the Beznau (15 and 172 to 175), Böttstein (301/2), and Riniken (303/2) boreholes are waters from the eastern Tabular Jura. The ^2H and ^{18}O contents of these waters fall along the meteoric water line (Figure 3.2.6). The very low ^{14}C contents of these samples yield model ages of > 17 ka, which is consistent with the ^{39}Ar results. The high helium contents of these samples are also consistent with a great water age, as is the increased ^{40}Ar content of the Böttstein water. Because of their age, the δ^2H and $\delta^{18}O$ values of these waters would not be expected to be like those of high-^3H waters. However, the pattern of the relationship between δ^2H and noble gas temperatures for the Böttstein and Riniken samples is like that of present Muschelkalk water. The ^{18}O and ^2H contents of the Riniken sample are like those of recent waters, however, in agreement with its noble gas recharge temperature;

- Samples 304/3 and 304/14 represent an Na-Cl-SO_4 water of relatively high-H_2S content from the Schafisheim borehole in the northern part of the Molasse Basin. No ^{14}C interpretation is possible for this sample, but its ^{39}Ar content suggests an admixture of young water. Its high He content, however, qualitatively suggests an old water component of great age. Although the noble gas recharge temperature of these samples is consistent with modern temperatures, their stable isotope contents are depleted relative to modern water. Because it is a mixture, the isotopic composition of this water cannot be interpreted unequivocally;

- The ^3H and ^{85}Kr contents of sample 302/10, from the Weiach borehole, suggests that it contains no more than a few per cent of young water (Table 4.3.1). Its ^{14}C content leads to a modern model age while its ^{39}Ar content is equivalent to a model age of about 800 years, which overlaps the ^{14}C results. This sample also has a relatively low ^{4}He content. Thus, the majority of the water comprising the Weiach sample cannot be more than one thousand years old, consistent with its noble gas temperature which is similar to that of the present climate. The ^{2}H and ^{18}O contents of sample 302/10 are lower than those of any other Muschelkalk sample. Modern waters as depleted as this sample are found only at high altitude in the Alps and in waters bodies such as the Rhine River and the Bodensee, which have their sources in the Alps. The relative hydraulic heads are such that the Rhine River between the Bodensee and the Rheinfall could be the source of water to the Muschelkalk at Weiach. The noble gas temperature of this sample corresponds to that of the Rhine River downstream of the Bodensee, while the isotopic composition of its dissolved sulphate is within the range of Keuper evaporites. This would be consistent with a flow path to the Muschelkalk from the Rhine through overlying Keuper deposits.

This explanation for the origin of this sample requires the existence of a zone of high flow within the Muschelkalk in the Weiach region. Another zone of high flow accounts for some of the characteristics of water found at Bad Lostorf (SCHMASSMANN, 1977).

- Samples 129 and 90, from Liel Neuer Brunnen and Krozingen 3, respectively, in the region of the Rhine Graben. These samples are from outside of the geographic area of northern Switzerland, and their isotopic composition probably reflects recharge at a higher elevation in the Black Forest (SCHMASSMANN AND OTHERS, 1984, p. 219, 253).

Several of the Magden boreholes (Weiere 271, Falke 272, and Eich 273) yield mixed waters. All were sampled more than once with different ^{14}C results and, for Weiere, different noble gas results. The ^{3}H contents of all samples from these boreholes were below detection, however. These samples are interpreted as the products of mixing between components with model ages of < 5 ka (272/102) and > 15 ka (271/101). The sample with the highest ^{14}C content from these three boreholes (272/102) has the highest noble gas temperature and is most enriched in ^{2}H of all the Magden samples. Similar enriched ^{2}H and ^{18}O contents are found in water from a fourth Magden borehole, Stockacher (274/102). This water contains ^{3}H and has the highest ^{14}C and ^{39}Ar contents of any Magden sample. The ^{39}Ar contents of all Magden samples are correlated with their ^{14}C contents (r = 0.8). The ^{14}C content of the old component of these waters is so low as to preclude the presence of ^{39}Ar, unless underground production is occurring. Thus, the measured ^{39}Ar contents can be attributed to the younger mixing component, which would suggest an age of c. 0.5 ka, based on a ^{39}Ar content of c. 25 per cent of modern in the samples from Eich (273). This is consistent with the ^{14}C model age (1.5 ± 0.9 ka) and ^{39}Ar age (0.4 to 0.6 ka) of the Stockacher sample (274/102). The young component is probably similar to water sampled elsewhere in the Folded and western Tabular Jura.

The sample from the Muschelkalk at Leuggern (306/1) is also a mixed water with high ^{39}Ar, relatively high ^{14}C, and low ^{4}He contents. This sample can be interpreted as a mixture of at least three water types, as discussed in Table 4.4.2.

The waters sampled at the northern border of the Molasse Basin at Schinznach (119 and 120) are mixtures of recent superficial and older, deeper groundwaters, as indicated by their high ^{3}H contents. The ^{14}C model age of their old component is > 0.6 ka. Lostorf 4 (124) contains no ^{3}H but has a ^{14}C model age of < 10 ka. The ^{2}H and ^{18}O contents of these samples also correspond to those of waters of the present climatic regime.

The samples from the thermal springs of Baden and Ennetbaden (201 through 219) have low, but measurable ^{14}C contents. These samples have ^{3}H concentrations which indicate mixing with modern waters in quantities large enough to provide all the measured ^{14}C. Thus, the model ages are reported as younger limiting ages of an old component. These waters are enriched in ^{18}O relative to meteoric water, which may indicate oxygen isotope exchange between water and rock as a result of high temperatures and (or) long residence times at depth. A relatively high He content was measured in one of these samples.

As discussed in Section 4.4 and shown in Table 8.3.1, the ^{39}Ar contents of all but one of the Muschelkalk samples are consistent with their ^{14}C contents. The sample from Pratteln (8/2) has a ^{39}Ar content of > 200 per cent modern. Such high ^{39}Ar contents are generally found only in waters from crystalline rock where they result from *in situ* production. The occurrence of such a high ^{39}Ar content in this sample requires that it contain some water with a crystalline source. This is supported by the increased ^{40}Ar/^{36}Ar ratio in this sample.

8.3.2 Geochemical Evolution

The sulphur and oxygen isotopic composition of sulphates from Muschelkalk rock and water is discussed in Section 7.1.4. The δ^{34}S values of six mineral sulphates from the upper and middle Muschelkalk of the Böttstein, Weiach and Schafisheim boreholes are within the range of minerals from the Muschelkalk of Middle Europe. The δ^{34}S values of dissolved sulphates are shown in Figure 7.1.7. Most Muschelkalk waters are within NIELSEN's (1979) range of Muschelkalk sulphates. Samples with ^{34}S contents lower than those of typical Muschelkalk sulphates include a number of high-^{3}H superficial waters of low dissolved sulphate contents including samples 65/2 and 66/2 from the Alter Hauensteintunnel and 70/2, 71/2, 72/2 and 74/2 from the Hauensteinbasistunnel. These waters probably contain sulphate dissolved from the overlying Keuper. The δ^{34}S values of minerals from the Keuper are between 13 and 16 per mil. Sample 306/1 from the Nagra Leuggern borehole has already been discussed as a mixed water. Its δ^{34}S value suggests that at least one component is of Keuper origin.

The only high-sulphate water with a δ^{34}S value lower than that of typical Muschelkalk sulphate is the sample from Weiach (302/10b). As discussed in the previous section, this is a young water which appears to have been recharged from the Rhine River downstream of the Bodensee. Its sulphate isotopic composition suggests that the flow during recharge was through the Keuper.

Samples from two locations, Schafisheim (304/3 and 304/14) and Schinznach Bad S2 (120/1), have δ^{34}S values enriched relative to the normal range of Muschelkalk minerals, and to the bulk of the water samples from northern Switzerland as well. These are waters with relatively high H_2S concentrations from the northern part of the Molasse Basin. H_2S production from dissolved sulphate can enrich the residual sulphate in the heavier isotope. This mechanism would be consistent with the H_2S contents of these waters, because the most highly enriched sample is from Schafisheim, which also has the highest total H_2S contents of any waters collected as part of this study.

The $^{87}Sr/^{86}Sr$ ratio of minerals and dissolved strontium are discussed in Section 7.2. Samples of anhydrite from a middle Muschelkalk marker horizon in the Weiach, Schafisheim and Leuggern boreholes all had $^{87}Sr/^{86}Sr$ ratios of 0.7078. This ratio is characteristic of Muschelkalk sea water.

The $^{87}Sr/^{86}Sr$ ratio of strontium dissolved in water from the Muschelkalk of the Nagra test boreholes is from 0.7087 to 0.7097. This dissolved strontium must be derived from the dissolution of other minerals in addition to anhydrite. The $^{87}Sr/^{86}Sr$ ratios of samples of calcite and dolomite from the Böttstein, Schafisheim and Leuggern boreholes are from 0.7081 to 0.7105. Thus, the strontium dissolved in Muschelkalk water appears to be a product of the dissolution of both sulphate and carbonate minerals from the aquifer itself.

The carbon and oxygen isotopic composition of carbonate minerals from the Muschelkalk, and their relationship to the oxygen and carbon isotopic composition of water and dissolved carbonate, are discussed in Section 5.3. One calcite each from the Leuggern and Schafisheim boreholes and three calcites and two dolomites from the Böttstein borehole were analysed. The $\delta^{13}C$ values of all samples are between -1.3 and +1.4 per mil, within the normal range of marine carbonate minerals. Only one sample, a dolomite from 137.2 m at Böttstein, has a $\delta^{18}O$ value typical of a marine carbonate mineral (+1.7 per mil). The $\delta^{18}O$ of the remaining carbonates is from -8.6 to -13.1 per mil.

The ^{18}O content of calcite precipitated from water depends upon the ^{18}O content of the water itself, and on the temperature. The $\delta^{18}O$ values of calcite in equilibrium with water presently in the Muschelkalk at Böttstein and Leuggern are -12.3 and -9.3 per mil, respectively. The $\delta^{18}O$ values measured on calcite from the Böttstein and Leuggern boreholes are between -9.9 and -13 per mil, overlapping the range of $\delta^{18}O$ values in equilibrium with present groundwater from those boreholes. Some of these calcites could well have precipitated from present groundwater. Calcite in equilibrium with waters from Weiach, Riniken and Schafisheim would have $\delta^{18}O$ values from -15.9 to -18.8 per mil. No calcite was measured from the Weiach and Riniken boreholes, and only one sample from Schafisheim, with a $\delta^{18}O$ value of -8.9 per mil. The Schafisheim calcite analysed is far from ^{18}O isotopic equilibrium with the water now present in the Muschelkalk of that borehole.

Samples 201 through 221, the thermal waters of Baden and Ennetbaden, plot to the right of the meteoric water line in Figure 3.2.6. Isotope exchange between water and rock at relatively high temperatures would lead to an enrichment in ^{18}O like that observed in these samples (see, for example, TRUESDELL AND HULSTON, 1980, Figure 5-1).

Muschelkalk water has low to moderate uranium concentrations from 0.06 to 1.4 ppb. There is no geographical pattern in the dissolved uranium data, but two groups of samples can be distinguished (Sections 7.3.3.2 and 7.3.4.2).

- One group, with low to medium $^{234}U/^{238}U$ ratios, includes samples from Weiach (302/10), Leuggern (306/1), Pratteln (8/2) and Densbüren (14/3). This suggests younger groundwaters dissolving uranium, consistent with their other isotopic properties (Table 8.3.1). The water from Schafisheim (304/3 and 14) also has a low $^{234}U/^{238}U$ ratio, but this sample also has a low uranium content suggesting it is an old water (Figure 7.3.5);

- The other group, with a high $^{234}U/^{238}U$ ratio, includes the samples from Böttstein (301/2), Riniken (303/2), Frenkendorf (9/2), Magden Weiere (271/101) and Magden Falke (272/101). These higher ratios suggest relatively older waters, and so are consistent with other isotope data for the Böttstein and Riniken samples and the older component of the mixed Magden samples. The Frenkendorf sample has a model ^{14}C age of 7.2 ka. This is younger than that of the old components of the other samples, but apparently is sufficient to have reached a high $^{234}U/^{238}U$ ratio.

Seven Muschelkalk rock samples were analysed for uranium- and thorium-series isotopes. Various isotope ratios of four samples from the shallow Muschelkalk in the Böttstein and Leuggern boreholes indicate uranium accumulation some $10^{5.5}$ to 10^6 years ago. The deeper samples from the Weiach, Riniken and Schafisheim boreholes have all activity ratios equal to one, indicating stability of these rocks for at least 10^6 years.

^{36}Cl measurements were made on groundwater samples from the Böttstein, Weiach and Schafisheim (two samples) boreholes (Section 6.2). The ^{36}Cl values of all samples are proportional to their chloride contents and apparently originate from the leaching of chloride and ^{36}Cl produced within the Muschelkalk itself, or in directly underlying formations.

$^3He/^4He$ ratios were measured on two samples from the Muschelkalk--301/2 from Böttstein and 302/10b from Weiach. Both are higher than could have been produced by underground production in the formation itself. They are the same as those measured and calculated for crystalline rock, and indicate that the origin of the helium in the Muschelkalk was in the crystalline. However, the helium concentration of the Weiach sample is low, and it would also be possible to explain the $^3He/^4He$ ratio of this water as the product of some atmospheric contamination.

Total helium concentrations are strongly correlated with the chloride concentrations of the various Muschelkalk samples. This suggests that both are related to the extent of water-rock interaction, and that the helium content cannot be interpreted in terms of water ages alone.

8.3.3 Conclusions about Muschelkalk Groundwater

The Muschelkalk is an extensive, continuous aquifer in northern Switzerland, but different factors dominate the type of flow and groundwater chemistry in different regions. In particular:

- The Böttstein (301/2), Beznau (15), and Riniken (303/2) samples represent an area of little groundwater circulation in the eastern Tabular Jura. These waters have high residence times and ^{2}H and ^{18}O contents (except Riniken) different from recent high-^{3}H water. The relationship between the ^{2}H values and noble gas temperatures of two of these samples is like of that of recent Muschelkalk water. The isotopic and noble gas properties of these waters are consistent with recharge in the region of northern Switzerland and southern Germany. They are so inconsistent with conditions in the Alps as to completely rule out recharge there;

- The locations sampled in the Folded and western Tabular Jura west of the Böttstein, Beznau and Riniken boreholes yielded groundwaters from much more active circulation systems. Most of these waters are from modern to not greater than 10 ka in age. Their δ^2H and δ^{18}O values and noble gases recharge temperatures are like those of the high-^{3}H water, suggesting recharge in the northeastern Jura and Black Forest area. The water sampled in some of the boreholes in this region is of a mixed origin. The wells in the Magden region, for example, yield water with varying proportions of the usual western Tabular and Folded Jura waters with an older component. The Pratteln borehole, although it yields dominantly water of western Muschelkalk origin, has a high ^{39}Ar content and ^{40}Ar/^{36}Ar ratio suggesting an influence by the crystalline;

- Samples from the northern edge of the Molasse Basin south and west of the sampled region of limited circulation are virtually all mixed waters. One end-member is a modern water while the other is of considerable age. The old water is probably not identical with the Böttstein-Beznau-Riniken type of water because the Baden springs show an oxygen isotope shift. Although the Lostorf 4 waters (124) are chemically similar to the sample from Weiach (302/10), their ^{2}H and ^{18}O contents suggest entirely separate sources of recharge water. The relationship of water from the Schafisheim borehole to other waters is unclear.

8.4 Buntsandstein, Permian and Crystalline

Groundwaters in these formations can be described in terms of waters with contrasting isotopic and chemical characteristics. One type has relatively low total mineralization, (< 5 g/l) and is found in the Black Forest and in the upper part of the crystalline in the Nagra boreholes in northern Switzerland. The second type is highly mineralized (37 to c. 100 g/l), and was sampled from the Permian sediments of the North-Swiss Permo-Carboniferous Trough in the Weiach and Riniken boreholes. The third type is of intermediate mineralization (< 16 g/l), and is found in the Buntsandstein and in the deep crystalline of the Böttstein and Weiach boreholes.

The hydrochemistry of these waters is discussed in Sections 1.5.6 and 1.5.7 and shown in Tables 1.5.4 and 1.5.5. Isotope data bearing on recharge conditions and model ages of waters from these units are given in Table 8.4.1.

To explore the evolution of these waters, it is important to consider the origin of the waters themselves separately from the origin of the salts dissolved in them. Isotopic data on the water isotopes ^{2}H and ^{18}O are useful indicators of water origin, while both chemical and isotopic data provide information on the origin of dissolved salts. One specific aspect of the chemistry of waters from the Buntsandstein, Permian and crystalline, the relationship between their bromide and chloride contents, is particularly useful in this connection. Bromide and chloride concentrations are widely used in studies of saline groundwaters (*e.g.*, CARPENTER, 1978; NORDSTROM AND OLSSON, 1987; BANNER AND OTHERS, 1989).

The relations of these waters can be illustrated graphically in three ways. The first is a plot of δ^2H values against total mineralization (as residue on evaporation) shown in Figure 3.2.9. The second, Figure 8.4.1, is a plot of the logarithms of the bromide concentrations of the waters against the logarithms of their chloride contents. The third illustration, Figure 8.4.2, shows the δ^{34}S and δ^{18}O of sulphate dissolved in waters from these formations as well as in two mineral samples from the Permian of Weiach. The sulphate data are discussed in detail in Sections 7.1.5 and 7.1.6, and provide information on the origin of the dissolved salts and on mixing among waters in these units.

The basis for the interpretation of Br/Cl ratios are the observations that waters leaching crystalline rock have Br/Cl ratios of 0.01 while sea water has a Br/Cl ratio of 0.003. These ratios are represented by the dashed lines in Figure 8.4.1. The lower line corresponds to a Br/Cl mass ratio of about 0.003, which is the ratio in sea water. A point representing sea water is shown on this figure. The upper line corresponds to a Br/Cl mass ratio of 0.01, which is the ratio found in saline waters in a number of crystalline rocks. For example, this is the ratio in the brines of the Canadian Shield (FRAPE AND FRITZ, 1987) and in Stripa groundwaters (NORDSTROM AND OTHERS, 1985, 1989a), as well as in leachates both of Stripa granite (NORDSTROM AND OTHERS, 1985, 1989b) and granite from the Böttstein core (PETERS, personal communication, August 1989).

Both the Cl and Br and, hence, their ratio are conserved in virtually all water-rock reactions, except those involving the precipitation or dissolution of chloride minerals. When halite, the principle chloride mineral, precipitates, the larger bromide ion is excluded from the chloride position in the crystal lattice. Thus, halite itself has much lower Br/Cl ratios than the water from which it precipitated, and the brines remaining after halite precipitation are enriched in bromide. Likewise, halite dissolution will lower the Br/Cl ratio of a water. Thus, the Br/Cl ratio of a water will not be higher than that of its crystalline or marine source, unless its salinity has increased (by evaporation, for example) to above halite saturation (*c.* 400, NaCl/l). The Br/Cl ratio of a water will be lowered only by the dissolution of chloride minerals. Intermediate Br/Cl ratios can result from mixing waters of various origins.

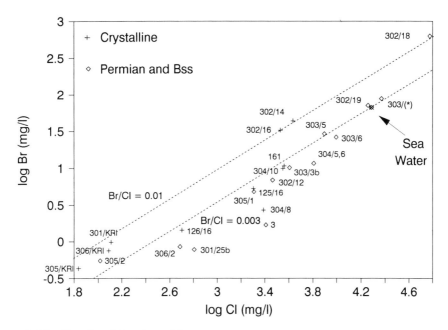

Figure 8.4.1: Graph of the logarithms of bromide concentrations against those of chloride concentrations in waters from the Buntsandstein, Permian and Crystalline.

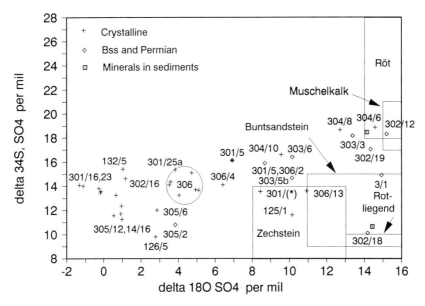

Figure 8.4.2: $\delta^{34}S$ and $\delta^{18}O$ values of sulphate dissolved in waters from the Buntsandstein, Permian and Crystalline and from minerals from the Permian of the Weiach borehole.

8.4.1 Waters of Low Total Mineralization from the Crystalline

Waters of low mineralization are found in the crystalline and can be discussed in terms of two subgroups.

The first is composed of four samples from locations in the Black Forest and includes the Nagra samples from Sulzburg Waldhotel (166) and Görwihl Bohrloch 30 (167) and two samples reported by RUDOLPH AND OTHERS (1983), and referred to here as B40 and B41. The ^{18}O and ^{2}H contents and the noble gas temperatures of the samples in this group are consistent with recharge under present climatic conditions. In addition, they have high ^{14}C contents which are consistent with young recharge. Their relatively high uranium concentrations also suggest oxidizing conditions and a low age. The characteristics of these samples are those of recently recharged water in the crystalline. Samples from several other locations in the Black Forest can also be considered as members of this subgroup. However, there is no noble gas or ^{14}C data for them, so they are not included in Table 8.4.1. These are the samples from Bürchau (110), Rothaus (150), Menzenschwand (162,163 and 164) and St. Blasien (165).

The second and much larger subgroup of waters of low mineralization is composed almost entirely of samples from the crystalline of the Kaisten (305), and Leuggern (306) boreholes, the thermal boreholes at Zurzach (131 and 132), and from the upper crystalline at Böttstein (301). The group also includes samples from the crystalline of the Black Forest at Waldkirch (95) and Freiburg-im-Breisgau (RUDOLPH AND OTHERS, 1983, designated here as sample B42), and from the Permian at Kaisten (305/2). The chemical character of sample B42 is unknown, but its isotopic characteristics place it in this group.

Isotopic and chemical characteristics of this second subgroup include:

- A narrow range of δ^2H and δ^{18}O values, centred on the meteoric water line and ranging from c. -72 to -75 per mil and c. -10.0 to -10.6 per mil, respectively (Figures 3.2.7 and Table 8.4.1);

- Noble gas analyses corresponding to temperatures between 2° and 6°C, with all but one result from northern Switzerland below 4.5°C. These samples make up a majority of the Group 1 samples discussed in the noble gas chapter, Section 3.3 and shown in Figure 3.3.8a;

- δ^{34}S and δ^{18}O values of dissolved sulphate ranging from 10 to 16 per mil and from -2 to +7 per mil, respectively (Figure 8.4.3). This is the range of δ^{34}S values found in what appears to be the younger of the two generations of barite in the crystalline (Figure 7.1.10);

- Little or no ^{3}H, and ^{14}C contents leading to model ages of > 12 ka for most samples (Table 8.4.1);

- Dissolved carbonate with $\delta^{13}C$ values of c. -9 to -10 per mil. Many calcites from core from the Nagra boreholes are at oxygen isotope equilibrium with the waters now present and at the present temperature of the crystalline (Figures 5.3.1, 5.3.4 and 5.3.6);

- Average Br/Cl ratios between 0.006 and 0.008 for the crystalline of the Böttstein, Kaisten, and Leuggern boreholes (Figure 8.4.1);

- Relatively high dissolved He concentrations in the narrow range of $5 \cdot 10^{-5}$ to $4 \cdot 10^{-4}$ cm^3/cm^3 H$_2$O. The $^{40}Ar/^{36}Ar$ ratios are less high, the maximum values being 309.

Nagra borehole samples fall in the lower left corner of Figure 3.2.9. Only samples from Waldkirch (95) and from the 923 m zone at Leuggern (306/11b, 12, 13) are distinguished in that figure. Waldkirch is in the Black Forest, from a flow system unrelated to that of northern Switzerland.

The ^{14}C contents of most samples in this group are either below detection or correspond to model ages of > 12 ka. The samples from the Buntsandstein-weathered crystalline horizon at Böttstein (301/5, 25b) contain ^{14}C giving a model age of 8 to 9 ka. (Table 8.4.2). ^{14}C is also present in several samples from the Böttstein crystalline, but its presence there may result from flow downward from the Buntsandstein-weathered crystalline horizon during drilling operations (PEARSON, 1985).

Waters in this subgroup from the upper crystalline in the Nagra boreholes and at Zurzach are chemically as well as isotopically similar. Their residues on evaporation range from 1.0 to 1.6 g/l, sodium is their dominant cation, and their major anions are present in sub-equal amounts with, generally, SO$_4$, ≥ HCO$_3$ ≥ Cl. Waters from the Kaisten boreholes (305) have a slightly higher total mineralization, due to relatively higher sodium and sulphate contents than waters from Böttstein (301), Leuggern (306) and Zurzach (131 and 132) (Table 1.5.5; SCHMASSMANN AND OTHERS, 1984, Figure 38).

The $^{87}Sr/^{86}Sr$ ratio of strontium dissolved in these waters ranges from 0.7166 to 0.7178. These ratios are lower than those of whole-rock samples. Two barite samples from Leuggern have $^{87}Sr/^{86}Sr$ ratios like those of the groundwaters, which is consistent with the similarity of $\delta^{34}S$ values in barite and dissolved sulphate in the crystalline.

The $^{87}Sr/^{86}Sr$ ratios of three samples of calcite were also measured. Only one had the same ratio as the dissolved strontium. It is not possible to examine the consistency of the $^{87}Sr/^{86}Sr$ and ^{18}O data on calcites from the crystalline because all determinations appear to have been made on different mineral samples. The availability of fluid-inclusion, chemical and isotopic analytical results on the same samples would add to the understanding of water-rock reaction processes in the crystalline. Collection of such data is a recommended future activity.

The Black Forest samples with relatively high ^3H and (or) ^{14}C contents can also be distinguished chemically from the larger group of samples mostly from the Nagra boreholes, as pointed out by SCHMASSMANN AND OTHERS (1984, Sections 6.1.1 and 6.1.2). The high-^3H Black Forest samples from Rothaus (150) and Bürchau (110), for example, are of the Ca-Na-HCO$_3$ type, while samples from northern Switzerland are Na-SO$_4$-HCO$_3$-Cl waters. The distinctive chemistry of the latter samples would be consistent with leaching and the occurrence of feldspar hydrolysis reactions. The greater extent of these reactions in the North-Swiss waters than in those from the Black Forest is consistent with the lower ^{14}C contents of the former group.

The conclusion from these data is that the waters in the crystalline of northern Switzerland were recharged in the Black Forest where the crystalline is exposed, and flowed southwards to their present location. They have relatively low ^{2}H and ^{18}O contents, and their noble gas temperatures are lower than those of modern water in the Black Forest, suggesting that all were recharged under cooler climatic conditions as would have existed at a time of glaciation. The low and unmeasurable ^{14}C contents correspond to model ages of > 12 ka, and their high ^{4}He contents also correspond to waters of some age.

The isotope characteristics of water in the crystalline from the Nagra boreholes and at Zurzach are so similar that they cannot be used to establish flow directions or age differences between the boreholes. The generally accepted conceptual model is of flow from the eastern slopes of the Black Forest to the south beneath the Rhine, and then to the west and northwest to discharge to the Rhine in the vicinity of Leuggern. This pattern is based on geographical considerations rather than the isotopic composition of the water samples themselves.

A preliminary interpretation of the samples from the Siblingen borehole (307) adds a complication to this simple picture. Siblingen is located in what would be interpreted as a region of flow from the flanks of the Black Forest recharge area to the northern Swiss crystalline region sampled by the Böttstein, Zurzach, Kaisten, and Leuggern boreholes. The salinity of the Siblingen samples is about half of that of water from the other boreholes, which is consistent with a younger age.

The ^{2}H and ^{18}O contents of the Siblingen waters, however, are more depleted than any of the north Swiss crystalline or Buntsandstein samples as shown in Figure 3.2.7. This suggests that the Siblingen waters would have been recharged at a time of even more extreme cold than prevailed at the time the waters further to the south were recharged.

The ^{14}C contents of the Siblingen samples are only partially consistent with this hypothesis. All have ^{14}C contents corresponding to model ages of about 16 ka and greater. However, while most of the more southerly waters have ^{14}C contents below detection, and possibly much greater ages, the Böttstein, Buntsandstein, and uppermost crystalline waters have ^{14}C model ages of about 10 ka, and the Zurzach waters have model ages of *c.* 18 to 21 ka.

8.4.2 **Waters of High Mineralization from Permian Sediments**

The group of samples with highest mineralization (37 to *c.* 100 g/l) comprises groundwaters from the Permian sediments at 1408 m and 1116 m at Weiach (302/17, 18 and 302/19, 20) and at 1362 m at Riniken (303/(*); Table 3.2.5). No uncontaminated sample of the 1362 m Riniken groundwater could be collected, so the values given here are based on extrapolations from measurements on samples 303/16 through 303/20 (WITTWER, 1986, Section 4.3.4; PEARSON AND OTHERS, 1989, Table 4.3). Because of uncertainty in the extrapolations, no conclusions are based solely on the Riniken sample.

These waters have the following isotopic and chemical characteristics:

- Their mineralization is dominated by sodium and chloride;

- They are enriched in ^{2}H and ^{18}O relative to the waters of the first group. The most positive values of δ^2H and δ^{18}O, -31.5 and -4.61 per mil, are found in the most highly mineralized sample, 302/18 (Figure 3.2.8);

- The most highly mineralized sample (302/18) has a relatively low dissolved sulphate content with an isotopic composition like that of an anhydrite nodule from 1022 m in the Weiach borehole, and typical of Rotliegend or Buntsandstein mineral sulphate (Figure 8.4.2) elsewhere in Middle Europe. Sample 302/19 from the Permian at 1116 m in the Weiach borehole and two samples of lower total mineralization from the Buntsandstein of the Weiach and Riniken boreholes (302/12 and 303/3) have much higher dissolved sulphate contents. Their isotopic compositions are similar to that of anhydrite cement from 1164 m in the Weiach borehole, and typical of mineral sulphate from the Muschelkalk or Röt evaporite with, perhaps, some Buntsandstein or Rotliegend contribution. (Figures 8.4.2 and 7.1.9);

- The ^{14}C content of the most highly mineralized sample is below detection. The other sample with high mineralization has a measurable ^{14}C content, but this may result from the oxidation of modern organic material introduced during drilling (Section 2.1.2.1). The highly negative δ^{13}C value of 302/19,20 (-25.4 per mil) is consistent with this explanation;

- A noble gas temperature is available only for the 1116 m Weiach sample (302/20), but it has the remarkably high value of 27.6 ± 6.2°C. Samples of intermediate mineralization from the Permian of Riniken (303/6) and the crystalline of Schafisheim (304/5-7 and 304/10) also have relatively high noble gas temperatures of 17.9 ± 7.8, 10.9 ± 0.8, and 10.4 ± 1.6°C, respectively.

The most saline sample (302/17, 18) is from the Rotliegend of the Weiach borehole at 1408 m. Its chloride content is 60 g/l, about three times that of normal sea water, but its ^{2}H and ^{18}O contents are those of a non-evaporated meteoric water and its Br/Cl ratio is 0.011, which is typical of crystalline rock. The sulphate content of this water is low (264 mg/l), far below saturation with anhydrite, and its ^{34}S and ^{18}O contents are consistent with the age of its source formation. The He content of this water, like its chloride content, is the highest of any sample from the Buntsandstein, Permian or crystalline (Figure 6.4.4). These properties are consistent with a water of very great age--perhaps approaching that of its source formation.

The dissolved solids in sample 302/18 could have formed by extensive leaching of the sediments in which the water is now found. The Rotliegend of northern Switzerland is non-marine, so the leached material would have been of a continental evaporite origin.

The fact that the sediments themselves are in large part crystalline-rock debris would account for the crystalline signature of the Br/Cl ratio. The ^{2}H and ^{18}O contents of this sample show that the water has not undergone extensive evaporation, and so it must represent a different water from the one associated with the deposition of the easily leachable minerals in the formation. However, the isotopic composition of the dissolved sulphate suggests that this water could not have been much younger than the deposition of the formation.

The waters from the 1116 m zone at Weiach (302/19, 20) and the Riniken sample 303/(*) may have a different origin than the deeper Weiach sample. This is suggested by their different Br/Cl ratios (Figure 8.4.1), different relationships between δ^2H and residues on evaporation (Figure 3.2.9), and different sulphate concentrations and isotopic compositions (Figure 8.4.2). The Br/Cl ratios can be interpreted as indicating that the mineralization in these waters is of marine origin, while the waters themselves have ^{2}H and ^{18}O contents suggesting continental meteoric waters. This could occur if a marine brine were diluted by continental surface or groundwater. The fact that the isotopic composition of the sulphate in these waters corresponds to an age younger than that of the host rock suggests that the diluted brines were of Röt evaporite or even Muschelkalk age.

In spite of uncertainties about the details of the origin of the mineralization of these three highly-saline waters, there is little doubt that they are very old. As Table 8.4.1 shows, these samples have high He contents, and the ^{40}Ar/^{36}Ar measurement on one of them (302/19) is the highest value reported in this study.

8.4.3 Waters of Mixed Origin

This section discusses the remaining samples from the Buntsandstein, Permian, and crystalline. As Figures 3.2.9, 8.4.1 and 8.4.2 show, some of these samples can be explained as mixtures between the waters of low and high mineralization discussed in the two preceding sections. The operation of additional processes, though, is required to account for features of other samples.

The samples from the crystalline at Weiach from 2218 m (302/16) and 2267 m (302/14) have lower salinities than many samples from the Permian, but have Br/Cl ratios of 0.01, suggesting that their mineralization results from leaching of their host rock. The ^{87}Sr/^{86}Sr ratio of the Sr dissolved in sample 302/16 is 0.7167, consistent with the value of 0.7144 for rock from the same interval and supporting a leaching origin. As Figure 3.2.8 shows, the δ^2H and δ^{18}O values of samples 302/16 and 302/14 are not along the meteoric water line, but are enriched in ^{18}O. This could also be a result of reaction between water and rock.

The most highly saline water from the crystalline was from 1326 m at Böttstein (301/(*)). The composition of this water was extrapolated from analyses of samples diluted by varying amounts of drilling fluid (PEARSON, 1985), and so must be used with caution. There are no bromide or strontium isotope data for this sample, but its δ^2H and δ^{18}O values, which are based on extrapolation of diluted samples and so subject to error, plot above the meteoric water line (Figure 3.2.8). Waters plotting above this line are unusual, but are known from the highly saline waters of the Canadian Shield. The isotopic composition of these shield waters has been attributed to oxygen isotope exchange between water and silicate minerals (PEARSON, 1987). At relatively low temperatures, the fractionation factor between water and silicates is large enough to deplete the water in ^{18}O relative to meteoric water of the same hydrogen isotopic composition. At higher temperatures, the fractionation factor would be smaller, so that exchange would drive the $\delta^{18}O$ value of a water to more positive values. This process could also account for the ^{18}O shift in the Weiach waters.

The Br/Cl ratio of waters of low mineralization from the crystalline at Böttstein (301), Kaisten (305), and Leuggern (306), and the Rotliegend of Kaisten (305/2) are intermediate between marine and crystalline rock values (Figure 8.4.1). This suggests sources of mineralization mostly from leaching of the crystalline rock, but with a marine contribution, possibly from marine rock overlying the Buntsandstein.

A number of samples have Br/Cl ratios below that of sea water. This indicates a source of bromide-depleted chloride to these waters, most probably the dissolution of evaporitic halite.

The sample with the lowest Br/Cl ratio is from the Buntsandstein-Rotliegend at Kaiseraugst (3). The isotopic composition of the sulphate dissolved in this sample is at the edge of the range typical of Buntsandstein sulphate elsewhere, and intermediate between ranges typical of Rotliegend, Röt or Muschelkalk sulphate (Figure 8.4.2). Halite is present in the Muschelkalk, and a Muschelkalk influence on the mineralization of the Kaiseraugst water would explain its Br/Cl ratio, and be consistent with its sulphate isotopic composition.

Samples from the Buntsandstein-weathered crystalline at Böttstein, (301/25), the Buntsandstein of Leuggern (306/2), and the Rotliegend at Kaisten (305/2) have bromide and chloride concentrations intermediate between those of the Kaiseraugst water (3) and those of water from the crystalline of the Böttstein (301), Kaisten (305) and Leuggern (306) boreholes. As Figure 8.4.2 shows, the isotopic composition of the sulphate in these three waters is also intermediate between that of the Kaiseraugst water and the range of compositions found in waters from the crystalline. The mineralization of these waters appears to have a mixed origin, including both a normal crystalline component and a component like that giving rise to the Kaiseraugst water.

The Br/Cl ratios of the remaining samples from the Permian and Buntsandstein are the same as, or lower than, that of sea water. The bromide and chloride contents (Figure 8.4.1), the total mineralization and δ^2H values (Figure 3.2.9), and the $\delta^{34}S$ and $\delta^{18}O$ values of the dissolved sulphate (Figure 8.4.2) of all samples are consistent with their being mixed waters. The end-members of the mixtures are waters of the range of types found in the Buntsandstein at Leuggern (306/2) and Böttstein (301/25b), or, at the extreme, at Kaiseraugst (3), and water from the Permian of the Permo-Carboniferous Trough, as sampled at Weiach (302/19) and Riniken (303/(*)). The Weiach, Riniken, and Schafisheim boreholes are within, or near, the Permo-Carboniferous Trough, while the Kaisten borehole is some distance from it.

The samples of low mineralization from the Buntsandstein at Böttstein (301/25b) and Leuggern (306/2) have ^{2}H and ^{18}O contents, which are virtually identical to those of water from the upper crystalline in the same boreholes. Their relatively higher chloride concentrations must result from the addition of bromine-free chloride.

Waters of intermediate salinity from the crystalline were sampled at Schafisheim (304/8, 9 and 304/10) and at Säckingen (125, 126, 161). The total mineralization of the Schafisheim and Säckingen Stammelhof (161) waters are only slightly lower than those from the crystalline at Weiach (302/14, 16), but they are, at least in part, of a different origin. The Br/Cl ratios of the Säckingen and one of the Schafisheim samples (304/10) are slightly lower than that of sea water, while that of the other Schafisheim sample (304/8) is much lower. The ^{34}S and ^{18}O contents of the Schafisheim samples are similar to Röt or Muschelkalk sulphate, or to a mixture of Röt or Muschelkalk and additional sulphate typical of the low-salinity waters of the upper crystalline. As discussed in Section 7.1.6, geophysical evidence suggests that the crystalline at Schafisheim is surrounded by sediments of the Permo-Carboniferous Trough. The similarity of the mineralization of the Schafisheim crystalline water to water from the Permian elsewhere (sample 302/12 from 985 m at Weiach, for example) is consistent with this structure.

No ^{34}S data are available on the sample from Säckingen Stammelhof (161), but such measurements were made on samples of water from the crystalline at other locations at Säckingen, the Badquelle (125) and the Margarethenquelle (126). Both these samples have tritium corresponding to at least 50 per cent young water. While the ^{34}S and ^{18}O contents of sulphate from the Margarethenquelle (126) are similar to those of the low-salinity waters from the upper crystalline, those from the Badquelle water (125), suggest it contains a strong admixture of Buntsandstein sulphate as found in the Kaiseraugst (3) sample. Thus, the mineralization of waters from the crystalline at Säckingen is more similar to that in the Buntsandstein at Kaiscraugst (3) or Kaisten (305/1) than to water found in the sediments of the Permo-Carboniferous Trough to the south.

8.4.4 Conclusions about Buntsandstein, Permian and Crystalline Water

Waters of relatively low mineralization (< c. 1.5 g/l residue on evaporation) are found in the crystalline of the Kaisten (305) and Leuggern (306) boreholes, in the upper part of the Böttstein (301) borehole and at Zurzach (131, 132). These waters are isotopically similar to young water from the Black Forest and modern precipitation in northern Switzerland, consistent with their recharge in that area. However, the recharge temperatures of these waters derived from their noble gas contents are much lower than the present mean annual temperatures in the region, suggesting that they were recharged at a time of cooler climate. This is in keeping with their model ^{14}C ages of > 10 ka. Waters in the Buntsandstein and Permian in these boreholes are similar to the water in the underlying crystalline, except for slightly higher salinities. The source of the chloride in the Buntsandstein of the Leuggern, Kaisten and Böttstein boreholes as well as at Kaiseraugst (3) appears to be related to the dissolution of halite.

The water from the deep crystalline at Böttstein and Weiach and from the Rotliegend at 1408 m at Weiach is highly mineralized. This mineralization developed by extensive *in situ* water-rock reactions.

Water of about the salinity of sea water is found in the upper part of the Rotliegend section at Weiach at 1116 m (302/19, 20), and in the Rotliegend of Riniken at 1362 m (303/(*)). These waters developed by dilution with continental meteoric water of a brine younger than the age of the formation itself. Water from the crystalline of Schafisheim and from the Buntsandstein and Rotliegend intervals in Schafisheim, Riniken and Weiach boreholes are influenced by mixing with water like that found in the upper crystalline as well as additional solution of evaporite minerals. The waters of intermediate mineralization from several locations at Säckingen (125, 126, 161) contains salts originally of Buntsandstein origin, like those in the Buntsandstein at Kaiseraugst (3), Böttstein (301), Leuggern (306) and Kaisten (305).

The flow path consistent with this distribution of chemistry and isotopic properties is one of virtually stagnant water in the Permo-Carboniferous Trough and deep crystalline of the Böttstein, Weiach and Schafisheim boreholes. For the most part, the salinity of the deep waters was developed from leaching of the present host rocks--*e.g.*, samples 302/14 and 302/16 from the crystalline rock at Weiach, and 302/18, 302/19, and 303/20, corrected, from Permian sediments of the Weiach and Riniken boreholes. The exception to this, samples 304/8 through 10 from Schafisheim, are from the crystalline in a region surrounded by sediments, and indicate some flow from the sediments to the crystalline.

There is increasing groundwater circulation at shallower depths and to the north of the Permo-Carboniferous Trough. Even in the more shallow crystalline waters, if the model ^{14}C ages represent real groundwater residence times and the noble gas temperatures truly reflect recharge under cooler climatic conditions, flow is sluggish compared with that in the overlying sedimentary aquifers such as the Keuper and the Muschelkalk in the Folded and western Tabular Jura.

8.5 Inter-Aquifer and Regional Flow

The isotope results describe active groundwater flow systems in the sedimentary aquifers of the Folded and western Tabular Jura, grading through regions of lesser activity in the sediments and upper crystalline rock of the eastern Tabular Jura and deeper Molasse Basin to the deep crystalline and sediments of the Permo-Carboniferous Trough, in which circulation is so limited that water ages may approach those of the formations themselves. The deep saline waters sampled directly, or present as mixtures in more dilute waters, have undergone such extensive water-rock interactions that evidence of their origin has been lost. The remaining waters were all recharged in the region of northern Switzerland or southern Germany. There is no evidence of recharge from such distant sources as the Alps.

There is little evidence of strong flow among the sedimentary aquifers above the Buntsandstein. Even in the eastern Tabular Jura, where intra-aquifer flow is sluggish, the isotope data do not indicate inter-aquifer flows. The isotope data are not precise enough to rule out all possibility of inter-aquifer flow, but will be useful to put upper limits on the amounts of such flow permissible in regional aquifer models. The isotope data do mark several areas of anomalous flow, however, including the presence of crystalline water in the Muschelkalk at Pratteln, the presence of older mixed water in the Muschelkalk at Magden and Baden, zones of very rapid flow in the Muschelkalk at Weiach and Lostorf, and the mixing of Molasse water types at Eglisau.

Water in the upper crystalline and associated Permian and Buntsandstein of northern Switzerland is recharged where these formations crop out in and around the Black Forest. However, these waters have such a homogenous character that the isotope data do not distinguish any flow patterns. South of the Rhine, these waters have ^{14}C model ages ranging from 13 to 21 ka to > 31 ka. The fact that the waters are in chemical and isotopic equilibrium with such fracture minerals as calcite and barite is consistent with these ages. The relationship of water in the crystalline at Siblingen to that south of the Rhine is not clear, but it is difficult to account for all the isotope data if Siblingen is considered to be on a flow path leading from the Black Forest to the crystalline south of the Rhine.

The isotopic and chemical character of water from the Permo-Carboniferous Trough and the deep crystalline is consistent with extensive water-rock reactions, and the presence of solutes, and perhaps water itself, of age close to that of the formations. There is enough difference between these waters to distinguish those formed by sediment-water reaction from those formed by crystalline-water reactions regardless of where they may have been sampled. Data on these waters will be particularly useful in calibrating models of water-rock geochemical evolution.

8.6 **Recommended Further Studies**

Throughout this report, and particularly in this Chapter 8, uncertainties in interpretation have been pointed out and suggestions about additional studies which could resolve the ambiguities have been made. These recommendations are repeated here in summary form.

- Continue time series analyses of mixed waters such as are found at Magden and Eglisau. Serial analyses make it possible to characterize end-members of mixtures;

- Perform fluid-inclusion temperature determinations, chemical analyses, and $^{87}Sr/^{86}Sr$ and $\delta^{18}O$ isotope measurements on the same samples of calcite from the crystalline;

- Sample a number of groundwaters from the Black Forest for determinations of:

    + Noble gas temperatures
    + ^2H, ^3H and ^{18}O contents
    + ^{13}C and ^{14}C contents
    + Chemistry

Present information about these entities is insufficient to explain the relationship between water in the Black Forest and the upper crystalline of northern Switzerland. For example, what noble gas temperatures are displayed by young waters of the Black Forest? Are they all as high as the two published values on samples from south of Freiburg-im-Breisgau (RUDOLPH AND OTHERS, 1983), or are some temperatures lower and closer to those of the Nagra borehole samples as a result of recharge only during cold seasons?

- Sample additional locations which may indicate interconnections between formations as do the samples from Pratteln, for example;

- The chemical differences between the superficial, clearly young waters of the Black Forest and the Nagra borehole waters may, if carefully modelled, provide order-of-magnitude indications of water ages.

For a better understanding of the hydrogeological history of the Molasse Basin, it would be helpful to complete the data set of the Eglisau boreholes with ^{39}Ar-, noble gas- and carbon isotope analyses. Noble gas analyses in modern deep groundwaters, *e.g.*, Beuren and Schönenbuch, would allow more accurate interpretations of the results in old waters. Measurements of the ^{40}Ar/^{36}Ar ratios in more sampling points of the regional programme would give additional information on groundwater evolution in the Molasse Basin. Finally, it is desirable to find Na-Cl waters closer to the assumed saline end-member than the investigated ones.

Table 8.1.1: Summary of isotope data on recharge temperatures and model ages of Tertiary and Malm aquifer waters. (Page 1 of 2)

Location Number	Name	Source Formation (1)	Noble Gas Temperatur °C ± 1SD (Tab. 3.3.1)	3H ± 2SD (TU) (Table 3.2.1)	Delta 2H (per mil)	Delta 18O (per mil)	Delta 13C (per mil) (Tables 5.2.1,	14C pmc 5.2.2)	14C "Ages" ka ± 1SD	39Ar % mdn (Tab. 4.4.1,	39Ar "Age" a 4.4.2)	4He cm3 STP/36Ar cm3 H2O (2)	40Ar
SHALLOW CA-MG-HCO3 WATERS													
42/1	Schwarzenberg	tOSM		86.8 ± 6.0	-71.6	-10.16							
130/2	Gränichen	tOSM		68.1 ± 4.5	-71.8	-10.11							
33/2	Fulenbach	tUSM		87.0 ± 5.7	-76.6	-10.55							
32/2	Attisholz	tUSM		68.1 ± 4.5	-72.6	-10.12							
7/1	Malleray	jo		93.3 ± 6.2	-74.1	-10.61							
99/1	Welschingen	jo(-q)		109.6 ± 7.3	-71.8	-9.89							
80/2	Durlinsdorf	jo		47.1 ± 3.2	-65.5	-9.26							
MIXED WATERS OF VARIOUS WATER TYPES													
30/1	Wildegg	jo		88.8 ± 5.9	-70.7	-10.01							
85/1	Steinenstadt	teo		95.1 ± 6.2	-65.6	-8.88							
501/102B	Reichenau	q	43 2.8 ± 1.3	24.0	-86.3	-12.09	-15.2	24.0	>6.5				
DEEP AND MIXED CA-MG-HCO3 WATERS													
96/1,2 *	Oberbergen	tvu		4.1 ± 0.7	-61.6	-8.81	-10.5	50.7	0.2 ± 1.0				
170/2	Dettenberg	tOMM		1.0 ± 0.7	-71.6	-10.01							
502/101B	Sauldorf	tOMM		4.0 ± 0.6		-9.94	-10.8	39.6	1.5 ± 0.8				
5/1,3 *	Schönenbuch	tUSM		1.0 ± 0.6	-63.8	-8.90	-14.2	68.0	Modern				
149/1 and 100	Beuren	jo		17.0 ± 1.3	-70.4	-9.80	-11.2	52.4	Modern				
NA-HCO3 WATERS													
98/1	Konstanz	tOMM		<0.7	-92.3	-12.71	-4.3	<1.3	>13				
98/102B				0.5	-90.9	-12.72	-4.0	0.6	>26				
128/1	Mainau	tOMM	45 2.1 ± 0.8	<0.9	-83.5	-11.62	-5.5	<2.2	>19				
128/101B				0.3	-83.4	-11.71	-5.7	0.7	>31				
508/101B	Birnau	tOMM	44 3.0 ± 0.2	0.5	-91.1	-12.40	-6.4	1.9	>22				
509/108B	Ravensburg	tOMM		0.4 ± 0.6		-12.22	-4.7	0.8	>17				
1/1,3 *	Zürich Aqui	tOMM	23 2.4 ± 1.3	<0.7	-88.0	-11.89	-3.0	<1.3	>19	<4.9	>1100	2.6E-07 295 1.3E-05 296	

Table 8.1.1: Summary of Isotope data on recharge temperatures and model ages of Tertiary and Malm aquifer waters. (Page 2 of 2)

Location Number	Name	Source Formation	(1)	Noble Gas Temperatur °C ± 1SD (Tab. 3.3.1)	3H ± 2SD (TU) (Table 3.2.1)	Delta 2H (per mil)	Delta 18O (per mil)	Delta 13C (per mil) (Tables 5.2.1, 5.2.2)	14C pmc	14C "Ages" ka ± 1SD	39Ar % mdn	39Ar "Age" a (Tab. 4.4.1, 4.4.2)	4He cm3 STP/cm3 H2O	40Ar/36Ar (2)
97/1	Singen	jO			<0.7	-90.7	-12.43	-5.2	<2.2	>18				
97/105B		jO	46	1.9 ± 1.7	0.6	-86.8	-12.30	-6.1	0.5	>31				
127/1-17 *	Lottsteten	jO			0.8	-73.7	-10.23	-3.8	2.0	>13				
127/104					0.8 ± 0.7	-72.0	-10.30	-3.9	3.5	>11				
77/1	Neuwiller	jmHR			<0.7	-75.8	-10.35	-4.4	1.3	>26				

NA-CL WATERS

2/1-3 *	Tiefenbrunnen	tOMM	24	9.7 ± 5.4	<0.8	-72.6	-9.09	-3.1	<2.2	>13	<16	>700	5.4E-05	302
141	Eglisau 1	tUSM			<0.7	-74.8	-9.89	-8.7	11.1	>12 +				
142/2-15 *	Eglisau 2	tUSM			<1.0	-72.4	-9.35							
143/2-17 *	Eglisau 3	tUSM			0.9	-70.4	-8.91							
144/2-16 *	Eglisau 4	tUSM			0.9	-67.0	-8.22							
304/1,2 *	Schafisheim	tUSM			<1.2	-61.7	-6.41	-8.5	40.5	Modern			5.6E-05	357
302/6,9 *	Welach	jOkI			<3.6	-61.8	-5.76	-4.2	12.5	>1			1.5E-04	

(1) Sample number used in text, tables and figures of Section 3.3.

(2) 4He and 40Ar/36Ar data from Tables 6.4.1 ad 3.3.2, respectively.

* Values given are averages of several analyses or combinations of data from different analyses.

\+ Mixed sample: Limiting age of old component given.

B Values from BERTLEFF (1986), noble gas temperatures recalculated (Section 3.3).

Table 8.2.1: Summary of isotope data on recharge temperatures and model ages of Dogger, Lias, and Keuper aquifer waters.

Location Number	Name	Source Formation	3H ± 2 SD TU -- (Table 3.2.1) --	Delta 2H (per mil)	Delta 18O (per mil)	Delta 13C (per mil)	14C pmc	14C "Age" ka ± 1 SD -- (Tables 5.2.1, 5.2.2) --	4He cm3 STP/ cm3 H2O
56/2	Weissensteintunnel 1.480	ju	<1.2	-75.3	-10.96				
171/101	Beznau 104.5-109.5 m	km	5.6 ± 3.9		-8.77	-6.4	<4.7	>12	
303/1	Riniken	km	<1.2	-41.2	-6.43	-8.2	3.35	>17	1.3E-04
	RHINE GRABEN								
109/1	Munzingen	jmHR	17.8 ± 1.5	-62.7	-8.88				
106/1	Riedlingen Subtherm	jmHR	10.2 ± 0.9	-66.3	9.43				
108/1	Müllheim	jmHR	18.1 ± 1.3	-64.0	-8.93				
77/1	Neuwiller	jmHR	<0.7	-75.8	-10.35	-4.4	1.20	>26	
81/1	Bellingen 1	jmHR	<1.0	-76.0	-9.99				
83/1	Bellingen 3	jmHR	<0.7	-75.1	-9.95				
84/1	Steinenstadt Thermal	jmHR	<0.7	-75.9	-10.14				

Table 8.3.1: Isotope data on recharge temperatures and model ages of Upper Muschelkalk aquifer waters. (Page 1 of 2)

Location Number	Name	Noble Gas Temperature °C ± 1 SD (Tab. 3.2.1)	3H ± 2SD TU (Table 3.2.4)	Delta 2H (per mil)	Delta 18O (per mil)	Delta 13C (per mil)	14C pmc (Tables 5.2.1, 5.2.2)	14C "Age" ka ± 1SD	39Ar % mdn (Tab. 4.4.1, 4.4.2)	39Ar "Age" a ± 1SD	4He cm3 STP/ cm3 H2O	40Ar/36Ar (2)
	FOLDED AND WESTERN TABULAR JURA											
8/2	Pratteln 26	9.7 ± 0.3	9.2 ± 0.8	-66.7	-9.29	-9.5	39.4	1.4 ± 1.6	293		1.2E-07	312
9/2	Frenken- 27 dorf	7.7 ± 0.9	<1.1	-64.4	-9.14	-9.8	19.0	7.2 ± 1.0	< 24	≥550	1.7E-07	313
12/1	Kaisten Felsbohrung		69.8 ± 4.8	-67.2	-9.48	-14.0	75.6	Modern				
14/1	Densbüren		5.2 ± 0.6	-72.3	-10.23	-8.9	33.3	2.0 ± 1.2				
14/3	" 29	7.5 ± 1.7	7.8 ± 0.9	-73.9	-10.24	-9.6	43.4	0.4 ± 1.1	61.5	190 ± 30 EM: c 220	1.4E-07	≥296
16/1	Windisch BT2		83.7 ± 5.6	-70.1	-9.52	-13.4	77.3	Modern				
123/110	Lostorf 3			-71.4	-10.15	-9.4	44.0	Modern	52.5	250 ± 30		
123/111	"		5.8 ± 0.7			-9.6	44.1	Modern				
123/113	"					-9.3	41.0	0.6 ± 1.3				
271/101	Magden Welere	37	7.6 ± 1.4	<0.8	-70.8	-10.14	-8.1	4.0	18 ± 2			7.9E-07	292
271/103	"	38	5.3 ± 1.5	<0.8	-74.6	-10.49	-6.8	8.5	10 ± 2	16.3	≥700	3.1E-07	
271/104	"			<1.1	-71.5	-9.84	-8.2	10.3	10 ± 1.0	10.8	≥700	7.9E-07	
272/102	Magden Falke	39	9.0 ± 0.5	<1.6	-66.2	-9.24	-9.6	28.1	3.3 ± 1.1	16.3	≥700	2.7E-07	305
272/103	"			<1.0	-72.3	-9.22	-9.3	5.5	16.5 ± 1.1	22.8	≥570		
273/102	Magden Eich			<1.9	-69.8	-9.83	-10.5	19.9	6.8 ± 1.0	23.4	≥570		
273/103	"			<0.9	-69.3	-9.50	-9.5	16.7	7.6 ± 1.2	39.1	c350		
274/102	Magden Stockacher			4.3 ± 0.7	-68.5	-9.30	-10.6	37.8	1.5 ± 0.9		EM: c 550		
306/1	Leuggern 53.5-96.4	a	9.0 ± 0.7	11.6 ± 1.1	-68.3	-9.96	-15.8	12.0	6	44.5	≥300	4.7E-06	297
	EASTERN TABULAR JURA											
15/1	Beznau	28	0.7 ± 5.5	<1.0	-81.2	-11.09	-5.4	0.4	29	<4.6	1200	4.0E-05	
301/2	Böttstein 123.2-202.5	1	6.3 ± 1.2	<1.0	-78.2	-11.16	-4.8	1.5	17	<6.3	1000	2.1E-04	≥310
303/2	Riniken 617.3-696.0	12	8.1 ± 2.5	<1.1	-71.6	-10.14	-4.8	0.9	19	<7.2	1000	6.8E-05	303

Table 8.3.1: Isotope data on recharge temperatures and model ages of Upper Muschelkalk aquifer waters. (Page 2 of 2)

Location Number	Name	Noble Gas Temperature °C ± 1 SD (Tab. 3.2.1)	3H ± 2SD TU (Table 3.2.4)	Delta 2H (per mil)	Delta 18O (per mil) (Table 3.2.4)	Delta 13C (per mil) (Tables 5.2.1, 5.2.2)	14C pmc	14C "Age" ka ± 1SD (Tab. 5.2.2)	39Ar % mdn	39Ar "Age" a ± 1SD (Tab. 4.4.1, 4.4.2)	4He cm3 STP/ cm3 H2O	40Ar/ 36Ar (2)
NORTHERN PART OF MOLASSE BASIN												
119/118	Schinznach alt		46.9 ± 5.1	-72.1	-10.19	-12.0	41.4	1.6 ± 0.9				
119/119	"					-10.4	42.0	0.4 ± 1.2				
120/1	Schinznach S2					-9.3	27.1	0.6				
120/118	"		18 ± 2.1	-72.7	-10.23	-10.4	24.5	2.3				
120/119	"					-9.4	28.0	Modern				
124/109	Lostorf 4		<0.5		-10.33	-6.8	14.8	5.7 ± 2.1				
124/110	"		<1.0	-70.2	-10.18	-6.8	15.5	5.3 ± 1.7				
124/111	"			-70.5	-8.99	-9.3	15.0	6.1 ± 1.8				
201/1 M	Ennetbaden Allgemeine		3.3 ± 0.8			-7.0	6.3	4			1.1E-04	301
203/112	Ennetbaden Schwanen		6.2 ± 0.6	-68.4	-9.01	-8.5	7.2	8				
206/102	Baden Limmatquelle		3.6 ± 0.5	-70.0	-8.99	-9.9	6.0	12				
201/1,2 M	Ennetbaden Allgemeine		4.8 ± 0.7	-69.5	-8.94						9.1E-05	3.2 ± 1
219/112	Baden Verenahof		5.8 ± 0.7			-8.0	7.0	7				
302/10	Welch 20 822-896.1	7.8 ± 0.5	3.2 ± 0.7	-86.6	-11.76	-4.6	13.3	Modern	12.8	800 ± 60	1.7E-06	≥294
304/3	Schafisheim 15 1227.8-12??	7.0 ± 1.7	<0.9	-80.0	-11.02		4.4	- -			2.4E-04	
304/14	Schafisheim 1240.7-1261.6		1.4 ± 0.7	-78.6	-10.81				12.3	800		

(1): Sample number used in text, tables and figures of Section 3.3.

(2): 4He and 40Ar/36Ar data from Table 6.4.1 and Table 3.2.2, respectively.

M: He concentrations and 40Ar/36Ar ratos from Weizmann Institute analyses, Table 2.3.1.

Table 8.4.1: Isotope data on recharge temperatures and model ages of Buntsandstein, Permian and Crystalline waters. (Page 1 of 4)

Location Number	Name	Source Formation (1)	Chloride (g/l)	Noble Gas Temperature °C ± 2SD (Tab. 3.3.1)	3H ± 2SD TU	Delta 2H (per mil) (Tab. 3.2.5)	Delta 18O (per mil)	Delta 13C (per mil)	14C pmc (Tab.5.2.1,5.2.2)	14C "Age" ka ± 1SD	36Cl/Cl x E-15 Meas. ± 1SD Calc. (Tab.6.2.1)	4He cm3 STP/ cm3 H2O (Tab.6.4.1,3.3.1)	40Ar/36Ar (2)
	WATERS OF LOW MINERALIZATION FROM THE CRYSTALLINE												
166/1	Sulzburg Waldhotel	KRI	0.0		23.8 ± 1.8	-66.3	-9.82	-13.2	45.0	1.5 ± 0.5			
167/1	Görwihl Bohrloch 30	KRI	0.0		57.3 ± 4.0	-69.0	-10.21	-21.2	107.8	Modern			
B40 (3)	Freiburg 1m Br.D	KRI 40		7.8		-63.9			82.5				
B41 (3)	Freiburg 1m Br.D	KRI 41		7.0		-64.9			68.4				
95/2	Waldkirch	KRI 30	0.2	4.5 ± 0.7	<0.9	-71.7	-10.26	-8.00	0.69	>31		3.5E-04	309
131/105 /115	Zurzach 1	KRI "	0.1		<0.5 1.0		-10.30 -10.30	-9.60 -9.50 -9.40	4.62 6.78 4.67	19 ± 1.0 15 ± 1.0 18 ± 1.0			
132/103 /131	Zurzach 2	KRI "	0.1		1.0	-73.7 -74.8	-10.09	-9.40	3.50 2.9	21 ± 1.0			
B43 (3)	Freiburg 1m Br.D	KRI 42		3.8									
301/8c, 9	Böttstein 393.9-405.1	KRI	0.1	2.8 ± 1.4	<1.0	-72.7	-10.05	-9.30	8.25	13.2 ± 1.0	13 ± 10	2.2E-04	27.5
301/16, 17	Böttstein 608.0-628.8	KRI	0.1	4.2 ± 1.5	1.0	-74.1	-10.01	-9.00	7.89	13.4 ± 1.0		1.4E-04	≥295
301/12b, 13	Böttstein 618.5-612.1	KRI	0.1	2.8 ± 1.2	1.0	-72.7	-10.04	-8.80	1.30	>27	43 ± 9	2.1E-04	≥303 34.4
301/23	Böttstein 640.8-657.2	KRI		4.2 ± 0.4	3.2 ± 0.7	-74.6	-9.97	-9.10	13.80	> 7	48 ± 19	1.9E-04	306 51.3
301/18	Böttstein 782.0-802.8	KRI	0.1	2.1 ± 0.9	0.9 ± 0.9	-73.7	-9.96	-9.10	7.64	13.8 ± 1.0	29 ± 10	1.5E-04	≥291 44.1
	Böttstein Crystalline	KRI 4		3.2 ± 0.8		-73.6	-10.01						
305/2	Kaisten 276.0-292.5	r 6	0.1	3.6 ± 0.6	0.7 ± 0.3	-73.5	-10.36	-17.80	0.80	>22	32 ± 6	9.3E-05	309 14.2
305/4	Kaisten 299.3-321.5	KRI	0.1	3.7 ± 1.5	1.1 ± 0.7	-73.5	-10.29	-9.80	0.98	>29		1.1E-04	289
305/6	Kaisten 475.5-489.8	KRI	0.1	2.4 ± 1.0	1.3 ± 0.7	-72.7	-10.34	-10.40	0.86	>31	27 ± 6	1.0E-04	296 34.9
305/9	Kaisten 816.0-822.9	KRI	0.1	2.3 ± 0.6	0.9 ± 0.7	-73.8	-10.32	-9.90				1.0E-04	298

405

Table 8.4.1: Isotope data on recharge temperatures and model ages of Buntsandstein, Permian and Crystalline waters. (Page 2 of 4)

Location Number	Name	Source Formation (1)	Chloride (g/l)	Noble Gas Temperature °C ± 2SD (Tab.3.3.1)	3H ± 2SD TU	Delta 2H (per mil) (Tab.3.2.5)	Delta 18O (per mil)	Delta 13C (per mil)	14C pmc (Tab.5.2.1,5.2.2)	14C "Age" ka (Tab.5.2.1,5.2.2)	36Cl/Cl x E-15 Meas. ± 1SD (Tab.6.2.1)	36Cl/Cl x E-15 Calc.	4He cm3 STP/ cm3 H2O (Tab.6.4.1,3.3.1)	40Ar/36Ar (2)
305/12	Kaisten 1021.0-1040.9	KRI	0.1	3.6 ± 2.2	<0.8	-74.1	-10.61	-9.20	0.84	>30	53 ± 8	36.3	5.3E-05	
305/14, 15	Kaisten 1140.8-1165.8	KRI	0.1	2.8 ± 1.7	<0.8	-71.7	-10.40						4.6E-05	
305/16, 17	Kaisten 1238.0-1305.8	KRI	0.1	3.5 ± 2.2	1.2 ± 0.7	-73.2	-10.46	-10.10	1.56	>26	54 ± 8	38.7	5.1E-05	298
	Kaisten Crystalline	KRI 7		3.1 ± 0.6		-73.2	-10.40							
306/4	Leuggern 235.1-267.5	KRI	0.2	3.6 ± 2.3	0.8 ± 0.7	-73.3	-10.49	-10.10	1.36	>27	73 ± 14	39	4.8E-05	≥296
306/5, 6	Leuggern 440.4-448.1	KRI	0.1	5.5 ± 1.4	<1.2	-73.0	-10.35	-10.10	3.19	>20			4.6E-05	*298
306/7	Leuggern 507.4-568.6	KRI	0.1	2.6 ± 0.8	<1.0	-71.9	-10.37	-10.20	2.41	>23			6.2E-05	300
306/9	Leuggern 702.0-709.5	KRI	0.1	2.4 ± 0.7	<1.3	-74.6	-10.52	-10.20	1.40	>27	170 ± 10	21.9	8.9E-05	301
306/17	Leuggern 834.5-859.5	KRI	0.1	4.3 ± 0.7	3.5 ± 0.7	-73.5	-10.48	-9.80	2.18	>23			6.0E-05	
306/11b, 13	Leuggern 916.2-919.7	KRI	0.2		8.1 ± 0.9	-71.8	-10.35				40 ± 6	32.8	2.2E-04	*
306/20	Leuggern 1179.3-1227.2	KRI	0.1		1.3 ± 0.7	-72.3	-10.42	-26.00	4.50	>16	35 ± 5	35.5		
306/26	Leuggern 1427.4-1439.4	KRI	0.4		1.4 ± 0.9	-72.1	-10.02	-23.00	11.70	> 8				
306/16	Leuggern 1637.4-1649.3	KRI	0.1	1.9 ± 1.4	0.7 ± 0.7	-71.9	-10.25	-9.30	1.19	>27	38 ± 6	62.3	1.1E-04	≥298
306/23	Leuggern 1642.2-1688.9	KRI	0.1	5.1 ± 0.9									5.2E-05	
	Leuggern Crystalline	KRI 11		3.6 ± 1.3		-72.7	-10.36							

Table 8.4.1: Isotope data on recharge temperatures and model ages of Buntsandstein, Permian and Crystalline waters. (Page 3 of 4)

Location Number	Name	Source Formation(1)	Chloride (g/l)	Noble Gas Temperature °C ± 2SD (Tab.3.3.1)	3H ± 2SD TU	Delta 2H (per mil) (Tab.3.2.5)	Delta 18O (per mil)	Delta 13C (per mil)	14C pmc (Tab.5.2.1,5.2.2)	14C "Age" ka ± 1SD	36Cl/Cl × E-15 Meas. ± 1SD Calc. (Tab.6.2.1)	4He cm3 STP/ cm3 H2O (Tab.6.4.1,3.3.1)	40Ar 36Ar (2)
WATERS OF HIGH MINERALIZATION FROM THE PERMIAN													
302/19 20	Weiach 1109.2-1123.8	r	22	18.4 27.6 ± 6.2	2.2 ± 0.6	-37.8	-5.32	-25.40 c. 13			20 ± 8	15.3 1.6E-03	395
302/18	Weiach 1401.1-1415.7	r	59.5		6.4 ± 0.9	-31.5	-4.61	-4.40	0.40		29 ± 4	26.9 4.5E-03	*
303/20, Corr	Riniken 1354.0-1369.0	r	23.4			-36.8	-3.98						
WATERS OF INTERMEDIATE MINERALIZATION FROM THE BUNTSANDSTEIN AND CRYSTALLINE													
125/1	Säckingen Badquelle	KRI	1.5		46.5 ± 2.0	-66.6	-8.89	-13.00	36.9	Modern		1.7E-04	
/108	"	"	1.6		39.4 ± 4.2	-66.4	-8.99	-14.20	34.0	>3.9			
/109	"	"	1.7		35.3 ± 4.3			-12.40	28.0	>4.3			
159/4	Rheinfelden	s- KRI	32	5.9 ± 1.1	<0.9	-70.7	-9.24	-7.10	2.2	>16		5.3E-04	317
/106	"	"	0.6										
161/1	Säckingen Stammelhof	KRI	33 3.6	9.4 ± 1.2	1.5 ± 0.7	-67.1		-10.90	2.9	>22		2.7E-04	350
301/25b	Böttstein 305.2-319.8	s- KRI	0.6	2.8 ± 0.4	0.2 ± 0.2	-73.2	-10.08	-9.30	14.03	9.0 ± 0.9		1.1E-04	314
/5,6	Böttstein 305.6-327.6	"	0.7	4.7 ± 0.7	<0.7	-73.7	-10.15	-6.30	12.13	7.6 ± 1.4	14 ± 7	8.7 1.3E-04	301
	Böttstein Bss-Cryst.	s- KRI	2	3.8 ± 1.0		-73.5	-10.12		13.08	8.3 ± 0.7			
302/12	Weiach 981.0-989.6	s	21 2.9	6.0 ± 0.5	<0.8	-60.5	-8.12	-8.30	3.63	>17	19 ± 3	8.7 1.8E-03	306
305/1	Kaisten 97.1-129.9	s	5 2.0	5.4 ± 1.3	0.4 ± 0.2	-68.3	-9.03	-17.20	1.60	>25		3.0E-04	328
306/2	Leuggern 208.2-227.5	s	9 0.5	4.5 ± 1.2	<0.8	-72.9	-10.39	-9.30	1.56	>26	69 ± 10	9.5 8.8E-05	

Table 8.4.1: Isotope data on recharge temperatures and model ages of Buntsandstein, Permian and Crystalline waters. (Page 4 of 4)

Location Number	Name	Source Formation(1)	Chloride (g/l)	Noble Gas Temperature °C ± 2SD (Tab. 3.3.1)	3H ± 2SD TU	Delta 2H (per mil) (Tab. 3.2.5)	Delta 18O (per mil)	Delta 13C (per mil)	14C pmc	14C "Age" ka ± 1SD (Tab.5.2.1,5.2.2)	36Cl/Cl x E-15 Meas. ± 1SD Calc. (Tab.6.2.1)	4He cm3 STP/ cm3 H2O (2) (Tab.6.4.1,3.3.1)	40Ar/ 36Ar (2)

WATERS OF INTERMEDIATE MINERALIZATION FROM THE PERMIAN AND CRYSTALLINE

Location Number	Name	Source Formation(1)	Chloride (g/l)	Noble Gas Temp °C ± 2SD	3H ± 2SD TU	Delta 2H	Delta 18O	Delta 13C	14C pmc	14C "Age" ka	36Cl/Cl x E-15 Meas. ± 1SD	Calc.	4He cm3 STP/ cm3 H2O	40Ar/ 36Ar
3/1	Kaiseraugst	s-r	2.6		<0.7	-72.5	-10.35	-8.30	2.85	>18				
3/3	Kaiseraugst	s-r	2.6		<0.7	-72.0	-10.28	-9.20	2.55	>20				
302/16	Weiach 2211.6-2224.6	KRI	3.4		2.7 ± 0.7	-62.0	-8.17	-26.70	10.10		40 ± 5	45.5	2.6E-03 *	
302/14	Weiach 2260.5-2273.5	KRI	4.3			-60.2	-7.65						2.5E-03 *	
303/3	Riniken 793.0-820.2	s-r 13	4.1	7.3 ± 0.3	0.7 ± 0.7	-65.4	-6.92	-9.20	0.55	>30	34 ± 7	12	1.1E-03	324
303/5b, c	Riniken 958.4-972.5	r	9.1		<1.1	-57.5	-6.73	-7.60	10.85	>6.8	34 ± 6	32.4		
303/6	Riniken 977.0-1010.0	r 14	4.9	17.9 ± 1.8	2.2 ± 1.0	-53.4	-6.65	-14.60	40.81	> 3			2.0E-03	357
304/5 -7	Schafisheim 1476.0-1500.4	s- 16 KRI	6.4	10.9 ± 0.8	1.8 ± 0.7	-58.0	-5.42	-21.20	1.40	>21	19 ± 4	16.4	1.4E-03	365
304/8, 9	Schafisheim 1564.5-1577.7	KRI 17	2.4	3.3 ± 4.8	2.5 ± 0.7	-64.8	-6.66				27 ± 5	84.3	2.7E-03	302
304/10	Schafisheim 1883.5-1892.3	KRI 18	3.6	10.4 ± 1.6	1.5 ± 0.9	-60.8	-5.55	-9.20	0.70	>31	38 ± 5	75.3	9.1E-04	349

(1): Sample number used in text, tables and figures of Section 3.3.

(2): 4He and 40Ar/36Ar data from Table 6.4.1 and Table 3.3.2, respectively.

(3): Data from RUDOLPH AND OTHERS (1983).

* : Indicates helium concentration from PSI analysis.

9. SUMMARY

Nagra, the Swiss National Cooperative for the storage of radioactive waste, has been carrying out a comprehensive field investigation programme to assess the performance of potential repositories for the final disposal of nuclear waste in northern Switzerland. As part of this study, water samples representing different aquifers were collected for hydrogeochemical and isotope analysis. About 140 existing wells and springs were sampled as part of the "regional programme" and about 40 useful samples were taken from zones in the seven boreholes drilled under the Nagra deep borehole programme. This report presents all isotope data collected as part of the Nagra programme in northern Switzerland together with supplementary literature data as appropriate. It also includes a first comprehensive synthesis and interpretation of the meaning of these data.

The isotopes studied have the potential to yield information about the flow rate, origin and geochemical evolution of groundwater. Isotopes addressing flow rates included ^{3}H, ^{3}He, ^{4}He, ^{14}C, ^{36}Cl, ^{36}Ar, ^{39}Ar, ^{40}Ar, ^{85}Kr, ^{234}U and ^{238}U. Data bearing on the location and conditions of recharge included the concentrations of dissolved noble gases (Ne, Ar, Kr, and Xe), and the ^{2}H/^{1}H and ^{18}O/^{16}O ratios of the water. The origin of solutes and the geochemical evolution of the waters were explored using ^{2}H/^{1}H ratios, ^{3}He and ^{4}He concentrations and ratios, the ^{18}O/^{16}O, ^{13}C/^{12}C and ^{34}S/^{32}S ratios in dissolved and mineral carbonates, sulphates, and sulphides, ^{87}Sr/^{86}Sr ratios in dissolved and mineral strontium, ^{36}Cl/Cl, ^{37}Ar/Ar, and ^{40}Ar/^{36}Ar ratios, the concentrations and ratios of various uranium and thorium isotopes and their daughter products in solution and in minerals, and the concentrations of other dissolved constituents.

9.1 Analytical Results and Interpretative Methods

9.1.1 Quality of Sampling and Analysis

Samples for isotopes and dissolved gases were collected and analysed using several techniques. Replicate results, comparison with chemical analyses and measurements of drilling fluid tracers, and consideration of internal consistency were used to evaluate the quality of the isotope and noble gas data. Results could be affected by mixing of waters of different history at or near the point of collection, contamination during sampling, difficulties with sample preservation during storage and shipment, and the sample preparation and analytical technique employed.

Water mixing and sample contamination was of particular concern for the interpretation radioisotopes. The presence of small amounts of admixed younger waters or sample contamination would have important consequences on the determination of groundwater residence times. Large gas samples were required for ^{39}Ar analyses and were also used for ^{85}Kr measurements. Contamination in these samples could be detected and the ^{39}Ar values corrected. ^{85}Kr alone and with ^{3}H values measured the proportion of young water in the sample. The smallest amount of admixed young water which could be detected is of the order of one to three per cent. The possible presence of modern ^{14}C in materials used in drilling the deep boreholes or entrained during sampling required that the ^{14}C results below a few per cent of modern carbon (pmc) be treated as limiting rather than as absolute concentrations.

Difficulties with sample preservation during collection, storage, and shipment affected some of the dissolved gas results. Sample degassing either in the subsurface or during collection can considerably change the gas composition. The quality of the noble gas results on individual samples was judged by the consistency of the recharge temperatures calculated from their Ar, Ne, Kr, and Xe contents. About 35 per cent of these samples could not be used for calculating recharge temperatures. Ar and He results on samples collected and analysed by different methods agreed with those of the noble gas samples which yielded consistent recharge temperatures.

Differences in sample preparation and analytical techniques affect the carbon and sulphur isotope results. Replicate carbon isotope samples collected by direct precipitation and analysed by conventional counting and mass spectroscopy agreed well. Equivalent samples analysed by accelerator mass spectrometry and extracted by the evolution of CO_2 gas from small water samples contained less ^{14}C and ^{13}C, due perhaps to fractionation during CO_2 extraction. Sulphur isotope analyses were made at two laboratories. Only three samples were analysed by both, and these agreed poorly. However, the considerable body of results from each lab is internally consistent and agrees well with the results from the other lab. Uranium concentrations were measured in three laboratories using different techniques with good agreement at all concentrations above the detection limit for each technique.

9.1.2   Water Origin and Recharge Conditions

The ^2H, ^{18}O, and noble gas contents of groundwaters provide information about the temperature and climatic regime under which the water was recharged, as well as on the evolution of the water in the subsurface. Correlations among δ^2H and $\delta^{18}O$ values of modern precipitation with altitude and temperature were developed from Nagra and other analyses of precipitation, surface water, and shallow groundwater, and from climatological records. The line describing the δ^2H and $\delta^{18}O$ values of these data is indistinguishable at the 95 per cent confidence limit from the global meteoric water line. The slopes and intercepts of lines describing δ^2H and $\delta^{18}O$ variations with altitude differ among various Swiss climatic regimes. The isotope-temperature relationships used to interpret the groundwater data were developed from these isotope-altitude equations and regional temperature-altitude relationships. This could be done reasonably well for the area of northern Switzerland and adjacent southern Germany, and the Swiss Mittelland, from which most of the Nagra groundwater samples were taken.

The concentrations of the noble gases Ar, Ne, Kr, and Xe in groundwaters reflect the temperature and atmospheric pressure (altitude) prevailing during recharge. These concentrations can also be affected by the entrainment of excess air during recharge, by subsurface-produced ^{40}Ar, and by gas losses either in the subsurface or during sampling. These processes will affect each noble gas to a different extent. Corrections for them were made for each sample to minimize the differences between the recharge temperatures calculated for each gas. After correction, about three quarters of all samples yielded recharge temperatures with standard deviations of less than 5°C from the mean of the temperatures calculated from the individual gases.

The relationship existing at the time of recharge between the temperature and stable isotopic composition of a groundwater can change as a result of groundwater mixing, isotope exchange before or after recharge, and gas losses or gains not accounted for in the corrections described above. In addition, this relationship for a groundwater may differ from that of present recharge water because of changes with time of the isotopic composition of the ocean source of all precipitation, in the paths of movement of the air masses bringing precipitation to the recharge area and in the seasonal patterns of recharge.

The relationship between δ^2H and noble gas temperatures for all samples from the Muschelkalk coincides with that calculated for recent water from northern Switzerland. The noble gas temperatures of Muschelkalk samples with ^{14}C contents greater than 10 pmc correspond to present mean annual temperatures. From this, it is evident both that noble gas correction procedures are correct, and that these Muschelkalk waters were recharged under modern climatic conditions.

Water from the crystalline and associated Buntsandstein and Permian of northern Switzerland, as sampled from the Leuggern, Kaisten, and Böttstein (above 900 m) boreholes, have closely similar isotopic compositions and noble gas temperatures. These waters are presumed to have recharged from the southeastern part of the Black Forest, but the relationship between their δ^2H and noble gas temperatures differs from that of recent waters from this region. The relationship is similar to that calculated from a few samples of recent water from the western Black Forest, but the noble gas temperatures of the crystalline waters are from 2° to 5°C, well below modern mean annual temperatures in the region. These waters were apparently recharged under colder and less continental climatic conditions than those of the present, as probably existed during times of increased glaciation. This is consistent with the absence of ^{14}C in most of the crystalline waters of northern Switzerland.

Samples from the Tertiary and Malm have a δ^2H-temperature relationship like that of recent water, but with lower noble gas temperatures, suggesting recharge at a time of lower temperature than at present, but with similar overall climatic conditions.

Various patterns are evident in the samples from the deep crystalline and in the Buntsandstein and Permian waters not associated with the upper crystalline. The δ^2H-temperature relationships of most are like that of recent water, but a number have lower temperatures suggesting recharge under colder climatic conditions. The data on many samples are consistent with the hypothesis that the Black Forest is the recharge area. It is of interest that a few samples of deep origin and high $^{40}Ar/^{36}Ar$ and ^4He contents, suggesting high residence times, have noble gas temperatures as high as or higher than those of the present time. This indicates recharge conditions similar to or warmer than the present.

The ^{18}O contents of water from the crystalline and associated Buntsandstein are consistent with their ^2H and noble gas contents. Water from the deep crystalline at Weiach is enriched in both ^{18}O and ^2H, but retains a meteoric signature. Waters from the Buntsandstein and Permian of the Riniken, Weiach, and Schafisheim boreholes are mixtures between meteoric water of low salinity and a highly saline, ^{18}O-enriched water, the origin of which is not clear. Water from the crystalline at Schafisheim resembles that from the Permian and is presumed to have its source in such a water.

In general, waters with ^{14}C contents above 5 pmc have recharge temperatures similar to the mean annual temperatures now prevailing in northern Switzerland. Those with lower ^{14}C contents have noble gas temperatures from nearly 0° to over 11°C suggesting that they represent recharge during a wide range of past climatic conditions.

The ^2H and ^{18}O contents of the north Swiss groundwaters confirm the noble gas evidence on their origins and conditions of recharge, and provide similar information for samples for which no noble gas data are available.

Most samples from the Quaternary, Tertiary, and Malm have δ^2H and $\delta^{18}O$ values like those of recent water. Several waters of the Na-HCO$_3$ type from intermediate depths are depleted in ^2H and ^{18}O, suggesting recharge under cooler conditions. The radioisotope model ages of these samples are consistent with recharge during glacial times. An additional group of samples, including those from the Malm at Weiach and the USM at Schafisheim, appear to be mixtures between recent waters of low salinity and a highly saline water enriched in ^{18}O. The origin of the latter water is not clear.

Most Muschelkalk samples also have ^2H and ^{18}O contents and noble gas temperatures like those in modern waters. Several, including those from the Weiach, Böttstein, Schafisheim and Beznau boreholes are depleted in the heavier isotopes. The radioisotopes of the samples from Böttstein and Beznau are consistent with recharge during a period of glaciation. The Weiach Muschelkalk has radioisotope model ages indicating recharge within the last millennium, yet it has ^2H and ^{18}O contents consistent with modern precipitation only in the Alps. The Rhine River and the Bodensee contain water with an Alpine isotopic signature, and are interpreted as the source of recharge to the Muschelkalk at Weiach. Water from the Baden springs is a mixture of a recent water with a more saline, ^{18}O enriched water. ^{18}O enrichment is common in thermal waters.

9.1.3 Water Ages

Several radioisotopes with a range of half lives were analysed to obtain quantitative and qualitative information on groundwater mixing and residence times. Radioisotope data require a conceptual model for groundwater flow if they are to be interpreted as ages. Short-lived isotopes and young groundwaters were interpreted using both the exponential and the piston flow models. Longer-lived isotopes and old water were usually interpreted using the piston flow model.

The short-lived isotopes ^3H ($t_{1/2}$ = 12.3 a) and ^{85}Kr ($t_{1/2}$ = 10.8 a), from the testing of thermonuclear devices and reactor fuel rod reprocessing, are present in the atmosphere and young groundwater. Concentrations and ratios of these isotopes can yield both model ages of young water and the proportion of such water present in mixed samples.

^{39}Ar ($t_{1/2}$ = 269 a) is produced both in the atmosphere and in the subsurface. Most samples from the Buntsandstein and all samples from stratigraphically lower units had ^{39}Ar contents above 100 per cent of modern (the atmospheric level) due to a high rate of production and release in crystalline rock. The ^{39}Ar activities in Buntsandstein waters confirm their direct connection with underlying crystalline or Permian waters. Most samples from sedimentary aquifers had measured ^{39}Ar values below 100 per cent, from which model ages were calculated assuming no underground production. These ages range from c. 200 to c. 800 years with limiting values of > 1000a.

^{14}C ($t_{1/2}$ = 5730 a) is present in the atmosphere both naturally and as a result of nuclear tests and reactor operations. The ^{14}C/C_{tot} ratio in groundwaters is strongly affected by geochemical reactions and its interpretation requires a model for the carbonate geochemical evolution of a sample. ^{14}C model ages range from c. 1 to c. 21 ka with limiting values of < 1 and > 32 ka. Only a few samples have non-limiting model ages from both ^{39}Ar and ^{14}C. However, the agreement between the ages of those few samples and of the additional results which can be interpreted as limiting ages is sufficiently good to corroborate the models used to develop groundwater ages from both isotopes.

Ages were calculated from measured ^{14}C values using a geochemical model of the evolution of the chemical and stable carbon isotopic compositions of the water. Many parameters are required for the model, so to ensure that uncertainties in their values were included in the uncertainty of the final model age, risk analysis techniques were used in the model calculations. Only limiting ages were reported from some samples, even though their measured ^{14}C contents were well above the analytical detection limit. These samples were either contaminated with drilling material, as indicated by their chemical quality, tracer content, or ^{3}H level, or were samples with admixed young components, as indicated by their ^{3}H and ^{85}Kr contents.

The activities of several of the isotopes measured were a result of underground production by radioactive decay or neutron, reactions rather than residual from activities present in the recharging groundwater. Underground production rates were calculated to support the interpretation of these isotopes using data on the chemical composition of rocks from the Nagra boreholes. For example, the measured high activities of ^{37}Ar and ^{39}Ar in water from the crystalline were consistent with calculated production rates.

All but possibly one of the ^{36}Cl ($t_{1/2}$ = 3 · 10^5 a) measurements are also above levels possible from atmospheric sources, and must, therefore, have been produced underground by neutron irradiation of ^{35}Cl. The ^{36}Cl/Cl ratios in many samples approach equilibrium with the production rate in the rock, which requires a minimum residence time of about 1.5 ma. There is a strong linear correlation between the ^{36}Cl and the total Cl contents of all water samples, indicating that both ^{36}Cl and common Cl have the same source, the minerals or highly-mineralized fluid inclusions of the water-bearing rock. Therefore, the high ^{36}Cl contents of the waters reflect the fact that the residence time of both ^{36}Cl and Cl in the rock is long, rather than that the water is necessarily very old.

Both ^{3}He and ^{4}He are stable isotopes. They are produced underground in such large quantities that groundwaters often have dissolved helium concentrations several orders of magnitude above those which result from the dissolution of atmospheric He. The calculated ^{3}He/^{4}He ratio is independent of rock porosity and ranges from $3 \cdot 10^{-7}$ in the crystalline to $< 1 \cdot 10^{-8}$ in the Muschelkalk. The measured ^{3}He/^{4}He ratio in all samples was about $2 \cdot 10^{-7}$, indicating a crystalline source for the helium in waters in all formations. The high absolute concentrations of helium in water, particularly from the sediments, do not necessarily represent long accumulation times of helium in a closed system, but rather indicate a transport of helium from depth. Diffusion is too slow to account for all the He transport required, so some convective transport must also be occurring. There is a linear relationship between the He and the Cl contents of samples from the crystalline, which suggests that both He and Cl are entering the water from the same source--presumably leaching of minerals or fluid inclusions of the rock itself.

The ^{40}Ar/^{36}Ar ratio in some samples is above that corresponding to the dissolution of atmospheric argon. This results from the underground production of ^{40}Ar. Measured ^{40}Ar contents, like measured He contents, lead to unrealistically long residence times if they are assumed to result from underground production in a closed system. High ^{40}Ar and He contents require the operation of long-range transport processes. ^{40}Ar, like He, is linearly correlated with Cl and its concentration also seems to reflect the extent of water-rock interactions.

The ^{4}He, ^{36}Cl, radiogenic ^{40}Ar and high chloride concentrations cannot be interpreted to yield absolute water ages. However, considered together, they indicate a very long residence time for some of the waters sampled.

9.1.4 Chemical Evolution

The stable isotopic compositions of carbon in dissolved and mineral carbonate, sulphur in dissolved and mineral sulphates and sulphides, and oxygen in water, mineral carbonate, and dissolved and mineral sulphate were measured. They provide information about the geochemical evolution of the water-rock system, and about the history and flow paths of groundwaters.

Many samples of calcite from the crystalline of the Böttstein, Kaisten, and Leuggern boreholes are in carbon and oxygen isotopic equilibrium with the groundwater. The δ^{34}S and δ^{18}O values of two generations of barite in the crystalline were measured. The older barites are similar to sedimentary sulphate minerals from the overlying formations, while the younger barites contain less ^{18}O. The ^{34}S and ^{18}O contents of sulphate in most waters from the crystalline were more similar to those of the younger barites than to those of the older ones, and were consistent with isotopic equilibrium between the dissolved sulphate and the younger barites. The isotopic equilibrium between dissolved carbonate and calcite, and between dissolved sulphate and barite in the crystalline is consistent with the fact that the waters are at chemical equilibrium with these minerals. The petrographic observation that the young barites and many calcite crystals in the crystalline are fresh, not eroded, suggests that they are forming from rather than dissolving in water now present in the crystalline.

The ^{34}S and, to the lesser extent, the ^{18}O content of sedimentary sulphate minerals has varied with time. Thus, the isotopic composition of dissolved sulphate can be used to characterize its mineral origin. Waters from the Keuper and all but those of lowest sulphate concentration from the Molasse and Malm have dissolved sulphate isotopes in the range typical of those of sedimentary minerals in their source formations. Most waters from the Muschelkalk also have dissolved sulphates similar to those of Muschelkalk mineral sulphate. The sulphate in waters from a few Muschelkalk locations more closely resembles Keuper or Malm mineral sulphate, and suggests flow through those formations to the Muschelkalk.

The isotopic composition of sulphate from waters from the Buntsandstein and Permian is similar to that of sedimentary sulphate of various ages. These include sulphate of the age of the formations themselves as well as those typical of overlying formations. Sulphate in some waters with low sulphate concentrations also resembles sulphate in waters from the crystalline.

Deep groundwaters from the crystalline, Permian and Buntsandstein, and Muschelkalk have distinctive dissolved strontium contents and $^{87}Sr/^{86}Sr$ ratios. The strontium isotopes of Muschelkalk groundwater are within the ranges of those of minerals in the formation, but are higher than those of Muschelkalk sea water and of primary sulphate minerals. Therefore, the strontium dissolved in Muschelkalk water must be a product of both sulphate and carbonate (principally dolomite) mineral dissolution.

The strontium isotope ratio of whole rock from the crystalline basement gives an isochron of 279 ± 5 ma corresponding to a Permian hydrothermal event known from K-Ar dates. The strontium isotope ratio of groundwater in the crystalline is inconsistent with whole rock dissolution, but suggests that the dissolved strontium results from plagioclase alteration.

The strontium isotope ratios of some Buntsandstein and Permian waters are like those of crystalline water, and are consistent with the fact that the mineralogy of their host rocks are similar to those of the crystalline, or that extensive exchange of waters between the several formations has occurred in some locations. Other waters from the Buntsandstein have strontium concentrations and isotope ratios requiring the additional solution of calcite and sulphate mineral cements.

Analysis of uranium concentrations and isotope ratios, and of the ratios of other isotopes in the uranium and thorium decay series were measured on a number of groundwater and rock samples. The groundwater data indicate sources and direction of flow and relative ages. The rock data provide evidence on water-rock reactions which have occurred during the last 10^6 a.

Water from the crystalline and associated Buntsandstein show zones of high and low uranium concentrations. High uranium concentrations are found in waters from the Black Forest, indicating recharge under oxidizing conditions. To the south, the uranium concentration decreases through a transition zone in the Rhine Valley where groundwaters become reducing. The $^{234}U/^{238}U$ ratio increases southward from the Black Forest, and reaches a relatively high ratio in the transition region in the Rhine Valley. South of this zone the ratio decreases as ^{234}U decays. Waters from the overlying aquifers show no pattern of uranium concentration or isotope ratios.

About half of the rock samples from all rock types show radioactive disequilibrium among various parent/daughter pairs, indicating rock-water interactions during the past 10^6 a. In crystalline rock, such disequilibria are concentrated in altered rock containing calcite. About one third of the samples also show $^{232}Th/^{228}Th$ disequilibria, indicating recent rock-water reactions with high ^{228}Ra mobility.

9.2 Hydrogeological and Hydrochemical Conclusions

Interpretation of the isotope and noble gas data of this report, supported by other chemical data from the Nagra programme, leads to conclusions about the origin, ages, and flow paths of water within the major aquifer groups, as well as about flow between these groups.

9.2.1 Tertiary and Malm

These formations contain water of three contrasting chemical types. In the Molasse Basin, these types overlie one another, but the boundary between them is not concordant with stratigraphic boundaries.

The uppermost is a Ca-Mg-HCO$_3$ type which is represented by a number of high-^{3}H young waters as well as by low-^{3}H older or mixed waters. These have water isotopes characteristic of recent recharge, modern ^{14}C model ages, and appear to be recent waters in rapid circulation.

The second water is an Na-HCO$_3$ type and is found at intermediate depths in the Molasse Basin. The noble gas contents and water isotopes of all except two samples of this water correspond to a much cooler climate than that of the present. Only limiting ^{14}C model ages can be determined for these waters, but the lowest is greater than 11 ka. These waters are older than the overlying Ca-Mg-HCO$_3$ waters, and have ages and isotopic compositions which are consistent with recharge during a period of cooler climate, such as would have prevailed during a time of glaciation. They are enriched in ^{13}C relative to the overlying waters, which is consistent with the development of their Na-HCO$_3$ character by reactions driven by ion exchange of Na for Ca.

The third water is a Na-Cl type and occurs beneath the Na-HCO$_3$ water or, at the northern border of the Molasse Basin, directly in contact with Ca-Mg-HCO$_3$ waters. Samples of the Na-Cl water have ^{2}H contents similar to those of the upper Ca-Mg-HCO$_3$ water, but are enriched in ^{18}O and plot to the right of the meteoric water line, by amounts that increase with their salinity. One saline ^{18}O-enriched water has a high $^{40}Ar/^{36}Ar$ ratio, and another has a high He content. The Na-Cl samples appear to be mixtures of waters of the Ca-Mg-HCO$_3$ or Na-HCO$_3$ types, with a water of very great age, perhaps even more enriched in ^{18}O and salinity than any of those sampled. The ^{18}O enrichment, salinity, and high ^{40}Ar and He contents suggest this end-member water has a very great, but unknown age.

9.2.2 Dogger, Lias and Keuper

The only low-^{3}H samples from these units outside of the Rhine Graben are samples from a tunnel in Lias and two Keuper samples from the Beznau and Riniken boreholes. The Keuper samples have limiting ^{14}C model ages > 12 ka and salinities > 14 g/l, and represent a region of little or no flow. The water isotopes of the Riniken sample fall on the meteoric water line, but are so enriched that the samples cannot have been formed as meteoric water in a region remotely resembling northern Switzerland at the present time.

9.2.3 Muschelkalk

More samples of Muschelkalk water were taken than of any other water-bearing unit or associated group. Most samples are from areas of relatively active circulation in the Folded and western Tabular Jura. Most of these waters are from modern to not greater than 10 ka in age, and their water isotopes and noble gas values suggest recharge in the Jura and Black Forest areas under climatic conditions like those of the present. A few boreholes in this region contain, in part, a much older water. The borehole of Pratteln has a high ^{39}Ar content and ^{40}Ar/^{36}Ar ratio suggesting an influence by crystalline water.

An area of little groundwater circulation in the eastern Tabular Jura is represented by samples from the Böttstein, Beznau, and Riniken boreholes. These waters have high residence times and different water isotopes than recent waters. However, the relationship between the ^{2}H contents and noble gas temperatures of the Böttstein and Riniken samples is consistent with recharge in the region of northern Switzerland and southern Germany under climatic conditions like those of the present.

Samples from the northern edge of the Molasse Basin, south and west of the region of limited circulation, are virtually all mixtures between a modern water and one of considerable age. The older water is probably not identical with that found in the Böttstein-Beznau-Riniken boreholes because the Baden springs show an oxygen isotope shift consistent with their high temperatures.

The Weiach sample has very low ^{2}H and ^{18}O contents typical of alpine precipitation or Rhine River water. As it has ^{39}Ar and ^{14}C contents consistent with a model age of c. 800 a and a very low ^{4}He content, the Rhine River must be taken as the zone of its recharge. This indicates a region of very rapid circulation in the Muschelkalk in the vicinity of Weiach. The chemistry and a ^{14}C model age of c. 5 ka for the Lostorf 4 sample suggest a similar area of high circulation in that region as well.

9.2.4 Buntsandstein, Permian and Crystalline

The crystalline as sampled at Kaisten, Leuggern, Zurzach, and the upper part of the Böttstein boreholes contains chemically homogeneous water of low mineralization, which is similar isotopically to young water in the Black Forest, but has noble gas temperatures corresponding to recharge at lower temperatures than prevail at the present. Most samples have limiting ^{14}C model ages of > 20 ka, but the Zurzach and uppermost

Böttstein samples have ages between 13 and 21 ka. These ages are consistent with the noble gas evidence that recharge occurred under cooler climatic conditions.

Preliminary results from the Siblingen borehole are not consistent with other upper crystalline samples. Siblingen waters have ^{14}C model ages similar to those of the Zurzach and upper Böttstein samples, but are very strongly depleted in ^{2}H and ^{18}O. If the Siblingen water was recharged during the last glaciation, the other crystalline samples with their higher ^{2}H and ^{18}O contents should represent recharge from a prior period of moderate climate. If this is the case, the model ^{14}C ages of the Zurzach and upper Böttstein samples would be too young.

Water in the Buntsandstein and Permian in the Böttstein, Kaisten, and Leuggern boreholes is similar to that in the underlying crystalline, except for slightly higher salinities which appear to be related to the dissolution of halite.

Mixtures of saline waters of different ages are found in the deeper Buntsandstein, Permian, and crystalline. As a group, these samples have ^{2}H, ^{18}O, and He contents, and ^{40}Ar/^{36}Ar ratios that qualitatively indicate great age. This is consistent with the development of their salinity by extensive water-rock interactions.

9.2.5 Inter-Aquifer and Regional Flow

The isotope results describe active groundwater flow systems in the sedimentary aquifers of the Folded and western Tabular Jura, grading through regions of lesser activity in the sediments and upper crystalline rock of the eastern Tabular Jura and deeper Molasse Basin to the deep crystalline and sediments of the Permo-Carboniferous Trough. Circulation in the deep crystalline and the Trough is so limited that water ages may approach those of the formations themselves. The deep saline waters sampled directly, or present as mixtures in more dilute waters, have undergone such extensive water-rock interactions that evidence of their origin has been lost. The remaining waters were all recharged in the region of northern Switzerland or southern Germany. There is no evidence of recharge from such distant sources as the Alps.

There is little evidence of strong flow among the sedimentary aquifers above the Buntsandstein. Even in the eastern Tabular Jura, where intra-aquifer flow is sluggish, the isotope data do not indicate inter-aquifer flows. The isotope data are not precise enough to rule out all possibility of inter-aquifer flow, but will be useful to put upper limits on the amounts of such flow permissible in regional aquifer models. The isotope data do mark several areas of anomalous flow, however, including the presence of crystalline water in the Muschelkalk at Pratteln, the presence of older mixed water in the Muschelkalk at Magden and Baden, zones of very rapid flow in the Muschelkalk at Weiach and Lostorf, and the mixing of Molasse water types at Eglisau.

Water in the upper crystalline and associated Permian and Buntsandstein of northern Switzerland is recharged where these formations crop out in and around the Black Forest. However, these waters have such a homogenous character that the isotope data do not distinguish any flow patterns. South of the Rhine these waters have ^{14}C model ages ranging from 13 to 21 ka to > 31 ka. The fact that the waters are in chemical and isotopic equilibrium with such fracture minerals as calcite and barite is consistent with these ages. The relationship of water in the crystalline at Siblingen to that south of the Rhine is not clear, but it is difficult to account for all the isotope data if Siblingen is considered to be on a flow path leading from the Black Forest to the crystalline south of the Rhine.

The isotopic and chemical character of water from the Permo-Carboniferous Trough and the deep crystalline is consistent with extensive water-rock reactions, and with the presence of solutes, and perhaps water itself, of an age close to that of the formations. There is enough difference between waters to distinguish those formed by sediment-water reaction from those formed by crystalline-water reactions regardless of where they may have been sampled. Data on these waters will be particularly useful in calibrating models of water-rock geochemical evolution.

10. LITERATURE CITED

Alexander, W. R., Scott, R. D., Mackenzie, A. B., and McKinley, I. G., 1988, A natural analogue study of radionuclide migration in a water conducting fracture in crystalline rocks: Radiochimica Acta, v. 44/45, p. 283-289.

Amberger, G., Siegenthaler, U., and Verstraete, P., 1981, Etudes en cours de la nappe souterraine de l'Arve: Eclogae geologicae Helvetiae, v. 74, p. 225-232.

Andrews, J. N., 1985, The isotopic composition of radiogenic helium and its use to study groundwater movement in confined aquifers: Chemical Geology, v. 49, p. 339-351.

Andrews, J. N., and Lee, D. J., 1979, Inert gases in groundwater from the Bunter Sandstone of England as Indicators of Age and palaeoclimatic trends: Journal of Hydrology, v. 41, p. 233-252.

Andrews, J. N., and Wood, D. F., 1972, Mechanism of radon release in rock matrices and entry into groundwaters: Transactions of the Institution of Mining and Metallurgy, v. B81, p. 198-209.

Andrews, J. N., Giles, I. S., Kay, R. L., Lee, D. J., Osmond, J. K., Cowart, J. B., Fritz, P., Barker, J. F., and Gale, J., 1982, Radioelements, radiogenic helium and age relationships for groundwaters from the granites at Stripa, Sweden: Geochimica et Cosmochimica Acta, v. 44, p. 201-206.

Andrews, J. N., Balderer, W., Bath, A. H., Clausen, H. B., Evans, G. V., Florkowski, T., Goldbrunner, J. E., Ivanovich, M., Loosli, H., and Zojer, H., 1983, Environmental Isotope Studies in two aquifer systems, *in* Isotope Hydrology 1983: Proceedings of a Symposium, Vienna, 12-16 September 1983: Vienna, International Atomic Energy Agency, p. 535-576.

Andrews, J. N., Fontes, J. Ch., Michelot, J. L., and Elmore, D., 1986, In Situ neutron-flux, ^{36}Cl-production and groundwater evolution in crystalline rocks at Stripa, Sweden: Earth and Planetary Science Letters, v. 77, p. 49.

Andrews, J. N., Davis, S. N., Fabryka-Martin, J., Fontes, J. Ch., Lehmann, B. E., Loosli, H. H., Michelot, J. L., Moser, H., Smith, B., and Wolf, M., 1989, The in-situ production of radioisotopes in rock matrices with particular reference to the Stripa granite: Geochimica et Cosmochimica Acta, v. 53, p. 1803-1815.

Andrews, J. N., Florkowski, T., Lehmann, B. E., and Loosli, H. H., 1990, Underground production of radionuclides in the Milk River aquifer, *(to be published in* Applied Geochemistry).

Andrews, R. W., and Pearson, F. J., Jr., 1984, Transport of ^{14}C and uranium in the Carrizo Aquifer of south Texas, *in* Proceedings of a Materials Research Society Symposium, Vol. 26: Amsterdam, Elsevier, p. 1085-1092.

Back, W., Hanshaw, B. B., Plummer, L. N., Rahn, P. H., Rightmire, C. T., and Rubin, M., 1983, Process and rate of dedolomitization: Mass transfer an ^{14}C dating in a regional carbonate aquifer: Geological Society of America Bulletin, v. 94, p. 1415-1429.

Balderer, W., 1983, Bedeutung der Isotopenmethoden bei der hydrogeologischen Charakterisierung potentieller Endlagerstandorte für hochradioaktive Abfälle: Baden, Switzerland, Nagra, Technischer Bericht 83-04, 158 p.

Balderer, W., 1985a, The Nagra Investigation Project for the Assessment of Repositories for High-Level Radioactive Wastes in Geologic Formations: Mineralogical Magazine, v. 49, p. 281-288.

Balderer, W., 1985b, Sondierbohrung Böttstein: Ergebnisse der Isotopenuntersuchungen zur hydrogeologischen Charakterisierung der Tiefengrundwässer: Baden, Switzerland, Nagra, Technischer Bericht 85-06, 260 p.

Balderer, W., 1990, Hydrogeologische Charakterisierung der Grundwasservorkommen innerhab der Mollasse der Nordost Schweiz aufgrund von Hydrochemischen und Isotopenuntersuchungen: Steirische Beiträge zur Hydrogeologie, v. 41, *(in press)*.

Balderer, W., Rauert, W., and Stichler, W., 1987a, Environmental isotope study of the deep groundwaters in northern Switzerland, *in* Isotope Techniques in Water Resources Development: Proceedings of a Symposium, Vienna, 30 March-3 April 1987: Vienna, International Atomic Energy Agency, p. 455-474.

Balderer, W., Fontes, J. Ch., Michelot, J. L., and Elmore, D., 1987b, Isotopic investigations of the water-rock system in the deep crystalline rock of northern Switzerland, *in* Fritz, P., and Frape, S. K., *eds.*, Saline Water and Gases in Crystalline Rocks: Geological Association of Canada, Special Paper 33, p. 175-195.

Banner, J. L., Wasserburg, G. J., Dobson, P. F., Carpenter, A. B., and Moore, C. H., 1989, Isotopic and trace element constraints on the origin and evolution of saline groundwaters from central Missouri: Geochimica et Cosmochimica Acta, v. 53, p. 383-398.

Bentley, H. W., 1978, Some comments on the use of Chlorine-36 for dating very old groundwaters, *in* Davis, S. N., *ed.*, Proceedings of a Workshop on Dating Old Groundwater, Tucson, Arizona, 16-18 March 1978: U.S. Department of Energy, 138 p.

Bentley, H. W., Phillips, F. M., and Davis, S. N., 1986, Chlorine-36 in the terrestrial environment, *in* Fritz, P., and Fontes, J. Ch., *eds.*, Handbook of Environmental Isotope Geochemistry, Volume 2, The Terrestrial Environment, B: Amsterdam, Elsevier, p. 427-480.

Bertleff, B., 1986, Das Strömungssystem der Grundwässer im Malm-Karst des WestTeils des süddeutschen Molassebeckens: Abhandlungen des Geologischen Landesamtes Baden-Württemberg, v. 12, p. 1-271.

Bertleff, B., Hammer, W., Joachim, H., Koziorowski, G., Stober, I., Strayle, G., Villinger, E., and Werner, J., 1987, Hydrogeothermiebohrungen in Baden-Württemberg - Eine Uebersicht: Z. dt. geol. Ges., v. 138, p. 411-423.

Bertleff, B., Koziorowski, G., Leiber, J., Ohmert, W., Prestel, R., Stober, I., Strayle, G., Villinger, E., and Werner, J., 1988, Ergebnisse der Hydrogeothermiebohrungen in Baden-Württemberg: Jh. geol. Landesamt Baden-Württemberg, v. 30, p. 27-116.

Bider, M., 1978, Nordöstlicher Jura und Juranordfuss, *in* Regionale Klimabeschreibungen - Klimatologie der Schweiz Band II, 1. Teil. Beiheft zu Annalen d. Schweiz. Meteorologische Zentralanst.: p. 115-178.

Bläsi, H.-R., 1987, Lithostratigraphie und Korrelation der Doggersedimente in den Bohrungen Weiach, Riniken und Schafisheim: Eclogae geologicae Helvetiae, v. 80, p. 415-430.

Bläsi, H.-R., Dronkert, H., Matter, A., and Ramseyer, K., 1989, Diagenese des Muschelkalk- und des Buntsandsteinaquifers in den NAGRA Tiefbohrungen: Baden, Switzerland, Nagra, Interner Bericht 89-54, 71 p.

Blaser, P., and others, *in preparation*, Sondierbohrung Siblingen: Documentation der Wasserprobenahmen und Hydrochemische Rohdaten der Sondierbohrung Siblingen: Baden, Switzerland, Nagra, Interner Bericht 89-29.

Blavoux, B., Burger, A., Chauve, P., and Mudry, J., 1979, Utilisation des isotopes du milieu à la prospection hydrogéologique de la chain karstique du Jura: Revue de géologie dynamique et géographie physique, v. 21, p. 295-306.

Brunner, R., 1973, Ein neues Massenspektrometer in der Schweiz: Der Elektroniker.

Burke, W. H., Denison, R. E., Hetherington, E. A., Koepenick, R. B., Nelson, H. F., and Oto, J. B., 1982, Variation of seawater $^{87}Sr/^{86}Sr$ throughout Phanerozoic time: Geology, v. 10, p. 516-519.

Carpenter, A. B., 1978, Origin and chemical evolution of brines in sedimentary basins: Oklahoma Geological Survey, Circular 79, p. 60-77.

Cherdynstev, V. V., Chalov, P. I., and Khaidarov, G. Z., 1955, Transactions of the 3rd Session of Committee for Determination Absolute of Ages of Geologic Formations: Izv. Akad. Nauk SSR, p. 175.

Claypool, G. E., Holser, W. T., Kaplan, I. R., Sakai, H., and Zak, I., 1980, The age of sulfur and oxygen isotopes in marine sulphates and their mutual interpretation: Chemical Geology, v. 28, p. 199-260.

Cuttell, J. C., 1983, The application of uranium and thorium series isotopes to the study of certain British aquifers: University of Birmingham, 269 p. *(unpublished)*.

Cuttell, J. C., Lloyd, J. W., and Ivanovich, M., 1986, A study of uranium and thorium series isotopes in chalk groundwaters of Lincolnshire, U.K.: Journal of Hydrology, v. 86, p. 343-365.

Czubek, J. A., 1988, SLOWN2.BAS program for calculation of the rock neutron slowing down parameters: Krakow, Institute of Nuclear Physics, Report No. 1397/AP.

Dearlove, J. P. L., 1989, Analogue studies in natural rock systems: Uranium series radionuclide transport and REE distribution and transport: Cambridgeshire College of Arts and Technology, PhD. Thesis.

Deines, P., 1980, The isotopic composition of reduced organic carbon, *in* Fritz, P., and Fontes, J. Ch., *eds.*, Handbook of Environmental Isotope Geochemistry, Volume 1, The Terrestrial Environment, A: Amsterdam, Elsevier, p. 329-406.

Deines, P. D., Langmuir, D., and Harmon, R. S., 1974, Stable carbon isotope ratios and the existence of a gas phase in the evolution of carbonate ground waters: Geochimica et Cosmochimica Acta, v. 38, p. 1147-1164.

Diebold, P., 1986, Erdwissenschaftliche Untersuchungen der Nagra in der Nordschweiz: Strömungsverhältnisse und Beschaffenheit der Tiefengrundwässer: Mitteilungen Aargauischen Naturforschenden Gessellschaft, v. 31, p. 11-52.

Diebold, P., and Müller, W. H., 1985, Szenarien der geologischen Langzeitsicherheit: Risiko analyse für ein Einlager für hochaktiv Abfälle in der Nordschweiz: Baden, Switzerland, Nagra, Technischer Bericht 84-26, 110 p.

Downing, R. A., Smith, D. B., Pearson, F. J., Jr., Monkhouse, R. A., and Otlet, R. L., 1977, The age of groundwater in the Lincolnshire Limestone, England, and its relevance to the flow mechanism: Journal of Hydrology, v. 33, p. 201-216.

Drever, J. I., 1982, The Geochemistry of Natural Waters: Englewood Cliffs, New Jersey, Prentice-Hall, 380 p.

Dronkert, H., Bläsi, H.-R., and Matter, A., 1989, Facies and Origin of Triassic Evaporites from the Nagra Boreholes, Northern Switzerland: Baden, Switzerland, Nagra, Technischer Bericht 87-02, *(in press)*.

Dubois, J. D., and Flück, J., 1983, Rapport préliminaire sur les eaux thermominerales profondes entre Baden et le Schwarzwald: Basel, Swiss National Energy Research Foundation (NEFF), NEFF 165 1B 019.

Dubois, J. D., and Flück, J., 1984, Geochemistry: Utilisation of geothermal resources of the Baden area: Basel, Swiss National Energy Research Foundation (NEFF), NEFF 165 1B 032, 129 p.

Fabryka-Martin, J. T., Davis, S. N., Elmore, D., and Kubik, P., 1989, In-situ production and migration of ^{129}I in the Stripa granite, Sweden: Geochimica et Cosmochimica Acta, v. 53, p. 1817-1823.

Faure, G., 1977, Principles of isotope geology: New York, John Wiley & Sons, 419 p.

Fauth, H., Hindel, R., Siewers, U., and Zinner, J., 1985, Geochemischer Atlas Bundesrepublik Deutschland: Bundesanstalt für Geowissenschaften und Rohstoffe.

Feige, Y., Oltman, B. G., and Kastner, J., 1968, Production rates of neutrons in soils due to natural radioactivity: Journal of Geophysical Research, v. 73, p. 3135-3142.

Fleischer, R. L., 1988, Alpha recoil damage: Relation to isotopic disequilibrium and leaching of radionuclides: Geochimica et Cosmochimica Acta, v. 52, p. 1459-1466.

Florkowski, T., 1989, Natural production of radioactive noble gases in the geosphere: IAEA Consultants meeting on Isotopes of Noble Gases as Tracers in Environmental Investigations, Vienna, 29 May - 2 June, 1989.

Fontes, J. Ch., 1983, Dating of Ground Water, *in* Guidebook on Nuclear Techniques in Hydrology, 1983 Edition: Vienna, International Atomic Energy Agency, Technical Report Series No. 91, p. 285-317.

Fontes, J. Ch., and Garnier, J. M., 1979, Determination of the Initial ^{14}C Activity of the Total Dissolved Carbon: A Review of the Existing Models and a New Approach: Water Resources Research, v. 15, p. 399-413.

Fontes, J. Ch., and Michelot, J. L., 1985, Stable isotope geochemistry of sulphur components at Stripa, *in* Nordstrom, D. K., and others, Hydrogeological and hydrochemical investigations in boreholes - Final report of the Phase I geochemical investigations of the Stripa groundwaters: Stockholm, Sweden, Swedish Nuclear Fuel and Waste Management Co., SKB Technical Report 85-06, p. 7:1-7:17.

Fontes, J. Ch., Brissaud, I., and Michelot, J. L., 1984, Hydrological implications of deep production of chlorine-36: Nuclear Instruments and Methods, v. B5, p. 303.

Fontes, J. Ch., Fritz, P., Louvat, D., and Michelot, J. L., 1989, Aqueous sulphates from the Stripa groundwater system: Geochimica et Cosmochimica Acta, v. 53, p. 1783-1789.

Forster, M., Moser, H., and Loosli, H. H., 1983, Isotope hydrological study with C-14 and Ar-39 in the Bunter Sandstones of the Saar Region, *in* Isotope Hydrology 1983: Proceedings of a Symposium, Vienna, 12-16 September 1983: Vienna, International Atomic Energy Agency, p. 515-533.

Frape, S. K., and Fritz, P., 1987, Geochemical trends for groundwaters from the Canadian Shield, *in* Fritz, P., and Frape, S. K., *eds.*, Saline Water and Gases in Crystalline Rocks: Geological Association of Canada, Special Paper 33, p. 19-38.

Friedman, I., and O'Neil, J. R., 1977, Compilation of stable isotope fractionation factors of geochemical interest: U.S. Geological Survey, Professional Paper 440-KK, p. KK1-KK12.

Fritz, B., Clauer, N., and Kam, M., 1986, Strontium isotopes and magnesium control as indicators of the origin of saline waters in crystalline rocks, *in* Fritz, P., and Frape, S. K., *eds.*, Saline Water and Gases in Crystalline Rocks: Geological Association of Canada, Special Paper 33, 252 p.

Fritz, P., 1982, Comments on isotope dating of ground waters in crystalline rocks, *in* Narasimhan, T. N., *ed.*, Recent Trends in Hydrogeology: Boulder, Colorado, Geological Society of America, Special Paper 189, p. 361-373.

Fritz, P., and Fontes, J. Ch., *eds.*, 1980, Handbook of Environmental Isotope Geochemistry, Volume 1, The Terrestrial Environment, A: Amsterdam, Elsevier, 545 p.

Fritz, P., and Fontes, J. Ch., *eds.*, 1986, Handbook of Environmental Isotope Geochemistry, Volume 2, The Terrestrial Environment, B: Amsterdam, Elsevier, 557 p.

Garcia, D., 1986, Etude isotopique et géochimique des eaux thermales des Vosges méridionales - application géothermique: Documents du B.R.G.M. No. 112.

Garrels, R. M., and Christ, C. L., 1965, Solutions, Minerals, and Equilibria: New York, Harper & Row, 450 p.

Gascoyne, M., 1987, The use of uranium series disequilibrium for site characterisation as an analogue for actinide migration, *in* Chapman, N. A., *ed.*, Proceedings of a Symposium on Natural Analogues in Radioactive Waste Disposal, Brussels, Belgium, 28-30 April 1987: Brussels, Committee of the European Communities, CEC-Report No. EUR 11037 EN, p. 356-373.

Geyh, M., and Mairhofer, J., 1970, Der natürliche Carbon-14- und Tritium-Gehalt der Wässer, *in* Batsche, H., and others, Kombinierte Karstwasseruntersuchungen im Gebiet der Donauversickerung (Baden-Württemberg) in den Jahren 1967-1969: Steirische Beiträge zur Hydrogeologie, v. 22.

Hadermann, J., and Jakob, A., 1987, Modelling small scale infiltration experiments into bore cores of crystalline rock and break-through curves: Baden, Switzerland, Nagra, Technical Report 87-07, 36 p.

Haug, A., 1985, Feldmethoden zur Grundwässerentnahme aus Tiefbohrungen und zur Hydrochemischen Überwachung der Bohrspülung: Baden, Switzerland, Nagra, Technischer Bericht 85-07, 71 p.

Herzberg, O., and Mazor, E., 1979, Hydrological Applications of Noble Gases and Temperature Measurements in Underground Water Systems: Examples from Israel: Journal of Hydrology, v. 41, p. 217-231.

Hofmann, B., 1989, Genese, Alteration und rezentes Fliess-System der Uranlagerstätte Krunkelbach (Menzenschwand, Südschwartzwald): Baden, Switzerland, Nagra, Technischer Bericht 88-30, *(in press)*.

Hofmann, B., Dearlove, J. P. L., Ivanovich, M., Lever, D. A., Green, D. C., Baertschi, P., and Peters, Tj., 1987, Evidence of fossil and recent diffusive element migration in reduction haloes from permian red-beds of northern Switzerland, *in* Chapman, N. A., *ed.*, Proceedings of a Symposium on Natural Analogues in Radioactive Waste Disposal, Brussels, Belgium, 28-30 April, 1987: Brussels, Committee of the European Communities, CEC-Report No. EUR 11037 EN, p. 217-238.

Hunziker, J. H., Steiner, H. R., and Hurford, A., 1987, Absolute Altersbestimmungen, Sondierbohrung Böttstein, *in* Peters, Tj., Matter, A., Bläsi, H.-R., and Gautschi, A., *eds.*, Sondierbohrung Böttstein - Geologie: Baden, Switzerland, Nagra, Technischer Bericht 85-02, p. 230.

Ingerson, E., and Pearson, F. J., Jr., 1964, Estimation of age and rate of motion of ground-water by the ^{14}C-method, *in* Miyake, Y., and Koyama, T., *eds.*, Recent Researches in the Fields of Hydrosphere Atmosphere and Nuclear Geochemistry, Ken Sugawara Festival Volume: Tokyo, Maruzen, p. 263-283.

Institut für Radiohydrometrie, 1981, Jahresbericht 1980: München, Gesellschaft für Strahlen-und Umweltforschung mbH, GSF-Bericht.

Institut für Radiohydrometrie, 1982, Jahresbericht 1981: München, Gesellschaft für Strahlen-und Umweltforschung mbH, GSF-Bericht, R 296, 154 p.

Institut für Radiohydrometrie, 1983, Jahresbericht 1982: München, Gesellschaft für Strahlen-und Umweltforschung mbH, GSF-Bericht, R 328, 193 p.

Institut für Radiohydrometrie, 1984, Jahresbericht 1983: München, Gesellschaft für Strahlen-und Umweltforschung mbH, GSF-Bericht, R 368, 197 p.

Institut für Radiohydrometrie, 1985, Jahresbericht 1984: München, Gesellschaft für Strahlen-und Umweltforschung mbH, GSF-Bericht.

Institut für Radiohydrometrie, 1986, Jahresbericht 1985: München, Gesellschaft für Strahlen-und Umweltforschung mbH, GSF-Bericht.

Institut für Radiohydrometrie, 1987, Jahresbericht 1986: München, Gesellschaft für Strahlen-und Umweltforschung mbH, GSF-Bericht.

International Atomic Energy Agency, 1981, Stable Isotope Hydrology: Deuterium and Oxygen-18 in the Water Cycle: Vienna, International Atomic Energy Agency, Technical Report Series No. 210, 334 p.

International Atomic Energy Agency, 1983a, Guidebook on Nuclear Techniques in Hydrology, 1983 Edition: Vienna, International Atomic Energy Agency, Technical Report Series 91, 439 p.

International Atomic Energy Agency, 1983b, Isotope techniques in the hydrogeological assessment of potential sites for the disposal of high-level radioactive wastes: Vienna, International Atomic Energy Agency, Technical Report Series 228, 151 p.

International Atomic Energy Agency, 1987, Studies on Sulphur Isotope Variations in Nature: Chemistry, Geology and Raw Materials/Hydrology: Proceedings of an Advisory Group Meeting, Vienna, 17-20 June 1985: Vienna, International Atomic Energy Agency.

Isenschmid, Ch., 1985/86, Nagra Hydrochemisches Untersuchungsprogramm. Inventar der Probeentnahmestellen das Regionalprogramms: Baden, Switzerland, Nagra, *(unpaged)*.

IUPAC, 1979, Solubility Data Series v. 1, 2 and 3: Pergamon Press.

Ivanovich, M., 1982, Uranium series disequilibria applications in geochronology, *in* Ivanovich, M., and Harmon, R. S., *eds.*, Uranium series disequilibrium: Applications to environmental problems: Oxford, Oxford University Press, p. 56-78.

Ivanovich, M., and Harmon, R. S., *eds.*, 1982, Uranium series disequilibrium: Applications to environmental problems: Oxford, Oxford University Press, 571 p.

Ivanovich, M., and Wilkins, M. A., 1984, Harwell Report on the Analysis of Nagra rock and fracture infilling samples from Böttstein core: Uranium-series disequilibria measurements: Harwell, U.K., Report AERE-G3371.

Ivanovich, M., Wilkins, M. A., and Dearlove, J., 1986, Analysis of uranium series disequilibrium data in Nagra's deep borehole solid and liquid samples: Harwell, U.K., Report AERE-G4193.

Jäckli, H., 1970, Kriterien zur Klassifikation von Grundwasservorkommen: Eclogae geologicae Helvetiae, v. 63, p. 389-434.

Jäger, E., 1962, Rb-Sr age determination on micas and total rocks from the Alps: Journal of Geophysical Research, v. 67, p. 13.

Jäger, E., 1979, The Rb-Sr method, *in* Jäger, E., and Hunziker, J. C., *eds.*, Lectures in Isotope Geology: Berlin, Springer.

Kanz, W., 1987, Grundwasserfliesswege und Hydrogeochemie in tiefen Graniten und Gneisen: Geologische Rundschau, v. 76, p. 265-283.

Kay, R. L. F., and Darbyshire, D. P. F., 1986, A strontium isotope study of groundwater-rock interaction in the Carnmenellis granite, *in* Proceeding of the 5th International Symposium on Water-Rock Interaction, Reykjavik, August 1986: p. 329-332.

Keil, R., 1979, Hochselektive spektralphotometrische Spurenbestimmung von Uran VI mit Arsenazo III nach Extraktionstrennung: Fres. Z. Anal. Chem., v. 297, p. 384-387.

Kempter, R., 1987, Fossile Maturität, Paläothermogradienten und Schichtlücken in der Bohrung Weiach im Lichte von Modellberechnungen der thermischen Maturität: Eclogae geologicae Helvetiae, v. 80, p. 543-552.

Kester, D. R., 1975, Dissolved Gases other than CO_2, *in* Riley, J. P., and Skirro, G., *eds.*, Chemical Oceanography, Second Edition: New York, Academic Press.

Kirchhofer, W., *ed.*, 1982, Klimaatlas der Schweiz: Bern, Schweiz. Meteorologische Anstalt.

Krouse, H. H., 1980, Sulphur isotopes in our environment, *in* Fritz, P., and Fontes, J. Ch., *eds.*, Handbook of Environmental Isotope Geochemistry, Volume 1, The Terrestrial Environment, A: Amsterdam, Elsevier, p. 435-471.

Kubik, P. W., Korschinek, G., Nolte, E., Ratzinger, U., Ernst, H., Teichmann, S., and Morinaga, H., 1984, Accelerator mass spectrometry of ^{36}Cl in limestone and some paleontological samples using completely stripped ions: Nuclear Instruments and Methods, v. B5, p. 326.

KUER, 1989, Report for the years 1985 and 1986: Federal Commission for the Survey of Radioactivity (KUER): Bern, Bundesamt für Gesundheitwesen.

Kussmaul, H., and Antonsen, O., 1985, Hydrochemische Labormethoden für das Nagra-Untersuchungsprogramm: Baden, Switzerland, Nagra, Technischer Bericht 85-04, 83 p.

Lally, A. E., 1982, Chemical procedures, *in* Ivanovich, M., and Harmon, R. S., *eds.*, Uranium series disequilibrium: Applications to environmental problems: Oxford, Oxford University Press, p. 79-106.

Lambert, S. J., 1987, Feasibility study: Applicability of geochronologic methods involving radiocarbon and other nuclides to the groundwater hydrology of the Rustler formation: Albuquerque, New Mexico, Sandia National Laboratories, SAND86-1054, 72 p.

Land, L. S., 1980, The isotopic and trace element geochemistry of dolomite: The state of the art: Society of Economic Paleontologists and Mineralogists, v. 28, p. 87-110.

Langmuir, D., 1978, Uranium solution-mineral equilibria at low temperatures with applications to sedimentary ore deposits: Geochimica et Cosmochimica Acta, v. 42, p. 547-569.

Laubscher, H., 1986, The eastern Jura: Relations between thinskinned basement tectonics, local and regional: Geologische Rundschau, v. 75, p. 535-553.

Laubscher, H., 1987, Die tektonische Entwicklung der Nordschweiz: Eclogae geologicae Helvetiae, v. 80, p. 287-303.

Lister, G. S., 1989, Reconstruction of palaeo air temperature changes from oxygen isotopic records in Lake Zürich: The significance of seasonality: Eclogae geologicae Helvetiae, v. 82, p. 219-234.

Loosli, H. H., 1983, A dating method with ^{39}Ar: Earth and Planetary Science Letters, v. 63, p. 51-62.

Loosli, H. H., Lehmann, B. E., and Balderer, W., 1989, Argon-39, argon-37 and krypton-85 isotopes in Stripa groundwaters: Geochimica et Cosmochimica Acta, v. 53, p. 1825-1829.

Ludin, A. I., 1989, Krypton-81 und Krypton-85, Bern, Lizentiatsarbeit, Universität Bern, *(unpublished)*.

Matter, A., 1987, Faciesanalyse und Ablagerungsmilieus des Permokarbons im Nordschweizer Trog: Eclogae geologicae Helvetiae, v. 80, p. 345-367.

Matter, A., Peters, Tj., Bläsi, H.-R., Meyer, J., Ischi, H., and Meyer, Ch., 1987, Sondierbohrung Weiach - Geologie: Baden, Switzerland, Nagra, Technischer Bericht 86-01, 470 p.

Matter, A., Peters, Tj., Bläsi, H.-R., Schenker, F., and Weiss, H. -P., 1988b, Sondierbohrung Schafisheim - Geologie: Baden, Switzerland, Nagra, Technischer Bericht 86-03, 350 p.

Matter, A., Peters, Tj., Isenschmid, Ch., Bläsi, H.-R., and Ziegler, H.-J., 1988a, Sondierbohrung Riniken - Geologie: Baden, Switzerland, Nagra, Technischer Bericht 86-02, 200 p.

Matthess, G., 1982, Properties of Groundwater: New York, John Wiley & Sons, 397 p.

McNutt, R. H., Frape, S. K., and Fritz, P., 1984, Strontium isotope composition of some brines from the Precambrian Shield of Canada: Isotope Geoscience, v. 2, p. 205-215.

Meyer, J., 1987, Die Kataklase im kristallinen Untergrund der Nordschweiz: Eclogae geologicae Helvetiae, v. 80, p. 323-334.

Michelot, J. L., Bentley, H. W., Brissaud, I., Elmore, D., and Fontes, J. Ch., 1984, Progress in environmental isotope studies (^{36}Cl, ^{34}S, ^{18}O) at the Stripa site, *in* Isotope Hydrology 1983: Proceedings of a Symposium, Vienna, 12-16 September 1983: Vienna, International Atomic Energy Agency, p. 207-229.

Milton, G., 1986, Palaeohydrological inferences from fracture calcite analysis, *in* Proceedings of Symposium on Isotope Geochemistry of Groundwater and Fracture Material in Plutonic Rock, Quebec, 1-3 Oct. 1986: Monte Ste. Marie, Quebec.

Mook, W. G., 1980, Carbon-14 in hydrogeological studies, *in* Fritz, P., and Fontes, J. Ch., *eds.*, Handbook of Environmental Isotope Geochemistry, Volume 1, The Terrestrial Environment, A: Amsterdam, Elsevier, p. 49-74.

Mook, W. G., 1983, Principles of Isotope Hydrology: Notes for Introductory Course on Isotope Hydrology: Amsterdam, The University, Department of Hydrogeology and Geographical Hydrology, 79 p.

Mook, W. G., 1986, ^{13}C in atmospheric CO_2: Netherlands Journal of Sea Research, v. 20, p. 211-223.

Moser, H., Stichler, W., Rank, D., and Rajner, V., 1981, Ergebnisse von Messungen des Gehalts an Deuterium, Sauerstoff-18 und Tritium in Wasserproben aus dem Einzugsgebiet der Langeten, *in* Leibundgut, Ch., and Harum, T., *eds.*, Tracerhydrologische Untersuchungen im Langetental (Schweiz): Steirische Beiträge zur Hydrogeologie, p. 5-124.

Mullis, J., 1987, Fluideinschluss-Untersuchungen in den Nagra Bohrungen der Nordschweiz: Eclogae geologicae Helvetiae, v. 80, p. 553-568.

Müller, G., 1980, Die Klimaregionen der Schweiz, *in* Die Beobachtungsnetze der Schweiz, Konzept 1980: Zürich, Meteorologische Anstalt.

Müller, W. H., Huber, M., Isler, A., and Kleboth, P., 1984, Erläuterung zur geologischen Karte der zentralen Nordschweiz 1:100'000: Baden, Switzerland, Nagra, Technischer Bericht 84-25, 234 p.

Münnich, F., 1958, Untersuchung der Energietönung und des Wirkungsquerschnitts einigen durch thermische Neutronen ausgelösten (n, α) - Prozesse: Zeitung für Physik, v. 153, p. 106-123.

Nagra, 1984, Die Kernbohrung Beznau: Baden, Switzerland, Nagra, Technischer Bericht 84-34, 111 p.

Nagra, 1985, Sondierbohrung Böttstein - Untersuchungsbericht: Baden, Switzerland, Nagra, Technischer Bericht 85-01, 190 p.

Nagra, 1988, Sedimentstuide-Zwischenbericht 1988: Möglichkeiten zur Endlagerung langlebiger radioaktiver Abfälle in den Sedimenten der Schweiz: Baden, Switzerland, Nagra, Technischer Bericht 88-25, 456 p.

Nagra, 1989, Hydrochemische Analysen, 9 vols: Isotopen- und Edelgas- Untersuchungen, 3 vols: Baden, Switzerland, Nagra, Interner Bericht 89-06. *(unpaged)*.

Neretnieks, I., 1980, Diffusion in the rock matrix: An important factor in radionuclide retardation?: Journal of Geophysical Research, v. 85, p. 4379-4397.

Neretnieks, I., 1981, Age dating of groundwater in fissured rock: Influence of water volume in micropores: Water Resources Research, v. 17, p. 421-422.

Nielsen, H., 1979, Sulfur isotopes, *in* Jäger, E., and Hunziker, J. C., *eds.*, Lectures in Isotope Geology: Berlin, Springer, p. 283-312.

Nordstrom, D. K., and Munoz, J. L., 1985, Geochemical Thermodynamics: Menlo Park, California, Benjamin/Cummings, 465 p.

Nordstrom, D. K., and Olsson T., 1987, Fluid inclusions as a source of dissolved salts in deep granitic groundwaters, *in* Fritz, P., and Frape, S. K., *eds.*, Saline Water and Gases in Crystalline Rocks: Geological Association of Canada, Special Paper 33, p. 111-119.

Nordstrom, D. K., Andrews, J. N., Carlsson, L., Fontes, J. Ch., Fritz, P., Moser, H., and Olsson, T., 1985, Hydrogeological and hydrochemical investigations in boreholes - Final report of the Phase I geochemical investigations of the Stripa groundwaters: Stockholm, Sweden, Swedish Nuclear Fuel and Waste Management Co., SKB Technical Report 85-06, 247 p.

Nordstrom, D. K., Ball, J. W., Donahoe, R. J., and Whittemore, D. 1989a, Groundwater chemistry and water-rock interactions at Stripa: Geochimica et Cosmochimica Acta, v. 53, p. 1727-1740.

Nordstrom, D. K., Lindblom, S., Donahoe, R. J., and Barton, C. C. 1989b, Fluid inclusions in the Stripa granite and their possible influence on the groundwater chemistry: Geochimica et Cosmochimica Acta, v. 53, p. 1741-1756.

O'Nions, R. K., and Oxburgh, E. R., 1983, Heat and helium in the earth: Nature, v. 306, p. 429-431.

Oeschger, H., 1974, Discussion, *in* Isotope Techniques in Groundwater Hydrology, 1974, Volume II: Vienna, International Atomic Energy Agency, p. 91.

Oeschger, H., and Siegenthaler, U., 1972, Umgebungsisotope im Dienste der Hydrologie und Ausblick auf neue Methoden: Gas-Wasser-Abwasser, v. 113, p. 501-508.

Ohmoto, H., and Lasaga, A. C., 1982, Kinetics of reactions between aqueous sulphates and sulphides in hydrothermal systems: Geochimica et Cosmochimica Acta, v. 46, p. 1727-1745.

Osmond, J. K., and Cowart, J. B., 1976, The theory and uses of natural uranium isotopic variations in hydrology: Atomic Energy Review, v. 144, p. 621-669.

Osmond, J. K., and Cowart, J. B., 1982, Ground Water, *in* Ivanovich, M., and Harmon, R. S., *eds.*, Uranium series disequilibrium: Applications to environmental problems: Oxford, Oxford University Press, p. 202-245.

Osmond, J. K., Cowart, J. B., and Ivanovich, M., 1983, Uranium isotopic disequilibrium in groundwater as an indicator of anomalies: International Journal of Applied Radiation Isotopes, v. 34, p. 283-308.

Pearson, F. J., Jr., 1985, Sondierbohrung Böttstein - Results of hydrochemical investigations: Analysis and interpretation: Baden, Switzerland, Nagra, Technischer Bericht 85-05, 131 p.

Pearson, F. J., Jr., 1987, Models of mineral controls on the composition of saline groundwaters of the Canadian Shield, *in* Fritz, P., and Frape, S. K., *eds.*, Saline Water and Gases in Crystalline Rocks: Geological Association of Canada, Special Paper 33, p. 39-51.

Pearson, F. J., Jr., 1989, Uncertainty analyses of models of ground-water carbonate isotope evolution, *in* Miles, D. L., *ed.*, Proceedings of the 6th International Symposium on Water-Rock Interaction, Malvern, U.K., 3-8 August 1989: Rotterdam, A. A. Balkema, p. 545-552.

Pearson, F. J., Jr., and Fisher, D. W., 1971, Chemical composition of atmospheric precipitation in the Northeastern United States: U.S. Geological Survey, Water-Supply Paper 1535-P, p. P1-P23.

Pearson, F. J., Jr., and Hanshaw, B. B., 1970, Sources of Dissolved Carbonate Species in Groundwater and Their Effects on Carbon-14 Dating, *in* Isotope Hydrology 1970: Proceedings of a Symposium, Vienna, 9-13 March 1970: Vienna, International Atomic Energy Agency, p. 217-286.

Pearson, F. J., Jr., and Rightmire, C. T., 1980, Sulphur and oxygen isotopes in aqueous sulphur compounds, *in* Fritz, P., and Fontes, J. Ch., *eds.*, Handbook of Environmental Isotope Geochemistry, Volume 1, The Terrestrial Environment, A: Amsterdam, Elsevier, p. 227-258.

Pearson, F. J., Jr., and White, D. E., 1967, Carbon-14 ages and flow rates of water in Carrizo sand, Atascosa County, Texas: Water Resources Research, v. 3, p. 151-161.

Pearson, F. J., Jr., Fisher, D. W., and Plummer, L. N., 1978, Correction of ground-water chemistry and carbon isotopic composition for effects of CO_2 outgassing: Geochimica et Cosmochimica Acta, v. 42, p. 1799-1807.

Pearson, F. J., Jr., Noronha, C. J., and Andrews, R. W., 1983, Mathematical modelling of the distribution of natural ^{14}C, ^{234}U and ^{238}U in a regional groundwater system: Radiocarbon, v. 25, p. 291-300.

Pearson, F. J., Jr., Lolcama, J. L., and Scholtis, A., 1989, Chemistry of groundwaters in the Böttstein, Weiach, Riniken, Schafisheim, Kaisten and Leuggern boreholes: A hydrochemically consistent data set: Baden, Switzerland, Nagra, Technischer Bericht 86-19, 102 p.

Pekdeger, A., and Balderer, W., 1987, The Occurrence of Saline Groundwaters and Gases in the Crystalline Rocks of Northern Switzerland, *in* Fritz, P., and Frape, S. K., *eds.*, Saline Water and Gases in Crystalline Rocks: Geological Association of Canada, Special Paper 33, p. 157-174.

Peters, Tj., 1986, Structurally incorporated and water extractable chlorine in the Böttstein granite (N. Switzerland): Contributions to Mineralogy and Petrology, v. 94, p. 272-273.

Peters, Tj., 1987, Das Kristallin der Nordschweiz: Petrographie und hydrothermale Umwandlungen: Eclogae geologicae Helvetiae, v. 80, p. 305-322.

Peters, Tj., Matter, A., Bläsi, H.-R., and Gautschi, A., 1986, Sondierbohrung Böttstein - Geologie: Baden, Switzerland, Nagra, Technischer Bericht 85-02, 230 p.

Peters, Tj., Matter, A., Isenschmid, Ch., Bläsi, H.-R., Meyer, J., and Ziegler, H.-J., 1988a, Sondierbohrung Kaisten - Geologie: Baden, Switzerland, Nagra, Technischer Bericht 86-04, *(in press)*.

Peters, Tj., Matter, A., Bläsi, H.-R., Isenschmid, Ch., Kleboth, P., Meyer, J., and Meyer, Ch., 1988b, Sondierbohrung Leuggern - Geologie: Baden, Switzerland, Nagra, Technischer Bericht 86-05, 250 p.

Pilot, J., Rosler, H. J., and Muller, P., 1972, Zur geochemischen Entwicklung des Meerwassers und mariner Sedimente im Phanerozoikum mittels Untersuchungen von S-, O- und C-Isotopen: Neue Bergbautechnik, v. 2, p. 161-168.

Plummer, L. N., 1977, Defining Reactions and Mass Transfer in Part of the Floridan Aquifer: Water Resources Research, v. 13, p. 801-812.

Plummer, L. N., and Back, W., 1980, The mass balance approach: Application to interpreting the chemical evolution of hydrologic systems: American Journal of Science, v. 280, p. 130-142.

Plummer, L. N., Parkhurst, D. L., and Thorstenson, D. C., 1983, Development of reaction models for ground-water systems: Geochimica et Cosmochimica Acta, v. 47, p. 665-686.

Ramseyer, K., 1987, Diagenese des Buntsandsteins und ihre Beziehung zur tektonischen Entwicklung der Nordschweiz: Eclogae geologicae Helvetiae, v. 80, p. 383-395.

Rauber, D., 1987, Edelgase im Grundwasser, Bern, Switzerland, Inaugural Dissertation, Universität Bern, *(unpublished)*.

Reardon, E. J., and Fritz, P., 1978, Computer Modelling of Groundwater ^{13}C and ^{14}C Isotope Compositions: Journal of Hydrology, v. 36, p. 201-224.

Rozanski, K., Sonntag, C., and Munnich, O., 1982, Factors Controlling Stable Isotope Composition of European Precipitation: Tellus, v. 34, p. 142-150.

Rudolph, J., Rath, H. K., and Sonntag, C., 1983, Noble Gases and Stable Isotopes in ^{14}C-dated Palaeowaters from Central Europe and The Sahara, in Isotope Hydrology 1983: Proceedings of a Symposium, Vienna, 12-16 September 1983: Vienna, International Atomic Energy Agency, p. 467-497.

Sauer, K., 1971, Herkunft und Zusammensetzung der Bad Krozinger Thermalsäurlinge: Heilbad und Kurort, v. 23, p. 82-92.

Savin, S. M., 1980, Oxygen and hydrogen isotope effects in low temperature mineral-water reactions, in Fritz, P., and Fontes, J. Ch., eds., Handbook of Environmental Isotope Geochemistry, Volume 1, The Terrestrial Environment, A: Amsterdam, Elsevier, p. 283-327.

Schmassmann, H., 1977, Die Mineral und Thermalwässer von Bad Lostorf: Mitteilungen Naturforschenden Gessellschaft Solothurn, v. 27, p. 150-290.

Schmassmann, H., 1987, Neue Erkenntnisse zur Beschaffenheit der Tiefengrundwässer der Nordschweiz: Eclogae geologicae Helvetiae, v. 80, p. 569-578.

Schmassmann, H., *in preparation*, Hydrochemische Synthese der Nordschweiz (Tertiar- und Malm-Aquifers): Baden, Switzerland, Nagra, Technischer Bericht 88-07.

Schmassmann, H., Balderer, W., Kanz, W., and Pekdeger, A., 1984, Beschaffenheit der Tiefengrundwässer in der zentralen Nordschweiz und angrenzenden Gebieten: Baden, Switzerland, Nagra, Technischer Bericht 84-21, 335 p.

Scholtis, A., 1988a, OBS-1: Hydrogeochemische Untersuchungen am Potentiellen Standort Oberbauenstock: Baden, Switzerland, Nagra, Interner Bericht 88-45, 45 p.

Scholtis, A., 1988b, PPG-1: Hydrogeochemische Untersuchungen am Potentiellen Standort Piz Pian Grand: Baden, Switzerland, Nagra, Interner Bericht 88-46, 19 p.

Schotterer, U., and Müller, I., 1982, Isotope und Chemie des tiefen Karstwassers im Laufental: Eclogae geologicae Helvetiae, v. 75, p. 601-606.

Schotterer, U., Felber, H. U., and Leibundgut, Ch., 1982, Tritium and oxygen-18 as natural tracers in the complex hydrology of the Alpine basin of Grindelwald (Switzerland): Beiträge zur Geologie der Schweiz - Hydrologie, v. 28, p. 435-444.

Schreiber, K. F., Kuhn, N., Hug, C., Schreiber, C., and Zeh, W., 1977, Wärmegliederung der Schweiz, Beilage 1:500'000: Gebiete unterschiedlichen Föhneinflusses: Bern, Eidgenössisches Justiz- und Polizeidepartement.

Schüepp, M., 1981, Klima und Wetter, *in* Spiess, E., Atlas der Schweiz.

Schwarcz, H. P., Gascoyne, M., and Ford, D. C., 1982, Uranium-series disequilibrium studies of granitic rocks: Chemical Geology, v. 36, p. 87-102.

Schweingruber, M., 1981, Löslichkeits- und Speziationsberechnungen für U, Pu, Np und Th in natürlichen Grundwässern. Theorie, thermodynamische Dateien und erste Anwendungen: Würenlingen, Eidgenössisches Institut für Reaktorforschung, EIR-Bericht 449.

Siegenthaler, U., 1972, Bestimmung der Verweildauer von Grundwasser im Boden mit radioaktiven Umweltisotopen (C-14, Tritium): Gas-Wasser-Abwasser, v. 52, p. 283-290.

Siegenthaler, U., 1980, Herkunft und Verweildauer der Mineralwässer aufgrund von Isotopenmessungen, *in* Högl, O., Die Mineral und Heilquellen der Schweiz: p. 101-108.

Siegenthaler, U., and Oeschger, H., 1980, Correlation of delta ^{18}O with temperature and altitude: Nature, v. 285, p. 314-317.

Siegenthaler, U., and Schotterer, U., 1977, Hydrologische Anwendungen von Isotopenmessungen in der Schweiz: Gas-Wasser-Abwasser, v. 57, p. 501-506.

Siegenthaler, U., Oeschger, H., and Tongiorgi, E., 1970, Tritium and oxygen-18 in natural water samples from Switzerland, *in* Isotope Hydrology 1970: Proceedings of a Symposium, Vienna, 9-13 March 1970: Vienna, International Atomic Energy Agency, p. 373-385.

Siegenthaler, U., Schotterer, U., and Oeschger, H., 1983, Sauerstoff-18 und Tritium als natürliche Tracer für Grundwasser: Gas-Wasser-Abwasser, v. 63, p. 477-483.

Smellie, J. A. T., MacKenzie, A. B., and Scott, R. D., 1986, An analogue validation study of natural radionuclide migration in crystalline rocks using U-series disequilibrium studies: Chemical Geology, v. 55, p. 233.

Sprecher, C., and Müller, W. H., 1986, Geophysikalisches Untersuchungsprogramm Nordschweiz: Reflexionsseismishe 82: Baden, Switzerland, Nagra, Technischer Bericht 84-15, 170 p.

Steiger, R. H., and Jäger, E., 1977, Subcommission on Geochronology: Convention on the use of decay constants in geo- and cosmochronology: Earth and Planetary Science Letters, v. 36, p. 359-362.

Stumm, W., and Morgan, J. J., 1981, Aquatic Chemistry, Second Edition: New York, John Wiley & Sons, 780 p.

Suter, M., Beer, J., Bonani, G., Hofmann, H. J., Michel, D., Oeschger H., Synal, H. A., and Wölfli, W., 1987, ^{36}Cl-studies at the ETH/SIN-AMS facility: Nuclear Instruments and Methods, v. B29, p. 211.

Szabo, B. J., and Kyser, T. K., 1985, Uranium, thorium isotopic analysis and uranium series ages of calcite and opal, and stable isotopic compositions of calcite from drill cores UE25al, USWG-2 and USWG3/GU-3, Yucca Mountain, Nevada: U.S. Geological Survey, Open File Report 85-1224, 25 p.

Taylor, B. E., Wheeler, M. C., and Nordstrom, D. K., 1984, Isotopic composition of sulphate in acid mine drainage as measure of bacterial oxidation: Nature, v. 308, p. 538-541.

Thiel, K., Vorwerk, R., Saager, R., and Stupp, H. D., 1983, ^{235}U fission tracks and ^{238}U-series disequilibria as a means to study recent mobilisation of uranium in Archaean pyritic conglomerates: Earth and Planetary Science Letters, v. 65, p. 249-262.

Thonnard, N., Willis, R. D., Wright, N. C., Davis, W. A., and Lehmann, B. E., 1987, Resonance ionization spectroscopy and the detection of ^{81}Kr: Nuclear Instruments and Methods, v. B29, p. 398-406.

Thorstenson, D. C., Fisher, D. W., and Croft, M. G., 1979, The geochemistry of the Fox Hills-Basal Hell Creek Aquifer in southwestern North Dakota and northwestern South Dakota: Water Resources Research, v. 15, p. 1479-1498.

Thury, M., and Diebold, P., 1987, Überlick über das geologische Untersuchungs-Programm der Nagra in der Nordschweiz: Eclogae geologicae Helvetiae, v. 80, p. 271-286.

Torgerson, T., 1989, Terrestrial helium degassing fluxes and the atmospheric helium budget: implications with respect to the degassing processes of continental crust: Chemical Geology, v. 79, p. 1-14.

Torgerson, T., and Clarke, W. B., 1985, Helium accumulation in groundwater, I: An evaluation of sources and the continental flux of crustal ^{4}He in the Great Artesian Basin, Australia: Geochimica et Cosmochimica Acta, v. 49, p. 1211-1218.

Torgerson, T, Kennedy, B. M., Hiyagon, H., Chiou, K. Y., Reynolds, J. H., and Clark, W. B., 1989, Argon accumulation and the crustal degassing flux of ^{40}Ar in the Great Artesian Basin, Australia: Earth and Planetary Science Letters, v. 92, p. 43-56.

Truesdell, A. H., and Hulston, J. R., 1980, Isotopic evidence on environments of geothermal systems, *in* Fritz, P., and Fontes, J. Ch., *eds.*, Handbook of Environmental Isotope Geochemistry, Volume 1, The Terrestrial Environment, A: Amsterdam, Elsevier, p. 179-226.

Tullborg, E.-L., 1989, $\delta^{18}O$ and $\delta^{13}C$ in fracture calcite used for interpretation of recent meteoric water circulation, *in* Miles, D. L, ed., Proceedings of the 6th International Symposium on Water-Rock Interaction, Malvern, U.K., 3-8 August 1989: Rotterdam, A. A. Balkema, p. 695-698.

Turcotte, D. L., and Schubert, G., 1987, Tectonic implications of radiogenic noble gases in planetary atmospheres, *in* Proceedings of Eighteenth Lunar and Planetary Science Conference, Houston, Texas, March 1987: Houston, p. 1028-1029.

Veizer, J., 1983, Chemical diagenesis of carbonates: Theory and application of trace element technique, *in* Arthur, M. A., Anderson, T. F., Kaplan, I. R., Veizer, J., and Land, L. S., eds., Stable isotopes in sedimentary geology: Society of Economic Paleontologists and Mineralogists, Short Course 10-3, p. 3:1-3:99.

Vogel, J. C., 1967, Investigation of groundwater flow with radiocarbon, *in* Isotopes in Hydrology: Proceedings of a Symposium, Vienna, 14-18 November 1966: Vienna, International Atomic Energy Agency, p. 355-369.

Vuataz, F., 1980, Prospection géothermique dans la zone Koblenz Wildegg-Dielsdorf: Basel, Swiss National Energy Research Foundation (NEFF), NEFF 026.

Vuataz, F., 1982, Hydrogéologie, géochimie et géothermie des eaux thermales de suisse et des regions alpines limitrophes: Matériaux pour la géologie de la Suisse - Hydrologie, v. 29.

Weber, H. P., Sattel G., and Sprecher, C., 1986, Sondierbohrungen Weiach, Riniken, Schafisheim, Kaisten, Leuggern - Geophysikalische Daten: Baden, Switzerland, Nagra, Technischer Bericht 85-50, 90 p.

Wexsteen, P., 1987, Hydrogéologie et géochimie des eaux minerales de la région de Scuol-Tarasp (Basse-Engadine, Grisons, Suisse), Genève, University of Genève, 190 p., *(unpublished)*.

Wexsteen, P., and Matter, A., 1989, Isotope geothermometry and flow regimes in the Muschelkalk aquifer from Northern Switzerland (Nagra Deep Drilling Program), *in* Miles, D. L., ed., Proceedings of the 6th International Symposium on Water-Rock Interaction, Malvern, U.K., 3-8 August 1989: Rotterdam, A. A. Balkema, p. 761-764.

Wexsteen, P., Jaffe, F. C., and Mazor, E., 1988, Geochemistry of cold CO_2-rich springs of the Scuol-Tarasp region, lower Engadine, Swiss Alps: Journal of Hydrology, v. 104, p. 77-92.

Wigley, T. M. L., 1975, Carbon-14 dating of groundwater from closed and open systems: Water Resources Research, v. 11, p. 324-328.

Wigley, T. M. L., Plummer, L. N., and Pearson, F. J., Jr., 1978, Mass transfer and carbon isotope evolution in natural water systems: Geochimica et Cosmochimica Acta, v. 42, p. 1117-1139.

Wittwer, C., 1986, Sondierbohrungen Böttstein, Weiach, Riniken, Schafisheim, Kaisten, Leuggern: Probenahmen und chemische Analysen von Grundwässern aus den Sondierbohrungen: Baden, Switzerland, Nagra, Technischer Bericht 85-49, 142 p.

Wolf, M., Rauert, W., and Weigl, F., 1981, Low-level measurement of tritium by hydrogenation of propadiene and gas counting of propane: International Journal of Applied Radiation and Isotopes, v. 32, p. 919-928.

York, D., 1969, Least squares fitting of a straight line with correlated errors: Earth and Planetary Science Letters, v. 5, p. 320-324.

Yurtsever, Y., and Gat, J. R., 1981, Atmospheric waters, in Stable Isotope Hydrology: Deuterium and Oxygen-18 in the Water Cycle: Vienna, International Atomic Energy Agency, Technical Report Series No. 210, p. 103-142.

Zielinski, R. A., Bush, C. A., Spengler, R. W., and Szabo, B. J., 1986, Rock water interaction in ash-flow tuffs, Yucca Mountain, Nevada, USA - The record from uranium studies: Uranium, v. 2, p. 361-386.

Zimmermann, P. H., Feichter, J., Rath, H. K., Crutzen, P. J., and Weiss, W., 1989, A global three-dimensional source-receptor model investigation using ^{85}Kr: Atmospheric Environment, v. 23, p. 25.

Zuber, A., 1986, Mathematical Models for the interpretation of environmental radioisotopes in groundwater systems, in Fritz, P., and Fontes, J. Ch., eds., Handbook of Environmental Isotope Geochemistry, Volume 2, The Terrestrial Environment, B: Amsterdam, Elsevier, p. 1-59.

APPENDIX

*Table of samples
discussed in this report*

Appendix: Table of samples discussed in this report. (Page 1 of 19)

Location Sample	Short Name	Name	Community	Formation	Coordinate y	Coordinate x	z	Date Range	Reference
1/1 1/3	Zürich Aqui	Thermalwasserbrunnen Aqui	Zürich ZH	tOMM	682125	245560	420	21-30JUL1981 04MAY1987	
2/1	Zürich Tiefenbrunnen	Thermalwasserbohrung Tiefenbrunnen	Zürich ZH	tOMM	684200	245350	408	21JUL1981	
2/3								21-30OCT1982	
3/1 3/3	Kaiseraugst	Bohrung WB 5	Kaiseraugst AG	s-r	622630	264726	300	16JUL1981 06FEB1984	
5/1	Schönenbuch	Gr. W. Fassung Kappelenmatt	Schönenbuch BL	tUSM	604579	265465	326	07-08SEP1981	
5/3								06FEB1984	
7/1	Malleray	Produktionsbohrung	Malleray BE	jo	585890	231670	735	09SEP1981	
8/2	Pratteln	Bohrung 41.J.8	Pratteln BL	mo	620736	264539	271	07-15FEB1984	
9/1 9/2	Frenkendorf	Bohrung 34.J.1	Frenkendorf BL	mo	621042	262365	305	26NOV1981 08FEB1984	
10/1	Oberdorf 1	Bohrung 92.J.1	Oberdorf BL	mo(+?Xm)	623248	249274	506	24NOV1981	
11/2	Oberdorf 2	Bohrung 92.J.2	Oberdorf BL	mo	623727	249323	500	09FEB1984	
12/1	Kaisten Felsbohrung	Felsbohrung Breitematt	Kaisten AG	mo	645988	264509	353	15JUL1981	
13/1	Oberhof	Felsbohrung Hurstet	Oberhof AG	mo	642814	254407	550	10SEP1981	
14/1	Densbüren Felsbohrung	Felsbohrung Asp	Densbüren AG	mo	646455	255054	517	14JUL1981	
14/3								10FEB1984	
15/1 15/3 15/4	Beznau	Bohrung NOK Nr.7904 Beznau	Döttingen AG	mo	659491	267242	326	22JUL1981 08FEB1984 13NOV1984	
16/1 16/101	Windisch BT 2	Bohrung BT 2 Mülligen	Windisch AG	mo	659490	257200	355	23JUL1981 12DEC1979	Vuataz (1980)

A1

A2

Appendix: Table of samples discussed in this report. (Page 2 of 19)

Location Sample	Short Name	Name	Community	Formation	Coordinate y	Coordinate x	z	Date Range	Reference
17/101	Birmenstorf BT 3	Bohrung BT 3	Birmenstorf AG	mo	660100	257650	341	05FEB1980	Vuataz (1980)
19/1	Meltingen 2	Mineralquelle 2	Meltingen SO	km(+?m)	611560	248300	605	08SEP1981	
21/1 21/2a	Lostorf Gips	Subthermale Gipsquelle	Lostorf SO	k	637521	243306	530	16AUG1985 26MAY1986	
22/1	Oberdorf Mineral	Mineralquelle	Oberdorf BL	mm	623494	248637	515	16JUL1981	
23/1 23/4	Eptingen	Mineralquelle Melsten	Eptingen BL	km	629310	248740	730	16JUL1981 16AUG1985	
24/1	Ramsach	Mineralquelle Ramsach Bad	Häfelfingen BL	mm	632924	250437	800	10SEP1981	
25/1	Sissach	Mineralquelle	Sissach BL	km	627670	258870	485	17JUL1981	
26/1	Wintersingen	Mineralquelle	Wintersingen BL	km	629740	260040	480	17JUL1981	
27/1	Magden Magdalena	Magdalenaquelle	Magden AG	km	628490	263865	335	25NOV1981	
28/2 28/3	Zeihen	Chillholzquelle B	Zeihen AG	km	649400	255850	610	15JUL1981 10FEB1984	
29/1 29/105 29/106 29/107	Schinznach Dorf	Mineralquelle	Schinznach Dorf AG	km	652695	255090	395	23JUL1981 15MAY1979 24JUL1979 31OCT1979	Vuataz (1980) Vuataz (1980) Vuataz (1980)
30/1	Wildegg	Jodquelle	Möriken-Wildegg AG	jo	655055	252150	357	21JUL1981	
31/2	Flüh	Subtherme Flüh	Hofstetten SO	JmHR	604490	259480	388	08SEP1981	
32/2	Attisholz	Badquelle Attisholz	Riedholz SO	tUSM	610350	230850	448	09SEP1981	
33/2	Fulenbach	Quelle Aarentränkl	Fulenbach SO	tUSM	630790	237400	410	13-14JUL1981	
34/2	Rohr	Quelle nw Dorf	Rohr SO	mo	638740	251360	620	13JUL1981	

Appendix: Table of samples discussed in this report. (Page 3 of 19)

Location Sample	Short Name	Name	Community	Formation	Coordinate y	Coordinate x	z	Date Range	Reference
35/2	Kaisten Tuttigraben	Tuttigrabenquelle	Kaisten AG	mo	646500	265530	365	15JUL1981	
36/2	Küttigen Fischbach	Fischbachquelle	Küttigen AG	mo	645040	253540	535	10SEP1981	
37/2	Küttigen Stäglimatt	Quelle Stäglimatt	Küttigen AG	ju	646210	253730	560	14JUL1981	
38/2	Lorenzenbad	Lorenzenbad-Quelle	Erlinsbach AG	mo	641930	252120	510	11SEP1981	
39/2	Densbüren Jura	Quelle beim Restaurant Jura Asp	Densbüren AG	mo	646030	254930	545	14JUL1981	
42/1	Schwarzenberg	Mineralquelle Schwarzenberg	Gontenschwil AG	tOSM	652910	234390	610	21JUL1982	
47/1 47/101 47/102 47/103	Warmbach West	Warmbachquelle West	Schinznach Dorf AG	mo	652635	255170	390	23JUL1981 15MAY1979 24JUL1979 01NOV1979	Vuataz (1980) Vuataz (1980) Vuataz (1980)
48/101	Windisch Reusufer	Reussuferquelle Gipsmühle	Windisch AG	mo	659530	257200	340	23JUL1979	Vuataz (1980)
50/101	Münzlishausen Gipsgrueb	Quelle Gipsgrueb Münzlishausen	Baden AG	ju?	662860	258900	682	23JUL1979	
56/2	Weissenstein 1.480	Weissensteintunnel, km 1.480 ab SP	Oberdorf SO	ju	603340	233010	682	21JUL1982	
65/2 65/101	Alter Hauenstein 0.975	Alter Hauenstein-tunnel, km 0.975 ab NP	Läufelfingen BL	mo	632035	248090	533	19JUL1982 11NOV1982	Dubois & Flück (1983)
66/2 66/103 66/104	Alter Hauenstein 1.080	Alter Hauenstein-tunnel, km 1.080 ab NP	Läufelfingen BL	mo	632085	247990	530	19JUL1982 11NOV1982 23FEB1983	Dubois & Flück (1983) Dubois & Flück (1983)

Appendix: Table of samples discussed in this report. (Page 4 of 19)

Location Sample	Short Name	Name	Community	Formation	Coordinate y	Coordinate x	z	Date Range	Reference
68/2	Hauenstein-basis 3.135	Hauensteinbasis-tunnel, km 3.135 ab NP	Zeglingen BL	jmHR	634830	251595	439	20JUL1982	
70/2	Hauenstein-basis 5.040	Hauensteinbasis-tunnel, km 5.040 ab NP	Wisen SO	mo	635295	249750	426	20JUL1982	
70/101								11NOV1982	Dubois & Flück (1983)
70/102								23FEB1983	Dubois & Flück (1983)
71/2	Hauenstein-basis 5.174	Hauensteinbasis-tunnel, km 5.174 ab NP	Wisen SO	mo	635330	249620	425	20JUL1982	
71/102								11NOV1982	Dubois & Flück (1983)
71/103								23FEB1983	Dubois & Flück (1983)
72/2	Hauenstein-basis 5.896	Hauensteinbasis-tunnel, km 5.896 ab NP	Lostorf SO	mo	635505	248920	420	20JUL1982	
73/1	Hauenstein-basis 5.930	Hauensteinbasis-tunnel, km 5.930 ab NP	Lostorf SO	mo	635510	248885	419	06JUN1984	
74/2	Hauenstein-basis 6.030	Hauensteinbasis-tunnel, km 6.030 ab NP	Lostorf SO	mo	635535	248790	419	20JUL1982	
76/2	Hauenstein-basis 6.785	Hauensteinbasis-tunnel, km 6.785 ab NP	Lostorf SO	jmHR	635720	248055	413	20JUL1982	
76/102								11NOV1982	Dubois & Flück (1983)
76/103								23FEB1983	Dubois & Flück (1983)
77/1	Neuwiller	Forage thermal	Neuwiller F	jmHR	605800	263150	360	07SEP1981	
80/2	Durlinsdorf	Subtherme	Durlinsdorf F	jo	585190	259345	462	22JUL1982	
81/1	Bellingen 1	Thermalbrunnen 1 (Markusquelle)	Bad Bellingen D	jmHR	608430	286500	225	17AUG1981	
83/1	Bellingen 3	Thermalbrunnen 3	Bad Bellingen D	jmHR	608410	285980	225	17AUG1981	
84/1	Steinenstadt Thermal	Thermalwasserbohrung Steinenstadt	Neuenburg am Rhein	jmHR	609130	289915	225	18AUG1981	
85/1	Steinenstadt Mark	Markgräfler Mineral-quelle Steinenstadt	Neuenburg am Rhein	teo	608990	290110	225	17-18AUG1981	

Appendix: Table of samples discussed in this report. (Page 5 of 19)

Location Sample	Short Name	Name	Community	Formation	Coordinate y	Coordinate x	z	Date Range	Reference
86/2	Badenweiler 1	Thermalquelle 1 (Stollenquelle)	Badenweiler D	mo	617580	294530	445	02MAR1982	
88/2	Badenweiler 3	Produktionsbohrung 3	Badenweiler D	s-r	617280	294290	440	02MAR1982	
90/2	Krozingen 3	Thermalwasserbohrung 3	Bad Krozingen D	mo	618980	307690	225	02MAR1982	
92/1	Zähringen	Thermalwasserbohrung Zähringen	Freiburg i. Br. D	mo	630500	319880	234	01MAR1982	
94/1	Freiburg 2	Thermalwasserbohrung 2	Freiburg i. Br. D	mo-s	624520	314870	223	01MAR1982	
95/1 95/2	Waldkirch	Thermalwasserbohrung	Waldkirch D	KRI	638310	326400	251	01MAR1982 11APR1984	
96/1 96/2	Oberbergen	Badquelle Oberbergen	Vogtsburg D	tvu	617530	327000	300	27NOV1981 11-12APR1984	
97/1 97/103 97/105	Singen	Thermalwasserbohrung	Singen D	jo	703990	289690	430	28JUL-10SEP81 01AUG1969 21MAR1983	Geyh & Mairhofer (1970) Bertleff (1986)
98/1 98/102	Konstanz	Thermalwasserbohrung	Konstanz D	tOMM	733270	280990	400	29JUL1981 21MAR1983	Bertleff (1986)
99/1	Welschingen	Brunnen Welschingen WV westl. Hegau	Engen D	jo(-q)	700480	298160	478	29JUL1981	
100/1 100/101	Beuren	Brun. Beuren a.d. Aach Stadtw. Singen 1	Singen D	jo(-q)	708050	295200	438	28JUL1981 01AUG1969	Geyh & Mairhofer (1970)
101/1	Grenzach 1	Mineralwasserbrunnen Grenzach 1	Grenzach-Wyhlen D	mu-so	616375	267220	262	25NOV1981	
105/1	Liel Schlossbrunnen	Schlossbrunnen Liel	Schliengen D	jmHR	612640	287610	270	18AUG1981	
106/1	Riedlingen Subtherm	Subtherme Riedlingen	Kandern D	jmHR	613820	283740	293	20AUG1981	

Appendix: Table of samples discussed in this report.

Location Sample	Short Name	Name	Community	Formation	Coordinate y	Coordinate x	z	Date Range	Reference
108/1	Müllheim	Subtherme	Müllheim D	jmHR	614650	294760	265	20AUG1981	
109/1	Munzingen	Subtherme Munzingen	Freiburg i. Br. D	jmHR	619540	313200	207	27NOV1981	
110/2	Bürchau	Subtherme	Bürchau D	KRI	629140	291660	625	20AUG1981	
110/4								20FEB1984	
112/101	Bözberg 39.498	Bözbergtunnel SBB, km 39.498	Schinznach Dorf AG	q	652020	256600		06DEC1982	Dubois & Flück (1983)
112/102								24FEB1983	Dubois & Flück (1983)
113/101	Bözberg 39.502	Bözbergtunnel SBB, km 39.502	Schinznach Dorf AG	km	652020	256605		06DEC1982	Dubois & Flück (1983)
113/102								24FEB1983	Dubois & Flück (1983)
114/101	Bözberg 39.647	Bözbergtunnel SBB, km 39.647	Schinznach Dorf AG	mo	651910	255695		06DEC1982	Dubois & Flück (1983)
114/102								24FEB1983	Dubois & Flück (1983)
115/101	Bözberg 39.725	Bözbergtunnel SBB, km 39.725	Schinznach Dorf AG	mo	651840	256750		06DEC1982	Dubois & Flück (1983)
115/102								24FEB1983	Dubois & Flück (1983)
119/2	Schinznach Bad alt	Thermalwasserbrunnen (alte Fassung)	Schinznach Bad AG	mo	654743	256594	340	30OCT1981	
119/5								17AUG1982	
119/8								18MAY1983	
119/11								01MAR1984	
119/14								14NOV1984	
119/17								27-28MAY1986	
119/106								16JUN1977	Vuataz (1982)
119/109								07MAR1978	Vuataz (1982)
119/113								15MAY1979	Vuataz (1982)
119/114								24JUL1979	Vuataz (1982)
119/115								31OCT1979	Vuataz (1982)
119/116								08JAN1980	Vuataz (1982)
119/118								22JUN1983	Dubois & Flück (1984)
119/119								13DEC1983	Dubois & Flück (1984)
120/1	Schinznach Bad S2	Bohrung S2 (neue Fassung)	Schinznach Bad AG	mo	654738	256518	342	07AUG-10SEP81	
120/2								30OCT1981	
120/5								17AUG1982	
120/8								18MAY1983	
120/11								01MAR1984	
120/14								14NOV1984	
120/17								28MAY1986	

Appendix: Table of samples discussed in this report. (Page 7 of 19)

Location Sample	Short Name	Name	Community	Formation	Coordinate y	Coordinate x	z	Date Range	Reference
120/102								25FEB1980	Vuataz (1982)
120/103								27FEB1980	Vuataz (1982)
120/104								28FEB1980	Vuataz (1982)
120/107								02MAR1980	Vuataz (1982)
120/116								11NOV1982	Dubois & Flück (1984)
120/117								23FEB1983	Dubois & Flück (1984)
120/118								22JUN1983	Dubois & Flück (1984)
120/119								14DEC1983	Dubois & Flück (1984)
121/1	Lostorf 1	Bohrung 1	Lostorf SO	mo(-q)	637729	249324	487	13JUL1981	Schmassmann (1977)
121/103								21JUN1972	Vuataz (1982)
121/104								16JUN1977	Vuataz (1982)
121/107								07MAR1978	
122/101	Lostorf 2	Bohrung 2	Lostorf SO	mo	637330	249234		19JUN1972	Schmassmann (1977)
123/1	Lostorf 3	Bohrung 3	Lostorf SO	mo	637326	249242	549	05AUG1981	
123/2								26OCT1981	
123/5								16AUG1982	
123/8								18MAY1983	
123/11								20FEB1984	
123/14								12NOV1984	
123/16								15AUG1985	
123/17								26MAY1986	
123/101								26JUN1972	Vuataz (1982)
123/103								16JUN1977	Vuataz (1982)
123/106								07MAR1978	Vuataz (1982)
123/110								09OCT1979	Vuataz (1982)
123/111								23FEB1982	Dubois & Flück (1984)
123/112								11NOV1982	Dubois & Flück (1984)
123/113								12DEC1983	Dubois & Flück (1984)
124/1	Lostorf 4	Bohrung 4	Lostorf SO	mo	637322	249235	549	05AUG1981	
124/2								26OCT1981	
124/5								16AUG1982	
124/8								18MAY1983	
124/11								20FEB1984	
124/14								12NOV1984	
124/16								15AUG1985	
124/17								26MAY1986	
124/102								16JUN1977	Vuataz (1982)
124/105								07MAR1978	Vuataz (1982)

Appendix: Table of samples discussed in this report. (Page 8 of 19)

Location Sample	Short Name	Name	Community	Formation	Coordinate y	Coordinate x	z	Date Range	Reference
124/109								10OCT1979	Vuataz (1982)
124/110								23JUN1983	Dubois & Flück (1984)
124/111								13DEC1983	Dubois & Flück (1984)
125/1	Säckingen	Badquelle	Bad Säckingen D	KRI	637970	267630	300	22-30JUL1981	
125/2	Badquelle							27-28OCT1981	
125/3								05MAR1982	
125/5								20AUG1982	
125/8								16MAY1983	
125/11								22FEB1984	
125/14								14NOV1984	
125/16								12AUG1985	
125/17								26MAY1986	
125/106								08DEC1982	Dubois & Flück (1984)
125/107								25FEB1983	Dubois & Flück (1984)
125/108								27JUN1983	Dubois & Flück (1984)
125/109								15DEC1983	Dubois & Flück (1984)
126/2	Säckingen	Margarethenquelle	Bad Säckingen D	KRI	638915	267645	287	27OCT1981	
126/3	Margarethen							04MAR1982	
126/5								20AUG1982	
126/8								16MAY1983	
126/11								22FEB1984	
126/14								14NOV1984	
126/16								12AUG1985	
126/17								26MAY1986	
126/103								08DEC1982	Dubois & Flück (1984)
126/104								25FEB1983	Dubois & Flück (1984)
126/105								27JUN1983	Dubois & Flück (1984)
127/1	Lottstetten	Thermalwasserbohrung Nack	Lottstetten D	jo	686170	273670	405	24-28JUL1981	
127/2								23NOV1981	
127/3								04MAR1982	
127/4,5								19AUG1982	
127/8								16-20MAY1983	
127/11								24FEB1984	
127/14								16NOV1984	
127/16								12AUG1985	
127/17								27MAY1986	
127/102								08DEC1982	Dubois & Flück (1984)
127/103								25FEB1983	Dubois & Flück (1984)
127/104								26JUN1983	Dubois & Flück (1984)

Appendix: Table of samples discussed in this report. (Page 9 of 19)

Location Sample	Short Name	Name	Community	Formation	Coordinate y	Coordinate x	z	Date Range	Reference
128/1	Mainau	Bohrung Insel Mainau	Konstanz D	tOMM	731610	285540	400	29JUL1981	
128/101								21MAR1983	Bertleff (1986)
129/1	Neuer Brunnen	N. Brunnen Fa. Lieler Schloss-brunnen, Liel	Schliengen D	mo-s	612640	287600	270	18AUG1981	
130/2	Gränichen	Badquelle I	Gränichen AG	tOMM-tUS	650300	245750	415	26NOV1981	
131/2	Zurzach 1	Thermalwasser-bohrung 1	Zurzach AG	KRI	663972	271224	337	27OCT1981	
131/5								17AUG1982	
131/8								19MAY1983	
131/11								23FEB1984	
131/14								15NOV1984	
131/16								15AUG1985	
131/17								27MAY1986	
131/105								01JAN1976	Loosli (1983)
131/107								16JUN1977	Vuataz (1982)
131/111								07MAR1978	Vuataz (1982)
131/115								1978	Loosli (1983)
131/116								18JUL1979	LLC pers. commun. (1980)
131/117								22OCT1979	LLC pers. commun. (1980)
131/118								20MAY1980	LLC pers. commun. (1980)
132/1	Zurzach 2	Thermalwasser-bohrung 2	Zurzach AG	KRI	664020	271325	337	06AUG1981	
132/2								27OCT1981	
132/5								17AUG1982	
132/7								25FEB-30MAR83	
132/8								19MAY1983	
132/11								23FEB1984	
132/14								15NOV1984	
132/16								15AUG1985	
132/17								27MAY1986	
132/103								01JAN1976	Loosli (1983)
132/105								16JUN1977	Vuataz (1982)
132/113								06DEC1977	Vuataz (1982)
132/130								11NOV1982	Dubois & Flück (1984)
132/131								24FEB1983	Dubois & Flück (1984)

A9

Appendix: Table of samples discussed in this report. (Page 10 of 19)

Location Sample	Short Name	Name	Community	Formation	Coordinate y	Coordinate x	z	Date Range	Reference
139/101	Warmbach Ost	Warmbachquelle Ost	Schinznach Dorf AG	mo	652645	255170	390	15MAY1979	Vuataz (1980)
139/102								24JUL1979	Vuataz (1980)
139/103								01NOV1979	Vuataz (1980)
141/1	Eglisau 1	Mineralwasserbohrung 1	Eglisau ZH	tUSM	681620	269995	345	24-31JUL1981	
141/2								10AUG1981	
142/2	Eglisau 2	Mineralwasserbohrung 2	Eglisau ZH	tUSM	680820	269865	382	04MAR1982	
142/4								19AUG1982	
142/7								20MAY1983	
142/10								23FEB1984	
142/13								15NOV1984	
142/15								14AUG1985	
143/2	Eglisau 3	Mineralwasserbohrung 3	Eglisau ZH	tUSM	680400	269910	377	28OCT1981	
143/5								19AUG1982	
143/8								20MAY1983	
143/11								24FEB1984	
143/14								15NOV1984	
143/16								14AUG1985	
143/17								27MAY1986	
144/2	Eglisau 4	Mineralwasserbohrung 4	Eglisau ZH	tUSM	680470	269820	383	28OCT1981	
144/5								19AUG1982	
144/8								20MAY1983	
144/11								23FEB1984	
144/13								15NOV1984	
144/15								14AUG1985	
144/16								27MAY1986	
148/101	Hauenstein-basis 5.085	Hauensteinbasistunnel, km 5.085 ab NP	Wisen SO	mo	635305	249710	426	11NOV1982	Dubois & Flück (1984)
148/102								23FEB1983	Dubois & Flück (1984)
149/1	Beuren F	Brunnen F (Stadt Singen D) Beuren a.d. Aach	Singen D	jo (-tUSM)	707840	294750	447	28JUL1981	
149/2								09FEB1984	
150/1	Rothaus	Badische Staatsbrauerei Rothaus	Grafenhausen D	KRI	659660	295000	1000	03MAR1982	

A10

Appendix: Table of samples discussed in this report. (Page 11 of 19)

Loca-tion Sample	Short Name	Name	Community	Forma-tion	Coordinate y	Coordinate x	z	Date Range	Reference
159/2	Rheinfelden	Bohrung Engerfeld	Rheinfelden AG	KRI	627650	256680	300	20APR1983	
159/3								09MAY1983	
159/4								30NOV-19DEC83	
159/101								24JUN1983	Dubois & Flück (1984)
159/102								15DEC1983	Dubois & Flück (1984)
159/103								09MAY1983	
159/106								30NOV1983	
160/1	Schinznach Badschachen	Brunnen Badschachen	Schinznach Bad AG	q	654713	257088	341	09FEB1984	
161/1	Säckinger Stammelhof	Bohrung Stammelhof	Bad Säckingen D	KRI	638650	267430	292	22FEB1984	
162/1	Menzen-schwand 1	Uranerzgrube Menzenschwand Stelle 1	St. Blasien D	KRI	657634	271208	700	19SEP1985	
163/1	Menzen-schwand 2	Uranerzgrube Menzenschwand Stelle 2	St. Blasien D	KRI	645320	299080	700	19SEP1985	
164/1	Menzen-schwand 2b	Uranerzgrube Menzenschwand Stelle 2b	St. Blasien D	KRI	657634	271208	700	19SEP1985	
165/1	St. Blasien Erzgruben	Erzgrubenquelle	St. Blasien D	KRI	650510	289170	850	19SEP1985	
166/1	Sulzburg Waldhotel	Bohrung unterhalb Waldhotel	Sulzburg D	KRI	622880	296830	450	18-20SEP1985	
167/1	Görwihl Bohrloch 30	Bohrloch 30 Schluchseewerke	Görwihl D	KRI	645220	282190	900	18-20SEP1985	
170/2	Dettenberg	Dettenbergtunnel	Rorbas ZH	tOMM	685535	263895	427	02JUL1986	
171/101	Beznau 104.5-109.5 m	Bohrung NOK 7904 Beznau 104.5-109.5 m	Döttingen AG	km3b	659491	267242	326	14JAN1980	NAGRA (1984) NTB 84-34
172/101	Beznau 230 m	Bohrung NOK 7904 Beznau 230 m	Döttingen AG	mo3	659491	267242	326	04FEB1980	NAGRA (1984) NTB 84-34

A11

A12

Appendix: Table of samples discussed in this report. (Page 12 of 19)

Location Sample	Short Name	Name	Community	Formation	Coordinate y	Coordinate x	Coordinate z	Date Range	Reference
173/101	Beznau 226.7-232.0 m	Bohrung NOK 7904 Beznau 226.7-232.0 m	Döttingen AG	mo3	659491	267242	326	05MAR1980	Nagra (1984) NTB 84-34
174/101	Beznau 285.7-291.0 m	Bohrung NOK 7904 Beznau 285.7-291.0 m	Döttingen AG	mo1	659491	267242	326	28FEB1980	Nagra (1984) NTB 84-34
175/101	Beznau 299.0-304.3 m	Bohrung NOK 7904 Beznau 299.0-304.3 m	Döttingen AG	mm4	659491	267242	326	26FEB1980	Nagra (1984) NTB 84-34
191/101 191/102	Häusern Sägtobel	Sägtobelquelle	Häusern D	KRI	655150	291250	1025	27OCT1982 27JUN1983	Dubois & Flück (1984) Dubois & Flück (1984)
192/101 192/102 192/103	Uehlingen Stollenmund	Stollenmundquelle	Uehlingen-Birkendorf/	KRI	659550	287650	800	27OCT1982 25FEB1983 27JUN1983	Dubois & Flück (1984) Dubois & Flück (1984) Dubois & Flück (1984)
193/101	Uehlingen Giessbach	Giessbachquelle	Uehlingen-Birkendorf/	KRI	658450	287600	850	27OCT1982	Dubois & Flück (1984)
194/101 194/102	Dachsberg Hierbach	Hierbachquelle	Dachsberg D	KRI	648825	282100	800	27OCT1982 27JUN1983	Dubois & Flück (1984) Dubois & Flück (1984)
196/101 196/102 196/103	Hausen HH 1	Bohrung HH 1	Hausen AG	mo-mm	657836	256939	380	12APR1983 13APR1983 14APR1983	Dubois & Flück (1984) Dubois & Flück (1984) Dubois & Flück (1984)
197/101 197/102 197/103	Hausen BT 4	Bohrung BT 4	Birmenstorf AG	mo	660049	257455	344	19APR1983 25APR1983 26APR1983	Dubois & Flück (1984) Dubois & Flück (1984) Dubois & Flück (1984)
198/101	Lostorf3 oberer Aquifer	Bohrung 3 oberer Aquifer	Lostorf SO	mo	637326	249242	549	04APR1972	Schmassmann (1977)
201/1 201/3 201/6 201/9 201/12 201/14 201/15	Ennetbaden Allgemeine	Allgemeine Quelle	Ennetbaden AG	mo	666074	259262	351	14-30JUL1981 18AUG1982 17MAY1983 21FEB1984 13NOV1984 13AUG1985 29MAY1986	

Appendix: Table of samples discussed in this report. (Page 13 of 19)

Location Sample	Short Name	Name	Community	Formation	Coordinate y	Coordinate x	z	Date Range	Reference
201/102								06DEC1982	Dubois & Flück (1984)
201/103								22FEB1983	
203/3	Ennetbaden Schwanen (inne)	Schwanenquelle (innen)	Ennetbaden AG	mo	666060	259212	352	18AUG1982	Vuataz (1982)
203/103								16JUN1977	Vuataz (1982)
203/107								07MAR1978	
203/111								08DEC1982	Dubois & Flück (1984)
203/112								23FEB1983	Dubois & Flück (1984)
204/2	Baden Gr Heisse Stein	Grosse Heisse Steinquelle	Baden AG	mo	665965	259237	354	29OCT1981	
204/5								17MAY1983	
204/8								21FEB1984	
204/11								13NOV1984	
204/13								13AUG1985	
204/14								29MAY1986	
206/102	Baden Limmat	Limmatquelle	Baden AG	mo	666015	259238	353	23JUN1983	Dubois & Flück (1984)
207/2	Baden Verena	St. Verenaquelle	Baden AG	tOMM	665950	259239	354	30MAY1986	
210/3	Baden Wälderhut	Wälderhutquelle	Baden AG	mo	665949	259250	354	18AUG1982	
210/6								17MAY1983	
210/9								21FEB1984	
210/12								13NOV1984	
210/14								13AUG1985	
210/15								29MAY1986	
219/1	Baden Verenahof	Verenahofquelle	Baden AG	mo	665934	259243	356	01JUN-08JUL81	
219/2								29OCT1981	
219/5								18AUG1982	
219/8								17MAY1983	
219/11								21FEB1984	
219/14								13NOV1984	
219/16								13AUG1985	
219/17								29MAY1986	
219/103								16JUN1977	Vuataz (1982)
219/107								07MAR1978	Vuataz (1982)
219/111								27MAR1983	Dubois & Flück (1984)
219/112								14DEC1983	Dubois & Flück (1984)
221/3	Baden Heisse Stein	Grosse und Kleine Heisse Steinquellen	Baden AG	mo	665965	259237	354	18AUG1982	

Appendix: Table of samples discussed in this report. (Page 14 of 19)

Location Sample	Short Name	Name	Community	Formation	Coordinate y	Coordinate x	z	Date Range	Reference
231/1	Amsteg	Sondierstollen KW SBB II km 0.157	Amsteg UR	KRI	694300	180160	540	11SEP1985	
232/1	Gotthard-strassent.	Gotthardstrassent. Sicherheitsst. km 2.8	Andermatt UR	KRI	657634	271208	1120	11SEP1985	
233/1	Grimsel Transitgas	Grimsel Transitgas-stollen km 2.480 ab NP	Guttannen BE	KRI	667240	162800	1426	12SEP1985	
234/1	Lötschberg-tunnel	Lötschbergtunnel km 6.020 ab NP	Kandersteg BE	KRI	621250	143310	1202	12SEP1985	
235/108 235/110	Pfäfers	Thermalquelle	Pfäfers SG	KroSE	755880	204040	693	01MAR1966 01AUG1975	Comm. Thermalbäder Bad Ragaz 1982
251/101 251/102 251/103	Aare Schinz-nach Bad	Aare	Schinznach Bad AG	Fluss-wasser	654800	255230	341	11NOV1982 23FEB1983 22JUN1983	Dubois & Flück (1984) Dubois & Flück (1984) Dubois & Flück (1984)
252/101 252/102 252/103	Limmat Baden	Limmat	Baden AG	Fluss-wasser	665200	259370	349	11NOV1982 23FEB1983 22JUN1983	Dubois & Flück (1984) Dubois & Flück (1984) Dubois & Flück (1984)
253/101 253/102 253/103	Reuss Mülligen	Reuss	Mülligen AG	Fluss-wasser	660840	256700	340	11NOV1982 23FEB1983 22JUN1983	Dubois & Flück (1984) Dubois & Flück (1984) Dubois & Flück (1984)
254/101 254/102 254/103	Alb Niedermühle	Alb Niedermühle	Höchenschwand D	Fluss-wasser	651000	280800	600	27OCT1982 25FEB1983 27JUN1983	Dubois & Flück (1984) Dubois & Flück (1984) Dubois & Flück (1984)
270/101	Olsberg Wangliste	Feldschlösschen Bohrung Wangliste	Olsberg AG	mo	626494	263748	408	09NOV1983	
271/101 271/102 271/103 271/104	Magden Weiere	Feldschlösschen Bohrung Weiere	Magden AG	mo	628724	262820	352	13-14SEP1983 09MAY1985 11-13JUN1985 28OCT1986	

Appendix: Table of samples discussed in this report. (Page 15 of 19)

Location Sample	Short Name	Name	Community	Formation	Coordinate y	Coordinate x	z	Date Range	Reference
272/101 272/102 272/103	Magden Falke	Feldschlösschen Bohrung Falke 2	Magden AG	mo	628092	263210	403	23OCT1985 22-25NOV1985 24-28OCT1986	
273/101 273/102 273/103	Magden Eich	Feldschlösschen Bohrung Eich	Magden AG	mo	627656	262757	413	10FEB1986 16MAY1986 27-28OCT1986	
274/102	Magden Stockacher	Feldschlösschen Bohrung Stockacher	Magden AG	mo	629177	263651	388	28OCT1986	
275/101	Arisdorf	Feldschlösschen Bohrung Arisdorf	Arisdorf BL	mo	624668	260888		25FEB1987	
276/101	Zeiningen	Feldschlösschen Bohrung Gründel	Zeiningen AG	mo	632760	266446		11NOV1986	
301/1 301/2	Böttstein 123.2-202.5	Sondierbohrung 123.2-202.5 m	Böttstein AG	mo	659340	268556	347	29OCT1982 01NOV1982	
301/5 301/6	Böttstein 305.6-327.6	Sondierbohrung 305.6-327.6 m	Böttstein AG	s-KRI	659340	268556	347	16NOV1982 16NOV1982	
301/8c 301/9	Böttstein 393.9-405.1	Sondierbohrung 393.9-405.1 m	Böttstein AG	KRI	659340	268556	347	14DEC1982 15DEC1982	
301/12b 301/13	Böttstein 618.5-624.1	Sondierbohrung 618.5-624.1 m	Böttstein AG	KRI	659340	268556	347	22JAN1983 22JAN1983	
301/16 301/17	Böttstein 608.0-628.8	Sondierbohrung 608.0-628.8 m	Böttstein AG	KRI	659340	268556	347	09AUG1983 09AUG1983	
301/18	Böttstein 782.0-802.8	Sondierbohrung 782.0-802.8 m	Böttstein AG	KRI	659340	268556	347	14-16AUG1983	
301/19 301/20 301/21,22	Böttstein 1321.0-1331.4	Sondierbohrung 1321.0-1331.4 m	Böttstein AG	KRI	659340	268556	347	30SEP1983 07OCT1983 10OCT1983	
301/23 301/24a 301/24b	Böttstein 640.8-657.2	Sondierbohrung 640.8-657.2 m	Böttstein AG	KRI	659340	268556	347	20OCT1983 04NOV1983 05NOV1983	

Appendix: Table of samples discussed in this report. (Page 16 of 19)

Location Sample	Short Name	Name	Community	Formation	Coordinate y	Coordinate x	z	Date Range	Reference
301/25a 301/25b	Böttstein 305.2-319.8	Sondierbohrung 305.2-319.8 m	Böttstein AG	s-KRI	659340	268556	347	20JAN1984 20JAN1984	
302/6-8 302/5,9	Weiach 242.9-267.0	Sondierbohrung 242.9-267.0 m	Weiach ZH	Jok1	676750	268620	369	06MAR1983 05MAR1983	
302/10a 302/10b	Weiach 822.0-896.1	Sondierbohrung 822.0-896.1 m	Weiach ZH	mo	676750	268620	369	03APR1983 04APR1983	
302/11	Weiach 983.0-985.3	Sondierbohrung 983.0-985.3	Weiach ZH	s	676750	268620	369	01MAY1983	
302/12 302/13	Weiach 981.0-989.6	Sondierbohrung 981.0-989.6 m	Weiach ZH	s	676750	268620	369	19JUL1983	
302/14 302/15	Weiach 2260.5-2273.5	Sondierbohrung 2260.5-2273.5 m	Weiach ZH	KRI	676750	268620	369	03-04APR1984 06APR1984	
302/16	Weiach 2211.6-2224.6	Sondierbohrung 2211.6-2224.6 m	Weiach ZH	KRI	676750	268620	369	27-28APR1984	
302/17 302/18	Weiach 1401.1-1415.7	Sondierbohrung 1401.1-1415.7 m	Weiach ZH	r	676750	268620	369	15JUN1984	
302/19 302/20	Weiach 1109.2-1123.8	Sondierbohrung 1109.2-1123.8 m	Weiach ZH	r	676750	268620	369	27-28JUN1984 29JUN1984	
303/1	Riniken 501.0-530.5	Sondierbohrung 501.0-530.5 m	Riniken AG	km	656604	261900	385	25JUL1983	
303/2	Riniken 617.3-696.0	Sondierbohrung 617.3-696.0 m	Riniken AG	mo	656604	261900	385	17AUG1983	
303/3 303/4	Riniken 793.0-820.2	Sondierbohrung 793.0-820.2 m	Riniken AG	s-r	656604	261900	385	16SEP1983 17SEP1983	
303/5b 303/5c	Riniken 958.4-972.5	Sondierbohrung 958.4-972.5 m	Riniken AG	r	656604	261900	385	04OCT1983 05OCT1983	
303/6 303/7-9	Riniken 977.0-1010.0	Sondierbohrung 977.0-1010.0 m	Riniken AG	r	656604	261900	385	02NOV1983 03NOV1983	

Appendix: Table of samples discussed in this report. (Page 17 of 19)

Location Sample	Short Name	Name	Community	Formation	Coordinate y	Coordinate x	z	Date Range	Reference
303/19 303/20	Riniken 1354.0-1369.0	Sondierbohrung 1354.0-1369.0 m	Riniken AG	r	656604	261900	385	11JUL1985 15JUL1985	
304/1 304/2	Schafisheim 553.0-563.0	Sondierbohrung 553.0-563.0 m	Schafisheim AG	tUSM	653632	246757	421	20DEC1983 20DEC1983	
304/3	Schafisheim 1227.8-1239.0	Sondierbohrung 1227.8-1239.0 m	Schafisheim AG	mo	653632	246757	421	17FEB1984	
304/4 304/5 304/6 304/7	Schafisheim 1476.0-1500.4	Sondierbohrung 1476.0-1500.4 m	Schafisheim AG	s-KRI	653632	246757	421	29MAR1984 01APR1984 01-02APR1984 01-02APR1984	
304/8 304/9	Schafisheim 1564.5-1577.7	Sondierbohrung 1564.5-1577.7 m	Schafisheim AG	KRI	653632	246757	421	01-02MAY1984 01-02MAY1984	
304/10 304/11	Schafisheim 1883.5-1892.3	Sondierbohrung 1883.5-1892.3 m	Schafisheim AG	KRI	653632	246757	421	17JUN1984 18JUN1984	
304/14	Schafisheim 1240.7-1261.6	Sondierbohrung 1240.7-1261.6 m	Schafisheim AG	mo	653632	246757	421	21JAN1985	
305/1	Kaisten 97.0-129.9	Sondierbohrung 97.0-129.9 m	Kaisten AG	s	644641	265624	320	22-23FEB1984	
305/2 305/3	Kaisten 276.0-292.5	Sondierbohrung 276.0-292.5 m	Kaisten AG	r	644641	265624	320	01MAR1984 01MAR1984	
305/4 305/5	Kaisten 299.3-321.5	Sondierbohrung 299.3-321.5 m	Kaisten AG	KRI	644641	265624	320	15MAR1984 16MAR1984	
305/6 305/7	Kaisten 475.5-489.8	Sondierbohrung 475.5-489.8 m	Kaisten AG	KRI	644641	265624	320	03APR1984 04APR1984	
305/9 305/10,11	Kaisten 816.0-822.9	Sondierbohrung 816.0-822.9 m	Kaisten AG	KRI	644641	265624	320	02MAY1984 03MAY1984	
305/12 305/13	Kaisten 1021.0-1040.9	Sondierbohrung 1021.0-1040.9 m	Kaisten AG	KRI	644641	265624	320	05-06JUN1984 06JUN1984	

Appendix: Table of samples discussed in this report. (Page 18 of 19)

Location Sample	Short Name	Name	Community	Formation	Coordinate y	Coordinate x	z	Date Range	Reference
305/14 305/15	Kaisten 1140.8-1165.8	Sondierbohrung 1140.8-1165.8 m	Kaisten AG	KRI	644641	265624	320	13AUG1984 14AUG1984	
305/16 305/17	Kaisten 1238.0-1305.8	Sondierbohrung 1238.0-1305.8 m	Kaisten AG	KRI	644641	265624	320	27AUG1984 28AUG1984	
306/1	Leuggern 53.5-96.4	Sondierbohrung 53.5-96.4 m	Leuggern AG	mo	657634	271208	359	18JUL1984	
306/2 306/3	Leuggern 208.2-227.5	Sondierbohrung 208.2-227.5 m	Leuggern AG	s	657634	271208	359	06-08AUG1984 08AUG1984	
306/4	Leuggern 235.1-267.5	Sondierbohrung 235.1-267.5 m	Leuggern AG	KRI	657634	271208	359	14AUG1984	
306/5 306/6	Leuggern 440.4-448.1	Sondierbohrung 440.4-448.1 m	Leuggern AG	KRI	657634	271208	359	14SEP1984 15SEP1984	
306/7 306/8	Leuggern 507.4-568.6	Sondierbohrung 507.4-568.6 m	Leuggern AG	KRI	657634	271208	359	27SEP1984 28SEP1984	
306/9 306/10	Leuggern 702.0-709.5	Sondierbohrung 702.0-709.5 m	Leuggern AG	KRI	657634	271208	359	17-18OCT1984 18OCT1984	
306/11b 306/12 306/13	Leuggern 916.2-929.7	Sondierbohrung 916.2-929.7 m	Leuggern AG	KRI	657634	271208	359	28NOV1984 30NOV1984 07DEC1984	
306/16	Leuggern 1637.4-1649.3	Sondierbohrung 1637.4-1649.3 m	Leuggern AG	KRI	657634	271208	359	15FEB1985	
306/17 306/18 306/19	Leuggern 834.5-859.5	Sondierbohrung 834.5-859.5 m	Leuggern AG	KRI	657634	271208	359	25MAR1985 26MAR1985 27MAR1985	
306/20 306/21	Leuggern 1179.3-1227.2	Sondierbohrung 1179.3-1227.2 m	Leuggern AG	KRI	657634	271208	359	22APR1985 23APR1985	
306/23	Leuggern 1642.2-1688.9	Sondierbohrung 1642.2-1688.9 m	Leuggern AG	KRI	657634	271208	359	30APR1985	

Appendix: Table of samples discussed in this report. (Page 19 of 19)

Location Sample	Short Name	Name	Formation	Coordinate y	Coordinate x	z	Date Range	Reference
306/24 306/25 306/26	Leuggern 1427.4-1439.4	Sondierbohrung 1427.4-1439.4 m	KRI	657634	271208	359	09MAY1985 11MAY1985 14-15MAY1985	
307/1b	Siblingen 171.1-196.0	Sondierbohrung 171.1-196.0	mo	680090	286693	574	26SEP1988	
307/2b	Siblingen 337.0-345.1	Sondierbohrung 337.0-345.1 m	s	680090	286693	574	17OCT1988	
307/3	Siblingen 467.0-490.9	Sondierbohrung 467.0-490.9 m	KRI	680090	286693	574	09NOV1988	
307/4b	Siblingen 490.9-564.4	Sondierbohrung 490.9-564.4 m	KRI	680090	286693	574	26NOV1988	
307/6b	Siblingen 1154.0-1165.2	Sondierbohrung 1154.0-1165.2 m	KRI	680090	286693	574	21FEB1989	
307/7b	Siblingen 1493.3-1498.6	Sondierbohrung 1493.3-1498.6 m	KRI	680090	286693	574	31MAR1989	
322/2	FLG EM 85.012	Felslabor Grimsel Bohrung EM 85.012	KRI	667491	158754	1734	02SEP1986	
501/102	Reichenau	Neuer Brunnen Reichenau	q	720030	284600	399	21MAR1983	Bertleff (1986)
502/101	Sauldorf	Neuer Brunnen Roth	tOMM	725000	308700		12OCT1982	Bertleff (1986)
508/101	Birnau	Brunnen Kloster Birnau	tOMM	733560	289750		22MAR1983	Bertleff (1986)
509/101	Ravensburg	Ravensburg TB-1	tOMM	759700	294800		07JUL1983	Bertleff (1986)